Late Glacial and Postglacial Environmental Changes

Late Glacial and Postglacial Environmental Changes

Quaternary, Carboniferous-Permian, and Proterozoic

Edited by

I. PETER MARTINI

New York Oxford
OXFORD UNIVERSITY PRESS
1997

Oxford University Press

Oxford New York
Athens Auckland Bangkok Bombay
Calcutta Cape Town Dar es Salaam Delhi
Florence Hong Kong Istanbul Karachi
Kuala Lumpur Madras Madrid Melbourne
Mexico City Nairobi Paris Singapore
Taipei Tokyo Toronto

and associated companies in
Berlin Ibadan

Published by Oxford University Press, Inc.,
198 Madison Avenue, New York, New York 10016

Library of Congress Cataloging-in-Publication Data
Late glacial and postglacial environmental changes:
Quaternary, Carboniferous-Permian, and Proterozoic/
edited by I. Peter Martini.
p. cm. Includes bibliographical references and index.
ISBN 0-19-508541-8
1. Glacial epoch. 2. Paleoecology. 3. Glacial climates.
I. Martini, I. P. (Ireneo Peter), 1935–
QE697.L2945 1996 551.7'9—dc20 95-51665

1 3 5 7 9 8 6 4 2

Printed in the United States of America
on acid-free paper

PREFACE

Modern technology allows researchers to probe the most remote parts of Earth. Much work and information is being gathered from the remaining ice sheets and their surrounding seas. The ancient glacial record is being studied in increasing detail with results that often challenge conventional wisdom, and climatic models are being created for different geological periods.

Other major preoccupations include balancing the need for development against the need to ensure that the natural environment is not disrupted further, and predicting whether the recent warming climatic trend is temporary or will lead to "permanent" warmer climates. In order to contribute to the understanding of some of these problems, this book analyzes through studies of landforms and sedimentary successions, the behavior of the major glaciers that covered Earth in the recent past as well as during Carboniferous, Permian and earlier times. It is useful to note that without present-day experiences we would not be able to interpret the ancient geological record. However, some events, such as catastrophic floods, occur so infrequently as not to be registered in human history. This does not make such events less real. Therefore, a few innovative approaches are illustrated in the book, as well as a few methods on how to recognize ancient, cold-climate, terrestrial deposits.

This book is not intended to be comprehensive. It is meant to complement the many other recent syntheses on glacial sedimentation and paleoenvironments. For example, we have not attempted to deal to any extent with the information growing out of continuing studies of deep sea basins and Antarctica. Rather, through the research of the foremost experts on the subject matter, the book focuses on terrestial and paralic glacial and postglacial environments. It is hoped that this collection of syntheses and analyses of Quaternary and older glacial products and events will foster a better understanding of those times when rapid climatic and environmental changes occurred and that it will constitute a ready source of information for analyzing other ancient geological records or constructing models for possible future happenings.

I would like to thank the contributors for their chapters, for the critical reviews they have offered of one other's chapters, and for having endured gracefully the nagging reminders from the editor. Individual acknowledgements in the chapters have not been included, but each author and I received help from colleagues, students, technicians, secretaries, and various organizations. In addition, numerous colleagues helped with the critical review of various chapters. The editorial staff of Oxford University Press helped immensely with their patient advice, in preparing the figures for final publication, and the wonderful desk copy editor in doing a fantastic job in improving, assuring consistency, and Americanizing everybody's prose. The color maps were produced by the Paleogeographic Atlas Project of Chicago and funded by its corporate sponsors: AGIP, Amoco, BHP Minerals, Mobil, Petrobras, Total and Union Texas. All this amounted to a small army of people who worked to improve this book. I thank them all very much on behalf of the authors and myself.

I hope the reader will find some stimulation in reading this book, and I look forward to discovering the results of this stimulation in future published research.

Guelph, Canada I. P. M.

ACKNOWLEDGMENTS

Rarely is a scientific idea completely new. The content of this book reflects the cumulative knowledge we have about the subject matter. Thus, we thank the innumerable colleagues and mentors who have toiled and published before us. Similarly we acknowledge the various journals and book publishers for the use of the figures and tables reproduced here with permission. Those presented as originally published are indicated as "from" in the captions, and those modified and updated are indicated as "after" in the captions.

Journals/*Publishers*

Academia Nacional de Ciencias, Córdoba • Figures 9.1, 9.8
America Geophysical Union • Figures 11.7, 11.8
American Association of Petroleum Geologists • Figures 10.1, 11.9
Arctic and Alpine Research • Figure 2.2
Australian Bureau of Mineral Resources • Figures 12.1, 12.3, 12.4, 12.5, 12.6, 12.7
Boreas • Figure 3.5
Cambridge University Press, Cambridge • Figure 10.7
Columbia University Press, New York • Figure 15.8
Comptes Rendus • Figures 18.3, 18.5
Cuencas Sedimenarias Argentinas. Serie Correlación Geológica No. 6, Universidad de Tucumán • Figure 9.5
Elsevier Publishing Company, Amsterdam • Figure 19.1
Géographic physique et Quaternaire • Figure 2.1, Table 17.1
Geologi • Figure 3.10
Geological Society • Figure 15.10
Geological Society London, Quarterly Journal • Figure 11.3
Geological Society of America Bulletin • Figures 3.3, 6.1, 6.3, 6.8, 7.3, 9.8, 13.3, 19.2
Geological Survey of Canada • Figures 6.4, 6.10, Table 18.1
Geological Survey of India • Figure 16.1
Geologische Rundschau • Figures 9.2, 16.2
Geologisches Institut der Universität zu Köln, Sonderveröffentlichung • Figure 11.4
Geology • Figures 7.1, 7.4, 15.8
International Journal of Coal Petrology • Figures 18.4, 18.7, 19.5A and B, 19.6B and C
John Wiley and Sons, New York • Figure 18.1
Journal of Australian Geology and Geophysics • Figure 11.9
Journal of Glaciology • Figure 6.2
Memorial Botanical Survey of South Africa • Figure 10.5
Methuen, London • Table 5.2
Naturc • Figurc 2.8, 15.9, 17.3
Nauka, Novosibirsk • Figures 4.2, 4.4
Norsk Geograpafisk Tidsskrift • Figure 3.9
Paleoceanography • Figure 2.11
Pesa Journal • Figure 12.9

Precambrian Research • Figure 15.3
Proceedings of the seventh International Gondwana Symposium, Universidade do Saõ Paulo • Figures 19.4A and B, 19.6A
Progress in Physical Geography • Table 7.3
Quaternary Research • Figure 6.11
Quaternary Science Reviews • Figure 2.3, Table 5.2
Reunión Grupo Argentino Proyecto Pangea, Comunicaciones • Figure 9.1
Revista Asociación Geológica Argentina • Figure 9.6
Science • Figures 7.2, 15.8, 17.3
Scottish Journal of Geology • Figures 15.9, 15.12
Sedimentology • Figure 9.6
Svensk geografisk årsbok • Figure 3.6
Sveriges Geologiska Undersökning • Figure 3.7

CONTENTS

Contributors xi

1. Introduction *I. Peter Martini* 3

PART I THE QUATERNARY:
THE LAST GLACIATION AND DEGLACIATION

2. Northern Hemisphere (Laurentide) Deglaciation: Processes and Responses of Ice Sheet/Ocean Interactions *John T. Andrews* 9

3. The Last Scandinavian Ice Sheet and Its Down-Wasting *Jan Lundqvist* 28

4. Environment and Climate of Sartan Maximum and Late Glacial in Siberia *Stanislav A. Arkhipov* 53

5. Termination of the Pleistocene and Holocene Changes in South America and Other Glaciated Parts of the Southern Hemisphere *Chalmers M. Clapperton* 61

PART II THE QUATERNARY: MACRO-PROCESSES

6. Sediment Deformation Beneath the Laurentide Ice Sheet *Peter U. Clark* 81

7. Megafloods and Glaciation *Victor R. Baker* 98

PART III THE CARBONIFEROUS-PERMIAN:
A PALEOGEOGRAPHIC RECONSTRUCTION

8. Permian World Topography and Climate *Alfred M. Ziegler, Michael L. Hulver and David B. Rowley* 111

9. Glacial-Postglacial Transition in the Late Paleozoic Basins of Southern South America *Oscar R. López-Gamundí* 147

10. A Review of the Permo-Carboniferous Glaciation in Africa Johan N. J. Visser 169

11. Transition from Freezing to Subtropical Climates in the Permo-Carboniferous of Afro-Arabia and India *Helmut Wopfner and Satyendra M. Casshyap* 192

12. Permian Postglacial Environments of the Australian Plate *John F. Lindsay* 213

13. Upper Paleozoic Glacial and Postglacial Deposits, Central Transantarctic Mountains, Antarctica *John L. Isbell, Gina M. Seegers and Greg A. Gelhar* 230

14. Some Problems of the Permian (Asselian) Glaciation and the Subsequent Climate in the Permian *J. M. Dickins* 243

PART IV AN OLDER PRECAMBRIAN GLACIATION: A COMPARATIVE ANALYSIS

15. Tectonic and Glacioeustatic Controls on Postglacial Stratigraphy: Proterozoic Examples
Grant M. Young 249

PART V THE QUATERNARY AND CARBONIFEROUS-PERMIAN: A FEW CONSIDERATIONS ON CONTINENTAL WEATHERING AND ORGANIC DEPOSITS

16. Interpreting Glacial Climate from Detrital Minerals in Sediments *Prodip Dutta* 271

17. Paleosols of the Northern Part of North America: Their Features and Significance as Indicators of Past Climates *Charles Tarnocai* 276

18. Cold-Temperate Peats and Coals: Their Sedimentology and Composition *R. Marc Bustin* 294

19. Recent Cold Climate Peats of Central Canada and Permian Coals of Gondwana
I. Peter Martini 311

Index 329

CONTRIBUTORS

John T. Andrews
Institute of Arctic and Alpine Research
and Department of Geological Sciences,
University of Colorado

Stanislav A. Arkhipov
United Institute of Geology, Geophysics
and Mineralogy, Russian Academy of Sciences,
Novorsibisk, Russia

Victor R. Baker
Department of Geosciences,
University of Arizona

R. Marc Bustin
Department of Geological Sciences,
University of British Columbia

Satyendra M. Casshyap
Department of Geology, Aligarh Muslim
University, India

Chalmers M. Clapperton
Department of Geology, University of Aberdeen,
Scotland

Peter U. Clark
Department of Geology, Oregon State
University

Jim M. Dickins
Bureau of Mineral Resources,
Canberra, Australia

Prodip Dutta
Department of Geography and Geology,
Indiana State University

Greg A. Gelhar
Department of Geosciences, University of
Wisconsin at Milwaukee

Michael L. Hulver
Department of Geophysical Sciences,
University of Chicago

John L. Isbell
Department of Geosciences,
University of Wisconsin at Milwaukee

John F. Lindsay
Australian Geological Survey,
Canberra, Australia

Oscar R. Lopéz-Gamundí
Texaco Overseas Holdings Inc.,
Bellaire, Texas

Jan Lundqvist
Kvartägeoloigska Institutionen,
Stockholm, Sweden

I. Peter Martini
Department of Land Resource Science,
University of Guelph, Ontario

David B. Rowley
Department of Geophysical Sciences,
University of Chicago

Gina M. Seegers
Department of Geosciences,
University of Wisconsin at Milwaukee

Charles Tarnocai
Agriculture Canada, Centre for Land
and Biological Resources, Ottawa

Johan N. J. Visser
Deapartment of Geology, University of the
Orange Free State, Republic of South Africa

Helmut Wopfner
Geologisches Institut, Universität zu Köln,
Germany

Grant M. Young
Department of Geology, University of Western Ontario, London, Ontario

Alfred M. Zeigler
Department of Geophysical Sciences, University of Chicago

Late Glacial and Postglacial Environmental Changes

1. INTRODUCTION

I. Peter Martini

We live during a time of rapid climatic and environmental change just after a major glaciation. The Pleistocene sedimentary record provides information on climatic fluctuations that have occurred throughout the most recent glacial epoch, including evidence of interglacial periods warmer than the present. There is, however, considerable uncertainty as to whether the future will bring a return to glaciation or will introduce permanent warm, arid conditions. The geological record of the unique Carboniferous-Permian illustrates a full glacial cycle, including a return to warm climatic conditions. The Carboniferous-Permian early postglacial settings were similar to the Quaternary ones, at least in regard to development of large peatlands. By comparing and contrasting what happened and what is happening during the Quaternary with the Carboniferous-Permian and older Precambrian glaciations, it may be possible to formulate various scenarios for future climatic and environmental change and hence, ultimately, for the fate of humankind.

After this brief introduction, Part I contains analyses of the major Pleistocene ice sheets. In the second chapter, Andrews has integrated available data on the Laurentide Ice Sheet of North America with new information obtained from the study of glaciomarine sediments of the Labrador Sea and the North Atlantic. Special emphasis is placed on examination of the ocean/ice interface. In the third chapter, Lundqvist takes up the difficult task of synthesizing the events of the last glaciation and deglaciation of northern Europe. The behavior of those ice sheets, during both inception and termination, is rather different from the North American equivalents. This is due both to different geomorphology and to the effect of the warm Gulf Current on Europe. His contention is that in order to understand the record of older glaciations, the features of the deposits themselves as well as "their context and relationship to each other" are important. Arkhipov briefly summarizes the deglaciation events in western Siberia in the fourth chapter. He subscribes to the widespread glaciation hypothesis for that region. It is of interest to note not only that a large ice dome formed on the Kara Sea shelf, but that the shelf itself constituted a center of glaciation as well. Furthermore, he describes the presence of large proglacial lakes and their drainage during early stages of the deglaciation to the south toward the Black, Caspian, and Mediterranean seas. Clapperton, in the fifth chapter, reports on the modern and Pleistocene glaciers of South America. He gives evidence of the growth and waning of these glaciers. He also indicates that although humankind has a short-term effect on the environment and the climate, as far as glaciation and deglaciation are concerned, that effect is slight compared to that of other terrestrial and extraterrestrial processes.

Part II of the book deals with few macroprocesses involving the glaciers and their sediments and meltwater. Clark, in Chapter 6, analyzes the effect of the substratum on glacier behavior. He indicates that the outer parts of the Laurentide Ice Sheet could have easily slipped over a deformable substratum to acquire a relatively thin, wide equilibrium form. A by-product of this analysis is the realization that eskers preferentially developed where the glacier was underlain by rigid substratum rather than soft deformable sediments. In the seventh chapter, Baker presents a convincing case for the occurrence of proglacial, catastrophic meltwater floods during the Pleistocene. We have no direct experience of such macro-floods, but their occurrence is well documented by the geological record on both Earth and Mars. After almost three quarters of a century of research, the existence of such catastrophic meltwater-floods is generally accepted by the scientific community. Their effects under the glaciers themselves remain somewhat controversial.

This knowledge of Quaternary events, landscape and sedimentary successions should help in the understanding of the ancient glacial geological record. Parts III and IV analyze the ancient record. The time scale for resolution of the ancient deposits is not as fine as that for the Quaternary one. However, various stages of glaciation and deglaciation can still be recognized. Ziegler, Hulver, and Rowley, in Chapter 8, set the scene with new paleogeographic maps of the Permian, including the extension of glaciers during the Sakmarian. They reconstruct the distribution of lands, mountains and seas. It is on this landscape that glaciers formed and waned. In Chapters 9–13 Lopéz-Gamundí, Visser, Wopfner and Casshyap, Lindsay, and Isbell, Seegers and Gelhar show the distribution and behavior of Carboniferous-Permian glaciers and their effects respectively on South America, South Africa, Africa-Arabia-India, Australia, and Antarctica. They all indicate that the ancient glacial record is biased toward basinal deposits, whereas in the Pleistocene much continental material of the last glaciation is still preserved. A characteristic of the Carboniferous-Permian glaciations of the Southern Hemisphere is that they were followed everywhere by vast peat-coal deposits, some of which still significantly fuel the economies of countries like South Africa and Australia. The analyses of the ancient glaciations indicate that the initiation of such events is controlled by several factors, such as tectonic uplift and fast rates of movement of crustal plates. It is clear that the climatic changes that occurred in the Quaternary are important for humankind, but similar major changes also occurred through previous geological times. During past glacial periods ("ice house" periods), variations between cold and warm conditions were similar to those of the present time, although some difference may have existed because of differences in the composition of the atmosphere and in the intensity of solar radiation. Even during particularly warm geological periods, cold-climate sediments were deposited in some polar regions, although glaciers may not have been extensive. Interestingly, Wopfner and Casshyap remind us that the major glaciations ("ice house" periods) may be related to astronomic events. For example, they suggest that long-term cycles of about 300 Ma may be forced on Earth by "changes in mass distribution within the Milky Way System in the course of the Galactic year." Perhaps if we can shift our attention from the ground under our feet to the sky, answers to some of these questions may come from observations made through the Hubble telescope.

Some concerns about following "accepted wisdom" are expressed by Dickins in a brief note in Chapter 14. Was the Gondwana supercontinent indeed as massive throughout the Permian as believed? How did the various ice sheets or glaciers relate to each other? Dickins also suggests that extraterrestrial influences my explain some the major climatic variations on Earth.

What about times older than the Carboniferous-Permian? Glaciations occurred during the Devonian and Ordovician and also in the Precambrian. Actually, it was in the upper Precambrian (Proterozoic) that some the largest glaciations enveloped the Earth's surface. This is the subject of Part V. Young, in Chapter 15, illustrates the occurrences of such older glaciations over most continents, pointing out their similarities to and differences from those of the Carboniferous-Permian and the Quaternary. For example, the Carboniferous-Permian and Proterozoic glaciations occurred during periods of continentality. The Tertiary and Quaternary glaciations, however, occurred when the continents were scattered, although the fact that the large continental mass of Antarctica is located at a pole may have been a factor. In every case, major glacial periods occurred during times of continental emergence and tectonic uplift, which may have contributed to a reduction in atmospheric CO_2.

In the last part of the book, a few methods are presented on how to recognize true, cold, terrestrial deposits. Disentangling climatic and tectonic effects on sediments and rocks is often difficult. Dutta, in Chapter 16, resurrects the concept that good mineralogical analyses can help in distinguishing cold climate conditions where minerals such as feldspar are little affected by chemical weathering. He presents convincing evidence for a very rapid transition from cold glacial and postglacial settings to warm conditions in India. Although affected by various factors, well developed paleosol record periods of stable land surface, little affected by erosion and sedimentation. Thus, Tarnocai, in Chapter 17, shows that paleosols can be powerful indicators of past climatic conditions. He illustrates this point with a series of examples from the Tertiary and the Quaternary, including examples from cold-climate settings. Bustin and Martini deal with peat and coal respectively in Chapters 18 and 19. The presence of such postglacial organic deposits is one of the features that distinguish Carboniferous-Permian and Quaternary glaciations from all others. One common effect of such deposits has been a sensible postglacial drawdown of CO_2, perhaps compensated for in part by emission of methane. Coals derive from peats. Modern peats form both in tropical and cold boreal to subpolar regions. Bustin, in Chapter 18, gives an overview on the geometry, evolution, and characteristics of modern peatlands and analyzes them as an analog for Carboniferous-Permian Gondwana coals. Can cold-climate peat precursors be differentiated from tropical precursors in Carboniferous-Permian Gondwana coals? It turns out that the obvious climatic information recorded in peat deposits may have been obscured not only by advanced organic metamorphism but indirectly by tectonic influences during the very early stages of formation of the peat deposits. Indeed, under rapid rates of basin subsidence, organic material that is rapidly inundated can be a precursor to vitrinite-rich coals. On the other hand, in slowly subsiding basins, peats may remain exposed for some time to oxidation and microbial decomposition before burial and thus, independent of the organic composition, become a precursor to inertinite-rich coal. Furthermore, Bustin suggests that little is known about the maceral precursors of cold-climate peats, so it is difficult to reconstruct the ancient peatlands settings from petrographic analysis of coal. Martini, in Chapter 19, starts from the same premises, but focuses on the fact that peat and coal are just features of the landscape or rock succession. Peat can be considered to be both a soil, and as such a carrier of considerable climatic information, and a sediment. In the latter case, a peatland is the terrestrial analog to a submarine reef. Peat changes into coal through a series of diagenetic and metamorphic processes. This metamorphism leads to transformation of organic matter into macerals (organic equivalent to inorganic metamorphic minerals) much more rapidly than other surrounding inorganic sediment and rocks. The higher the degree of metamorphism (rank of coal), the less information is retained of the original deposits. However, the associated rocks provide additional climatic and tectonic information such that paleoenviromental reconstructions can be attempted. This reveals that the Gondwana Carboniferous-Permian coals do not constitute a homogeneous population; instead, a few Carboniferous and the lowermost Permian coals derive from true cold-climate peatlands and all the others from temperate and warmer settings.

What can be learned from all this? For one thing, a synthesis has been presented of the two more recent, major glaciations on Earth. In terms of processes and methods, only few examples have been given to indicate that much remains to be learned about the behavior of glaciers and about how to recognize true cold deposits. Furthermore, the Carboniferous-Permian postglacial settings are more like present ones than any other. Much can be learned from this, although, as indicated by various authors in this book and elsewhere, it remains difficult to compare in detail the Quaternary events with the Carboniferous-Permian ones because of the longer duration of the Paleozoic glaciations, the different stratigraphic records biased toward basinal deposits in the ancient, different tectonic settings, imprecise dating of the ancient rock, and the different organisms. In any case, when the analyses of these periods are made and we extrapolate to ponder expectations for the future, it looks as though, in the short run, Earth may experience various climatic cycles and perhaps even warmer conditions than the present, but the crustal and tectonic elements of Earth and the extraterrestrial influences are still such that in the long run, in spite of human activities, a new glaciation is more likely than a change toward warm and arid conditions similar to those that developed during the Late Permian.

PART I

THE QUATERNARY: The Last Glaciation and Deglaciation

2. NORTHERN HEMISPHERE (LAURENTIDE) DEGLACIATION: PROCESSES AND RESPONSES OF ICE SHEET/OCEAN INTERACTIONS

John T. Andrews

Research done over the past one to two decades on the late Quaternary deglaciation of Northern Hemisphere ice sheets has produced a wealth of data that now enables glacial geologists to address some of the underlying events which have characterized glacial processes and responses during the last great Termination (Duplessy et al. 1981; Bard et al., 1987; Ruddiman 1987; Fairbanks 1989; Fairbanks et al. 1991). A considerable increase in our knowledge has come from traditional glacial geological and geomorphological studies based on terrestrial glacial and nonglacial sequences (Denton and Hughes 1981; Dyke and Prest 1987; Fulton, 1989). Critical insights into the history of the Northern Hemisphere late Quaternary ice sheets have also been gained from several other sources, including the study of marine sediment and fossil sequences (Bond et al. 1992; Broecker et al. 1992), and the forward and inverse modeling of ice sheet/isostatic interrelationships (Tushingham and Peltier 1991). By and large, terrestrial mapping enables researchers to delimit the former margins of ice sheets whereas research on the continental shelves and the deep sea, as well as on glacial isostasy, adds information on the thickness and the volume of the ice sheets and the effect of ice-sheet history on global climate (Broecker and Denton, 1989).

From the perspective of the geological record of glaciations, one of the crucial lessons that might be taken from late Quaternary studies is the speed at which the glacial/ocean/atmosphere system can interact. There is also an increasing body of evidence that suggests that certain Northern Hemisphere ice sheets were extremely dynamic, not only in terms of Milankovitch time scales (20, 40, and 100 ka periodicities) (Imbrie et al. 1992) but on shorter time scales (ca. ≤10 ka) (Bond et al. 1992).

This chapter presents an overview of the late Quaternary deglaciation of the Northern Hemisphere, paying particular attention to the rates of change and to the nature of the physical evidence that supports the notion of a highly dynamic cryosphere/atmosphere/ocean linked system. These lessons from the recent past have relevance for the older records of continental-scale glaciation.

Nature of the Northern Hemisphere Ice Sheets

Our knowledge of the ice-sheet systems in the Northern Hemisphere was initially based on terrestrial glacial geological studies. However, we now increasingly rely on information and evidence from the glaciated continental margins and from the deep sea to understand better and date the ice/ocean/atmosphere interactions that appear to typify the Quaternary. This section outlines a number of important concepts that need to be kept in mind.

Denton and Hughes (1981), among others, drew attention to the potential differences in response to climate forcing between terrestrial ice sheets and marine-based ice sheets (Fig. 2.1). The present Greenland Ice Sheet is a model of the former system and is represented in the late Quaternary by the Cordilleran Ice Sheet of North America (Clague, 1989). Such ice sheets have beds, even allowing for glacial isostatic compensation, are primarily above sea level and/or have only limited connections between areas below sea level and the perimeter of the ice sheet. In contrast, marine-based ice sheets have large areas that are below sea level; these areas are linked via large channels to surrounding seaways. A present example is the West Antarctic Ice Sheet (Bindschadler, 1991), whereas late Quaternary

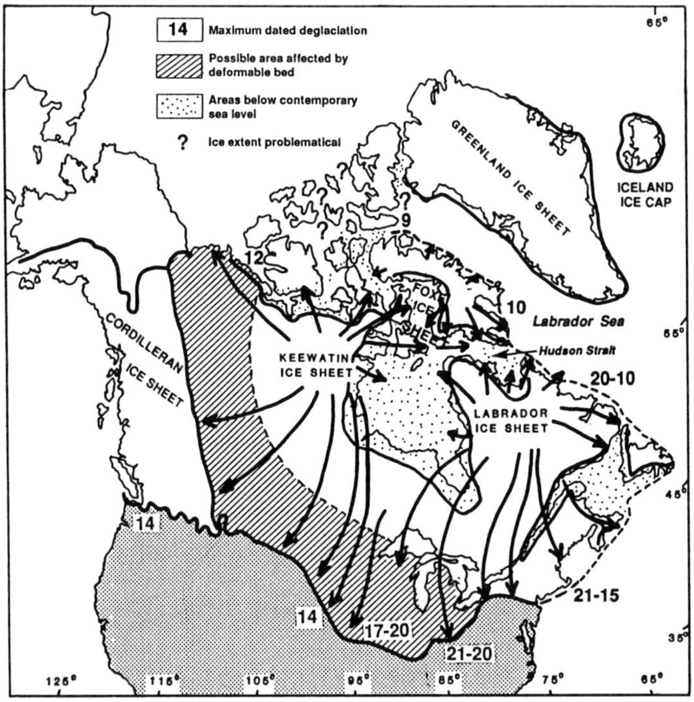

Fig. 2.1. Extent of ice sheets in northern North America during the last glaciation showing probable dispersal centers, major directions of ice flow, marine-based portions, areas with possible soft-bed deformation, and maximum dated evidence on deglaciation (after Dyke and Prest, 1987).

examples include the Laurentide Ice Sheet ofNorth America and the Fennoscandian Ice Sheet (Denton and Hughes, 1981). Whereas terrestrial ice sheets respond to changes in air temperature and the amount of precipitation falling on the ice sheet; that is, changes in their mass balances are associated directly with climatic controls, the dynamics of marine-based ice sheets are controlled by processes that affect the rate of calving of their ice fronts, which terminate in water (Thomas 1977; Hughes 1987, 1992). These can be associated with climatic factors, but they are also strongly affected by changes in relative sea level, physical oceanography, basal thermal regime of the ice sheet (MacAyeal 1992), and the nature of the bed (Boulton et al., 1985; Fisher et al., 1985; Alley 1991; Clark 1994).

These in turn may not be simple processes. For example, an increase in relative sea level at the front of an ice-sheet outlet might be attributed to a global rise in sea level associated with the melting of other ice sheets, but it might also be caused by delayed glacial isostatic depression associated with the mass growth of the ice sheet itself (Andrews, 1987a). If a grounded ice sheet advances across a continental shelf, the visco-elastic response to this increase in loading will take hundreds to thousands of years to reach equilibrium, since the rate of calving is a power function of water depth (Brown et al. 1982). This might be an example of a self-limiting processes unassociated with climate.

The apparently clear indications from the deep-sea faunal and geochemical records that the Quaternary ice ages are paced by variations in the Milankovitch insolation rhythms has, I think, been too easily accepted by those of us who deal with parts of specific ice sheets. The deep-sea records are, by and large, globally integrated responses that hide, or may hide, important differences in the responses of different ice sheets or even sectors of the same ice sheet (Fig. 2.1).

Yet geologists working on the old record are dealing with segments of former ice sheets. If the recent past is indeed the key to the past, then we must be aware of the pitfalls involved in the temporal and spatial scales of glacial response. Examples of specific questions might be (1) How globally significant is the Younger Dryas interval of glacial re-advance, an event so well documented from northwest Europe, and (2) does the late Quaternary global late glacial maximum of marine isotope stage 2 imply that this was the glacial maximum for all ice sheets or for all parts of an ice sheet? Remember here that the Laurentide Ice Sheet stretched from around 45° N to 75° N, or a distance of about 3500 km, encompassing a present difference in climate (mean annual temperature) from ca. +15°C to -15°C.

Differences in the surface climate of an ice sheet affect the dynamics of ice sheet behavior and also exercise a control on the thermal regime of the ice/bed interface (Sugden 1977). It is the nature of this interface is of greatest concern for the glacial stratigrapher. By and large, many existing models of ice sheets are in reality drawn from studies on valley glaciers, mainly from temperate latitudes, where glacially transported sediment lies both supraglacially (from rockfall) and subglacially (from glacial erosion). However, polythermal bed conditions (Blatter and Hutter 1991) indicate that sediment supply, transportation, and deposition are intimately associated with changes in the thermal regime of the bed. Indeed, the presence of water at the bed of an ice sheet or glacier is clearly one of the most fundamental controls on glacier behavior (Boulton and Dobbie 1993; Iken et al. 1993).

Most terrestrial reconstructions of ice sheets have considered ice sheet beds to be relatively dry, but radar surveys of the Antarctic Ice Sheet (Drewry 1981) suggest that large areas of that ice sheet are underlain by large subglacial lakes. Their presence has led to suggestions that this ice sheet may respond rapidly to a change in basal thermal regime, for example, if a series of subglacial lakes became interconnected and the ice sheet thus increase its flow rate catastrophically the equilibrium profile of an ice sheet is maintained by the balance between friction at the bed and the weight of the overlying ice (Paterson 1981). Observations and theory have recently been advanced in support of a hypothesis that states that large subglacial lakes also existed beneath the North American Laurentide Ice Sheet (Shoemaker 1992a, 1992b). Although the interpretation of the field evidence is controversial, it does appear to be gaining converts and must be seriously considered. In addition, there is increasing evidence that profiles of the southern and western fringes of the Laurentide Ice Sheet were extremely low and flat, raising the possibility that the low basal shear stresses required to drive such lobes (Fig. 2.2) were the product of subglacial sediment deforming at stresses significantly lower than those needed to deform ice (Matthews 1974; Boulton and Jones 1979; Beget 1986, 1987; Alley 1991; Clark, 1994; Clark and Walder, 1994). The differences are typically of the order of 50–90 kPa (Fig. 2.2).

The present and late Quaternary distribution of large ice masses on the surface of the earth is also of importance for the older glacial records. On late Cenozoic time scales it is obvious that we have two different sensitivities to climate change. Although arguments continue on the presence or absence of a Greenland or West Antarctic Ice Sheet over the last interglaciation (ca. 125 ka), today we have a world dominated by two large ice sheets appear to have existed in some form or another for the last 1 Ma or longer. In contrast, there are many other ice sheets of comparable size (Table 2.1), that have waxed and waned during this same interval. Is there anything in the respective stratigraphic records of these various ice sheets that allows us to differentiate between these different styles of behavior? Note here that the entire duration of the late Cenozoic ice age in the Northern Hemisphere is within one standard deviation of dating errors for much of the older glacial records!

One additional point must be made in this section, and that is that the bulk of the late Quaternary glacial sediment record resides not on land but in the nearshore basins, shelf troughs, and adjacent deep-sea basins. The dominant product of glacial erosion appears to be silt- and clay-size sediments (Andrews 1990; Andrews and Syvitski in press). On land, this fraction of the glacially derived sediments are dispersed by wind from the ice sheet and glacier margins to form thick and extensive loess deposits, whereas

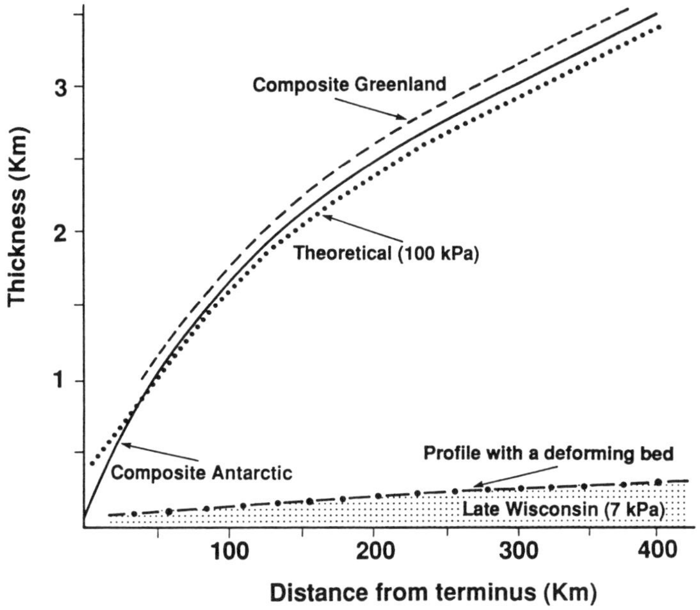

Fig. 2.2. Profiles of present, past, and a theoretical ice sheet with different basal shear stresses, showing the inferred thickness of some of the lobes around the southern margin of the Laurentide Ice Sheet that may have lain on deforming beds (after Beget, 1987).

around margins that terminate in the sea they are represented by thick sequences of fine-grained laminated to massive pebbly mudstones.

One final point concerns our ability to estimate accurately rates of change during deglaciation. Most of the chronology of the last termination can be assessed by radiocarbon dating. Recent comparisons of the association between radiocarbon years and those derived from U-series dating indicate that the discrepancy that existed between tree-ring (calendar) years and radiocarbon years up to about 9,000 ^{14}C years ago continued throughout marine isotope stage 2 between 13,000 and around 30,000 ^{14}C years ago (Fairbanks 1989; Bard, et al., 1990, 1993); thus radiocarbon years are "shorter" than calendar years, so rates of glacial retreat are compressed in the radiocarbon time scale. The late glacial maximum, for example, is radiocarbon dated at at around 18 ka, whereas on the insolation scale for 65°N (Fig. 2.3) the late glacial maximum occurred around 22 ka. With the calibration between ^{14}C and U-series ages this appears now to be the case (Fig. 2.3).

The Last Deglaciation

Our knowledge of the deglaciation history of the world's ice sheets has gained much from detailed mapping of continental glacial deposits (Fulton 1989), but it is equally true that a study of glacial marine facies on continental shelves (Syvitski 1993) and in the deep sea (Heinrich 1988; Piper et al. 1991) has added considerably to our insights into the timing and mechanisms of deglaciation. This essay discusses the evidence for the last deglaciation in a general sense but focuses specifically on the area of the ocean/ice sheet interface between the eastern sector of the Laurentide Ice Sheet and records from the Labrador Sea (Fig. 2.4) and the North Atlantic Ocean.

Last Glacial Maximum

The late glacial maximum based on the summer insolation curve for 65°N should have occurred at 22 cal ka (the term "cal" will be used to

TABLE 2.1. *Possible ice sheet volumes (x 10^6 km^3) during the late Quaternary (from Denton and Hughes 1981).*

Ice Sheet	Minimum ice volume	Maximum ice volume
Laurentide	30.5	34.8
Cordillera	0.26	1.84
Greenland	0.28	2.55[a]
Antarctica	9.81	9.81[a]
Scandinavian	7.06	7.32
Barents-Kara	0.87	6.25
Total (includes smaller ice masses not listed)	51.3	65.3
Sea level influence	127 m	163 m[b]

[a] Ice involved in deglaciation.
[b] To obtain equivalent sea level divide by area of the world oceans (316 x 10^6 km^3).
NB: Change of sea level based on Barbados data = 105 m (Tushingham and Peltier, 1991)denote "true" dates

derived from calibration with the U-series (Fairbanks, 1989; Bard et al. 1990, 1993), whereas a notation without "cal" refers to the radiocarbon age). The radiocarbon dates from around the southern (terrestrial) margin of the Laurentide Ice Sheet generally show that the late glacial maximum occurred between 18 and 20 ka (Mickelson et al.1983; Dyke and Prest 1987), but in areas associated with the marine portions of the Laurentide Ice Sheet (in the Canadian Arctic Channels, Arctic coast, and the eastern Canadian Arctic) (Fig. 2.1) dated retreat commenced significantly later, with maximum dates along the northern margin of the ice sheet around 12 ka (Fulton 1989). It is not known whether it is the marine or the "northern" element that is important in this difference in response.

The terrestrial evidence from the eastern Canadian Arctic of late glacial ice extent and for the timing of the late glacial maximum has been filled in since the 1981 synthesis of Denton and Hughes (Denton and Hughes 1981; Klassen and Shilts 1987; Fisher and Klassen 1988; Stravers, et al., 1992), but the attendant questions of minimum and maximum ice sheet extents still persist and have not yet been resolved. It is known from offshore marine evidence from some of the major troughs and channels that the extent of ice was probably greater than once thought (MacLean et al. 1989; Jennings 1993), but there is still little factual evidence to limit the extent of grounded ice along the eastern margin of the Laurentide Ice Sheet (Josenhans et al. 1986; Andrews et al., 1991; Piper et al. 1991). What is known is that the Laurentide Ice Sheet was still active along its eastern and northern margins until around 8 ka, well after the peak of the northern summer insolation maximum (Fig. 2.3). These late glacial episodes of pronounced glacial advances, even surges, have been ascribed to the effect of increased solid precipitation (Dyke and Prest 1987; Miller and Kaufman 1990).

Ice Sheet Instability: Detrital Carbonate (Heinrich) Events

The generally accepted view is that Quaternary ice sheet advances and retreats are driven by climatic forcings that were embedded within the changes in solar insolation on orbital time scales (Imbrie et al. 1992). However, the study of the ice core record from the Greenland Ice Sheet (Broecker et al. 1988; Dansgaard et al. 1989; Taylor et al. 1993) has demonstrated that atmospheric changes during the glacial/interglacial transition were extremely rapid and occurred at much higher frequencies than expected by insolation forcing.

Hughes and MacAyeal suggested (Hughes 1992; MacAyeal 1992) that there is an inherent instability in ice sheets associated with basal bed

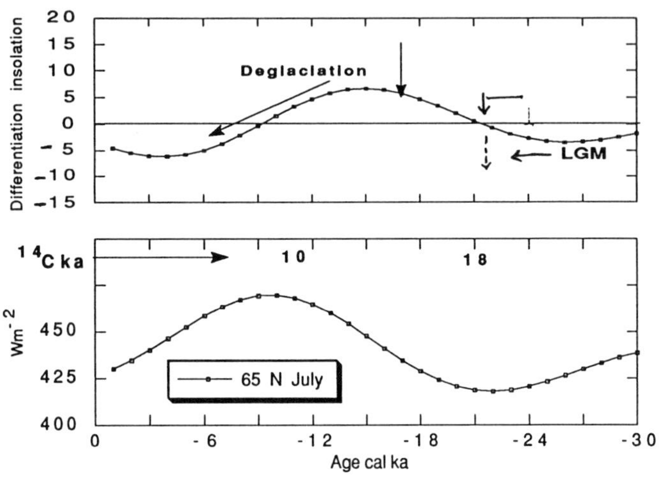

Fig. 2.3. (Top) Rate of change of incoming solar radiation at 65°N (Wm^2 ka). Note that on two radiocarbon dates (ka) are shown in comparison to cal ka; (Bottom) Insolation receipts at 65° N in July (Wm^{-2})(Berger and Loutre, 1991).

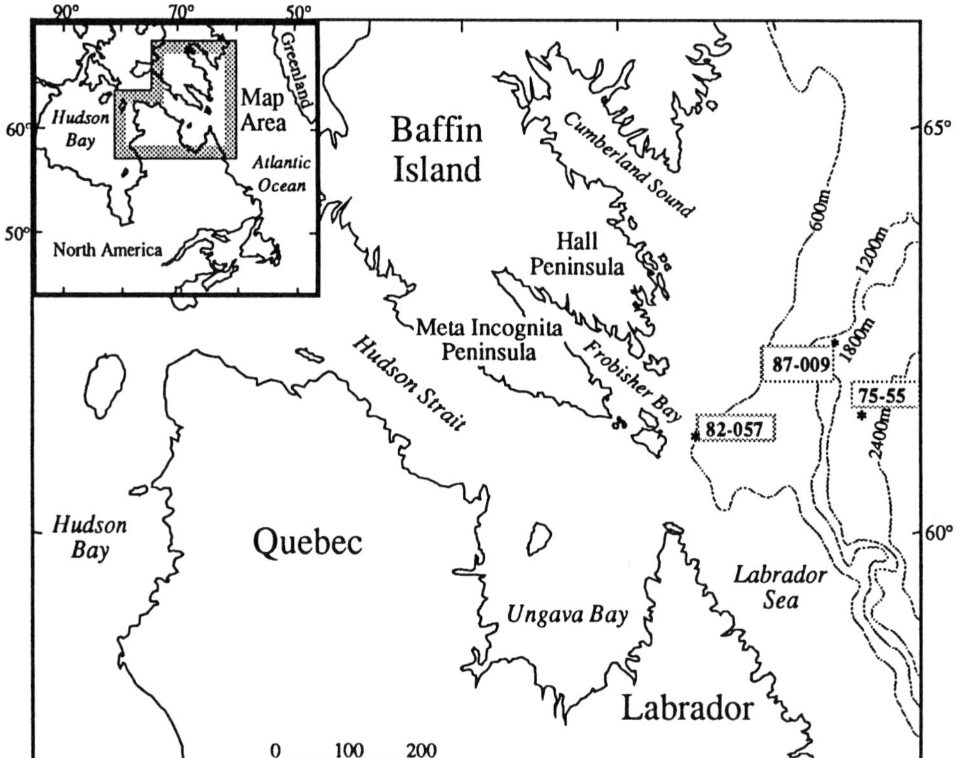

Fig. 2.4. Bathymetry and location of the cores outside the Hudson Strait trough discussed in this chapter. Data from these cores are illustrated in Figures 2.7, 2.9, and 2.11.

conditions underlying ice streams; that is, zones of fast-moving ice (km/a) embedded within a more normally deforming ice mass (Figs. 2.5, 2.6). The ice streams' high velocities might be associated with deformation of underlying "soft" sediments (Alley 1991; Clark 1991), or they might represent enhanced velocities caused by high water pressures at the bed of a rigid ice sheet. MacAyeal (1992) suggested that a quasi-periodicity for ice sheet collapse could be forecast for a model that allowed for changes in basal temperatures associated with periods of ice sheet growth and collapse. During intervals when the ice sheet was thickening the bed was cold and frozen, but at some critical juncture temperatures reached the pressure melting point and a short interval of rapid ice stream motion ensued (Figs. 2.5, 2.6). It is not yet clear which of the two alternative scenarios for the behavior of the ice sheet accurately describes the glaciological conditions during periods of ice-sheet instability (Figs. 2.5, 2.6), and indeed other scenarios might be applicable.

The physical evidence for some kind of high-frequency (≤10± ka) behavior of the Laurentide Ice Sheet has recently been detected from suites of cores stretching from just east of Hudson Strait (Fig. 2.4) all the way south and east to off the coasts of Newfoundland and Spain (Heinrich 1988; Andrews and Tedesco 1992; Bond et al. 1992; Grousset et al. 1993). "Heinrich-events" (H-) are marked by intervals of increased lithic abundances, an increase in sand-size carbonate fragments, and a marked decrease in planktonic foraminifera (Ruddiman and McIntyre 1981; Broecker et al. 1992). In the northwestern Labrador Sea they appear as distinct creamy-yellow layers composed of up to 60 percent detrital carbonate by weight (Andrews and Tedesco 1992). They appear to represent intervals when large amounts of Paleozoic carbonate-rich sediments from the floors of Hudson Strait and portions of the southeastern Baffin Shelf were stripped by glacial action. Because these sites are distal to the shelf edge and the maximum possible limit of glaciation, it is thought that Heinrich events represent periods of production of an enormous number of icebergs (Broecker et al. 1992; Andrews et al. in press).

Figure 2.7 shows the detrital carbonate and <125 μm weight percents for core HU87-033009 from the slope off southeastern Baffin Island in 1.5 km water depth (Tedesco 1993). Radiocarbon dates are on mg samples of foraminifera; they define four detrital carbonate (DC-) events which have been named DC-0,-1,-2, and -3. Andrews and Tedesco (1992) showed that DC-1 and -2 were coeval with H-1 and H-2 in the North Atlantic (Bond et al. 1992). These two events are ^{14}C dated at around 14 and 20 ka; DC-0 occurred during the latter part of the Younger Dryas chronozone, and DC-3 occurred shortly before 33 ka. In the type section for Heinrich events in the eastern North Atlantic (Heinrich 1988), H-4, -5, and -6 are not accurately dated but have approximate dates of 40, 50, and 60 ka (Bond et al. 1992; Broecker et al. 1992). There is at the moment a possible discrepancy in either the age and/or mineralogy of H-3/DC-3. H-3 is dated close to 28 ka in the North Atlantic (Bond et al. 1993), but Grousset et al. (1993) presented evidence that this unit lacked carbonate; hence the correlation with DC-3 is unclear.

What is not yet clear is whether the Heinrich events that are associated with discharge of icebergs, sediment, and meltwater from Hudson Strait (Figs. 2.1, 2.4) are associated with a massive calving event and a dramatic retreat and drawdown of the ice sheet (Fig. 2.6A,B) or whether they represent a surge of an ice stream across the shelf to the shelf break (Fig. 2.6C,D). The concept of the ice sheet being gutted by calving (Denton and Hughes 1981) appears to be a realistic model for the final collapse of the Laurentide Ice Sheet ca. 8 ka (Andrews 1987b), although this may have been driven by the northward expansion of proglacial lakes in the southern portion of Hudson Bay. A possible test for these alternative models lies in the bedrock geology that crops out from Hudson Strait to the shelf break (Fig. 2.5). Notice that the Paleozoic carbonate outcrop occurs west of the Hudson Strait sill; thus the DC-events in cores from the northwestern Labrador Sea probably represent erosion of this outcrop. However, in the first scenario (Fig. 2.5) the ice extends to the shelf break where the underlying bedrock is Tertiary offshore sediments. Smectite is a tracer clay mineral for this outcrop (Andrews et al. 1989), but this species is virtually absent from the DC-

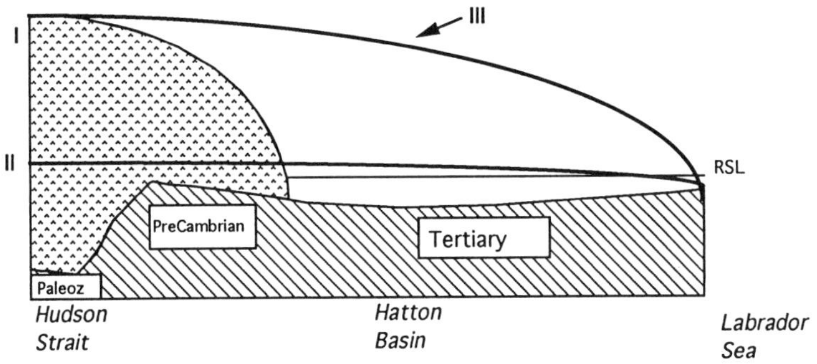

Fig. 2.5. Schematic diagram showing the ice sheet filling Hudson Strait and grounded on the inner shelf (Position I). Two possible alternatives for the form of the ice sheet during a subsequent advance are shown (see Fig. 2.6 for plan view). In Position II the ice has surged across the shelf as a thin ice lobe lying on a deforming bed, whereas in Position III the ice has advanced across the shelf to the shelf break, keeping the original (Position I) basal shear stress.

units at HU75-55 and HU87-009, at least when analyzed without removal of the carbonate. Smectite is, however, a major component of the clay-mineral composition in the inter-DC units. The presence or absence of smectite in offshore cores tells us something about the viability of the two scenarios (Figs. 2.5, 2.6), but exactly what is implied is still being actively investigated.

The probable link between these intervals of massive iceberg discharge and climate is the effect that meltwater has on the thermohaline circulation of the North Atlantic (Broecker and Denton 1989; Stocker and Wright 1991; Stocker et al. 1992). A cap of low-density water on the North Atlantic would curtail or displace the formation of North Atlantic Deep Water (NADW) to the south. This would dramatically cool temperatures in the North Atlantic and have a dramatic climatic impact on northwestern Europe. There is an analog for these conditions in the "Great Salinity Anomaly" of the 1960s on which reduced deep-water formation in the Labrador Sea (Dickenson et al. 1988).

Heinrich events probably stemmed from the marine-based portion of the Laurentide Ice Sheet, which has an area of around 3 x 10^6 km^2. However, the concept of deformable beds and a more dynamic Laurentide Ice Sheet has become much more popular recently and thus there is now serious doubt as to whether the Greenland Ice Sheet and the East Antarctic ice sheets are suitable analogs for the Laurentide Ice Sheet in terms of their volume/area ratio. Stratigraphic and dating studies around portions of the southern margin of the Laurentide Ice Sheet suggest that this margin was highly mobile during deglaciation, with short-lived intervals of retreat followed by short intervals of extensive re-advance (Mickelson et al. 1983; Dyke and Prest 1987). In the region of the southern and northern Great Lakes, Mickelson et al. (1983) document significant glacial advances about once every thousand years between 21 and ca. 10 ka. Various lines of evidence indicate that many of these ice lobes were thin, with low basal shear stresses (Matthews 1974; Beget 1987; Clark 1992), thus supporting various hypotheses suggesting that the Laurentide Ice Sheet had a southern fringe of ice which was thin and that was controlled by the strength of easily deformed basal sediments (Boulton et al. 1985; Fisher et al. 1985; Andrews 1987; Clark, 1994a,b).

Furthermore, it should not be forgotten that large sections of the southern and western margins of the Laurentide Ice Sheet terminated in proglacial lakes (Teller 1987), which probably had a profound influence on deglacial processes and weakened some of the distinctions between marine-based and terrestrial portions of ice sheets (Andrews 1973).

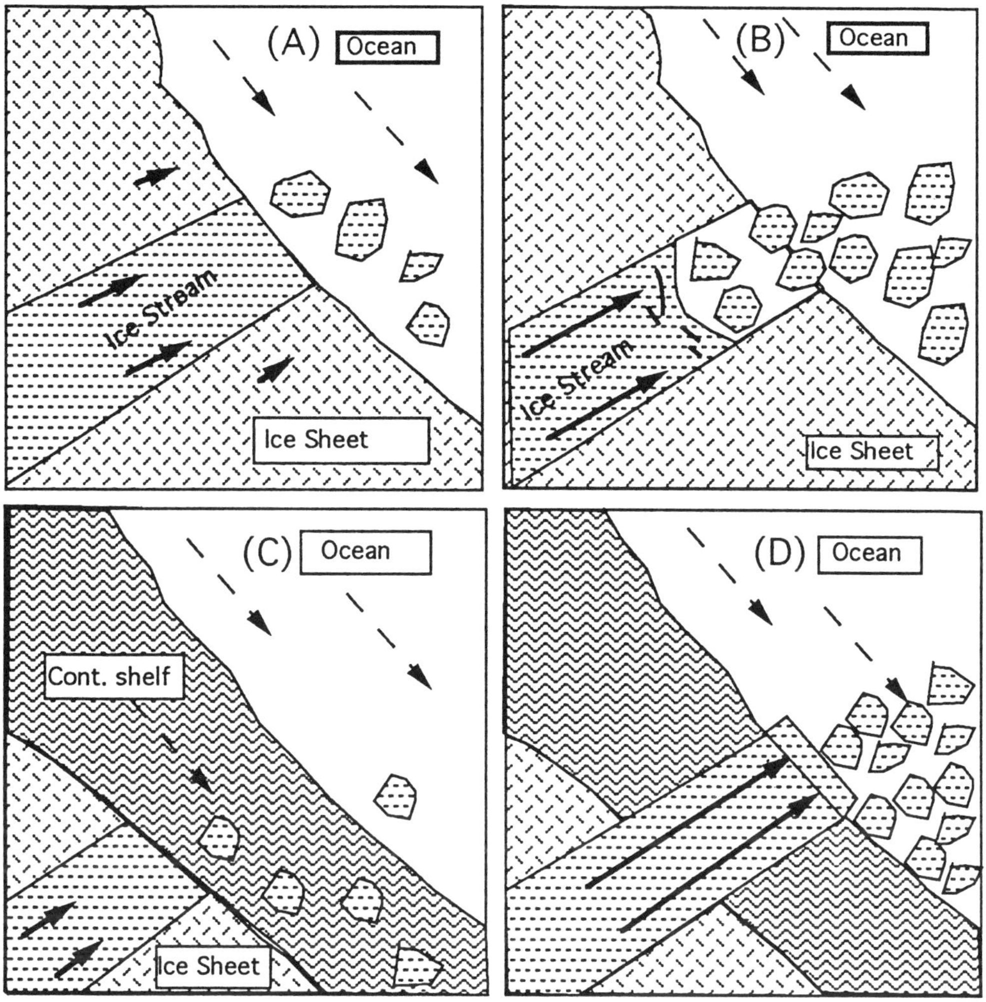

Fig. 2.6. Two possible scenarios in Hudson Strait and adjacent shelf leading to iceberg outbursts from an ice stream within Hudson Strait (see Figure. 2.5 for cross-section view). In the upper panels (A, B), the before-and-after sketches show ice retreat via a massive calving bay. In the lower panels (C, D) the ice margin is situated on the inner-shelf, and shelf and during Heinrich events the ice advances as a tongue across the shelf to the shelf break.

Deglaciation: View from Ice-Proximal/Ice-Distal Marine Locations

Terrestrial sites adjacent to ice margins can reflect details of the pattern of deglaciation, but such sites are frequently characterized by episodic sedimentation, so it is difficult to develop a complete time-series of environmental changes during the deglacial cycle. However, the Northern Hemisphere ice sheets occupied all or parts of the surrounding continental shelves during the maximum of the last glaciation (Denton and Hughes 1981; Fulton 1989; Piper et al. 1991); basins and troughs on these glaciated continental margins have thick sediment sequences (Josenhans et al. 1986; Praeg et al. 1986; Andrews et al. 1991) that retain much information on changes in ice extent and environment during deglaciation. In the last decade such evidence has been extracted from the vicinity of the northwestern European ice sheets, the Greenland Ice Sheet, and sections of the North American Ice Sheet (Vorren et

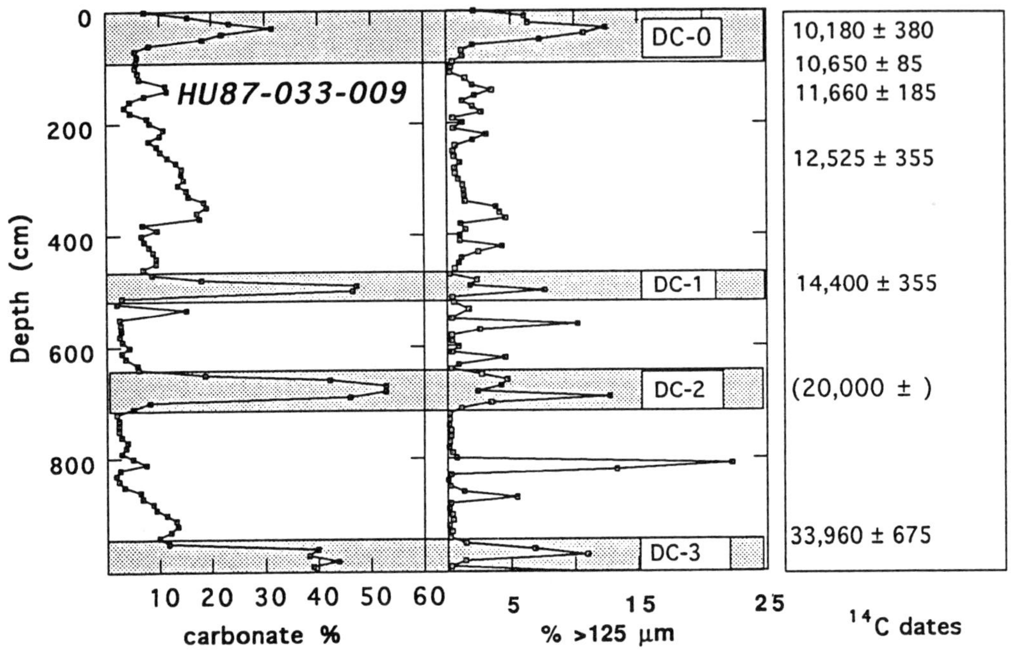

Fig. 2.7. Core HU87-009 from the slope off SE Baffin Island. The downcore plots for carbonate % (=total carbonate weight percent), %> 125± m (weight percent), and AMS ^{14}C dates (right column). The date in brackets is estimated from correlations with HU75-55. Shaded areas represent detrital carbonate events.

al.1983; Hald and Vorren 1987; Andrews, et al., 1990, 1991; Hein and Mudie 1991; Hald and Steinsund, 1992; Mienert et al. 1992). These investigations have used a multivariate approach to describing the environment during deglaciation, including sediment- and bio-facies and stable isotopic analyses of both benthic and planktonic foraminifera (Hald and Vorren 1987; Lehman et al. 1991).

As we shall see, these studies give much detail about regional changes in ice-sheet history, but it is important to compare this information with a global integration (if one is available) of the addition of meltwater to the world's oceans. In the past this has been attempted changes in the δ^{18}O of planktonic and benthic foraminifera (Shackleton and Opdyke 1973; Mix and Ruddiman 1985; Mix 1987) from sites in the subtropical or tropical open ocean; that is, sites well removed from the direct influence of glacial meltwaters. However, such studies had trouble deriving an absolute transfer coefficient for the conversion of a change in the δ^{18}O to an equivalent rise in sea level. However, a generally agreed on figure was around 0.11‰ δ^{18}O = 10 m change in sea level.

Studies of high-resolution deep-sea cores with adequate chronological control suggested that the globally integrated process of deglaciation did not occur uniformly but was marked by rapid changes in the flux of meltwater to the oceans (Duplessy et al. 1981; Bard et al. 1987; Ruddiman, 1987). Two specific steps or increases in the rate of deglaciation were deduced from the stable isotopic records on foraminifera (Duplessy et al. 1981) that were bracketed by AMS ^{14}C dates as occurring between 14.8 and 13.6 ka, Termination IA, and around 8.96 ka, Termination IB. Because the oceans tend to mix the surface isotopic signal, there has been discussion on the location(s) of the meltwater sources that contributed to these sharp changes in the near-surface water isotopic composition. The 14.5± ka event was until recently attributed to the disintegration of the Barents and Fennoscandinavian ice sheets (Jones and Keigwin 1988), but it has also been shown to be a major event in the Labrador Sea (Andrews et al. 1994) and hence associated with deglaciation of the Laurentide Ice Sheet as well.

Fairbanks (1989) has been able to measure directly the rise in world sea level at Barbados, a Caribbean island somewhat removed from the influence of glacial isostatic recovery (Tushingham and Peltier 1991). By coring through the coral reefs and then dating the samples by both ^{14}C and U-series, Fairbanks was able to break the circularity of the previous arguments and show that the mass of ice on the face of the planet over the last 17 ka was equivalent to a 125 ± 5 m change in sea level (Fairbanks 1989). When allowance is made for hydro-isostatic loading, the appropriate global sea level is around 105 m (Tushingham and Peltier 1991) (Table 2.1). This figure falls close to the minimum estimate calculated by Denton and Hughes (1981) (Table 2.1). However, these calculations were made using an ice sheet with a high basal yield stress (Fig. 2.2); as pointed out by Fisher et al.(1985), the influence of a deforming bed (Fig. 2.2) would be to reduce the volume of ice contained within the Laurentide Ice Sheet (Table 2.1).

The sea-level curve produced by Fairbanks (1989) shows two intervals of rapid rise (Fig. 2.8A); if this curve is differentiated to obtain a graph of the rates of change, then it is clear that the globally integrated process of deglaciation was not linear but could be described by two intervals of rapid change in sea level (Fig. 2.8B; Fairbanks 1989). Because these figures relate to changes in sea level, they must translate to massive and rapid changes in the volumes of the ice sheets, not just a retreat of the margins of ice sheets. As noted by Fairbanks (1989), the pattern of global sea level gives significant insights into the globally integrated response of the ice sheets, but it does not *a priori* mean that all ice sheets, or all margins of an ice sheet, were involved in both major intervals of melting. There is also some discrepancy between the timing of the two-step model of Duplessy et al. (1981) and the changes in sea level derived by Fairbanks (1989).

Thus, the next important question is: What technique or method allows the regional-scale responses to be extracted from the global integration? An integrated measure of the history of the eastern, marine-based sector of the Laurentide Ice Sheet can be gained from an examination of the sediment and meltwater fluxes at sites close to the margin of the ice sheet in the vicinity of the northwestern Labrador Sea (Figs. 2.4, 2.7, 2.9). Many of the cores that had been used previously to decipher and interpret the glacial record of the Northern Hemisphere had come from deep-sea abyssal sites and from locations well removed from the glaciated continental margins. More recently, attention has shifted to records from the glaciated continental margins and from adjacent slope/deep-sea sites (Hald and Vorren, 1987; Lehman et al. 1991; Andrews et al. 1991, 1994; Hillaire-Marcel et al. in press; Mienert et al. 1992).

Figure 2.9 shows the stable $\delta^{18}O$ from two cores (HU75-55 and HU87-009) from sites close to the major outlet of the Laurentide Ice Sheet into the North Atlantic; that is, the large channel of Hudson Strait that connected the interior of the Laurentide Ice Sheet with the Labrador Sea and the North Atlantic. Most researchers agree that this route drained about one quarter to one third of the basal area of the Laurentide Ice Sheet, some 3 to 4 x 10^6 km^2, with a steady-state discharge of 500–1000 km^3 of ice per year, about two to four times the present iceberg flux from the entire Greenland Ice Sheet (Reeh 1984, 1985). Thus, sites close to the margin of the Laurentide Ice Sheet, such as HU75-55 and HU87-009 (Fig. 2.4), would be well placed to record changes in sediment and meltwater fluxes and in sediment provenance during the deglacial cycle. Over the last 30 cal ka, summer solar insolation near the latitude of Hudson Strait has fallen to the late glacial maximum (solar minimum) values and then increased (Fig. 2.3). The long-term trend embodied in the $\delta^{18}O$ records follows this trend but with significant higher frequency events superimposed on this Milankovitch forcing. Isotopically the late glacial maximum occurs between 18 and 22 cal ka (Fig. 2.9), although this interval is broken by a pronounced light isotopic event. However, at about 17 cal ka there is a very pronounced change in both cores, with a rapid decrease in the $\delta^{18}O$ of near-surface planktonic foraminifera (Andrews et al. 1994) from about 4.6‰ to 2.4‰. A sharp, light isotopic event is delimited that subsequently varies around 3‰. I interpret this as a pronounced decrease in the surface salinity of the northwestern Labrador Sea, although it is not certain

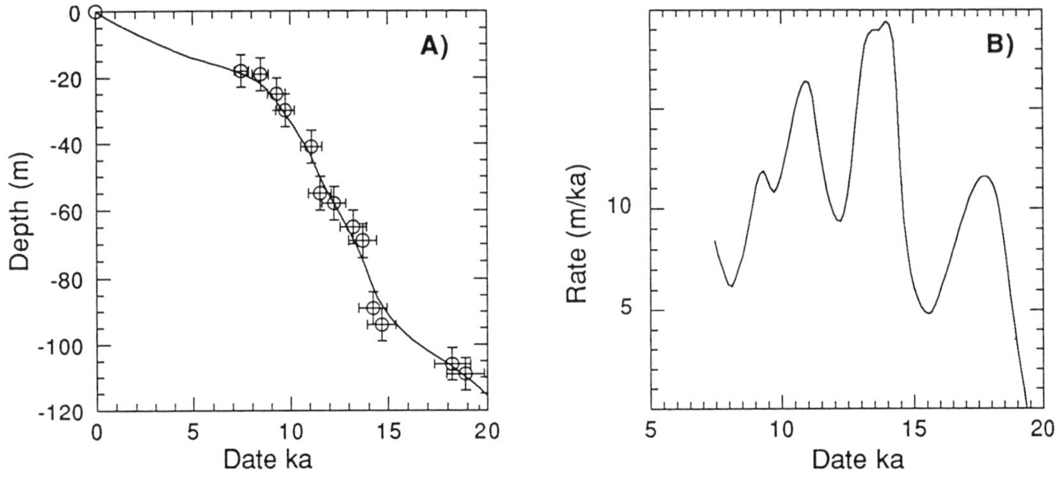

Fig. 2.8. A. Global sea-level rise from Barbados (Fairbanks, 1989), plotted on the U-series time scale; B. 1st derivative of the 3-point smoothed curve showing changes in the rate in rise of sea level during the last deglaciation (m/ka).

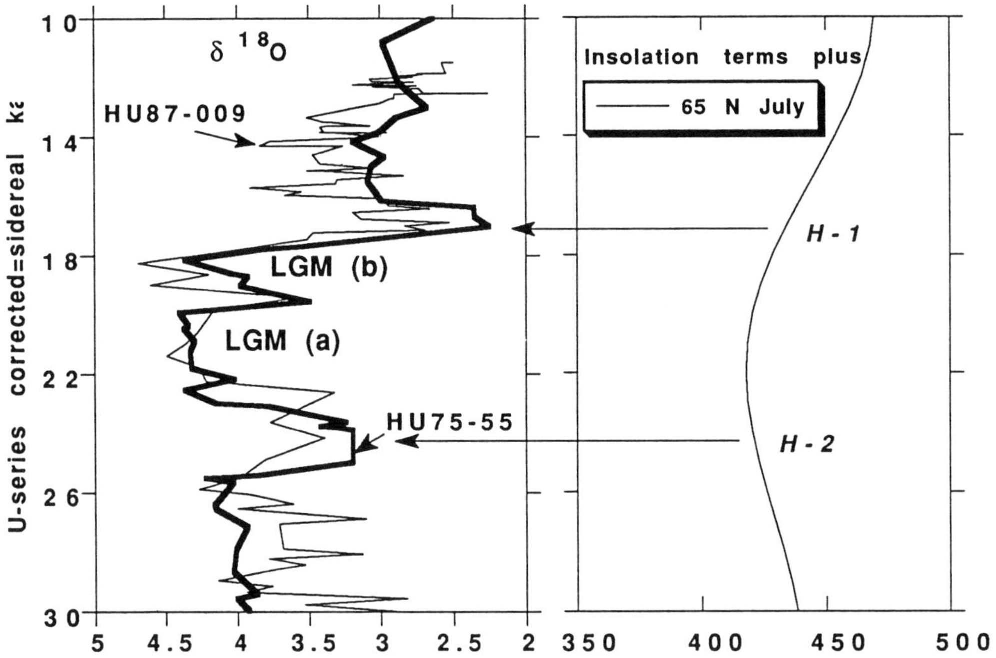

Fig. 2.9. Stable isotopic records from the eastern margin of the Laurentide Ice Sheet (see Figure 2.4 for location) showing changes in $\delta^{18}O$ of planktonic foraminifera against a U-series corrected time scale (cal ka). These records are plotted against the 65° N insolation curve, and the cal ka ages of H-1 and H-2 (DC-1 and DC 2, Fig. 2.7) are shown.

whether this is caused by the release of meltwater from the ice sheet *per se*, or whether it is related to the melting of large quantities of icebergs (Jenkins, in press; Labeyrie, et al., 1986) related to Heinrich event H-1 (Fig. 2.7). No datable terrestrial evidence from southeastern Baffin Island is available for this interval.

Isotopic studies on other cores from the eastern North Atlantic had previously detected a major light isotopic event close to 14 ka; this had been attributed to deglaciation from the Norwegian and Barents/Spitsbergen margins (Jones and Keigwin 1988; Keigwin et al. 1991; Weinelt et al., 1991), the data from close to the margin of the Laurentide Ice Sheet (Fig. 2.9) show that the Laurentide Ice Sheet also underwent significant retreat at the same time. AMS ^{14}C dated cores from the southeastern Baffin Shelf indicate that the ice sheet retreated rapidly to the inner shelf between 14 and 12 ka (Andrews et al. 1991, 1994).

Figure 2.10 compares the high-resolution stable isotopic records from the Canadian and Scandinavian margins (Lehman et al. 1991; Andrews et al. 1994) on a common time scale and compares these with a measure of the globally integrated change in global sea level (Mix and Ruddiman 1985). However, when these curves are compared to the Barbados sea level record the 14± ka melting record is not noted as a major shift in sea level. This suggests that the 14± ka event (ca. 17 cal ka) reflects a change in area but not in volume; this might indicate that the deglaciation was primarily of shelf-based ice, which would have had a minimal effect on global sea level. This despite the fact that it coincided with, or was caused by, a massive iceberg-rafting event H-1 (Fig. 2.7) (Andrews and Tedesco 1992; Bond et al. 1992). It thus appears that the 14 ka light isotope event (H-1) reflects the simultaneous deglaciation of shelf-based, ice sheet margins around the North Atlantic. What is also interesting about this comparison is the fact that the two isotopic curves converge toward the start of the Younger Dryas cool interval, thus implying that greater cooling took place in the NE North Atlantic then in the northwestern North Atlantic (Fig. 2.10).

What triggered this event is not yet clear. It may have been a strictly climatic response, or it may reflect an uncoupling of grounded ice-sheet margins from their beds associated with changes in relative sea level at the grounding line. Although the term "grounding line" is used, it appears probable that Northern Hemisphere ice sheets were not fringed by large ice shelves such as those typical of the current Antarctic continent (Oldale et al. 1990).

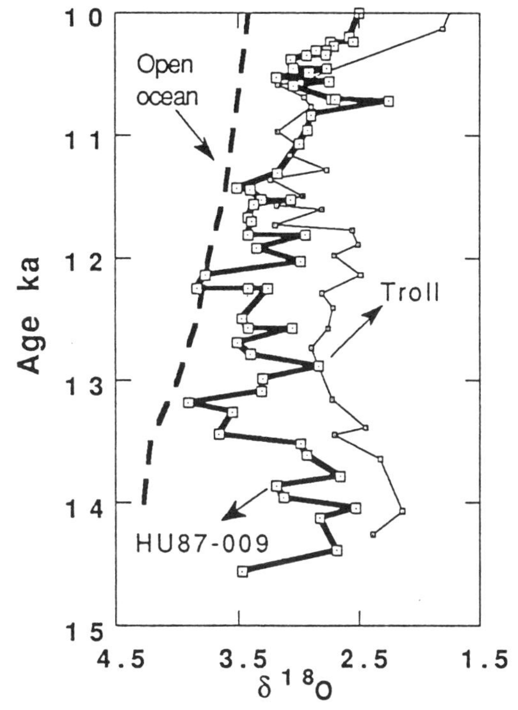

Fig. 2.10. Stable $\delta^{18}O$ records from planktonic foraminifera plotted from a core off Norway (Troll, Lehman et al., 1991) and for HU87-009 off Hudson Strait (Hillaire-Marcel et al., in press; Andrews et al. 1994). Both data sets have been interpolated between radiocarbon dates for a time scale. Note that HU87-009 is usually colder than Troll (heavier $\delta^{18}O$), but the records converge between 11.5 and 10.5 ka.

Studies on the deglaciation of the northeastern Canadian shelf must be combined with those from the adjacent slope and deep sea to judge properly the history of late Quaternary glacial conditions. Cores from basins on the southeastern Baffin Shelf, affected by ice streaming seaward from Hudson Strait, record a complex late glacial record (Praeg et al. 1986; Josenhans et al. 1986; Andrews et al., 1990, 1991). As an example, I show data from HU87-057 from Resolution Basin (Fig. 2.11), a small enclosed basin on the southeastern Baffin Shelf, NE of Hudson Strait at about 580 m water depth. Sedimentation on the shelf ceased during the early Holocene, since deglaciation of Hudson Bay and Hudson Strait was completed by about 8 ka. A series of AMS ^{14}C dates (Fig. 2.11) reveals that the core extends from about 7.8 to 24.7 ± 0.42 ka. There is one dating reversal which probably represents reworking of glacial marine sediments during a glacial advance. The total carbonate curve shows large-scale variations, which represents primarily changes in the calcite fraction, since dolomite remains nearly constant (Fig. 2.11). At 325 cm there is evidence of overriding ice in the form of a loading cast; dates either side of this unconformity differ by around 800 yrs. One explanation for the interval of nondeposition bracketed by these dates is that this unconformity may represent ice advance across Resolution Basin during DC-1/H-1. Toward the base of the core HU82-057 (Fig. 2.11) there are two dates that differ by nearly 9 ka over 10–15 cm, and I have tentatively indicated that this might represent a unconformity associated with DC-2 (Figs. 2.4, 2.11). The AMS ^{14}C date at 325 cm indicates that ice had retreated landward of Resolution Basin ≤13.2 ka; thus, during the Bölling/Alleröd interval the ice was distal to the core site and may indeed have retreated some way into Hudson Strait. However, sometime close to 11.2 ka the re-advance of the ice is recorded by an increase in the detrital carbonate, a decrease in the magnetic susceptibility, and an increase in the rate of sediment accumulation (Fig. 2.11). The Younger Dryas chronozone at this site is marked by a two-pronged carbonate spike with an intervening interval when total carbonate was only about 20 percent by weight. During this inter-Younger Dryas interval, laminated clays were deposited in Resolution Basin, possibly reflecting deposition beneath an ice shelf or near-permanent ice cover (Andrews et al. 1990). The carbonate peak in HU82-057 is approximately coeval with DC-0 in HU87-009 (Fig. 2.7), but mineralogically they represent different sources or processes, as DC-0 in HU87-009 is primarily dolomitic. The hypothesis that changes in total carbonate weight by percent are associated with glacial activity and ocean surface productivity is supported by the close relationship between this curve and that of changes in the $\delta^{18}O$ of planktonic foraminifera, with the $\delta^{18}O$ records showing lighter values (more iceproximal conditions) during the interval of ice advance ≤11 ka (Fig. 2.11).

Conclusion

Our insights into the deglacial processes of the last glacial/interglacial cycle have increased markedly over the last one to two decades. In no small measure, these increases have been associated with technological advances (for example, AMS ^{14}C dating), with the wealth of data being provided from deep-sea and shelf sediment cores and from a better understanding of how large ice sheets respond to changes in both surface and bed conditions. Modeling of both complex glaciological and ice sheet/ocean interaction systems have also been important in providing concepts and theories that can be tested against real field data. It is clear that the relatively simple models of deglacial processes that existed twenty years ago are now no longer valid. During the last deglaciation, former Northern Hemisphere ice sheets were much more dynamic than hitherto realized. Even during the height of the glaciation the occurrence of H-2 at about 202 ka, suggests that the ice sheet did not build slowly to an equilibrium profile. During deglaciation the ice sheets did not simply retreat in response to the increase in solar insolation but had intervals of retreat punctuated by major intervals of re-advance. It is not yet clear how far these were synchronous on the hemisphere or global levels. There are for example, questions about the global significance of H-1 or the Younger Dryas event and the synchroneity of the maximum extent of ice sheet margins.

It is also worth noting that despite all our efforts, we still do not know unequivocally the distribution of ice on the planet during the last glaciation! This is best brought out by the work on Tushingham and Peltier (Tushingham and Peltier 1991), who call for a 25 m sea level input into the world oceans from Antarctica. However, the evidence for this on the Antarctic continent is ambiguous (Andrews 1992; Colhoun et al. 1992); meanwhile, the extent and the chronology of ice sheets on the Eurasian shelf are still not resolved

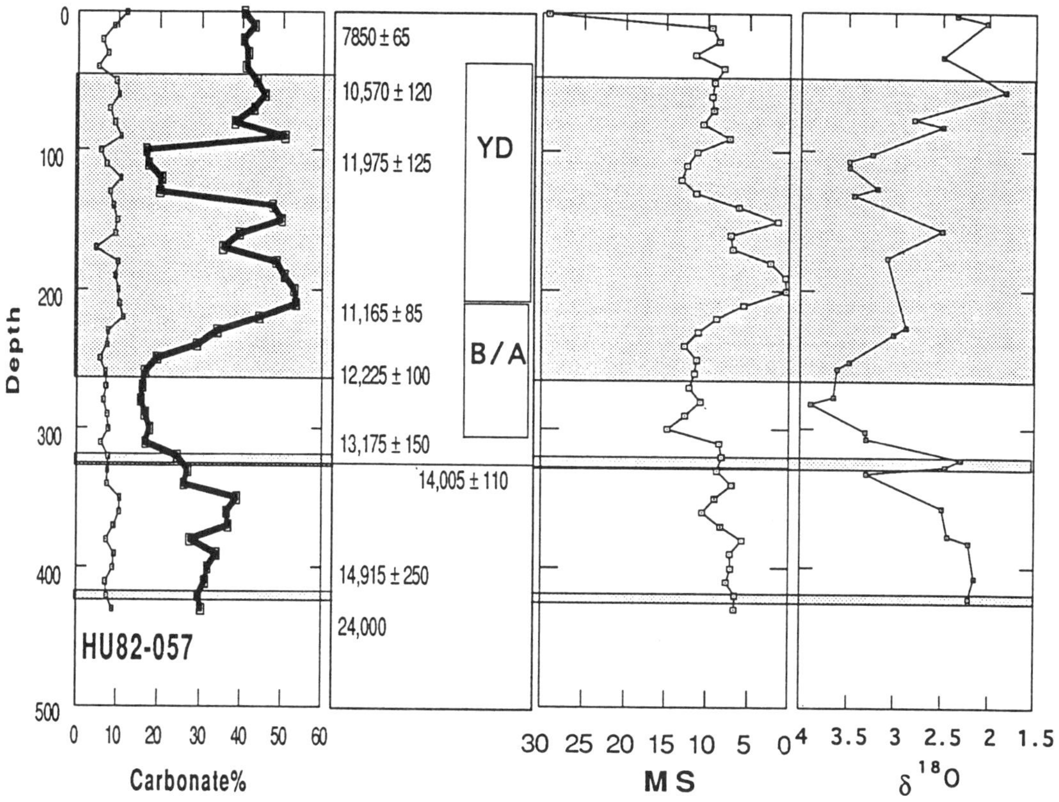

Fig. 2.11. Records from HU82-057, Resolution Basin, SE Baffin Shelf. Downcore changes in carbonate % (total carbonate weight %) (heavy line) and dolomite weight % (lighter line), ^{14}C dates, magnetic susceptibility (MS, 10^{-7} m^3/kg), and $\delta^{18}O$ of planktonic foraminifera. The boxes with YD and B/O represent the Younger Dryas and Bölling/Alleröd chronozones. The shaded area represents an interval of ice advance, as indicated by the increase in detrital carbonate (Andrews et al., 1990).

(Grosswald 1980; Denton and Hughes 1981), especially for those proposed for the East Siberian Shelf and the Kara Sea (Astakhov 1992).

When viewing the Pre-Quaternary glacial records, or even the early glacial events of the Late Cenozoic Ice Age, from the perspective of the last deglaciation, one is forced to wonder whether the stratigraphic record *per se* is able to distinguish the nuances of unstable ice sheet behavior that now appears to have been a feature of the last deglaciation. Without the chronological control that has been provided by radiocarbon dating, and more specifically by AMS ^{14}C dating, it is doubtful that the correct inferences on the rates of deglaciation could be assessed or that minor changes in core properties could be interpreted as glacial advances across sedimentary basins (the disconformities in HU82-057, Fig. 2.11). One of the single most important issues to arise from a survey of the last deglaciation is how far individual sectors of ice sheets adequately represent the average global behavior of the cryosphere system. We need to establish, on the basis of field observations and measurements, each regional history, in my view, it is incorrect simply to infer correlations between regional glacial histories and the ice volume record. As shown on Table 2.1, even an ice sheet the size of the Cordilleran Ice Sheet is within the noise of determining ice sheet/ocean volume interchanges.

References

Alley, R.B., 1991. Deforming bed origin for southern Laurentide till sheets. Journal of Glaciology, 37: 67–76.

Andrews, J.T., 1973. The Wisconsin Laurentide Ice Sheet: Dispersal centers, problems of retreat, and climatic implications. Arctic Alpine Research, 5: 185–199.

Andrews, J.T., 1987a. Glaciation and sea level: a case study. In: R.J.N.Devoy (Editor), Sea surface studies. A global view. Croom Helm, London, pp. 95–126.

Andrews, J.T., 1987b. The Late Wisconsin Glaciation and deglaciation of the Laurentide Ice Sheet. In: W.F. Ruddiman, H.E. Wright Jr. (Editors), North America and adjacent oceans during the last deglaciation. Geological Society of America, Boulder, CO, K-3, pp. 13–37.

Andrews, J.T., 1990. Fiord to deep-sea sediment transfers along the Northeastern Canadian Continental Margin: Models and Data. Géographie physique et Quaternaire, 44: 55–70.

Andrews, J.T., 1992. A case of missing water. Nature, 358: 281.

Andrews, J.T., Erlenkeuser, H., Evans, L., Briggs, W. and Jull, A.J.T., 1991. Meltwater and deglaciation SE Baffin Shelf (NE margin Laurentide Ice Sheet) between 13.5 and 8 ka: From stable O and C data. Paleoceanography, 6: 621–637.

Andrews, J.T., Erlenkeuser, H., Tedesco, K., Aksu, A. and Jull, A.J.T., 1994. Late Quaternary (Stage 2 and 3) Meltwater and Heinrich events, NW Labrador Sea. Quaternary Research, 41: 26–34.

Andrews, J.T., Evans, L.W., Williams, K.M., Briggs, W.M., Erlenkeuser, H., Hardy, I. and Jull, A.J.T., 1990. Cryosphere/ocean interactions at the margin of the Laurentide Ice Sheet during the Younger Dryas Chron: SE Baffin Shelf, Northwest Territories. Paleoceanography, 5: 921–935.

Andrews, J.T., Geirsdottir, A. and Jennings, A.E., 1989. Spatial and temporal variations in clay-and silt-size mineralogies of shelf and fiord cores, Baffin Island. Continental Shelf Research, 9: 445–463.

Andrews, J.T., Jennings, A.E., MacLean, B., Mudie, P., Praeg, D. and Vilks, G., 1991. The surficial geology of Canadian eastern Arctic and Polar continental shelves. Continental Shelf Research, 11: 791–819.

Andrews, J.T. and Syvistki, J.P.M., 1994. Sediment fluxes along high latitude continental margins (NE Canada and E. Greenland). In: W. Hay (Editor), National Academy of Sciences, Washington, D.C.,

Andrews, J.T. and Tedesco, K., 1992. Detrital carbonate-rich sediments, northwestern Labrador Sea: Implications for ice-sheet dynamics and iceberg rafting (Heinrich) events in the North Atlantic. Geology, 20: 1087–1090.

Andrews, J.T., Tedesco, K., Briggs, W.M. and Evans, L.W., 1994. Sediments, sedimentation rates, and environments, SE Baffin Shelf and NW Labrador Sea 8 to 26 KA. Canadian Journal of Earth Sciences.

Astakhov, V., 1992. The last glaciation in west Siberia. Sveriges Geologiska Undersokning, 81: 21–30.

Bard, E., Arnold, M., Duprat, J. and Duplessy, J.-C., 1987. Reconstruction of the last deglaciation: deconvolved records of del 18-O profiles, micropaleontological variations and accelerator mass spectrometric ^{14}C dating. Climate Dynamics, 1: 101–112.

Bard, E., Arnold, M., Fairbanks, R.G. and Hamelin, B., 1993. 230Th–234U and ^{14}C ages obtained by mass spectrometry on corals. Radiocarbon, 35: 191–200.

Bard, E., Fairbanks, R.G., Hamelin, B. and Zindler, A., 1990. Calibration of the ^{14}C timescale over the past 30,000 years using mass spectrometric U-Th ages from Barbados corals. Nature, 345: 405–410.

Beget, J., 1986. Influence of till rheology on Pleistocene glacier flow in the southern Great Lakes area, USA. Journal of Glaciology, 32: 235–241.

Beget, J., 1987. Low profile of the Northwest Laurentide Ice Sheet. Arctic and Alpine Research, 19: 81–88.

Berger, A. and Loutre, M.F., 1991. Insolation values for the climate of the last 10 million years. Quaternary Science Reviews, 10: 297–318.

Bindschadler, R.A. (Compiler), 1991. West Antarctic Ice Sheet Initiative, vol. 1, Science and implementation plan. NASA, Conference Publication 3115. pp. 58.

Blatter, H. and Hutter, K., 1991. Polythermal conditions in arctic glaciers. Journal of Glaciology, 37: 261–269.

Bond, G., Heinrich, H., Broecker, W.S., Labeyrie, L., McManus, J., Andrews, J.T., Huon, S., Jantschik, R., Clasen, S., Simet, C., Tedesco, K., Klas, M., Bonani, G. and Ivy, S., 1992. Evidence for massive discharges of icebergs into the glacial Northern Atlantic. Nature, 360: 245–249.

Bond, G., Broecker, W.S., Johnson, S., McManus, J., Labeyrie, L., Jouzel, J., and Bonani, G., 1993. Correlations between climate records from North Atlantic sediments and Greenland ice. Nature, 365: 143–147.

Boulton, G.S. and Dobbie, K.E., 1993. Consolidation of sediments by glaciers: relations between sediment geotechnics, soft-bed glacier dynamics and subglacial groundwater flow. Journal of Glaciology, 39: 26–44.

Boulton, G.S. and Jones, A.S., 1979. Stability of temperate ice caps and ice sheets resting on beds of deformable sediment. Journal of Glaciology, 24: 29–43.

Boulton, G.S., Smith, G.D., Jones, A.S. and Newsome, J., 1985. Glacial geology and glaciology of the last mid-latitude ice sheets. Journal of the Geological Society of London, 142: 447–474.

Broecker, W.S., Andree, G., Bonani, G., Wolfi, W., Oeschger, H. and Klas, M., 1988. Can the Greenland climatic jumps be identified in records from ocean and land? Quaternary Research, 30: 1–6.

Broecker, W.S., Bond, G., McManus, J., Klas, M. and Clark, E., 1992. Origin of the Northern Atlantic's Heinrich events. Climate Dynamics, 6: 265–273.

Broecker, W.S. and Denton, G.H., 1989. The role of ocean-atmosphere reorganizations in glacial cycles. Geochimica et Cosmochimica Acta, 53: 2465–2501.

Brown, C.S., Meier, M.F., and Post, A., 1982. Calving speed of Alaska tidewater glaciers, with application to Columbia Glacier. United States Geological Survey, Professional Paper 1258-C, 13 pp.

Clague, J., 1989. Quaternary geology of the Canadian Cordillera. In: R. J. Fulton (Editor), Quaternary geology of Canada and Greenland. Government of Canada, Ottawa, K-1, Decade of North American Geology, pp. 15–98.

Clark, P.U., 1991. Straited clast pavements: Products of deforming subglacial sediment? Geology, 19: 530–533.

Clark, P.U., 1992. Surface form of the southern Laurentide Ice Sheet and its implications to ice sheet dynamics. Geological Society American Bulletin: 104: 595–605.

Clark, P.U., 1994. Unstable behaviour of the Laurentide Ice Sheet over deforming sediment and its implications for climate change. Quaternary Research, 41: 19–25.

Clark, P.U., and Walder, J.S., 1994. Subglacial drainage, eskers, and deforming beds beneath the Laurentide and Eurasian ice sheets. Geological Society of America Bulletin, 106: 304–314.

Colhoun, E.E., Mabin, M.C.G., Adamson, D.A., and Kirks, R.M., 1992. Antarctic ice volume and contribution to sea-level fall at 20,000 yr BP from raised beaches. Nature, 358: 316–319.

Dansgaard, W., White, J.W.C. and Johnson, S.J., 1989. The abrupt termination of the Younger Dryas climate event. Nature, 339: 532–534.

Denton, G.H. and Hughes, T.J. 1981. The last great ice sheets. John Wiley & Sons, New York, 484 pp.

Dickenson, R.R., Meincke, J., Malmberg, S. and Lee, A., 1988. The "Great Salinity Anomaly" in the northern North Atlantic 1968–1982. Progress in Oceanography, 20: 103–151.

Drewry, D.J., 1981. Radio echo sounding of ice masses: principles and applications. In: A.P. Cracknell (Editor), Remote sensing in meteorology, oceanography, and hydrology. Ellis Horwood, Chicester, pp. 270–284.

Duplessy, J.-C., Delibrias, G., Turon, J.L., Pujol, C. and Duprat, J., 1981. Deglacial warming of the northeastern Atlantic Ocean: Correlation with the paleoclimate evolution of the European continent. Palaeogeography, Palaeoclimatology, Palaeoecology, 35: 121–144.

Dyke, A.S. and Prest, V.K., 1987. Late Wisconsinan and Holocene history of the Laurentide Ice Sheet. Géographie physique et Quaternaire, 41: 237–263.

Fairbanks, R.G., 1989. A 17,000-year glacio-eustatic sea level record: influence of glacial melting rates on the Younger Dryas event and deep-ocean circulation. Nature, 342: 637–642.

Fairbanks, R.G., Charles, C.D. and Wright, J.D., 1991. Origin of global meltwater pulses. In: R.E. Taylor et al. (Editors), Radiocarbon after four decades, Springer-Verlag, Berlin, pp. 473–500.

Fisher, D.A. and Klassen, R.A., 1988. Basal-flow conditions at the northeastern margin of the Laurentide Ice Sheet, Lancaster Sound. Canadian Journal of Earth Sciences 25: 1740–1750.

Fisher, D.A., Reeh, N. and Langley, K., 1985. Objective reconstruction of the late Wisconsinan Laurentide Ice Sheet and the significance of deformable beds. Géographie physique et Quaternaire, 39: 229–238.

Fulton, R.J. (Editor), 1989. Quaternary geology of Canada and Greenland. Geological Society of America, K-1, Boulder, CO, 839 pp.

Grosswald, M.G., 1980. Late Weischselian Ice Sheet of northern Eurasia. Quaternary Research, 13: 1–32.

Grousset, F.E., Labeyrie, L., Sinko, J.A., Cremer, M., Bond, G., Duprat, J., Cortijo, E. and Huon, S., 1993. Patterns of ice-rafted detritus in the glacial North Atlantic (40–55°N). Paleoceanography, 8: 175–192.

Hald, M. and Steinsund, P.I., 1992. Recent and Late Quaternary benthic foraminiferal assemblages in the southwestern Barents Sea. Studies in benthic foraminifera. Benthos 90, Tokai University, pp. 255–264.

Hald, M. and Vorren, T.O., 1987. Stable isotope stratigraphy and paleoceanography during the last deglaciation on the continental shelf of Troms, northern Norway. Paleoceanography 2: 583–599.

Hein, F.J. and Mudie, P.J., 1991. Glacial-marine sedimentation, Canadian polar margin, north of Axel

Heiberg Island. Géographie physique et Quaternaire, 45: 213–227.
Heinrich, H., 1988. Origin and consequences of cyclic ice rafting in the Northeast Atlantic Ocean during the past 130,000 years. Quaternary Research, 29: 143–152.
Hillaire-Marcel, C., de Vernal, A., Bilodeau, G. and Wu, G., in press. Isotope stratigraphy, sedimentation rates, deep circulation and carbonate events in the Labrador Sea during the last 200 kyr. Canadian Journal of Earth Sciences.
Hughes, T., 1987. Ice dynamics and deglaciation models when ice sheets collapsed. In: W.F. Ruddiman and H.E. Wright Jr. (Editors), North America and adjacent oceans during the last deglaciation, Geology of North America, K-3, Geological Society of America, Boulder, pp. 183–220.
Hughes, T., 1992. Abrupt climatic change related to unstable ice-sheet dynamics: toward a new paradigm. Palaeogeography, Palaeoclimatology, Palaeoecology, 97: 203–234.
Iken, A., Echelmeyer, K., Harrison, W. and Funk, M., 1993. Mechanisms of fast flow in Jakobshavns Isbrae, West Greenland: Part I. Measurements of temperature and water level in deep boreholes. Journal of Glaciology, 39: 15–15.
Imbrie, J., Boyle, E.A., Clemens, S.C., Duffy, A., Howard, W. R., Kukla, G., Kutzbach, J. et al., 1992. On the structure and origin of major glaciation cycles, 1, Linear responses to Milankovitch forcing. Paleoceanography, 7: 701–738.
Jenkins, A., in press. The melting of continental ice in the ocean and its impact on surface and bottom waters. In: W.R. Peltier (Editor), Ice in the climate system. NATO Advanced Workshop.
Jennings, A.E., 1993. Late Quaternary chronology of deglaciation, Cumberland Sound, N.W.T. Géographie physique et Quaternaire, 47: 21–42.
Jones, G.A. and Keigwin, L.D., 1988. Evidence from the Fram Strait (78 N) for early deglaciation. Nature, 336: 56–59.
Josenhans, H.W., Zevenhuizen, J. and Klassen, R.A., 1986. The Quaternary geology of the Labrador Shelf. Canadian Journal of Earth Sciences, 23: 1190–1214.
Keigwin, L.D., Jones, G.A. and Lehman, S.J., 1991. Deglacial meltwater discharge, North Atlantic deep circulation and abrupt climatic change. Journal of Geophysical Research, 96: 16811–16826.
Klassen, R.A. and Shilts, W.W., 1987. Bylot Island-Eastern Canadian Arctic. Excursion Guide Book A-1, XII International INQUA Congress, Ottawa, 54 pp.
Labeyrie, L.D., Pichon, J.J., Labracherie, M., Ippolito, P., Duprat, J. and Duplessy, J.-C., 1986. Melting history of Antarctica during the past 60,000 years. Nature, 322: 701–706.
Lehman, S.J., Jones, G.A., Keigwin, L.D., Andersen, E.S., Butenko, G. and Ostmo, S.-R., 1991. Initiation of Fennoscandian ice-sheet retreat during the last deglaciation. Nature, 349: 513–515.
MacAyeal, D.R., 1992. Irregular oscillations of the West Antarctic ice sheet. Nature, 359: 29–32.
MacLean, B., Sonnichson, G., Vilks, G., Powell, C., Moran, K., Jennings, A., Hodgson, D. and Deonarine, B., 1989. Marine geological and geotechnical investigations in Wellington, Byam Martin, Austin and adjacent channels, Canadian Arctic Archipelago. Geological Survey of Canada, Paper 89–11, 69 pp.
Matthews, W.H., 1974. Surface profiles of the Laurentide Ice Sheet in its marginal areas. Journal of Glaciology. 13: 37–43.
Mickelson, D.M., Clayton, L., Fullerton, D.S. and Borns, H.W.J., 1983. The Late Wisconsin glacial record of the Laurentide Ice Sheet in the United States. In: S.C.Porter (Editor), Late-Quaternary environments of the United States. University of Minnesota Press, Minneapolis, vol. 1, pp. 3–37.
Mienert, J., Andrews, J.T. and Milliman, J.D., 1992. The East Greenland continental margin (65 N) since the last deglaciation: changes in sea floor properties and ocean circulation. Marine Geology, 106: 217–238.
Miller, G.H. and Kaufman, D.S., 1990. Rapid fluctuations of the Laurentide Ice Sheet at the mouth of Hudson Strait: New evidence for ocean/ice-sheet interactions as control on the Younger Dryas. Paleoceanography, 5: 907–919.
Mix, A.C., 1987. The oxygen-isotope record of glaciation. In: W.F. Ruddiman and H.E.Wright, Jr. (Editors), North America and adjacent oceans during the last deglaciation. Geological Society of America, Boulder, CO, K-3, pp. 111–136.
Mix, A.C. and Ruddiman, A.F., 1985. Structure and timing of the last deglaciation: oxygen isotope evidence. Quaternary Science Reviews, 4: 59–108.
Oldale, R.N., Williams, R.S.J. and Coleman, S.M., 1990. Evidence against a Late Wisconsin Ice Shelf in the Gulf of Maine. Quaternary Science Reviews, 9: 1–13.
Paterson, W.S.B., 1981. The physics of glaciers. Pergamon Press, Oxford, 380 pp.
Piper, D.J.W., Mudie, P.J., Fader, G.B., Josenhans, H.W., MacLean, B. and Vilks, G., 1991. Quaternary Geology. In: M.J. Keen and G.L. Williams (Editors), Geology of the continental margin of eastern Canada. Geological Society of America, Boulder, CO, I-1, pp. 475–607.

Praeg, D.B., MacLean, B., Hardy, I.A. and Mudie, P.J., 1986. Quaternary geology of the southeast Baffin Island continental shelf, Geological Survey of Canada, Paper 85-14, 13 pp.

Reeh, N., 1984. Reconstruction of the glacial ice covers of Greenland and the Canadian arctic islands by three-dimensional, perfectly plastic ice-sheet modelling. Annals of Glaciology, 5: 115–121.

Reeh, N., 1985. Greenland Ice-Sheet mass balance and sea-level change. United States Department of Energy, Seattle, WA, pp. 155–171.

Ruddiman, W.E., 1987. Northern oceans. In: W.F. Ruddiman and H.E. Wright Jr. (Editors), North America and adjacent oceans during the last deglaciation. Geological Society of America, Boulder, CO, K-3, pp. 137–154.

Ruddiman, W.F. and McIntyre, A., 1981. The mode and mechanism of the last deglaciation: Oceanic evidence. Quaternary Research, 16: 125–134.

Shackelton, N.J. and Opdyke, N.D., 1973. Oxygen isotope and paleomagnetic stratigraphy of equatorial Pacific core V28-238: Oxygen isotope temperatures and ice volumes on 10^3 and 10^6 year scale. Quaternary Research, 3: 39–55.

Shoemaker, E.M., 1992a. Subglacial floods and the origin of low-relief ice-sheet lobes. Journal of Glaciology, 38: 105–112.

Shoemaker, E.M., 1992b. Water sheet outburst floods from the Laurentide Ice Sheet. Canadian Journal of Earth Sciences, 29: 1250–1264.

Stocker, T. and Wright, D., 1991. Rapid transitions of the ocean's deep circulation induced by changes in surface water fluxes. Nature, 351: 729–732.

Stocker, T.F., Wright, D.G. and Broecker, W.S., 1992. The influence of high-latitude surface forcing on the global thermohaline circulation. Paleoceanography, 7: 529–541.

Stravers, J.A., Miller, G.H. and Kaufman, D.S., 1992. Late glacial ice margins and deglacial chronology for Hudson Strait, Eastern Canadian Arctic. Canadian Journal of Earth Sciences, 29: 1000–1017.

Sugden, D.E., 1977. Reconstruction of the morphology, dynamics and thermal characteristics of the Laurentide Ice Sheet at its maximum. Arctic and Alpine Research, 9: 21–47.

Syvitski, J.P.M., 1993. Glaciomarine environments in Canada: an overview. Canadian Journal of Earth Sciences, 30: 354–371.

Taylor, K.C., Lamorey, G.W., Doyle, G.A., Alley, R.B., Grootes, P.M., Mayewski, P.A., White, J.W.C. and Barlow, L.K., 1993. The "flickering switch" of late Pleistocene climate change. Nature, 361: 432–436.

Tedesco, K., 1993. Late Quaternary paleoceanography of the Northwest Labrador Sea. University of Colorado, Boulder, 190 pp.

Teller, J.T., 1987. Proglacial lakes and the southern margin of the Laurentide Ice Sheet. In: W.F. Ruddiman and H.E. Wright Jr. (Editors), North America and adjacent oceans during the last deglaciation. Geology of North America, K-3, Geological Society of America, Boulder, pp. 39–70.

Thomas, R.H., 1977. Calving bay dynamics and ice sheet retreat up the St. Lawrence Valley system. Géographie physique et Quaternaire, 31: 347–356.

Tushingham, A.M. and Peltier, W.R., 1991. Ice 3G: A new global model of late Pleistocene deglaciation based upon geophysical predictions of post-glacial relative sea level change. Journal of Geophysical Research, 96: 4497–4523.

Vorren, T.O., Hald, M., Edvardsen, M. and Lind-Hansen, O.-W., 1983. Glacigenic sediments and sedimentary environments on continental shelves: General principles with a case study from the Norwegian shelf. In: J. Ehlers (Editor), Glacial deposits in North-West Europe, A.A. Balkema, Rotterdam, pp. 61–73.

Weinelt, M.S., Sarnthein, M., Vogelsand, E. and Erlenkeuser, H., 1991. Early decay of the Barents Shelf Ice Sheet - spread of stable isotope signals across the eastern Norwegian Sea. Norsk Geologisk Tidsskrift, 71: 137–140.

3. THE LAST SCANDINAVIAN ICE SHEET AND ITS DOWN-WASTING

Jan Lundqvist

Continental Ice Sheets and their Effect upon the Landscape

The waxing and waning of continental ice sheets, the pre-Quaternary as well as the Quaternary, had a considerable effect on the landscape, locally as well as regionally. When we try to interpret deposits from older glaciations, which are rather fragmentary, a comparison with modern or more recent glaciers and ice sheets is useful. The Pleistocene Scandinavian ice sheets and their down-wasting (Fig. 3.1) are favorable in this respect, being comparatively well known. In many respects they differ from the ice sheets on the North American continent with regard to both size and general climatic conditions. A study of the character and geographical distribution of the glacial deposits can give valuable knowledge to scientists dealing with older glacial deposits and help them to understand pre-Quaternary environments.

Knowledge about the glacial deposits themselves is essential, but the deglaciation history is almost as important. Scandinavia and its surroundings have been glaciated repeatedly during the Pleistocene. Since the glacial histories of these events were all different, it is appropriate to review them briefly. The detailed story of the waning of an ice sheet gives an important clue to the climatic situation at the time under consideration and can even give information about the conditions at the onset of the glaciation. The down-wasting of an ice sheet is controlled by a climatic amelioration and several other conditions, most important of which are the topography and the physical properties of the substratum of the ice. The latter are determined by the geological deposits and the mode of formation of the ice sheet. The start of the glaciation in turn is controlled by the climatic conditions. I will therefore review these factors briefly and then consider the visible direct effects of the Scandinavian inland ice on the landscape.

When an inland glacier moves over a terrain it erodes the substratum, especially in valleys and bedrock depressions. Soft rocks and sediments are deformed and partly disintegrated. Some of the released material is transported with the ice. When the ice melts down this material is left behind as a number of different glacial deposits (Fig. 3.2; Dreimanis, 1989). The most widespread deposit is till. At the base of the moving ice, lodgement till accumulates. This is mostly a diamicton characterized by dense texture, often strong fissility, and well-developed fabric, the preferred orientation of clasts and grains being parallel to the ice movement. By passive melting from the base of stagnant ice, basal melt-out till is formed, similar to lodgement till but usually less compact, often with numerous inclusions and laminae of sorted sediments and a veneer of silt or fine sand on the lower side of clasts. Striated stones are indicative of glacial movement.

During transport, debris move upward within the glacier and become concentrated on the surface where they are released by passive melt-out. Supraglacial melt-out till, characterized by loose texture, absence of clear fabric, and presence of some sorted sediment layers and inclusions is formed. Because of irregular topography on the ice surface, this debris often is remobilized for some distance before it comes to rest as flow till. The flow till has many inclusions of sorted sediments. Its fabric is often pronounced, with preferred orientation following the various flow directions. It is thus similar to other mass flow deposits not necessarily related to glacier ice, and by many specialists it is not regarded as a true till.

Under floating ice, such as ice shelves, or locally at ice margins in open water, mixtures of glacial debris and ordinary waterlaid sediments are formed. Such poorly sorted deposits are called waterlain till (Dreimanis, 1989). Where only scattered debris is found in the marine or lacustrine sediments we talk about ice-rafted

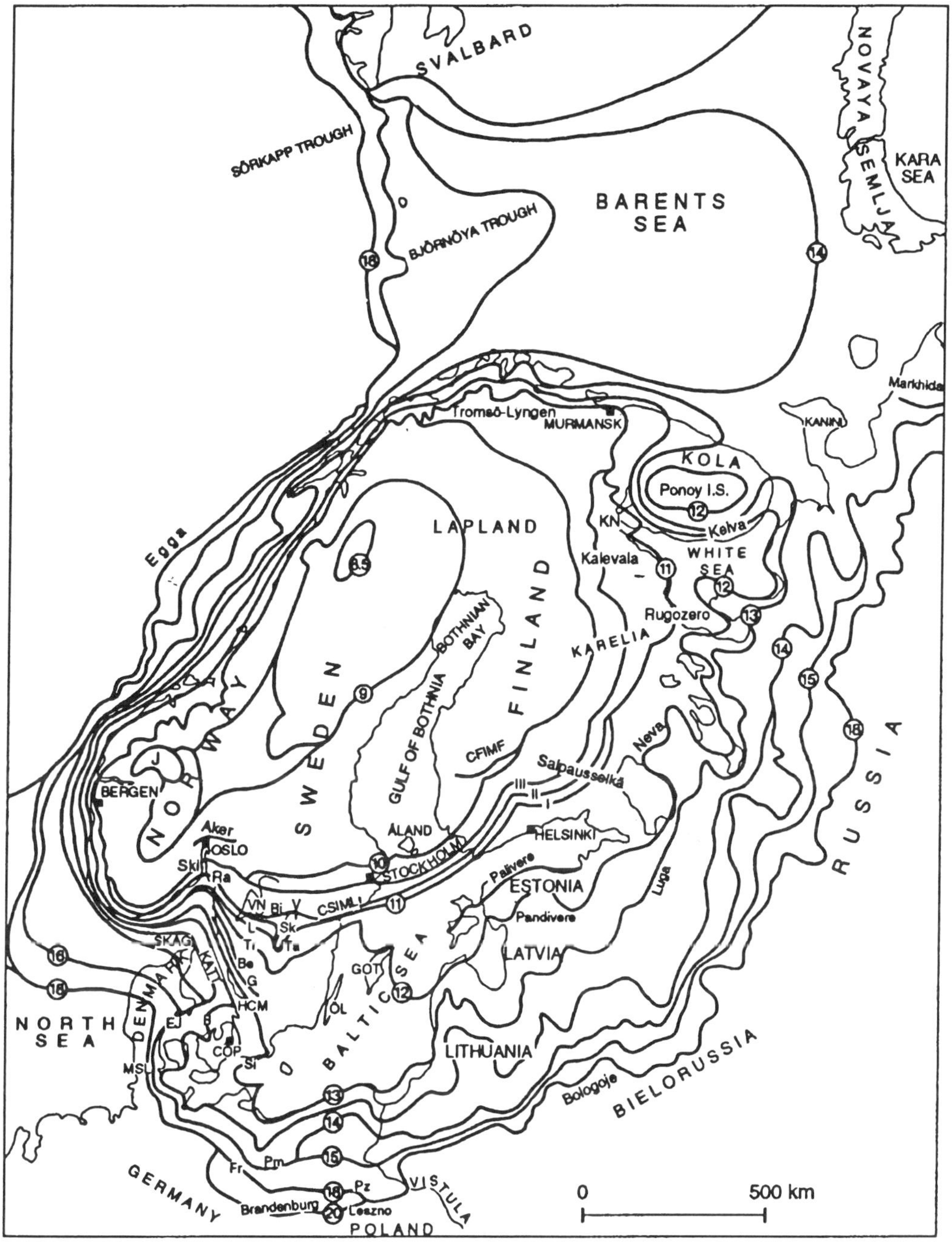

Fig. 3.1. Location map with schematic lines of ice recession. Numbers in circles show radiocarbon years (ka BP). Geographical names are in capital letters, names of moraines in lower-case letters. *Abbreviations*: Moraines: B = Belt; Be = Berghem; Bi = Billingen; CFIMF = Central Finnish Ice-Marginal Formation; CSIML = Central Swedish Ice-Marginal Line; EJ = East Jutland; Fr = Frankfurt; G = Gothenburg; HCM = Halland Coastal Moraines; L = Levene; MSL = Main Stationary Line; Pm = Pommeranian; Pz = Poznan; Si = Simrishamn; Sk = Skövde; Ta = Taberg; Tr = Trollhättan. I, II, and III refer to the three Salpausselkäs. *Place names*: GOT = Gotland; J = Jotunheimen; KN = Kandalaksha; KATT = Kattegat; ÖL = Öland; SKAG = Skagerrak; V = Lake Vättern; VN = Lake Vänern.

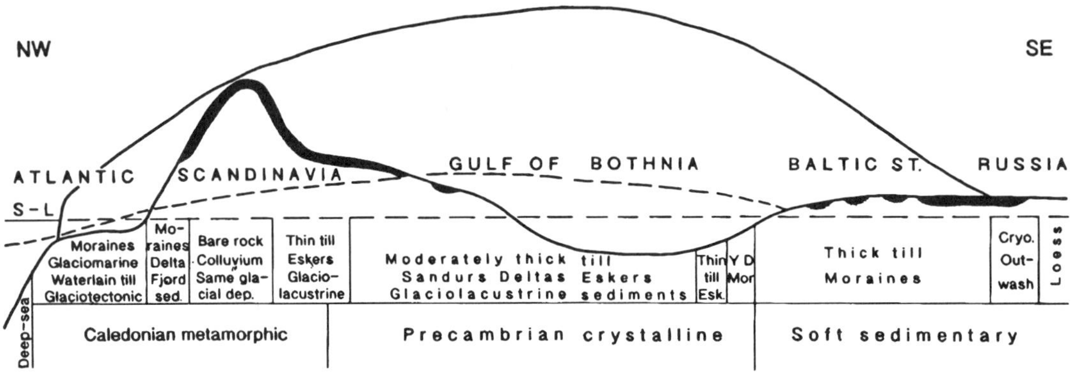

Fig. 3.2. Schematic profile through the Scandinavian Ice Sheet, the substratum, and the partly interpolated and hypothetical levels of the present and contemporaneous sea (S-L). The shadowing indicates cold-based ice and permafrost. The section, which is not to scale, also shows the bedrock and glacial and periglacial deposits. Esk = eskers; Y D Mor = Younger Dryas Moraines; Cryo = cryogenic features.

material or isolated dropstones. Such deposits are an indication of the presence of ice. An important grounding-line sediment system associated with a floating terminus consists of till or till-like sediments interfingering with purely waterlaid sediments (Fig. 3.3; King et al., 1991).

Solifluction and slumping are common processes on the glacier as well as in its surroundings, favoured by the presence of seasonal or permanent ground frost and lack of vegetation. In these environments numerous different deposits are formed, mostly diamictons with various structures and degree of sorting. In these settings we also find patterned ground and similar features caused by frost action.

Debris in glaciers can be partly sorted by meltwater activity. Well-sorted glaciofluvial deposits accumulate in different environments, more or less directly controlled by the presence of ice. Eskers and crevasse fillings are formed in tunnels and narrow conduits. Hummocky kame fields form at the ice margin. Some distance away from it, glaciofluvial deltas accumulate where the meltwater reaches the sea or a lake. In those water bodies the finest particles form glaciomarine or glaciolacustrine sediments. The silty or clayey sediments are influenced by the annual variations in melting of the ice. Rhythmites, varved clays and silts, which for the most part are indicative of a glacial environment, are formed.

The glacial deposits are influenced by a number of different factors. Among the most important are the conditions at the base of the ice, the bedrock or types of sediment over which the ice passed (Lundqvist, 1940), and the environment at the ice margin (Dreimanis, 1989). The thermal regime and the glacial dynamics are other influencing factors. In general, soft rocks and unconsolidated Quaternary sediments yield large amounts of sediments. Crystalline bedrock yields a more limited amount, but some rocks, such as coarse granites and veined gneisses, may give numerous large boulders. When the origin of the till is the same, it is often found that the till is more coarse-textured in valleys and low grounds due to shorter transport (Lundqvist, 1940). On the other hand, in such positions older fine-textured sediments may occur and, through reworking, affect the till correspondingly.

Glacials and Interglacials in Europe

We usually identify three or four different glaciations in Europe, separated by interglacials. Each of the glaciations (glacials) has been divided into cooler stadials and essentially ice-free interstadials (Fig. 3.4). At the beginning of each glaciation the landscape was probably similar to what has been described here. Due to different climatic and oceanographic conditions, however,

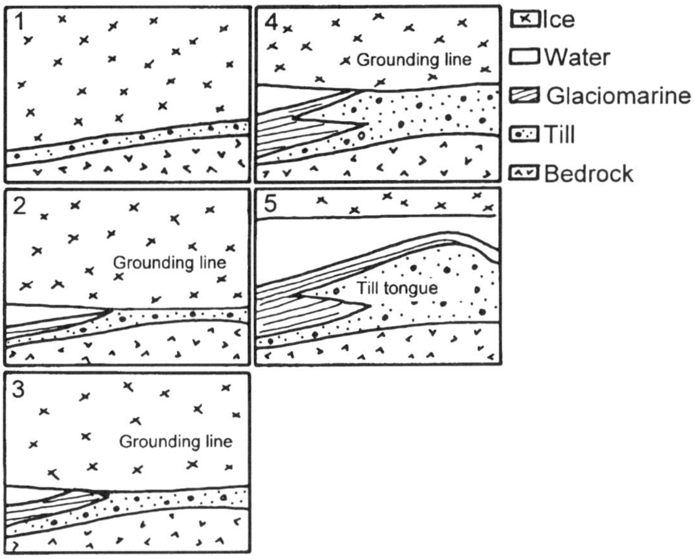

Fig. 3.3. Formation of till-tongue stratigraphy on the continental shelf. The formation follows the migration of a floating terminus of the ice (after King et al., 1991).

the glaciations started in different ways. This is important for the ice sheets themselves, for their down-wasting, and for the total amount of deposits remaining in different regions, the thickest being in the central area affected by repeated glaciations.

The first Weichselian glacial substage started according to the classical model; that is, that glaciation began as mountain glaciers, which gradually extended to form piedmont glaciers and later coalesced as ice caps and an ice sheet. It is possible that this first ice sheet never reached farther south than the central part of the Scandinavian Peninsula (Lundqvist, 1992). Within Scandinavia and Finland there are no indications of the presence of glacier ice during this substage outside that limit. In northwestern Siberia this glaciation was the most extensive during the Weichselian (Astakhov, 1992). The first Weichselian ice was centered in the mountains in the west but extended also to the north and the far east; that is, into coastal areas. The main part of the young Pleistocene morphology and deposits in northern Scandinavia is a result of this glaciation.

The second Weichselian glaciation is poorly known, but it was fairly similar to the first one (Lagerbäck, 1988a; Lundqvist, 1992). In Finland it is not identified (Hirvas, 1991), and its effect upon the substratum was weak (Lagerbäck, 1988a,b).

The last Weichselian glaciation may have started in the mountains, like the two earlier ones (Ljungner, 1949), but the evidence is weak. More important, that glaciation seems to have started simultaneously in lower areas outside the mountain range (instantaneous glacerization; Ives et al., 1975). Vast areas in northernmost Sweden do not show traces of this ice (Lagerbäck, 1988a,b). This fact has two important implications. First, the ground must have been frozen when it became covered by ice and the ice sheet was cold-based, at least at an early stage. Second, the ice sheet became centered east of the mountains. According to Ljungner's (1949) and other models (Denton and Hughes, 1981; Boulton et al., 1985) its center was situated over the innermost part of the Gulf of Bothnia (Bothnian Bay).

In western Siberia the extension of this last ice sheet was much more limited than that of earlier glaciers (Astakhov, 1992). The reason is probably the different conditions prevailing in the ocean adjacent to the ice sheet at each location. During the onset of the first glaciation at the end of the Eemian Interglacial, the sea temperature must have been higher than it was during the last glacial stage. This implies a more efficient source

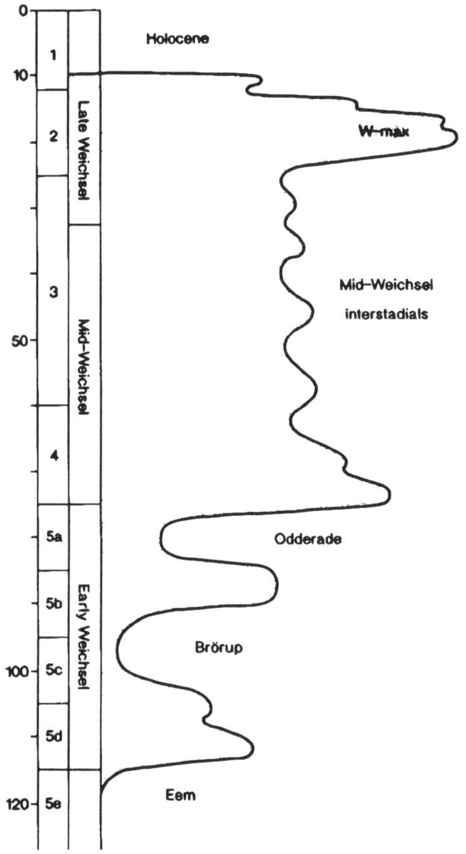

Fig. 3.4. Glaciation curve through the Weichselian. The curve indicates roughly the extent of the Scandinavian inland ice from north (left) to south. Names refer to the north European interstadials, and the scales show thousands of (uncalibrated) radiocarbon years BP (left) and the marine oxygen isotope stages. The shape of this curve agrees well with the shape of the isotope curve.

of moisture, favouring near-coast glaciation. At the beginning of the last glacial the sea temperature was lower, possibly even with frozen sea surface, resulting in a different type of nourishment of the glacier, with moisture coming from a more distant source area.

Conditions at the Beginning of the Last Glaciation

The central area of the glaciation is the Scandinavian Peninsula, the backbone of which is the Caledonian mountain range, whose highest peaks reach more than 2,000 meters above the present sea level. The bedrock consists of schists and some soft sedimentary rocks and volcanics. Outside the mountain range the peninsula belongs to the Baltic Shield, consisting of Precambrian crystalline rocks. They form a smooth landscape, with local hilly relief. The same applies to Finland and adjacent parts of Russia, namely Karelia and the Kola Peninsula.

Outside the region with crystalline bedrock, the Precambrian is covered by Paleozoic to Tertiary sedimentary rocks. Such rocks also form the main part of the sea floor fringing the Scandinavian Peninsula. These seas are generally shallow, although an exception important for glaciation is the 800m deep Norwegian Trench in the Skagerrak south of Norway. Barents Sea is part of the continental shelf. Its western part is intersected by the east-west-trending Björnöya Trough.

Soft Quaternary sediments cover the older formations. They are locally up to 200 m thick, but usually only of the order of 10 m. On the sea floor the thickness increases, as it does in the peripheral areas. In the central area of glaciation, the thickness is considerable as a result of several glaciations.

In the central area of glaciation, the Earth's crust was isostatically depressed by the ice sheet. During the deglaciation that area was covered by the sea. Because of postglacial rebound there has been a continuous lowering of the sea level. In the marginal areas the process was complex, with an initial sea-level rise followed by regression.

The Ice Sheet at its Maximum

The last Weichselian ice sheet, which grew after the Odderade Interstadial (Fig. 3.4), is fairly well known, although its early history is obscure. There are no deposits known from early stages of this glaciation in central Scandinavia or in Finland. There is stratigraphic evidence that parts of southernmost Sweden were not covered by an ice sheet after the Eemian until the latest phase of the glaciation, about 20,000 years ago (Lundqvist, 1992). On the other hand, the oldest Weichselian till in southeastern Denmark has been shown to be younger than the Odderade Interstadial (Houmark-Nielsen, 1987), indicating an early advance in the Baltic basin. Also in Poland, according to Drozdowski and Fedorowicz (1987), deposits earlier referred to the older Weichselian glaciations derived from the same ice advance, reaching into the lower Vistula Valley. Probably this advance corresponds to marine isotope stage 4.

During the long interval between marine isotope stage 4 and stage 2 (Martinson et al., 1987; Mangerud, 1991), when the ice reached its maximum extension, all of Scandinavia and Finland were probably ice-covered. Some evidence of a partial deglaciation during this phase is found in the western fringe areas of the glaciation (Hillefors, 1969; Miller, 1977; Mangerud, 1981). According to the model proposed by Lundqvist (1992), this means that the ice margin was fairly stationary along the western coasts of Scandinavia from isotope stage 4 until stage 2, but showed minor oscillations corresponding to the series of cool interstadials (Hengelo, Denekamp, and others) that have been identified on the European continent (Behre, 1989). Similar oscillations of the ice margin continued until the ice reached its maximum extension (Houmark-Nielsen, 1987, 1989).

The extension of the last Weichselian ice is best known on the European continent, where it is marked by extensive end moraines. These are the Brandenburg moraines in Germany, continuing in Poland as the Leszno moraines. The Brandenburg moraines have been dated with radiocarbon to 20,000–18,000 BP (Cepek, 1965). The Brandenburg moraines have in some areas been overridden by later ice and intersected by the Frankfurt (in Poland, the Poznan) moraines, which thus may form the outermost deposits, dated at 18,600–19,700 BP (Kozarski, 1988).

The Frankfurt moraines can be traced northwest into Denmark, where they turn west with a sharp angle. The position of the ice margin is seen as large moraines and as vast outwash plains extending to the bottom of the North Sea. The sharp corner can be ascribed to different ice masses that flowed in from Norway in the north and from the southern part of Scandinavia and the Baltic basin in the east. Lagerlund and Houmark-Nielsen (1993) have shown that there was a considerable fluctuation of the Baltic ice margin, between 15,000 and 13,200 BP. A transgression followed after the retreat of this ice, inundating the coastal areas of northern Denmark and southwestern Sweden. During this stage, debris from floating icebergs were deposited to form a thin diamicton.

The extension of the ice margin west from Denmark is controversial. Some authors think that the Weichselian ice, like the pre-Weichselian one, covered the North Sea. There was thus a confluence between the Scandinavian ice and the ice sheet over the British Isles. The marine sediments are sometimes overconsolidated and often contain ice-rafted material. It is not clear whether they were covered by a partly floating ice sheet, by sea ice, or just by icebergs. On the basis of the morphology of the sea floor, Wingfield (1989) proposed a glacier confluence in the northernmost part of the North Sea. Sejrup et al. (1994) postulated coalescence between the Scandinavian and the British ice sheets in a period between 29,400 and 22,000 years BP. The maximum extent in this area was prior to 22 ka BP. When the ice reached its maximum limit on the European continent, the confluence in the North Sea area was broken. Then the ice margin turned north in the middle of the North Sea, creating an ice-free corridor between Scandinavia and Scotland - Shetland Isles (Sejrup et al., 1987). Because of the eustatically lowered sea level, the main part of the North Sea floor was dry land, dissected by large drainage systems from the European continent. The ice must have traversed the deep Norwegian Trench in the east.

Off the Norwegian west coast, the glacier covered the continental shelf to its edge. The complex Egga moraines along the edge mark the outermost ice margin. Radiocarbon dates suggest an age of about 19,000 BP for the last glacial maximum here (Holtedahl, 1993).

In the north, the ice margin extended not very far off the present-day coast and followed the edge of the continental shelf toward Svalbard (Vorren, 1992). The main outflow followed the Björnöya Trough, where a convergent flow of ice and meltwater formed a huge delta.

The glaciation of Barents Sea has been subject to intense discussions. The present-day isostatic rebound may be interpreted as an indication of the existence of a separate ice dome over it (Hoppe, 1972; Forman, 1990). However, current opinion favors an ice cover of the Barents Sea that did not form a separate dome, but was the result of an inflow of ice from the adjacent higher areas, Scandinavia, Svalbard, Franz Josef Land and Novaya Semlja. Grosswald (1980, 1988, 1993) and Hughes et al. (1977) emphasize the importance of a Kara Sea ice dome, with ice flow across Novaya Semlja. Others even deny the existence of a continuous ice sheet over Barents Sea (Faustova and Velichko, 1992). Research in progress favors an ice center in the region around northeastern Svalbard (Salvigsen et al., 1992). This glacier probably covered the entire Barents Sea, although there may be some uncertainty concerning the southern part (Elverhøi and Solheim, 1983) and the precise age of its maximum.

The continuation of the ice margin to the east is even more controversial. According to one extreme opinion presented by Faustova and Velichko (1992), the margin of the Scandinavian Ice Sheet followed the coast only past the Kola Peninsula, where it turned south through northwestern Russia, with some interaction only in the region around the Kanin Peninsula with various glaciers covering Novaya Semlja, the northern Ural mountains, and adjacent areas. According to the opposite extreme view of Grosswald (1988, 1993) the margin passing around Svalbard and Franz Josef Land continued east along the edge of the Siberian shelf. He even postulated a floating ice shelf covering the entire polar basin. General opinion favors a confluence of the Scandinavian Ice Sheet and other glaciers in the north and east, centered over the shelf or over Novaya Semlja and other highlands. Whatever interpretation is true, it is clear that the connection between the Scandinavian Ice Sheet proper and other glaciers was broken fairly early. In northernmost Scandinavia there are no traces of the action of a Barents Sea glacier except for a few boulders from a northerly provenance (Holtedahl, 1960).

From the northeast, the glacier margin extended southwest toward the Leszno zone in Poland (Raukas and Gaigalas, 1993). Along this line the definition of the margin offers some problems. Well-developed moraines occur, mostly in lower areas where the glacier formed lobes. In other areas there is no morphological boundary, and the definition has to be made on the basis of the distribution of till units. This has generated different hypotheses on the extension of the ice (Obrutschew, 1930). In many places the marginal zone is marked by hummocky moraines, which form uplands as in Lithuania. In Poland, as in Germany, the southern ice margin is marked by moraines and by very large marginal drainage channels (pradolinas, Urstromtäler) towards the west. The entire drainage that now flows into the Baltic and the North seas was deflected to traverse the exposed bed of the North Sea and deposited extensive deltaic sediments.

There are different opinions about the conditions within the ice sheet during its maximum extension. In the model of Torell (1872, 1873) and Ljungner (1949), the ice divide, or center of spreading, gradually moved from the mountains to somewhere east of them. There is a consensus of opinion that the center of movement reached the region of maximum rebound on the coast of the Gulf of Bothnia, a model supported by drumlins radiating northwards from the Gulf. Most models indicate a unicentered ice sheet, although it is clear that for long times the high mountains in southern Norway served as a second center.

Possibly the shift from mountain-centered to east-centered ice sheet took place very early. There is evidence in the form of striae indicating a Late Weichselian ice divide in the mountains (Ljungner, 1949), but the dating of these striae is uncertain. In northernmost Scandinavia no such evidence exists. Even at the earliest stage, the ice divide was situated east of the mountain range (Lagerbäck, 1988a,b). In large areas the glacial landforms derive from the earlier substages of the Weichselian. The reason must be that the ice was primarily cold-based, which supports the idea of instantaneous glacierization east of the mountains. According to the "outrageous" model of Denton and Hughes (1981), even the Bothnian Bay served as a center of glaciation.

Several ice streams can be identified within the ice sheet. The most conspicuous one followed the Baltic depression. The low shear strength of the soft sediments in combination with the low terrain position favored a rapid outflow of ice, with dislocations taking place in the soft substratum (Boulton and Jones, 1979). According to the model of Lagerlund (1980, 1987) the outflow took place intermittently as surges. Local ice domes were formed on these surged masses due to greater precipitation (Liljequist, 1974) along the ice margin.

Lagerlund (1980, 1987) extended these ideas to a model for the glaciodynamics in southernmost Scandinavia. According to the classical model (Fig. 3.5) there was a broad ice stream following the Baltic depression south, turning west around the southern end of the Scandinavian Peninsula and then north through the sounds between the peninsula and Denmark (Lagerlund, 1977). This theory is supported by Ringberg (1989). Lagerlund's (1980, 1987) model (Fig. 3.5B), in contrast, implies the existence of marginal domes in the ice sheet. There were radial movements in the ice from these domes, and the glacial geology is the result of the interaction and displacements of them. The dome theory is based on interpretation of the sedimentology of till beds, but models generally favor the classical theory (Boulton et al., 1985; Holmlund and Fastook, 1993; Boulton and Payne, 1994).

There are also different opinions about the basal conditions in the ice sheet during its maximum. According to both the maximum and minimum models of Denton and Hughes (1981) there was basal melting below the entire central part. The region with warm-based ice was surrounded by a marginal zone of cold-based ice. However, field evidence from parts of Swedish Lapland and adjacent areas (Lagerbäck, 1988a,b; Kleman, 1992) demonstrates that the ice sheet in that area must have been frozen to its base to allow preservation even of fragile landforms. The same applies to regions in the central part of Scandinavia (Kleman et al., 1992; Lundqvist and Miller, 1992).

According to calculations by Hooke (1977) for the Laurentide Ice Sheet, basal melting occurred in a zone a few hundred kilometers wide near the ice margin, except for the outermost few hundred meters where the ice was frozen to its bed. Considering the general temperature conditions at the beginning of the glaciation, it is probable that the ice was cold-based in the central area. This line of evidence also applies to the Scandinavian Ice Sheet.

Deglaciation

Principles

The deglaciation of Scandinavia and adjacent areas was influenced by the varying topographic conditions. Where the glacier was active, marginal features like push moraines could be formed, mainly in the areas close to the maximum extension of the ice. Because of the cold-base conditions near the front, a high pore pressure built up behind it. If the shear stress of the underlying strata was exceeded, the effect was thrusting and other large-scale glaciotectonic disturbances. The end moraines are often formed by pushing, folding, or upthrusting in front of the glacier or by a combination of these features and contain till, meltwater sediments, slump, and flow deposits as well.

Push and glaciotectonic structures develop preferentially in areas with soft rocks or unconsolidated Quaternary sediments, that is, outside the Precambrian Shield. Glaciotectonic features are mainly restricted to the marginal regions, but similar structures could also form under the central part of the ice sheet (Aber and Lundqvist, 1988).

In the supra-aquatic areas within the Shield, surface melting was the dominant factor for the down-wasting of the ice sheet. End moraines and marginal formations of active ice are mostly missing. There is, however, an increasing tendency to formation of such features to the west in the maritime areas. Along the western coasts of Scandinavia, end moraines are well developed.

During the deglaciation, the ice divide gradually shifted back toward the mountain range. There was also a separation between two centers of the ice sheet, one over the South-Norwegian high

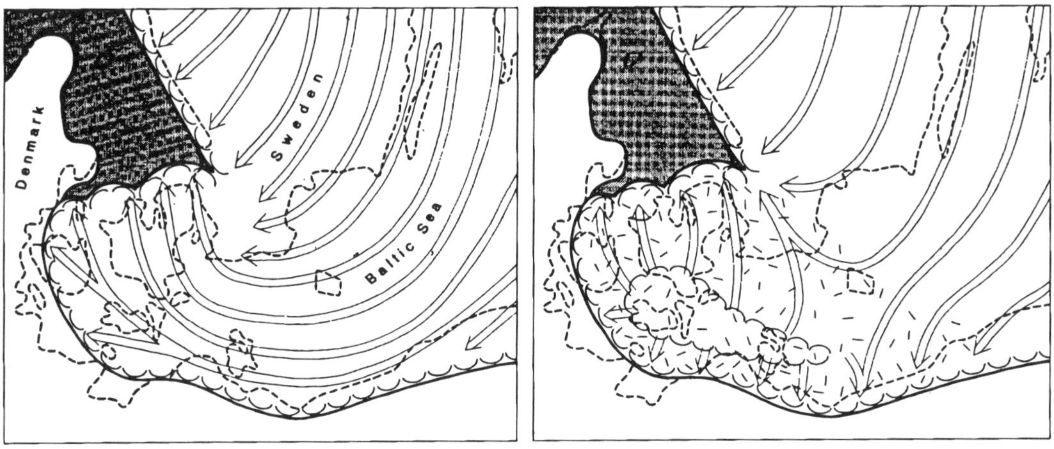

Fig. 3.5. Sketch map showing the two alternative models for the glaciation in southernmost Scandinavia. On the left is the classical ice-stream model; on the right is Lagerlund's (1987) marginal-dome model. Shaded area = ocean.

mountains and one more easterly along the range. Toward the end of the glaciation, the latter split into several centers. In the central part of the range these were situated near the eastern border of the mountains, while farther to the north the last centers occurred within the highest massifs.

Where the ice-sheet front was situated in water, calving contributed to the down-wasting. Calving took place partly as slumping from the steep margin, partly by upbreaking. Where ablation had made the ice thin, floating large sheets of ice could be separated, causing a rapid recession of the margin.

For these reasons, ice domes in deep basins were unstable. Calving could result in an increased outflow, partly as surges, of ice towards the basins. The outflow had an effect also upon the supra-aquatic areas, where the ice surface became lowered. This downdraw, emphasized by Denton and Hughes (1981), was an important factor for the deglaciation of some areas. The opposite effect, a buttressing, would be the case where thick lowland ice or an ice shelf blocked the outflow.

In the following section the process of deglaciation in different sectors of the ice sheet is analyzed. The correlation between different sectors is important, since it gives information about the behavior of the ice sheet and the climatic conditions.

Retreat of the Ice Margin before the Younger Dryas Event

In Germany and Poland, the ice reached its outermost limit at the Brandenburg-Leszno moraines 20,000–18,000 BP (Fig. 3.1; Cepek, 1965). The activity of the ice was great, and large moraines were formed. In deep basins the ice protruded as tongues, in places forming end-moraine ridges by push or glaciotectonics. While recession proceeded in one area, the ice front could advance in another. Consequently, the Brandenburg moraines were in some areas overridden by the ice, and the outer younger Frankfurt moraines formed. Recession at this early stage was slow in the southern sector. The front receded from the outer moraines past the Pomeranian moraines and others only to the present-day coast, a distance between 150 and 200 km in about 7,000 years. At the same time the North Sea area with Skagerrak and Kattegat, as well as northern and western Denmark, became free of ice, although affected by a temporary re-advance (Lagerlund and Houmark-Nielsen, 1993).

An effect of the difference in deglaciation rate was a broad ice lobe in the southern part of the Baltic, turning north over eastern Denmark. This lobe, like the glacier deriving from the South Norwegian high-mountain region, was very active. Retreats and re-advances alternated, as they did like before the glacial maximum (Kozarski, 1988; Houmark-Nielsen, 1989). In Denmark these shifting glacier movements were recorded by widespread glaciotectonic structures (Berthelsen, 1979). Many moraines were once defined in Denmark as recessional (Hansen, 1965), but most of them are poorly developed and have been reinterpreted as irregular hummocky moraine landscapes. According to modern opinion, only three of them are real moraines, namely, the outermost Main Stationary Line, the East Jutland, and the Belt Lines (Houmark-Nielsen, 1981, 1987).

There is a controversy concerning the glaciodynamics in and around the southern part of the Baltic (Fig. 3.5). According to the classical model, which still enjoy the most general support, there was a final re-advance of an ice lobe in the Baltic depression-the Young (or Low) Baltic re-advance- shortly after 13,500 BP (Lagerlund, 1977; Ringberg, 1989). The alternative model of Lagerlund (1980, 1987) questions whether this event occurred. Deposits of the so-called Low Baltic Ice, earlier interpreted as till, are considered to be mainly glaciolacustrine, formed by sedimentation in water, partly filled with drift ice that dropped boulders and finer debris. The essential difference between the two models concern to the definition of ice lobe versus glacier domes, and till versus glaciolacustrine sediments.

The importance of areal deglaciation has been emphasized by Marcussen (1977). Whether areal down-wasting or glacier margin retreat dominated is still an open question. Lodgement till (Dreimanis, 1989), push and glaciotectonic structures, drumlins and end moraines suggest the latter process. Large areas of hummocky ablation moraine and flow till (Dreimanis, 1989) indicate stagnant ice. A rapid deglaciation in the southern part of the Baltic, as shown by clay-varve chronology (Stay, 1979), suggests a similar stagnant ice condition.

The retreat of the ice margin in the Kattegat area in combination with the rapid breakup in the Baltic basin resulted in the lobe-shaped contour of the glacier on the South Swedish upland between those two areas. As a consequence over the southernmost part of the Scandinavian Peninsula the glacier margin has an east-west convex extension. There Lagerlund et al. (1983) and Möller (1987) identified a periodicity in the deglaciation. Before the Younger Dryas cold event there were five phases, with cool climate during what has been considered the warm Bölling and Alleröd phases separated only by the insignificant cool Older Dryas. The cool phases are marked by zones rich in hummocky moraine formed by bodies of cold-based ice left behind when the active ice margin retreated during the warmer phases.

Toward the west, the moraine zones concentrate to form end moraines. Along the Swedish west coast these are well developed (Fig. 3.6), consisting of high ridges traversing both valleys and higher ground. The outermost moraines are the Halland Coastal Moraines, a series of low ridges and deltas. According to Fernlund (1993), these ridges are glaciotectonically formed.

Most of the west Swedish moraines are well developed in the west and become discontinuous in the southeast (Hillefors, 1979; Johansson, 1982; Ronnert, 1989). They can be morphostratigraphically correlated with moraines in the Oslofjord area (Sørensen, 1992) and represent stagnations of the retreat, in some cases caused mainly by topographic conditions.

The Levene moraine is a distinct moraine with glaciotectonic structures indicating an advance of the ice margin. Its continuation east of Lake Vättern is problematic. Persson (1983) correlates a distinctive zone of thicker till and partly till-covered large deltas between Lake Vättern and the Baltic with the Levene line. This is the only clearly marked ice-marginal line in eastern Sweden.

In the Oslofjord region, ice-marginal deposits are very well developed (Fig. 3.7). Large moraines there indicate a deep estuary in the margin (Sørensen, 1979, 1992). The outermost moraines indicate the beginning of the Younger Dryas climatic deterioration. Corresponding moraines can be traced along the Norwegian coast and on the adjacent shelf (Andersen, 1979).

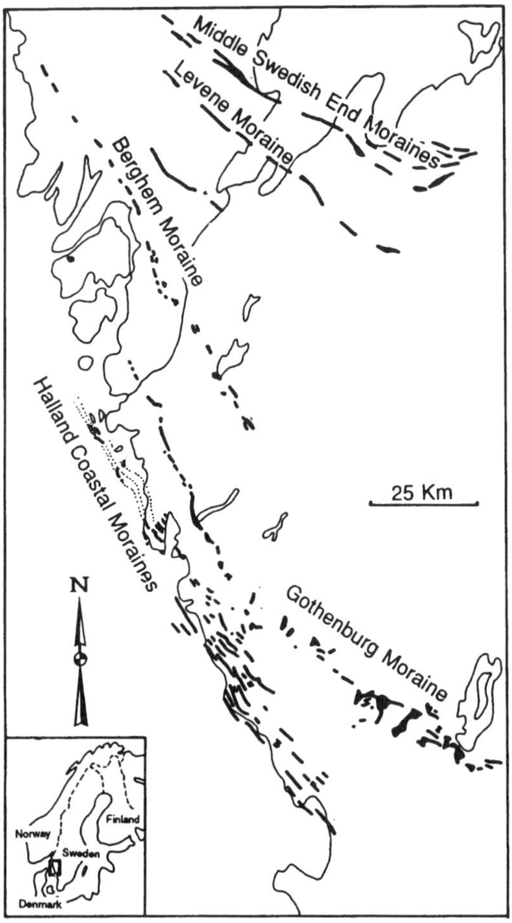

Fig. 3.6. The principal moraines along the Swedish west coast. The scattering and disappearance of the moraines in the southeast is notable. The Middle Swedish End Moraines are referred to in the text as the Skövde and the Billingen moraines (after Hillefors, 1975).

In general, the ice margin retreated from its outermost position at the Egga moraines to the coastal areas of today during the time span from the glacial maximum to the Younger Dryas cold event, that is, in about 7,000 years. To some extent this retreat may have been influenced by a downdraw effect, starting when the buttressing effect of the marine part of the ice sheet ceased. According to Anundsen (1985), the ice margin reached the coast of southwestern Norway at a very early stage. The outermost coastal area became deglaciated well before 15,000 BP.

According to Vorren et al. (1988) and Alm (1993), the maximum extent of the ice sheet on the northern shelf occurred prior to 22,000 BP. The retreat started early, but a climatic deterioration 19,000–18,500 BP caused a re-advance to the maximum limit. This implied a total covering of the Barents Sea shelf by a grounded ice sheet moving mainly from the area northeast of Svalbard. The movement converged toward the Björnöya and Sörkapp Troughs, where thick delta deposits accumulated. Ice moving into these areas originated from a considerable part of the ice sheet, comprising the Scandinavian part as well as Svalbard and Novaya Semlja (Kara Sea). According to Vorren (1992) the ice margin was situated close to the shelf edge until 16,000 BP. The retreat took place essentially between 16,000 and 13,000 BP (Vorren et al., 1988); downdraw effect caused a fairly rapid collapse of the entire Barents Sea ice sheet (Ruddiman and McIntyre, 1981), which could have had a triggering effect upon the North Atlantic current system, hence also upon the climate (Mercer, 1969).

East of Scandinavia the northern coast of the Kola Peninsula became free of ice in the time span 14,000–10,000 BP (Ekman and Iljin, 1991). Farther east, the Scandinavian Ice Sheet was isolated from ice flowing from the Novaya Semlja-Kara Sea region at an early stage (Faustova and Velichko, 1992).

The deglaciation of the Kola Peninsula and adjacent areas is subject to much debate (Ekman and Iljin, 1991). One reason is to be found in the very weak activity of an ice dome centered over the highland of interior Kola (Evzerov and Koshechkin, 1981). Along the southern coast of the peninsula the very large and well-developed Keiva moraines were formed by the ice lobe protruding into the White Sea, that is, from the southwest, or by the radial movement in the Ponoy Ice Cap in the interior of the peninsula. The moraines slope from the mouth of the White Sea basin into the bay, which has been used as an argument by Grosswald (1980) for a Kara Sea ice lobe extending southwest into the basin.

The older (Keiva II) moraine was probably formed in the contact zone between the White Sea lobe and the Ponoy Ice Cap. The Ponoy cap became isolated from the Scandinavian Ice Sheet prior to the Younger Dryas cold phase.

In the Alleröd warm phase (11,800–11,000 BP)

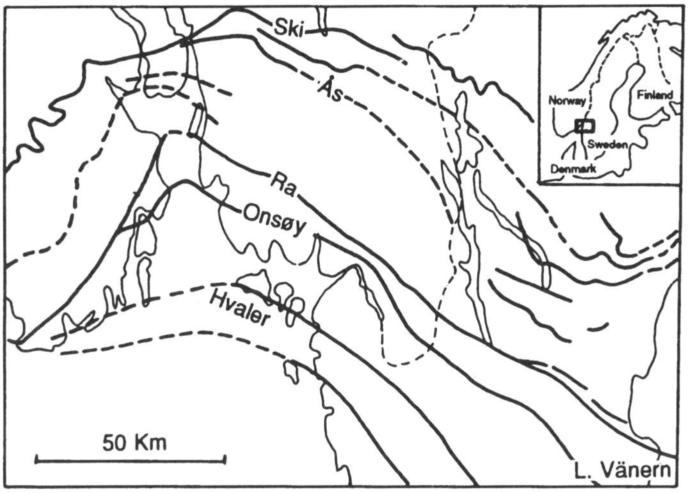

Fig. 3.7. Ice-marginal lines in the Oslofjord area. The southernmost line corresponds to the Berghem Moraine in Figure 3.6; the Ra is the continuation of the Middle Swedish End Moraines in Figure 3.6 (after Sørensen, 1992).

the main ice sheet in the eastern sector was active and characterized by formation of lobes extending into the White Sea, around the mountain areas in interior Kola, and into the Russian Plain southeast of the White Sea (Ekman and Iljin, 1991). The lobe-like behavior characterized the ice margin from the Barents Sea coast across the Russian Plain down into Poland and Germany. Over flat ground there was an irregular flow of ice lobes, while in more broken terrain the lobes followed the lower areas (Raukas, 1977). The lobes were thin and controlled mainly by the pre-existing relief. In some areas the lobes were thicker and more active, deepening and widening existing valleys to form tongue basins.

Along the entire ice margin from the White Sea to Germany, climatic fluctuations or other factors caused a stepwise retreat of the margin. Moraines and other marginal formations are in some places well defined, but since they are less well developed on higher ground, the connection between different parts of the ice margin is not clear everywhere. The dating of these lines is problematic, and the difficulties in correlating even neighboring moraine lobes makes a correlation across the Baltic Sea hypothetical. Correlations east of the Baltic Sea are based mainly on the assumption that different till units do correspond to certain moraines.

Younger Dryas Event

From its outer positions the ice margin retreated to a very well-defined position, corresponding to the Younger Dryas cold event (approximately the first 500 years of the Younger Dryas substage, 11,000–10,000 BP). In some areas the retreat proceeded beyond this position. Local re-advance resulted in an accelerated erosion behind the Younger Dryas line, demonstrated by a zone where bedrock is exposed (Kujansuu and Niemelä, 1984; Lundqvist, 1989a).

Due to eustatic lowering of sea level, the sounds between Scandinavia and Denmark were dry. South from Mount Billingen in southcentral Sweden, therefore, the Baltic Basin was isolated from the ocean in the west by the South Swedish highland. When the receding ice margin passed Billingen into the lower ground north of it, the Baltic Ice Lake, dammed up east of the highland, was catastrophically drained (Fig. 3.8). A direct connection between the basin and the ocean was opened. A huge delta of coarse sediments was deposited outside the drainage area (Strömberg, 1992). According to Björck and Digerfeldt (1984) drainage occurred twice, due to a Younger Dryas re-advance at 10,300–10,200 BP.

West of the Billingen area the Younger Dryas line continues as large moraines and marginal deltas (De Geer, 1909; Johansson, 1982; Lund-

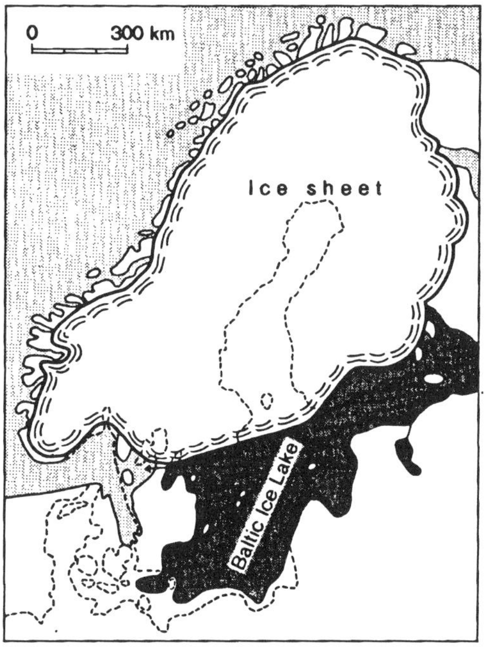

Fig. 3.8. The situation at the catastrophic drainage of the Baltic Ice Lake (dark grey), which was ponded east of Mt. Billingen (at the arrow) 26 m above sea-level (light grey) to the west. When the ice margin retreated north of the mountain, a connection between the ocean and the Baltic Basin was opened.

qvist, 1995). The stratigraphy indicates a re-advance of the glacier. The extent of the preceding retreat is unknown, but can be estimated to a few tens of kilometers.

The moraines can be morphologically followed westward. The Skövde and the Billingen lines converge into the huge Ra moraine, traversing the Oslo Fjord. It is dated at 11,000–10,700 BP (Gjessing, 1966; Sørensen, 1979, 1992). There seem to have been repeated stillstands of the ice margin in that area during the Younger Dryas. To the east, these moraines, like all other moraines in the Skagerrak region, become gradually more discontinuous and disappear. Scattered marginal deposits indicate a fairly dramatic history of retreat, implying strange ice movements, even toward the north-northwest (Larsson, 1945).

The moraines in the Oslofjord area converge west of the fjord to form a Younger Dryas moraine which can be followed more or less continuously (Fig. 3.9) along the entire coast of Norway and into the Kola Peninsula (Andersen, 1979; Andersen et al., 1995). The moraines are found on high ground as well as in valleys. In the west they indicate a considerable re-advance of the glacier, dated to late Younger Dryas. In the Bergen area this was almost at the end of the chronozone (Aarseth and Mangerud, 1974; Mangerud et al., 1979), which is much later than in the eastern part of the ice sheet.

One explanation for the different behavior of the glacier along the coast can be found in the bordering ocean. When the receding ice front on the shelf approached the coast, the buttressing effect of the shelf ice ceased, resulting in a downdraw from the mountain area and a retreat at least 50 km beyond the position of the Younger Dryas moraines (Mangerud et al., 1979). A longer time was consequently needed for the buildup of ice thick enough to cause a re-advance.

The maritime climate in combination with the topography explain for the existence of local glaciers today, and even more during the deglaciation. During warm episodes many of the numerous cirques may have been empty, but in cool stages, especially in the Younger Dryas, glaciers were more abundant (Andersen, 1968, 1975).

From the northern coast of Scandinavia the Younger Dryas (Tromsö-Lyngen) moraines continue along the coast of the Kola Peninsula, and become more discontinuous. The Ponoy Ice Cap in the central part of the peninsula became isolated from the main inland ice before the Younger Dryas time. The Younger Dryas moraines therefore turn south east of Murmansk toward the Kandalaksha Bay of the White Sea.

South of the White Sea there are almost continuous huge moraines from the Younger Dryas stage throughout Karelia. The oldest one of those, the Rugozero moraine, extends from the Kandalaksha Bay into Finland with lobe-shaped contours. West of it, the Kalevala moraine also runs into Finland. These moraines are several tens of meters high, sometimes complex, and accompanied by outwash deposits (Rainio et al., 1995).

One major problem concerns the relation between these two moraines and the Salpausselkäs in Finland (Fig. 3.10; Rainio 1985b, 1991). The two major moraines in Finland do not corre-

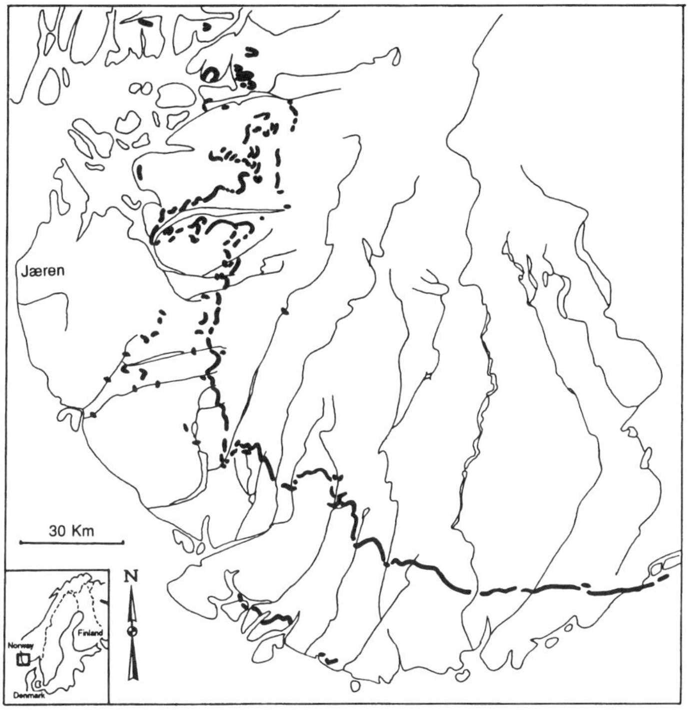

Fig. 3.9. The almost continuous Younger Dryas moraines in southwestern Norway continuation of the Ra in Figure 3.7 (after Andersen, 1954).

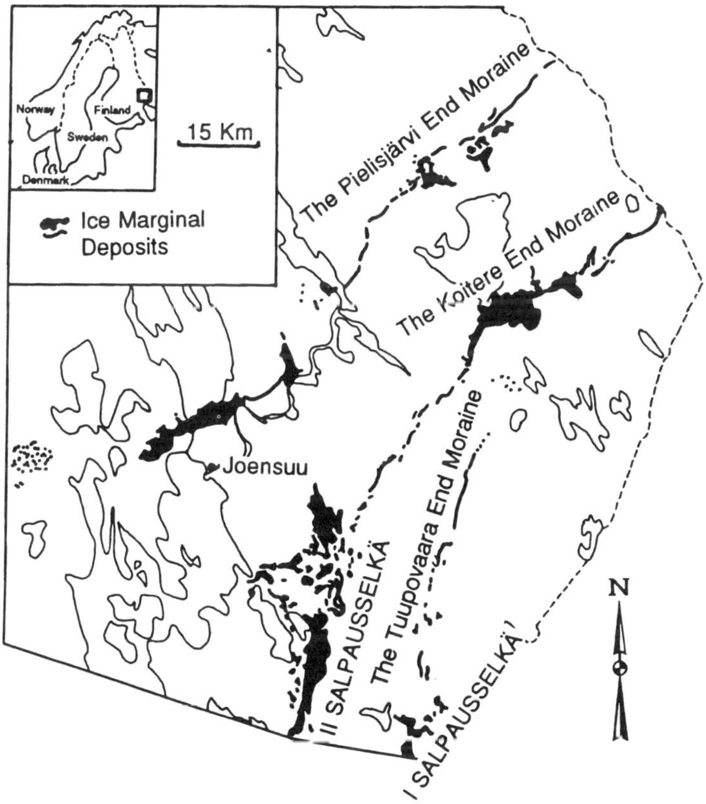

Fig. 3.10. Ice-marginal formations in Finnish Karelia. The very large First Salpausselkä disappears northeastward becoming the small Tuupovaara end moraine (after Rainio, 1985b).

spond to the two major moraines in Russian Karelia. This is of principal importance because it illustrates how an ice sheet may react differently in different sectors.

The large Younger Dryas moraines in Finland, the three Salpausselkäs, form two broad arcs of ridges (Ignatius et al., 1980). The eastern arc, the eastern end of which was described above, consists of the First (outer) and the Second Salpausselkä, while the Third Salpausselkä is in the western arc. The two older moraines are formed by very large ridges and deltas of mainly glaciofluvial sediments, to some extent alternating with till (Fig. 3.11). The Third Salpausselkä is more indistinct, consisting of shorter ridges alternating with small moraines of till. Like the other two moraines, it extends westward into the Baltic Sea, where it becomes much more concentrated, forming one single large ridge on the sea floor.

From the lobate contour, occurrences of till-covered sediments and the change in direction of glacial striae (Rainio, 1985a), it is evident that these moraines represent one or more advances of the ice margin. In general, this activity is correlated with the onset of the Younger Dryas climatic event around 11,000 BP. Although the two arcs pass directly into each other, the retreat of the ice margin from them was not synchronous (Donner, 1978). In addition, the advance was probably not synchronous along the entire front. This lack of synchronity between different sectors, illustrating the balance within the ice sheet, is of great climatic importance.

The precise dating of these moraines is an important issue, but presents great problems. It is based partly on a local clay-varve chronology (Fig. 3.12; Niemelä, 1971), and partly on the relation to the catastrophic drainage of the Baltic Ice Lake. Within the Salpausselkäs there are numerous deltas, the surfaces of which can be related to different stages of this ice-dammed lake and its successor, the Yoldia Sea. At the Second Salpausselkä there are deltas corresponding to the levels both before and after the drainage. This gives a date of the retreat from the Second Salpausselkä of 10,300–10,200 radiocarbon years BP.

In eastern Scandinavia the zone is marked by small moraines, deltas, and other marginal deposits, in some areas just by a zone of thicker till cover (Lundqvist and Saarnisto, 1990). The correlation across the Baltic between the Salpausselkäs and these formations is not clear. The clay-varve time scales are correlated by Strömberg (1990), but according to underwater seismics there is no direct morphological connection (Söderberg, 1988). There is also no indication whether there was a re-advance in the sea between Finland and the Scandinavian Peninsula.

Deglaciation after the Younger Dryas Event

The warm Holocene climate following the Younger Dryas cold event resulted in a rather quick collapse of the Scandinavian Ice Sheet as the ice margin retreated toward the high-mountain region.

From the Atlantic coast the deglaciation proceeded east. The ice margin receding from the coast was more active than in the east due to the topographic conditions and the influence of the maritime climate. End moraines were formed inside the Younger Dryas line from the Oslo Fjord in the south to the Kola Peninsula in the north. They were often accompanied by large deltas at sea level. In the Oslofjord area the Aker moraines (Gjessing, 1980) are conspicuous. Even at this late stage (9800 BP; Sørensen, 1979) the glacier was active enough to make a re-advance (Gjessing and Spjeldnaes, 1979). Like most of the older ones, these moraines disappear eastward (Lundqvist, 1988). Large deltas north of the Aker moraines (Holtedahl, 1974) are often considered as marginal formations indicating steps in the retreat of the ice margin, but this may be questioned. They may be compatible with accumulations in the Stockholm area, representing increased meltwater flow.

Due to the topographic conditions the ice sheet separated into two parts, one in the high-mountain region in southwestern Norway and one along the main part of the range. The partition of the ice took place in the large valleys north of the Oslo Fjord, through the catastrophic outbreak of large ice-dammed lakes (Holmsen, 1915). Garnes and Bergersen (1980) have demonstrated that the movements of the western ice became more depending on the local topography, and its retreat took place toward the Jotunheimen area,

Fig. 3.11. Section in Salpausselkä I in Finland showing alternating beds of diamicton and meltwater sediments. where the last remnants of this ice cap were situated (photo, J. Lundqvist, 1993).

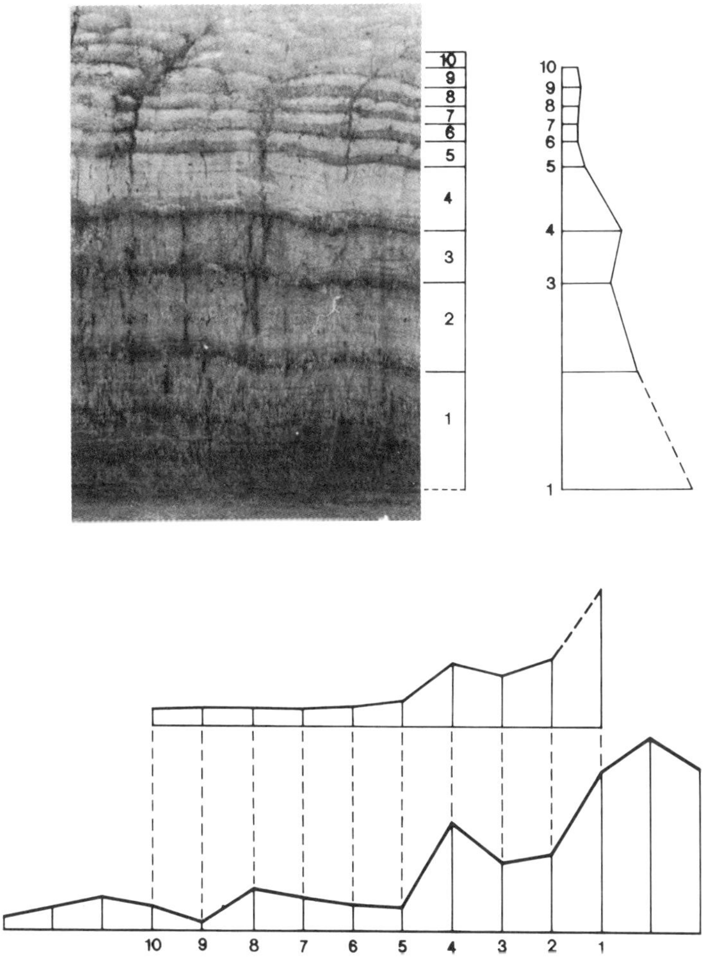

Fig. 3.12. Principles of the clay-varve dating method. The annual varves of the clay are measured. Their thicknesses are drawn perpendicular to a base line and combined to form a curve. This curve can be correlated with similar curves from other sites. Since the lowermost varve was deposited at the ice margin, the combination shows the time difference between different positions of the margin. This gives the possibility for absolute dating.

The eastern part of the ice sheet was the largest, extending from the area of bipartition to the Arctic sea. Its western margin retreated across the mountain range. In the high-mountain area in northern Scandinavia, recession was directed to these mountains (Kleman, 1990) but north and south of this region the receding front passed the range. It proceeded from higher to lower ground east of the mountains in the northernmost and southernmost areas. Large glacial lakes became ponded (Fig. 3.13; Lundqvist, 1972; Soyez, 1974). Between these areas the recession proceeded toward centers located along the eastern margin of the range (Ulfstedt, 1980). These centers formed culminations in the ice sheet, from which the flow radiated.

At the end of Younger Dryas an ice front extended in an almost east-west direction across the Baltic Sea south of the Åland Archipelago. From this position there was a rapid retreat north. A series of very small and low De Geer moraines occurs in a zone across Sweden, approximately on the latitude of Stockholm. They run subparallel to the ice margin and were earlier (De Geer, 1940) interpreted as annual end moraines, but are today considered as having formed in basal crevasses in the ice, although near the front (Hoppe, 1957). Their chronological value is debated (Möller, 1962). Large delta-like accumulations in the Stockholm area were formely interpreted as indications of a retardation of the ice recession, but have been shown (Brunnberg, in Risberg et al., 1991) to be a result of an increased rate of retreat.

In Sweden below the coastline of that time the deglaciation has been dated very precisely by means of clay-varve chronology (Fig. 3.12). It must be observed that for the time period under discussion there is a significant difference between radiocarbon and clay-varve (siderial, or sedimentological) years (Olsson, 1974; Berglund, 1979; Lundqvist, 1986).

In this whole area there was a continuous regular retreat of the ice margin (Strömberg, 1989). There is no indication of any halt in the recession corresponding to the Norwegian Preboreal stages. In the deep main part of the Gulf of Bothnia, the ice broke up quickly due to the calving effect. This resulted in a turning of the ice margin from an east-westerly to a more north-south orientation. The formation of calving bays is typical for subaquatic areas, in contrast to supra-aquatic areas, where the ice formed lobes on lower ground (Strömberg 1981).

Across the Bothnian Bay the ice on the Scandinavian Peninsula extended as a lobe to Finland (Sauramo, 1929). This implies some isolation from the accumulation area. The consequence was a rapid breakup in the bay (Andrén, 1990). Thinning of the ice caused changes in meltwater discharge, well-defined conduits being replaced by sheet flow (Hoppe, 1948). Broader deposits of glaciofluvial sediments mixed with till were formed instead of eskers.

A retreat from the Bothnian coast toward the mountain range was soon established. The receding eastern flank of the ice sheet approached the front receding east from the Atlantic. The shrinking ice wasted down, with isolation of dead-ice bodies in low terrain. The final retreat approached a center in the area of the highest mountains. Recent research (Lundqvist, 1973; Kleman, 1990) has shown that the glacier must have been active. Hummocky moraine was formed stepwise during the retreat of the ice margin, which itself requires active ice before the stagnation (Boulton, 1972).

In central Finland there is the distinct Central Finland Ice-Marginal Formation. This moraine shows the same arcuate form as the Salpausselkä moraines. It indicates a stagnation or even re-advance of the ice margin and corresponds to the Aker moraines in Norway. These moraines correspond to the Preboreal cooling (YD II) identified in the North Atlantic (Koç Karpuz and Jansen, 1992). There is thus evidence for this retardation or re-advance in the west and east, while in the intermediate region clay-varve chronology and the absence of moraines indicate that the recession was continuous (Strömberg, 1989).

The decrease of ice volume is not well known. Models have been constructed for the decay (Boulton et al., 1984; Holmlund and Fastook, 1993), but field observations are lacking or ambiguous. Curves of the eustatic sea-level changes give a general idea, but they are the combined effect from several ice sheets and their interpretation is debatable. The curve of Fairbanks (1989), supporting a two-step de-

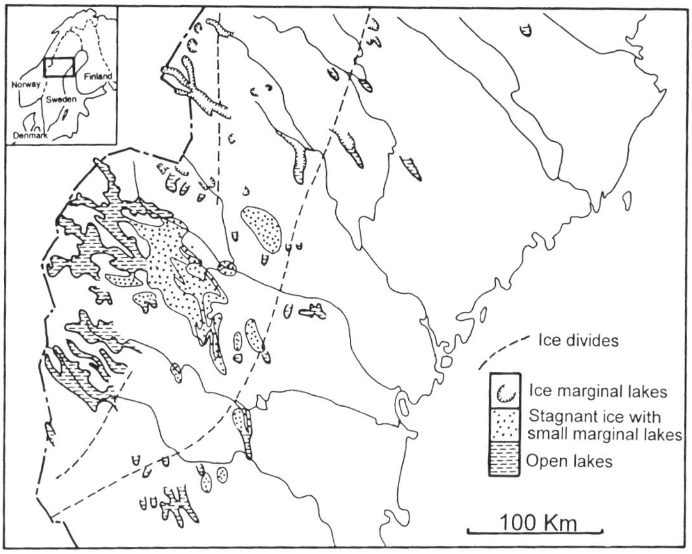

Fig. 3.13. Ice-dammed lakes in a part of central Sweden (after Lundqvist, 1972).

glaciation model, probably gives the best picture, but it is to the greatest extent influenced by the Laurentide Ice Sheet of North America. If applicable to the Scandinavian Ice Sheet, this curve implies maximum melting 14,000–12,000 BP and 10,000–7,000 BP, with a mid-deglacial pause centered around the Younger Dryas.

Deposits of the Scandinavian Ice Sheet

The continental shelf west and north of Scandinavia is covered by thick deposits that can be interpreted as waterlain till or as glaciomarine or marine sediments deposited by overriding ice sheets (Holtedahl, 1993). The character of these deposits makes it difficult to define the maximum extent of the ice sheet. In the Barents and Baltic seas, subject to greater isostatic down-warping, these conditions are less prominent. The glacial sediments are dominated by more regular tills and outwash or marine deposits. The Baltic has experienced marine stages as well as lacustrine conditions. The stratigraphy therefore shows changes between marine, brackish-water, and lacustrine sediments.

In the supra-aquatic areas, glacial deposits are dominated by till and meltwater sediments. Due to different basal conditions in the ice, the type and the amount of deposits are different in different zones of the ice sheet. In the marginal zone the total amount of sediment is large because of the soft bedrock in most of these areas, and the basal conditions. This is most pronounced in the moraines formed during deglaciation. Because of the thinness of the ice sheet in zones corresponding to the maximum extension, moraines are sometimes less conspicuous there (Faustova and Chebotareva, 1979). The cold base of the ice at its margin forced upward the debris eroded some distance behind the margin, resulting in a concentration near the ice front. Till of subglacial as well as supraglacial origin was deposited. The buildup of a high pore-water pressure behind the frozen margin, together with the pressure of the active glacier, gave rise to large-scale glaciotectonic features in this zone. These deposits form landscapes of end moraines or hummocky dead-ice moraine. Strong glacial erosion has created tongue basins with the eroded material redeposited as moraines at the distal end.

Large thrust masses consisting of rafts of bedrock and older Quaternary deposits are abundant in the outer zone of glaciation. Sediments from previous stages of the Quaternary are found to be redeposited in this way, which may cause misinterpretation of the preglacial and interglacial environment (Liivrand, 1991).

Outwash sediments are widespread outside the ice margin. On the European continent the blocking by the ice in the north resulted in westward linking to the North Sea region of the drainage from the entire southern sector of the ice sheet, including part of the eastern sector. This gave rise to a conspicuous system of drainage channels along the southern glacier margin at its earlier stages. To the west and in the North Sea, corresponding thick sediments accumulated. Wind action on the barren landscape gave rise to vast areas of loess fringing the glaciated area.

Between the broad marginal zone and the Younger Dryas moraines there is a zone with moderate amount of glacial deposits. This is most clearly seen in Scandinavia, where the bedrock is crystalline. The till cover is fairly thin, forming drumlinoid terrain or, in depressions, hummocky moraine. Coarse-grained outwash sediments are abundant. Older deposits are usually not preserved, indicating strong erosional activity. Only in protected positions, such as ancient valleys, are old deposits present in large amounts (Tavast et al., 1983).

The entire Younger Dryas zone is characterized by thick accumulations of till and other glacial sediments. Glaciotectonic structures are common. As a whole, this zone is similar to the outermost zone of glaciation, but narrower.

The zone of strong erosion behind the Younger Dryas moraines is conspicuous (Lundqvist, 1987). Inside it, the till cover is moderately thick. Glaciofluvial deposits, often forming large eskers, are abundant. The spacing of esker systems gives important information about the hydraulic conditions at the base of the waning ice sheet. Along the highest coastline the eskers widen and form extensive deltas.

In flat terrain like Finland the glacial morphology shows conspicuous lobate patterns (Kujansuu and Niemelä, 1984). Behind marginal formations there are drumlins or drumlinoid forms radiating toward the lobe front. Eskers and other glaciofluvial features follow the same pattern. In the more broken landscape of the Scandinavian Peninsula the pattern is distorted and obscured by the local topography, but traces of it can be observed. In the mountainous western part of the peninsula, no such pattern can be seen.

In central Scandinavia outside the mountain range, the till cover is thick, sometimes more than 50 meters. The thickness is generally the effect of more than one glacial event. The till shows conspicuous landforms, such as Rogen (Fig. 3.14; Lundqvist, 1989b) and Veiki (Hoppe, 1952) moraines. As has been shown by Lagerbäck (1988a), some of these forms are old, with insignificant or no reshaping by the last ice sheet. Where sediments are exposed and can be dated, it is clear that the effect of the last glaciation was small. The till from the last Weichselian stage is thin, in some areas even lacking (Lagerbäck, 1988a,b). The thickest till cover may correspond to the extension of the earlier Weichselian ice sheets, and perhaps to previous glacial stages (Lundqvist, 1992).

Till from the last glaciation may be entirely affected by glaciotectonic structures under the central part of the ice sheet (Lundqvist and Miller, 1992). These structures indicate significant internal movements in the ice during its down-wasting (Aber and Lundqvist, 1988).

The areal distribution of different till types and moraine forms illustrates the varying basal conditions in the ice sheet. In the central part of the Scandinavian Peninsula, Rogen moraine occurs approximately between lat. 61° N and 66° N. To the north it is replaced by Veiki moraine, which passes east into a special type of hummocky moraine called Pulju moraine (Aario, 1992). These zones may correspond to changes between cold-based and temperate ice in the southern area and dominantly cold-based ice in the north. The Veiki area is a region of cold base and insignificant accumulation, while in the Pulju area there was thicker accumulation.

In the high-mountain region, till is lacking in vast areas. In other areas there is a thin cover of diamicton, which is partly till, partly frost-shattered debris. The till is also in part reworked by frost action. Here it is often difficult to distinguish between till and weathered material.

In the lower part of the mountains and in their fringe areas, till and other glacial deposits are more abundant. In such regions, morphological traces of the deglaciation are often conspicuous. Lateral drainage channels, subglacially formed chutes, small eskers, and many other types occur (Mannerfelt, 1945). Together these forms indicate the areal down-wasting of the ice sheet, and have

Fig. 3.14. Rogen moraine landscape, consisting of ridges at right angles to the ice flow. Each ridge is made up of short crescentic elements, horns pointing down-ice (to the left). The area shown on the picture is about 2 km wide (photo, J. Lundqvist, 1963).

even been interpreted as evidence for stagnant conditions at that time.

According to Kleman (1990), the glacier showed activity even until its last remnants wasted down. However, vast areas of stagnant ice were certainly left behind, isolated from the accumulation area. The effect has been extensive occurrences of hummocky dead-ice moraine, especially on lower ground, alternating with more drumlinoid features, often on higher ground. The deposits in the hummocky moraine are usually referred to as till, but actually true basal till is rare. Flow till, slope, and slump deposits alternating with sorted meltwater sediments are most common. In drumlins there is often a surface layer of basal till, but in the interior we can find almost any kind of sediment. This demonstrates that the drumlins to a great extent are, or could be, remnants of older deposits, reshaped by overriding ice. Even when they are made up entirely of till, the till may partly derive from previous glacial stages. This is one of a number of facts that complicate the interpretation of the extensive glacial deposits of the Scandinavian Ice Sheet.

In the study of the more scattered deposits from pre-Quaternary ice sheets one should look for all the features described here. It is not only the deposits themselves that are important but also their context and relationship to one another. Tills are important and can often be recognized by characteristic texture or content of striated clasts, but if these do not occur, identification can be more difficult. The combination with other sediments, such as meltwater deposits and rhythmites can be informative. Wherever a typical glacial morphology can be traced, it is also indicative. The combination of regionally scattered features must be considered because of their importance for the general picture of the environment; attention must always be paid to the regional distribution of the deposits.

References

Aario, R., 1992. Puljo moraines and Sevetti moraines. Sveriges Geologiska Undersökning, Ca 81: 7–14.
Aarseth, I. and Mangerud, J., 1974. Younger Dryas

end moraines between Hardangerfjorden and Sognefjorden, western Norway. Boreas, 3: 3–22.

Aber, J.S. and Lundqvist, J., 1988. Glaciotectonic structures in central Sweden and their significance for glacial theory. Géographie physique et Quaternaire, 42: 315–323.

Alm, T., 1993. Øvre Æråvatn - palynostratigraphy of a 22,000 to 10,000 BP lacustrine record on Andøya, northern Norway. Boreas, 22: 171–188.

Andersen, B.G., 1954. Randmorener i Sørvest–Norge. Norsk Geografisk Tidsskrift, 14: 273–342.

Andersen, B.G., 1968. Glacial geology of western Troms, north Norway. Norges Geologiske Undersøkelse, 256, 160 pp.

Andersen, B.G., 1975. Glacial geology of northern Nordland, north Norway. Norges Geologiske Undersøkelse 320, 74 pp.

Andersen, B.G., 1979. The deglaciation of Norway 15,000–10,000 B.P. Boreas, 8: 79–87.

Andersen, B.G., 1980. The deglaciation of Norway after 10,000 B.P. Boreas, 9: 211–216.

Andersen, B.G., Mangerud, J., Sørensen, R., Reite, A., Sveian, H., Thoresen, M. and Bergström, B., 1995. Younger Dryas ice-marginal deposits in Norway. Quaternary International, 28: 147–169.

Andrén, T., 1990. Till stratigraphy and ice recession in the Bothnian Bay. University of Stockholm, Department of Quaternary Research, Report 18, 59 pp.

Anundsen, K., 1985. Changes in shore-level and ice-front position in Late Weichsel and Holocene, southern Norway. Norsk geografisk Tidsskrift, 39: 205–225.

Astakhov, V., 1992. The last glaciation in west Siberia. Sveriges Geologiska Undersökning, Ca 81: 21–30.

Behre, K.-E., 1989. Biostratigraphy of the last glacial period. Quaternary Science Reviews, 8: 25–44.

Berglund, B.E., 1979. The deglaciation of southern Sweden 13,500–10,000 B.P. Boreas, 8: 89–117.

Bergsten, H. and Nordberg, K., 1993. Middle Weichselian marine sediments in eastern Kattegat, Scandinavia. Boreas, 22: 311–318.

Bergström, R., 1992. The Fennoscandian End Moraine Zone in eastern Sweden. Sveriges Geologiska Undersökning, Ca 81: 43–50.

Berthelsen, A., 1979. Contrasting views on the Weichselian glaciation and deglaciation of Denmark. Boreas, 8: 124–132.

Björck, S. and Digerfeldt, G., 1984. Climatic changes at Pleistocene/Holocene boundary in the Middle Swedish Endmoraine zone, mainly inferred from stratigraphic indications. In: N-A. Mörner and W. Karlén (Editors), Climatic Changes on a Yearly to Millennial Basis. D.Reidel, Dordrecht, pp. 37–56.

Boulton, G.S., 1972. Modern Arctic glaciers as depositional models for former ice sheets. Journal of the Geological Society of London, 128: 361–388.

Boulton, G.S. and Jones, A.S., 1979. Stability of temperate ice caps and ice sheets resting on beds of deformable sediment. Journal of Glaciology, 24: 29–43.

Boulton, G.S., Smith, G.D. and Morland, L.W., 1984. The reconstruction of former ice sheets and their mass balance characteristics using a non-linearly viscous flow model. Journal of Glaciology, 30: 140–162.

Boulton, G.S., Smith, G.D., Jones, A.S. and Newsome, J., 1985. Glacial geology and glaciology of the last mid-latitude ice sheets. Journal of the Geological Society of London, 142: 447–474.

Boulton, G.S. and Payne, T., 1994. Mid-latitude ice sheets through the last glacial cycle: glaciological and geological reconstructions. NATO ASI Series I 22: 177–212.

Cepek, A.G., 1965. Geologische Ergebnisse der ersten Radiokarbondatierungen von Interstadialen im Lausitzer Urstromtal. Geologie, 14: 625–657.

De Geer, G., 1909. Dal's Ed. Some stationary ice-borders of the last Glaciation. Geologiska Föreningens i Stockholm Förhandlingar, 31: 511–556.

De Geer, G., 1940. Geochronologia Suecica Principles. Kungl. Svenska Vetenskapsakademien III:18:6, 367 pp.

Denton, G.H. and Hughes, T.J. (Editors), 1981. The last great ice sheets. John Wiley & Sons, New York, 484 pp.

Donner, J., 1978. The dating of the levels of the Baltic Ice Lake and the Salpausselkä moraines in South Finland. Commentationes Physico-Mathematicae, 48: 11–38.

Dreimanis, A., 1989. Tills: their genetic terminology and classification. In: R.P. Goldthwait and C.L. Matsch (Editors), Genetic classification of glacigenic deposits. A.A. Balkema, Rotterdam, pp. 17–83.

Drozdowski, E. and Fedorowicz, S., 1987. Stratigraphy of Vistulian glaciogenic deposits and corresponding thermoluminescence dates in the lower Vistula region, northern Poland. Boreas, 16: 139–153.

Ekman, I. and Iljin, V., 1991. Deglaciation, the Younger Dryas end moraines and their correlation in the Karelian A.S.S.R. and adjacent areas. Geological Survey of Finland, Guide 32: 73–99.

Elverhøi, A. and Solheim, A., 1983. The Barents Sea ice sheet-a sedimentological discussion. Polar Research 1 n.s.: 23–42.

Evzerov, V. and Koshechkin, B., 1981. Glaciation flow stages in the western part of the Kola Peninsula. In: G.I. Gorbunov, B.I. Koshechkin, and A.I. Lisitsyn (Editors), Glacial deposits and glacial history in eastern Fennoskandia. Academy of Sciences of the USSR, Apatity, pp. 38–47.

Fairbanks, R.G., 1989. A 17,000-year glacio-eustatic sea level record: influence of glacial melting rates on the Younger Dryas event and deep-ocean circulation. Nature, 342: 637–642.

Faustova, M.A. and Chebotareva, N.S., 1979. Structure and dynamics of the last European ice sheet. IGCP Project 73/1/24 Quaternary Glaciations in the Northern Hemisphere. Report No 5 on the session in Novosibirsk /USSR/, July, 17–28, 1978. Prague, pp. 99–106.

Faustova, M.A. and Velichko, A.A., 1992. Dynamics of the last glaciation in northern Eurasia. Sveriges Geologiska Undersökning, Ca 81: 113–118.

Fernlund, J.M.R., 1993. The long-singular ridges of the Halland Coastal Moraines, south-western Sweden. Journal of Quaternary Science, 8: 67–78.

Forman, S.L., 1990. Svalbard radiocarbon date list 1. University of Colorado. Institute of Arctic and Alpine Research, Occasional Paper 47, 48 pp.

Garnes, K. and Bergersen, O.F., 1980. Wastage features of the inland ice sheet in central South Norway. Boreas, 9: 251–269.

Gjessing, J., 1966. Deglaciation of southeast and east-central South Norway. Norsk geografisk Tidsskrift, 20: 133–149.

Gjessing, J., 1980. The Aker Moraines in southeast Norway. Norsk geografisk Tidsskrift, 34: 9–34.

Gjessing, J. and Spjeldnaes, N., 1979 Dating of the Grefsen moraine and remarks on the deglaciation of southeast Norway. Norsk geografisk Tidsskrift, 33: 71–81.

Grosswald, M.G., 1980. Late Weichselian Ice Sheet of northern Eurasia. Quaternary Research, 13: 1–32.

Grosswald, M.G., 1988. An Antarctic-style ice sheet in the Northern Hemisphere: toward a new global glacial theory. Polar Geography and Geology, 12: 239–267.

Grosswald, M.G., 1993. Extent and melting history of the Late Weichselian ice sheet, the Barents-Kara continental margin. In: W.R. Peltier (Editor), Ice in the climate system. NATO ASI Series I, 12. Springer Verlag, Berlin, Heidelberg, pp. 1–20.

Hansen, S., 1965. The Quaternary of Denmark. In: K. Rankama (Editor), The Quaternary 1. Interscience Publishers, New York, pp. 1–90.

Hillefors, Å., 1969. Västsveriges glaciala historia och morfologi. Naturgeografiska studier. Meddelanden från Lunds Universitets Geografiska Institution, Avhandlingar 60, 319 pp.

Hillefors, Å., 1975. Contribution to the knowledge of the chronology of the deglaciation of western Sweden with special reference to the Gothenburg moraine. Svensk geografisk årsbok, 51: 70–81.

Hillefors, Å., 1979. Deglaciation models from the Swedish west coast. Boreas, 8: 153–169.

Hirvas, H., 1991. Pleistocene stratigraphy of Finnish Lapland. Geological Survey of Finland, Bulletin 354, 127 pp.

Holmlund, P. and Fastook, J., 1993. Numerical modelling provides evidence of a Baltic Ice Stream during the Younger Dryas in Scandinavia. Boreas, 22: 77–86.

Holmsen, G., 1915. Brædæmte sjøar i nordre Østerdalen. Norges geologiske Undersøkelse 73, 211 pp.

Holtedahl, H. (Editor), 1993. Marine geology of the Norwegian continental margin. Norges geologiske Undersøkelse, Special Publication 6, 150 pp.

Holtedahl, O. (Editor), 1960. Geology of Norway. Norges Geologiske Undersøkelse 208, 540 pp.

Holtedahl, O., 1974. Noen glasifluviale israndavsetninger i den sydlige del av Glomma-vassdragets (nåværende) dreneringsområde. Med en del berggeologisk-morfologiske data fra de nordlige Øyerentrakter. Norges geologiske Undersøkelse, 306: 1–85.

Hooke, R. LeB., 1977. Basal temperatures in Polar Ice Sheets: a quantitative review. Quaternary Research, 7: 1–13.

Hoppe, G., 1948. Isrecessionen från Norrbottens kustland i belysning av de glaciala formelementen. Geographica, 20, 112 pp.

Hoppe, G., 1952. Hummocky moraine regions with special reference to the interior of Norrbotten. Geografiska Annaler, 34: 1–72.

Hoppe, G., 1957. Problems of glacial morphology and the Ice Age. Geografiska Annaler, 39: 1–18.

Hoppe, G., 1972. Ice sheets around the Norwegian Sea during the Würm glaciation. Ambio, Special Report, 2: 25–29.

Houmark-Nielsen, M., 1981. Glacialstratigrafi i Danmark øst for Hovedopholdslinien. Dansk geologisk Forening Årsskrift for 1980: 61–76.

Houmark-Nielsen, M., 1987. Pleistocene stratigraphy and glacial history of the central part of Denmark. Bulletin of the Geological Society of Denmark, 36: 1–189.

Houmark-Nielsen, M., 1989. The last interglacial-glacial cycle in Denmark. Quaternary International 3/4: 31–39.

Hughes, T., Denton, G.H. and Grosswald, M.G., 1977. Was there a late-Würm Arctic Ice Sheet? Nature, 266: 596–602.

Ignatius, H., Korpela, K. and Kujansuu, R., 1980. The deglaciation of Finland after 10,000 B.P. Boreas, 9: 217–228.

Ives, J.D., Andrews, J.T. and Barry, R.G., 1975. Growth and decay of the Laurentide Ice Sheet and comparisons with Fenno-Scandinavia. Naturwissenschaften, 62: 118–125.

Johansson, B.T., 1982. Deglaciationen av norra Bohuslän och södra Dalsland. Chalmers Tekniska Högskola-Göteborgs Universitet, Geologiska Institutionen A 38, 180 pp.

King, L.H., Rokoengen, K., Fader, G.B.J. and Gunleiksrud, T., 1991. Till-tongue stratigraphy. Geological Society of America Bulletin, 103: 637–659.

Kleman, J., 1990. On the use of glacial striae for reconstruction of paleo-ice sheet flow patterns. With application to the Scandinavian ice sheet. Geografiska Annaler, 72A: 217–236.

Kleman, J., 1992. The palimpsest glacial landscape in northwestern Sweden-Late Weichselian deglaciation landforms and traces of older west-centered ice sheets. Geografiska Annaler, 74A: 305–325.

Kleman, J., Borgström, I., Robertsson, A.-M. and Lilliesköld, M., 1992. Morphology and stratigraphy from several deglaciations in the Transtrand Mountains, western Sweden. Journal of Quaternary Science, 7: 1–17.

Koç Karpuz, N. and Jansen, E., 1992. A high-resolution diatom record of the last deglaciation from the SE Norwegian Sea: documentation of rapid climatic changes. Paleoceanography, 7: 499–520.

Kozarski, S., 1988. Time and dynamics of the last Scandinavian ice-sheet retreat from northwestern Poland. Geographica Polonica, 55: 91–101.

Kujansuu, R. and Niemelä, J. (Editors), 1984. Suomen maaperä. Geologian Tutkimuskeskus, Helsinki. Map.

Lagerbäck, R., 1988a. The Veiki moraines in northern Sweden-widespread evidence of an Early Weichselian deglaciation. Boreas, 17: 469–486.

Lagerbäck, R., 1988b. Periglacial phenomena in the wooded areas of northern Sweden-relicts from the Tärendö Interstadial. Boreas, 17: 487–499.

Lagerlund, E., 1977. Förutsättningar för moränstratigrafiska undersökningar på Kullen i Nordvästskåneteoriutveckling och neotektonik. University of Lund, Department of Quaternary Geology, Thesis 5, 106 pp.

Lagerlund, E., 1980. Litostratigrafisk indelning av Västskånes Pleistocen och en ny glaciationsmodell för Weichsel. University of Lund, Department of Quaternary Geology, Report 21, 120 pp.

Lagerlund, E., 1987. An alternative Weichselian glaciation model, with special reference to the glacial history of Skåne, South Sweden. Boreas, 16: 433–459.

Lagerlund, E., Knutsson, G., Åmark, M., Hebrand, M., Jönsson, L.-O., Karlgren, B., Kristiansson, J., Möller, P., Robison, J.M., Sandgren, P., Terne, T. and Waldemarsson, D., 1983. The deglaciation pattern and dynamics in South Sweden, a preliminary report. LUNDQUA Report 24, 7 pp.

Lagerlund, E. and Houmark-Nielsen, M., 1993. Timing and pattern of the last deglaciation in the Kattegat region, southwest Scandinavia. Boreas, 22: 337–347.

Larsson, W., 1945. Zur Kenntnis der spätglazialen Eisbewegungen westlich des Wenersees, Schweden. Sveriges Geologiska Undersökning, C 466, 13 pp.

Liivrand, E., 1991. Biostratigraphy of the Pleistocene deposits in Estonia and correlations in the Baltic region. Stockholm University, Department of Quaternary Research, Report 19, 114 pp.

Liljequist, G.H., 1974. Notes on meteorological conditions in connection with continental land-ice in the Pleistocene. Geologiska Föreningens i Stockholm Förhandlingar, 96: 293–298.

Ljungner, E., 1949. East-west balance of the Quaternary ice caps in Patagonia and Scandinavia. Bulletin of the Geological Institutions of the University of Upsala, 33: 11–96.

Lundqvist, G., 1940. Bergslagens minerogena jordarter. Sveriges Geologiska Undersökning, C 433, 87 pp.

Lundqvist, G., 1942. Var äro issjöarnas sediment? Geologiska Föreningens i Stockholm Förhandlingar, 64: 160–162.

Lundqvist, J., 1972. Ice-lake types and deglaciation pattern along the Scandinavian mountain range. Boreas, 1: 27–54.

Lundqvist, J., 1973. Isavsmältningens förlopp i Jämtlands län. Sveriges Geologiska Undersökning, C 681, 187 pp.

Lundqvist, J., 1986. Late Weichselian glaciation and deglaciation in Scandinavia. Quaternary Science Reviews, 5: 269–292.

Lundqvist, J., 1987. Glaciodynamics of the Younger Dryas marginal zone in Scandinavia. Implications of a revised glaciation model. Geografiska Annaler, 69A: 305–319.

Lundqvist, J., 1988. Younger Dryas-Preboreal moraines and deglaciation in southwestern Värmland, Sweden. Boreas, 17: 301–316.

Lundqvist, J., 1989a. Late glacial ice lobes and glacial landforms in Scandinavia. In: R.P. Goldthwait and C.L. Matsch (Editors), Genetic classification of glacigenic deposits. A.A. Balkema, Rotterdam/Brookfield, pp. 217–225.

Lundqvist, J., 1989b. Rogen (ribbed) moraine-identification and possible origin. Sedimentary Geology, 62: 281–292.

Lundqvist, J., 1992. Glacial stratigraphy in Sweden. Geological Survey of Finland, Special Paper 15, pp. 43–59.

Lundqvist, J., 1995. The Younger Dryas ice-marginal zone in Sweden. Quaternary International, 28: 171–176.

Lundqvist, J. and Saarnisto, M. (Editors), 1990. Termination of the Pleistocene. Field Conference Norway-Sweden-Finland May 9–16, 1990. Excursion and Abstracts. Geological Survey of Finland, Guide 31, 163 pp.

Lundqvist, J. and Miller, U., 1992. Weichselian stratigraphy and glaciations in the Tåsjö-Hoting area, central Sweden. Sveriges Geologiska Undersökning, C 826, 35 pp.

Mangerud, J., 1981. The Early and Middle Weichselian in Norway: a review. Boreas, 10: 381–393.

Mangerud, J., 1991. The Scandinavian Ice Sheet through the last interglacial/glacial cycle. In: B. Frenzel (Editor), Klimageschichtliche probleme der letzten 130 000 Jahre. G.Fischer, Stuttgart, New York, pp. 307–330.

Mangerud, J., Larsen, E., Longva, O. and Sønstegaard, E., 1979. Glacial history of western Norway 15,000–10,000 B.P. Boreas, 8: 179–187.

Mannerfelt, C. M., 1945. Några glacialgeologiska formelement och deras vittnesbörd om inlandsisens avsmältningsmekanik i svensk och norsk fjällterräng. Geografiska Annaler, 27: 1–239.

Marcussen, I., 1977. Deglaciation landscapes formed during the wasting of the late Middle Weichselian ice sheet in Denmark. Danmarks Geologiske Undersøgelse II: 110, 72 pp.

Martinsson, D.G., Pisias, N.G., Hays, J.D., Imbrie, J., Moore, T.C. and Shackleton, N.J., 1987. Age dating and orbital theory of the ice ages: development of a high resolution 0 to 300 000 year chronostratigraphy. Quaternary Research, 27: 1–29.

Mercer, J.H., 1969. The Allerød oscillation: an European climatic anomaly? Arctic and Alpine Research, 1: 227–234.

Miller, U., 1977. Pleistocene deposits of the Alnarp Valley, southern Sweden-Microfossils and their stratigraphical application. University of Lund, Department of Quaternary Geology, Thesis 4, 125 pp.

Möller, H., 1962. Annuella och interannuella ändmoräner. Geologiska Föreningens i Stockholm Förhandlingar, 84: 134–143.

Möller, P., 1987. Moraine morphology, till genesis, and deglaciation pattern in the Åsnen area, south-central Småland, Sweden. LUNDQUA Thesis 20, 146 pp.

Niemelä, J., 1971. Die quartäre Stratigraphie von Tonablagerungen und der Rückzug des Inlandeises zwischen Helsinki und Hämeenlinna in Südfinnland. Geological Survey of Finland, Bulletin 253, 79 pp.

Obrutschew, W.A., 1930. Die Verbreitung der Eiszeitspuren in Nord-und Zentralasien. Geologische Rundschau, 21: 243–283.

Olsson, I.U., 1974. The eighth International Conference on Radiocarbon Dating. Geologiska Föreningens i Stockholm Förhandlingar, 96: 37–44.

Persson, C., 1983. Glacial deposits and the central Swedish end moraine zone in eastern Sweden. In: J. Ehlers (Editor), Glacial deposits in north-west Europe. A.A. Balkema, Rotterdam, pp. 131–139.

Rainio, H., 1985a. Första Salpausselkä utgör randzonen för en landis som avancerat på nytt. Geologi, 37: 70–77.

Rainio, H., 1985b. Pohjois-Karjalan Salpausselkien aikaiset reunamuodostumajaksot tarvitsevat nimet [Names needed for north Karelian marginal formations contemporaneous with the Salpausselkä ridges]. Geologi, 37: 48–50.

Rainio, H., 1991. The Younger Dryas ice-marginal formations of southern Finland. Geological Survey of Finland, Guide, 32: 25–41.

Rainio, H., Saarnisto, M. and Ekman, I., 1995. Younger Dryas end moraines in Finland and NW Russia. Quaternary International, 28: 179–192.

Raukas, A., 1977. Ice-marginal formations and the main regularities of the deglaciation in Estonia. Zeitschrift für Geomorphologie, N.F., Suppl.-Bd 27: 68–78.

Raukas, A. and Gaigalas, A., 1993. Pleistocene glacial deposits along the eastern periphery of the Scandinavian ice sheets-an overview. Boreas, 22: 214–222.

Ringberg, B., 1989. Upper Late Weichselian lithostratigraphy in western Skåne, southernmost Sweden. Geologiska Föreningens i Stockholm Förhandlingar, 111: 319–337

Risberg, J., Miller, U. and Brunnberg, L., 1991. Deglaciation, Holocene shore displacement and coastal settlements in eastern Svealand, Sweden. Quaternary International, 9: 33–37.

Rokoengen, K., Bugge, T. and Løfaldli, M., 1979. Quaternary geology and deglaciation of the continental shelf off Troms, north Norway. Boreas, 8: 217–227.

Ronnert, L., 1989. The deglaciation of the Berghem area, south-western Sweden. Chalmers Tekniska Högskola-Göteborgs Universitet, Geologiska Institutionen A 64, 46 pp.

Ruddiman, W.F. and McIntyre, A., 1981. The North Atlantic during the last deglaciation. Palaeogeography, Palaeoclimatology, Palaeoecology, 35: 145–214.

Salvigsen, O., Adrielsson, L., Hjort, C., Johansson, K., Kelly, M., Landvik, J.Y. and Ronnert, L., 1992. Ice movements in eastern Svalbard. LUNDQUA Report, 35: 9–16.

Sauramo, M., 1929. The Quaternary geology of Finland. Bulletin de la Commission Géologique de Finlande, 86, 110 pp.

Sejrup, H.P., Aarseth, I., Ellingsen, K.L., Reither, E. and Jansen, E., 1987. Quaternary stratigraphy of the Fladen area, central North Sea: a multidisciplinary study. Journal of Quaternary Science, 2: 35–58.

Sejrup, H.P., Haflidason, H., Aarseth, I., King, E., Forsberg, C.F., Long, D. and Rokoengen, K., 1994. Late Weichselian glaciation history of the northern North Sea. Boreas, 23: 1–13.

Söderberg, P., 1988. Notes on the continuation of the Salpausselkä ice marginal zone in the northern Baltic Proper. Geological Survey of Finland, Special Paper 6, pp. 69–72.

Sørensen, R., 1979. Late Weichselian deglaciation in the Oslofjord area, south Norway. Boreas, 8: 241–246.

Sørensen, R., 1992. The physical environment of Late Weichselian deglaciation of the Oslofjord region, southeastern Norway. Sveriges Geologiska Undersökning, Ca 81: 339–346.

Soyez, D., 1974. Studien zur Geomorphologie und zum letztglazialen Eisrückzug in den Gebirgen Süd-Lapplands/Schweden. Geografiska Annaler, 56A: 1–71.

Stay, B., 1979. Warwenstratigraphische Untersuchungen an Bändertonen des Bornholm-Beckens und des pleistozänen Eisstausee-Beckens von Bara (bei Malmö). Thesis. Christian-Albrecht-Universität zu Kiel, 92 pp.

Strömberg, B., 1981. Calving bays, striae and moraines at Gysinge-Hedesunda, central Sweden. Geografiska Annaler, 63A: 149–154.

Strömberg, B., 1989. Last Weichselian deglaciation and clay varve chronology in East-Central Sweden. Sveriges Geologiska Undersökning, Ca 73, 70 pp.

Strömberg, B., 1990. A connection between the clay varve chronologies in Sweden and Finland. Annales Academiæ Scientiarum Fennicæ A III, 154: 1–32.

Strömberg, B., 1992. The final stage of the Baltic Ice Lake. Sveriges Geologiska Undersökning, Ca 81: 347–353.

Tavast, E., Raukas, A. and Künnapuu, 1983. O drevnikh pogrebennikh vrezakh v Tallinne [Ancient buried valleys in Tallinn]. Eesti NSV teaduste akadeemia tometised, 32: 79–85.

Torell, O., 1872. Undersökningar öfver istiden. I. Öfversigt af Kongl. Vetenskaps-Akademiens Förhandlingar 1872, 10: 25–66.

Torell, O., 1873. Undersökningar öfver istiden. II. Öfversigt af Kongl. Vetenskaps-Akademiens Förhandlingar 1873:1: 47–64.

Ulfstedt, A.-C., 1980. Isrecessionen i Västerbottens och södra Norrbottens fjälltrakter. University of Stockholm, Department of Physical Geography, Forskningsrapport 43, 106 pp.

Wingfield, R.T.R., 1989. Glacial incisions indicating Middle and Upper Pleistocene ice limits off Britain. Terra Nova, 1: 538–548.

Vorren, T., 1992. Glaciations of the Barents Sea-an overview. Sveriges Geologiska Undersökning, Ca 81: 367–372.

Vorren, T.O., Hald, M. and Lebesbye, E., 1988. Late Cenozoic environments in the Barrents Sea. Palaeoocenography, 3: 601–612.

Vorren, T.O., Vorren, K.-D., Alm, T., Gulliksen, S.S. and Løvlie, R., 1988. The last deglaciation (20,000 to 11,000 BP) on Andøya, northern Norway. Boreas, 17: 41–77.

4. ENVIRONMENT AND CLIMATE OF SARTAN MAXIMUM AND LATE GLACIAL IN SIBERIA

Stanislav A. Arkhipov

During the last Sartan glacial maximum (20–18 ka) the northwest sector of the Siberian subcontinent, including the Taimyr Peninsula, was covered by a large glacial sheet whose main centre was located on the Kara Sea shelf. This was the Siberian portion of the vast Euroasian Ice Sheet which covered both the continent and the shelf of the adjacent Arctic seas. Ice-dammed basins formed all along the ice front, and glacial waters flowed southwest through the Turgay Pass into the Aral, Kaspian, and Black seas and eventually into the Mediterranean Sea. Loess accumulated along the southern periphery of the West Siberian Plain beyond the ice-dammed basins. Treeless landscapes dominated in the periglacial regions of Siberia.

After the glacial maximum (18 to 14–13 ka), global warming and aridization led to a progressive disruption of the Sartan glaciers. During deglaciation (14–13 to 10 ka) dramatic environmental changes occurred. Transgression of a late glacial sea led to a rapid collapse of the Kara-centered glacier and to the disappearance of ice-dammed lakes. Rivers discharged north. Rapid and contrasting climate changes were reflected in flora and landscape changes.

The Sartan glaciation was the ninth and last in a series of Quaternary glaciations culminating 20–18 ka around the time of the North American late Wisconsinan and the European late Weichselian glaciations. The effects of the Sartan glaciation have been most extensively studied in western Siberia.

Two major hypotheses have been debated over the past century. One contends that the glacier had a restricted distribution on the continent around the North Urals, Byrranga on Taimyr, and the Putorana plateau. The other visualizes a widespread glaciation that included the continental part of north Siberia and the shelves of the adjacent Arctic seas. The restricted glacier distribution hypothesis was proposed by Cherskiy (1882) who thought that the extreme aridity of the Siberian climate precluded the development of widespread glaciations. This idea greatly influenced later authors, some of whom still accept the concept of a glaciation limited to the Siberian continent (Strelkov, 1965; Velichko, 1987, 1993; Astakhov, 1992). Other authors, however, gathered evidence to show that the glaciers were much more widespread (Kropotkin, 1876; Sax, 1953; Kind, 1974). In the 1950s and 1960s glacial features and deposits were discovered in northwestern Siberia and on the Taimyr Peninsula that seem to confirm widespread glaciation, a conclusion that was further confirmed by analysis of satellite images from Meteor and Landsat (Astakhov, 1976; Arkhipov et al. 1980; Arkhipov, 1984). The existence of a widespread Euroasian Ice Sheet that covered northwest Siberia, the Taimyr Peninsula, and the Kara shelf seems to be proven and is accepted in this essay (Grosswald, 1977, 1980, 1994; Hughes et al., 1977).

It is believed that this huge Euroasian Ice Sheet developed in response to the cooling of the arctic areas and the regression of the world oceans. The effect of the Gulf Stream current was cut off. Glaciation began on the continent, in the Urals, Byrranga Mountains and Putorana Plateau, and quickly spread to northwest Siberia and then to the adjacent Kara Sea shelf. Temperatures dropped drastically. During the Sartan thermal minimum the annual average temperature in the arctic high latitudes decreased, according to Chizhov (1970), to -33°C, and summer temperature to 12°C. Ice thickness increased progressively on the Kara Sea shelf, initially through freezing of the water from below and, after it grounded, from continuous snow precipitation whose ablation was minimal due to very low summer temperature.

It is believed that the glacier reached thicknesses of about 1.5–2.0 km in the ice divide of the Kara Sea shelf. From there the glacier flowed in various directions. The southward movement is well documented by an eastwest-oriented series of end moraines in the northern part of west Siberia. Such moraines could not have been formed by glaciers flowing from the Urals or the Byrranga and Putorana mountains (Fig. 4.1). The most western Bajdarateskij ice lobe flowed around the small Arctic Urals and formed well-developed end moraines on adjacent plains. Its traces along the flanks of the mountains occur to about 700 m asl. This suggests that the thickness of the glacier was in excess of 500 m. On the Taimyr Peninsula, the ice lobes spread out from the Kara ice center over the low Byrranga Mountains (Kind and Leonov, 1982). In north Enisei they collided with glaciers flowing from the Putorana Plateau, deflecting them to the south (Figs. 4.1, 4.2). The margins of the glaciers are clearly defined by various glacial landforms and sediments such as moraines, kames, glaciolacustrine deposits and eskers. These features were partially mapped for the first time in this region between the 1930s and the 1950s, and more accurate determination of their distribution became possible through analysis of remote sensing images (Astakhov, 1976). Three major Sartan marginal belts of glacial deposit were distinguished in this way. They are the Khaduttei, corresponding to the maximum stage at latitude 66–67°N, and two recessional ones; Tanama and Jamal-Gydan, traced to the north between 67–71°N (Fig. 4.2). The maximal Khaduttei stage is dated at about 20–18 ka (Fig. 4.3).

An important component of the Sartan environment was the ice-dammed lakes. In western Siberia a system of interconnected dammed basins appeared in front of the ice sheet (Fig. 4.2) and formed the so-called Mansy Lake Sea (Volkov, 1964, 1975). To the east, in the middle of Siberia, in the basins of Enisei, Lena, Angara, and Low Tunguska, there were also ice-dammed lakes, though smaller in size (Grosswald, 1977; Volkov et al., 1978). All of them drained through the Turgay Pass in the southwestern outskirts of western Siberia. Traces of glacier-dammed lakes and spillways of the southwestern Turgay drainage area are represented by the Urtam terrace (60–80 m above sea level), which is synchronous with the Khadattei marginal glacial belt. Radiocarbon age of the terrace ranges between 23–22 to 12.8–12.3 ka (Arkhipov et. al., 1980).

During glacial and stadial stages, the climate of Siberia was much colder than at present. At middle latitudes (70–50°N), vast periglacial areas experienced average annual temperatures 8–10°C lower than those of the present day. This Siberian periglacial area was characterized by a treeless landscape: tundra near the margin of the ice sheet and in the mountains, and tundrasteppe associations on the plains (Grichuk, 1970). The study of phytofossils (fossil pollen and spores) shows a dominance of spore (mainly green mosses), and herb and shrub species: herbage, absinth, cereals, arctic and hyperarctic lycopodia, shrub betula (*Betula nana*). This flora type developed during each glacial epoch and resulted in the recurrent disappearance of taiga forest and the southward expansion of tundra into the area now occupied by steppe (Arkhipov et. al., 1980). Thus, the paleogeography of the northwest of the Siberian subcontinent during the Sartan glacial maximum (20–18 ka) was characterized by glaciers and vast periglacial treeless areas covered by tundra-steppe landscapes.

Deglaciation

After the glacial maximum, deglaciation occurred relatively rapidly.

The first phase, 18 to 14–13 ka, produced aridization. This was caused by a sharp decrease in annual precipitation, and resulted in massive reduction of glacial sheet size. The Kara ice center received less snow, and the flow of glacial streams from it decreased; as a result, glacial margins on the continent began to recede to the north. Temporal stages of this retreat are marked by the recession of the Tanama and Jamal-Gydan ice margins north of the Khaduttei belt (Fig. 4.2). During the early stages, however, changes were not drastic. The main features of glacial sheets and dammed basins remained. Tundra and tundra-steppe dominated in the periglacial area (Arkhipov et. al.,1980).

The final wasting of the Sartan glaciers occurred between 14–13 and 10 ka. This de

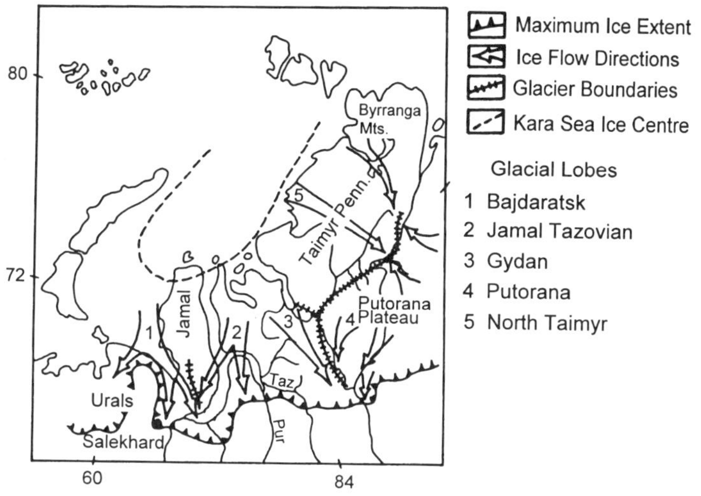

Fig. 4.1. Sartan ice sheet in western Siberia and Taimyr.

glaciation phase was part of a global event and was caused by changes in Earth's orbital parameters, according to Milankovich's theory (Imbrie and Imbrie, 1988). Global warming occurred, glaciers melted, and sea level rose. The sea covered the Kara shelf, which was still isostatically depressed by the weight of the glacier. The Kara glacier collapsed gradually, the western part of the Kara shelf was probably first freed of glacier ice due to the influence of warmer waters of the Atlantic Ocean. The ice cover of the eastern and central sectors of the shelf degraded more slowly, and remnants are still preserved on the North Land islands. On the continent, glaciers remained as scattered stagnant ice masses. This led to the formation of the last set of Jamal-Gydan recessional moraines in the northern part of the West Siberian Plain (Figs. 4.2, 4.3). The glaciers also retreated into the Arctic Urals and the Putorana Plateau, as indicated by some recessional moraines in the area.

The sea transgressed inland from the Kara Sea shelf. The area of Jamal, Gydan, and west Taimyr were flooded rapidly, and ice remnants formed icy islands, as indicated by a series of waterlaid moraines in the Jamal-Gydan peninsulas (Arslanov et al., 1983). Maximum sea level is recorded at a height of 50–60 m asl. This led to the burial of glacial sediments under marine deposits and to the drowning of the lower reaches of the Ob' and Irtysh rivers forming large estuaries. The evolution of these estuaries is well documented by a 55–60 m highest-sea-level mark and a series of three other terraces that, near Salekhard, are found respectively at 30–40 to 55–60 m asl (terrace III), 20–30 m asl (terrace II), and 15–18 m asl (terrace I). Similar terraces are also present in the Jamal and Gydan peninsulas and in western Taimyr. The terraces also cut into the Khaduttei moraines and can be traced to the north up to the Kara Sea, where the two upper terraces merge into a late glacial marine transgression plain. Marine microfauna (foraminifera) are found in deposits of terrace III. The lower terraces instead have alluvial and marine deposits (terrace II) and alluvial sediments (terrace I). This sequence indicates that the formation of terraces started during the maximum early late glacial marine transgression and proceeded after emersion due to isostatic uplift (Arkhipov et. al., 1977, 1980). In the Ob' and Irtysh valleys, south of Salekhard, the two upper terraces covered by estuarine sediments merge into Kolpashevo terrace, whose height is within the range of the transgression plains in the north. A series of ^{14}C dates from 12.8(12.3) to 10.6(10.2) ka obtained from the Kolpashevo terrace also indicates their similar age and correct late-glacial age.

The postglacial river drainage was reestablished in the north. Vast coastal plains emerged in eastern and northern Siberia (67–72°N), with cold and by sparse vegetation. Sediments were reworked by wind and developed permafrost and

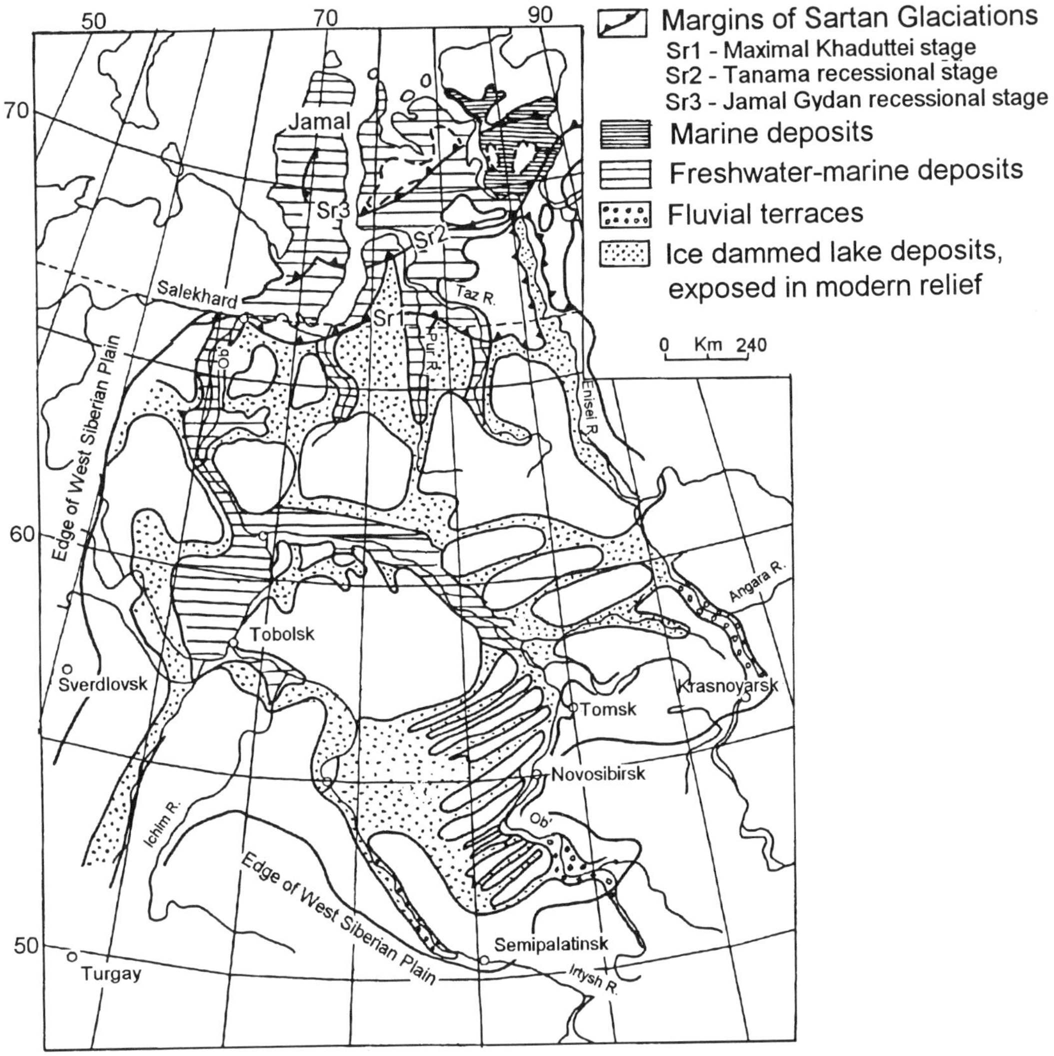

Fig. 4.2. Paleogeography of the West Siberian Plain during the Sartan glaciation (after Arkhipov et. al., 1980).

patterned grounds. In eastern Siberia the aeolian loess cover containing ice veins is called "edoma" (loess-ice formation; Tomirdiaro, 1980; Bolikhovsky, 1987). The wind-blown sediments locally reach significant thicknesses (15–20 m) and cover marine sediments of terrace III and alluvial deposits of terrace II. Similar loess deposits occur south of the permafrost zone (south of 60°N) in the inner regions of Siberia. This loess cover mantles the Kolpashevo terraces and underlies the modern river floodplain deposits. This loess has been dated from 10.6–10.2 to 9.8–9.2 ka and is therefore related to the Late Dryas event and records the late glacial time of the region.

Climatic and environmental changes during deglaciation have been recorded by variations in vegetation. A major change occurred during the early stages of late glacial marine transgression at the end of The Early Dryas and in Bölling (14–12.3 ka) times as indicated by phytofossils. The early arctic tundra and tundra-steppe were replaced by typical tundra and, in more southern regions, by forest tundra. During Bölling times (about 13–12.3 ka) an increased amelioration of the climate is evidenced along river valleys by the occurrence of spruce, pine, and betula forests and by local Siberian cedar (*Pinus siberica*). This resembles the modern northern taiga. Renewed

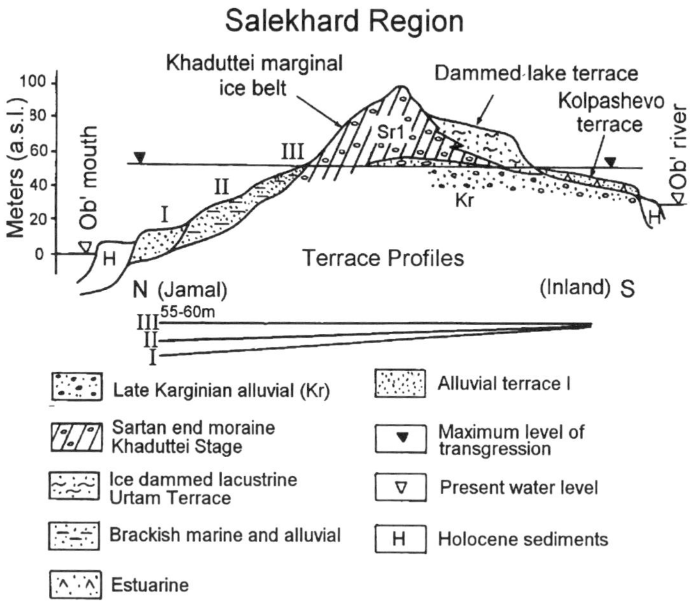

Fig. 4.3. Geological cross-section of the northwestern part of the West Siberian Plain.

cooling occurred in the Middle Dryas (about 12.3–11.8 ka), woody plants disappeared, and the tundra-steppe was restored. Subsequent warming during the 1,000-year-long Alleröd (11.8–10.8 ka) was accompanied first by wet and then by arid climatic conditions. Among xerophytes, cereal-absent assemblages and betula-spruce forests similar to the modern northern and middle taiga appeared for a short time, they then died and gave way to open betula woodlands. The last severe cooling occurred during the Late Dryas (about 10.8 - 10 ka). It led to the total disappearance of woody plants and the wide occurrence of xerophyte plant assemblages with elements of arctic flora. Winters became particularly cold. In southwestern Siberia, January average temperatures were 10°C lower than at present.

Modern Siberian landscapes began to develop in the pre-boreal period (10—9 ka) during the early Holocene. Carbon 14 dating indicates that the vast (about 10 million hectares) peatlands of west Siberia started developing about 9.1–9.2 ka (Nejshtadt, 1969; Arkhipov ct. al., 1980). Of particular interest is the Bol'shoe Vasjuganskoe Boloto (Big Vasjugan peat-bog), which occupies the whole northern part of Ob'-Irtysh interfluve and has an average peat thickness of about 5–6 m. During the hypsithermal (warmest Holocene time; 6–5 ka) in west Siberia, the flora zones shifted 500–600 km farther north than they are today (Fig. 4.4). The arctic tundra disappeared, forests of pine, *Abies* and cedar moved north. Lime, oak, and elm grew in the south-taiga zone among betula and pine woods. All this indicates that the annual temperatures in the present sub-arctic zone were 3–4°C higher than the modern ones during the hypsithermal time (Levina et al., 1989; Zakharov, 1991; Levina and Orlova, 1993; Arkhipov et al, 1994).

Conclusion

1. Several Pleistocene glaciations affected Siberia. The last Sartan glaciation (23–10 ka) is generally accepted to have been synchronous with the late Wisconsinan and the late Weichselian glaciations of, respectively, North America and Europe.

2. Major geological events characterized the end of each glaciation. However, each glacial epoch in Siberia differed from the others in size and volume of ice cover, spacial distribution of

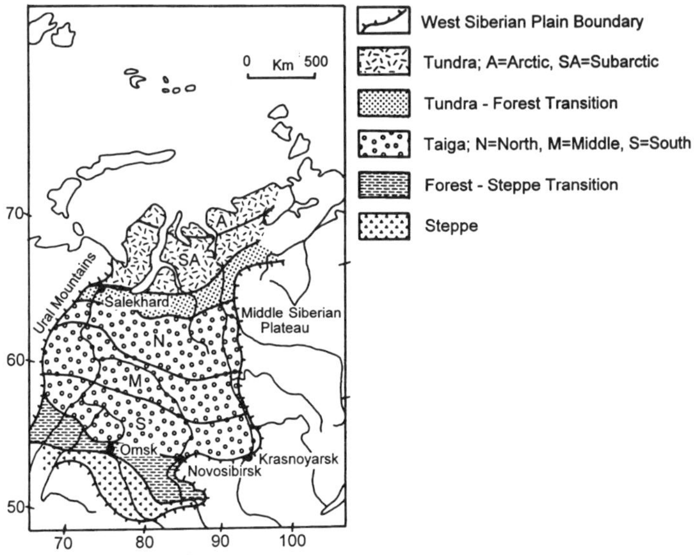

Fig. 4.4. Flora zonality and peatland of the West Siberian Plain according to Ilyina et al. (1985) and Nejshtadt (1969). The peatlands cover the taiga and the tundra-forest zones.

marginal glacial belts, and duration and number of stages and sizes of dammed basins and glacio-eustatic transgressions. Although similar in overall pattern, each deglaciation thus developed its peculiar features. The Sartan deglaciation, being the last, is the best preserved, and it provides an opportunity to determine in some detail the dramatic changes that occurred in the environments and vegetation from glacial to early postglacial times.

3. The Sartan glaciation, like other more ancient ones, probably started in the north Siberian mountains and spread to the Kara Sea shelf. A thick ice divide developed on the Kara Sea shelf. This is confirmed by the orientation and the space distribution of marginal ice features such as terminal and recessional moraines.

4. The ice sheet blocked the discharge of west Siberian rivers to the north. Therefore, numerous, vast ice-dammed lakes formed on the inner west Siberian regions and discharged southwest through the Turgay Pass.

5. The late Sartan deglaciation was accompanied by rapid and contrasting climate changes. These are manifested in wide-ranging, shifting floral zones. In the central regions of the western Siberian plain, the early postglacial tundra gave place to taiga and then again to tundra. Although there were several climatic reversals, with the appearance and disruption of woody plants, the overall postglacial trend from 14 to 9 ka was one of progressive climatic warming.

6. The deglaciation led, at the beginning of the Holocene (10 Ka), to the development of floral zones similar to the present ones. During the remainder of the Holocene, the changes were less dramatic and occurred gradually. Floral zones migrated farther to the north. During the climatic optimum (6–5 ka) the floral zones migrated farthest to the north by 500–600 km, suggesting the existence of an average temperature 3–4°C warmer than present-day temperature in what is now the subarctic zone.

7. At the beginning of the Holocene interglacial, peat started developing on the vast west Siberian wetlands.

References

Arkhipov, S.A. 1984. Late Pleistocene glaciation of western Siberia. In: A.A. Velichko A.A. (Editor), Late Quaternary environments of the Soviet Union. University of Minnesota Press. Minneapolis, pp. 13–19.

Arkhipov, S.A., Astakhov, V.I., Volkov, I.A., Volkova, V.S. and Panychev, V.A., 1980. Paleogeography of West Siberian Plain in the late Zyryanka glacial maximum. Nauka, Novosibirsk, 107 pp. (in Russian).

Arkhipov, S.A., Votakh, M.R., Golbert, A.V., Gudina, V.I., Dovgal, L.A. and Yudkevich, A.I., 1977. Late glaciation in lower Ob' river region. Nauka, Novosibirsk, 214 pp. (in Russian).

Arkhipov, S.A., Volkova, V.S., Bakhareva, V.A., Votakh, M.R., Levina, T.P., Krivonogov, S.K. and Orlova L.A., 1994. Nature climatic changes in west Siberia to 2000. Geology and Geophysics, 1: 3–21 (in Russian).

Arslanov, Kh.A., Lavrov, A.S. and Potapenko, L.M., 1983. New data on the late Pleistocene glaciation of the north of west Siberia. In: S.A. Arkhipov (Editor), Glaciations and paleoclimates in Siberia during the Pleistocene. Institute Geology and Geophysics Siberian Branch of Academy of Science USSR, Novosibirsk, pp. 27–35 (in Russian).

Astakhov, V.I., 1976. Geological evidence of the Pleistocene ice dispersal centre on the Kara Shelf. Doklady Acad. Sci. USSR 23: 1178–1181 (in Russian).

Astakhov, V.I., 1992: The last glaciation in west Siberia. Sveriges Geologiska Undersöking. Ser. Ca 81: 21–30.

Bolikhovsky, V.F., 1987. Edome deposits of west Siberia. In: J.M.K.Punning, I.K. Ivanova, N.V. Kind and O.A. Chichagova (Editors), New data on geochronology of the Quaternary Period. Nauka, Moscow, pp. 128–135 (in Russian).

Cherskiy, I.D., 1882. To the problem of traces of ancient glacials in east Siberia. Izvestia East-Siberian branch of Russian geographical society, 12: 4–5 (in Russian).

Chizhov O.P., 1970 On the changes of the arctic basin status from the last glacial maximum. In: A. I. Tolmachev (Editor), Arctic Ocean and its coast in the Cenozoic. Gydrometioizdat, Leningrad, pp. 71–76 (in Russian).

Grichuk, M.P., 1971. Principles of formation of present spore-pollen spectra as the basis of interpretation of fossil spectra. In: V. N. Saks (Editor), History of vegetation evolution in the extraglacial region during the late Pliocene and Quaternary. Nauka, Moscow, pp. 12–19 (in Russian).

Grosswald, M.G. 1977. Last Eurasian ice cover. Materials of glaciological studies, chronicle, discussion. Moscow, 30: 45–60 (in Russian).

Grosswald, M.G., 1980. Late Weichselian ice sheet of northern Eurasia. Quaternary Research, 13: 1–32.

Grosswald, M.G., 1994. Drumlin fields of the Novaja Zemlja Uralian Upland and their bearing on the Kara Sea center of glaciation. Geomorphology, 1: 40–53 (in Russian).

Hughes, T, Denton, G.H. and Grosswald, M.G., 1977. Was there a late-Wurm Arctic Ice Sheet? Nature 266: 596–602.

Ilyina, N.S., Lapshina, E.I. and Lavrenko, I.,N., 1985. Flora cover of West Siberian Plain. Novosibirsk, Nauka, 251 pp. (in Russian).

Imbrie, J. and Imbrie K.P., 1988. Ice ages-solving the mystery. Hillside, NJ, 1979. Translation from English. Progress, Moscow, 262 pp.

Kind, N.V., 1974. Late Quaternary geochronology according to isotopes data. Nauka, Moscow, 254 pp. (in Russian).

Kind, N.V., Leonov, B.N. (Editors), 1982: The anthropogene of the Taimyr Peninsula. Nauka, Moscow, 181 pp. (in Russian).

Kropotkin, P.A., 1876. Studies on glaciations. Notes of Irkutsk Branch of Russian Geographical Society, 1, 717 pp. (in Russian).

Levina, T.P. and Orlova, L.A., 1993. Holocene climatic rhythms of the south of west Siberia. Geology and Geophysics 3: 38–55 (in Russian).

Levina, T.P., Orlova, L.A., Panychev, V.A. and Skabichevshaja, N.A., 1989. Paleogeography and ^{14}C chronology at the Pleistocene-Holocene boundary in the pre-Altai Plain. In: V.Z. Zykina (Editor), Cenozoic of Siberia and USSR North East. Nauka, Novosibirsk, pp. 128–129 (in Russian).

Nejshtadt, M.I., 1969. Peat. In: M. I. Nejshtadt (Editor), The Holocene. Nauka, Moscow, pp. 206–215 (in Russian).

Saks, V.N., 1953. The Quaternary in the Soviet Arctic. Trudy Inst Geol. of the Arctic 77, Leningrad, Moscow, 627 pp. (in Russian).

Strelkov, S.A., 1965. The north of Siberia. Nauka, Moscow, 256 pp. (in Russian).

Tomirdiardo, S.V., 1980. Loess glacial formation of east Siberia during Late Pleistocene and Holocene. Nauka, Moscow, 183 pp. (in Russian).

Velichko, A.A., 1987: Modern status of the concept of Earth glaciations. Izvestya Academy Science USSR, Seria geograph., 3: 21–34 (in Russian).

Velichko, A.A., 1993: Introduction by the editor. In: A.A. Velichko (Editor), Evolution of landscapes and climates of the northern Eurasia. Late Pleistocene - Holocene elements of prognosis. Nauka, Moscow, 36–37 pp. (in Russian).

Volkov, I.A. and Volkova, V.S., 1964. On the late Pleistocene Lake Sea in the south of West Siberian Plain. In: V.N. Saks (Editor), Quaternary geology, geomorphology and palaeogegraphy of Siberia. Institute Geology and Geophysics, Siberian Branch of Academy of Science, USSR, Novosibirsk, pp. 109–129 (in Russian).

Volkov, I.A. and Volkova, V.S., 1975. Great

periglacial drainage system in Siberia. In: D.D. Kwasov (Editor), Lake History in Pleistocene. vol. 2., pp. 133–140 (in Russian).

Volkov, I.A., Grosswald, M.G. and Troitsky, S. 1978. On the drainage of periglacial waters during the last glaciation in west Siberia. Izvestya Academy of Sci. USSR, Seria geograph., 4: 25–35 (in Russian).

Zakharova V.A. (Editor), 1991. Evolution of the climate, biota and primitive-man habitat in the Late Cenozoic in Siberia. United Institute Geology, Geophysics, Mineralogy, Siberian Branch R.A.N., Novosibirsk, 89 pp. (in Russian).

5. TERMINATION OF THE PLEISTOCENE AND HOLOCENE CHANGES IN SOUTH AMERICA AND OTHER GLACIATED PARTS OF THE SOUTHERN HEMISPHERE

Chalmers M. Clapperton

The cyclical downswing in the amount of solar radiation received by the middle to high latitudes in the Northern Hemisphere led to the most recent glacial cycle which began about 125,000 years ago (Dawson, 1992). For the next 115,000 years, conditions on Earth were cooler for most of the time than they were in the 10,000 years of the Holocene. Large ice sheets developed over vast areas of the Northern Hemisphere, and montane glaciers advanced worldwide. In Antarctica, the smaller West Antarctic Ice Sheet expanded, but the giant East Antarctic Ice sheet (which has existed since the mid-Miocene some 14 million years ago) changed very little because of a reduction in the already low precipitation experienced on this highest and coldest continent. Evidence from the analysis of oxygen isotopes in cores from polar ice sheets and deep ocean sediments confirms that climate changed periodically during the glacial cycle, with intervals of severe cold giving way to warmer (but still colder than now) conditions every few thousand years. However, an overall trend toward conditions of maximum cold culminated in the last global glaciation maximum (LGGM) at 21,000–18,000 years ago, when the volume of glacier ice on the planet was highest and global (eustatic) sea level was lowest (by ca. 120 m). A consequence of this extreme environmental situation was that atmospheric and sea surface temperatures were also at their lowest (CLIMAP, 1981), atmospheric circulation was violently turbulent and widespread arid areas developed on most continents (Goudie, 1992; Clapperton, 1993a; Williams et al., 1993).

An upswing in solar radiation after 21,000 yr BP gradually initiated the collapse of Northern Hemisphere ice sheets and caused glacier recession worldwide, but this took several thousand years to happen, and the main explosion of warmth did not occur until about 14,000–13,000 yr BP (Broecker and Denton, 1990). It can be identified in most paleo-environmental records and has been labeled Termination 1. Despite this fundamental change toward global warming, relatively small-scale but distinctive reversals of the trend occurred ca. 12,500–10,000 yr BP, before true postglacial conditions pervaded the globe. For the past 10,000 years, Earth has enjoyed an interglacial climate, generally defined by an absence of large ice sheets over mid- to high latitudes in the Northern Hemisphere. However, it is clear that climate has continued to fluctuate on a small but significant scale, with important consequences for the biosphere in some places.

This chapter outlines the estimated extent of glaciers in South America and some other parts of the Southern Hemisphere prior to Termination 1, and traces glacial fluctuations during both the transition to interglacial conditions and the Holocene. The discussion focuses on regions for which the most data are available, and includes the Andes of South America, the Southern Alps of New Zealand, islands in the Southern Ocean and the sub-Antarctic, and West Antarctica (Fig. 5.1). The framework for this discussion is temporal rather than regional; we attempt to paint a broad canvas of environmental conditions in the Southern Hemisphere during the last glacial-interglacial transition.

The Modern Glaciers

It is probably helpful to give a general description of modern glaciers and ice fields in the Southern Hemisphere and some adjacent tropical mountains to assist a full appreciation of the changes that occurred during the past ca. 30,000 years as the Earth's climate moved to a state of

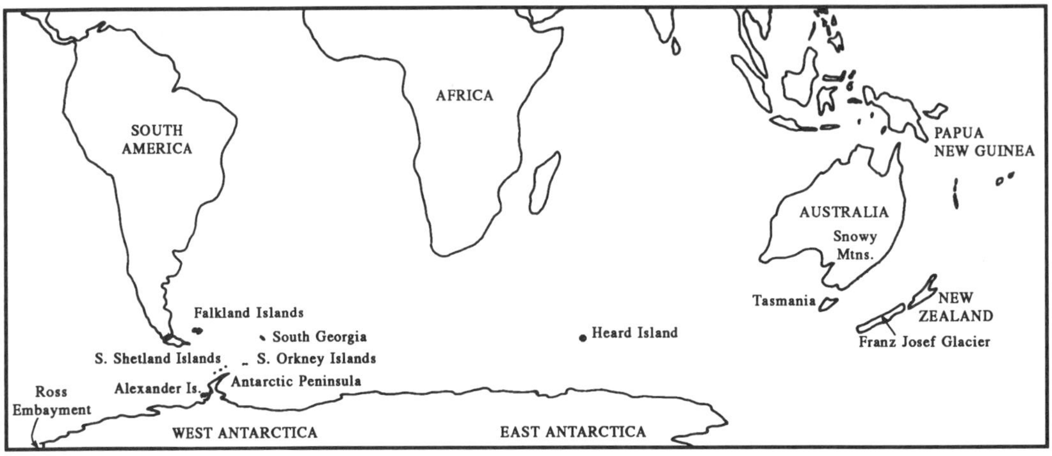

Fig. 5.1. Map of the Southern Hemisphere.

extreme cold before rapidly changing to the more equable conditions of the Holocene. This will be done on a regional basis, bearing in mind that the primary controls on glacier extent in most areas are temperature and precipitation. At any locality, these depend on factors such as altitude, latitude, proximity to the ocean, and circulation changes in the interlinked ocean-atmosphere system.

Antarctica

Most of the world's glacier ice is locked up in the East Antarctic Ice Sheet. This immense feature could be regarded as one of the wonders of the world; it blankets an entire continent with more than 2–3 km of ice thickness in places (Fig. 5.2). The most remarkable aspect of the ice cover over East Antarctica is its age, a full-blown ice sheet having formed by 14 million years ago in the geological period known as the Miocene (Kennett, 1977). It developed as plate tectonic movements isolated this high continent in the south polar region, where low temperatures combined with moisture from the surrounding Southern Ocean led to the progressive buildup of a continental ice sheet. It is known that the ice sheet was even larger than it is now sometime before about 11 million years ago, since volcanic ashes of this age lie on slopes previously covered by moving ice (Marchant et al., 1993), but subsequent topographical changes resulting from tectonic earth movements and increased aridity (resulting from altitude, extreme cold, and distance from the open ocean) caused the ice sheet to shrink roughly to its present proportions. While local glaciers adjacent to the ice sheet have fluctuated over short distances during the past 3 million years, expanding when the Earth's climate was warmer and producing increased precipitation over the maritime fringe of Antarctica, the main ice sheet has changed very little. Since mean annual temperature over the polar plateau is below -50°C, it would take a rise in temperature of at least 25°C to cause widespread meltdown of the ice sheet, according to glaciological modeling studies (Huybrechts, 1993). Thus, in terms of recent global warming scenarios, catastrophic deglaciation of East Antarctica is inconceivable. The East Antarctic Ice Sheet is also buffered from climatic change because it terminates mostly on land and has a relatively small proportion of its volume draining to the open sea through ice streams. This contrasts with the ice sheets that covered large parts of the Northern Hemisphere during the Pleistocene. As large areas of the Laurentide and Scandinavian ice sheets drained to the Atlantic through ice-streams, they were vulnerable to rapid disintegration when sea surface temperature and global sea level rose at the end of the glacial cycle (Denton and Hughes, 1986).

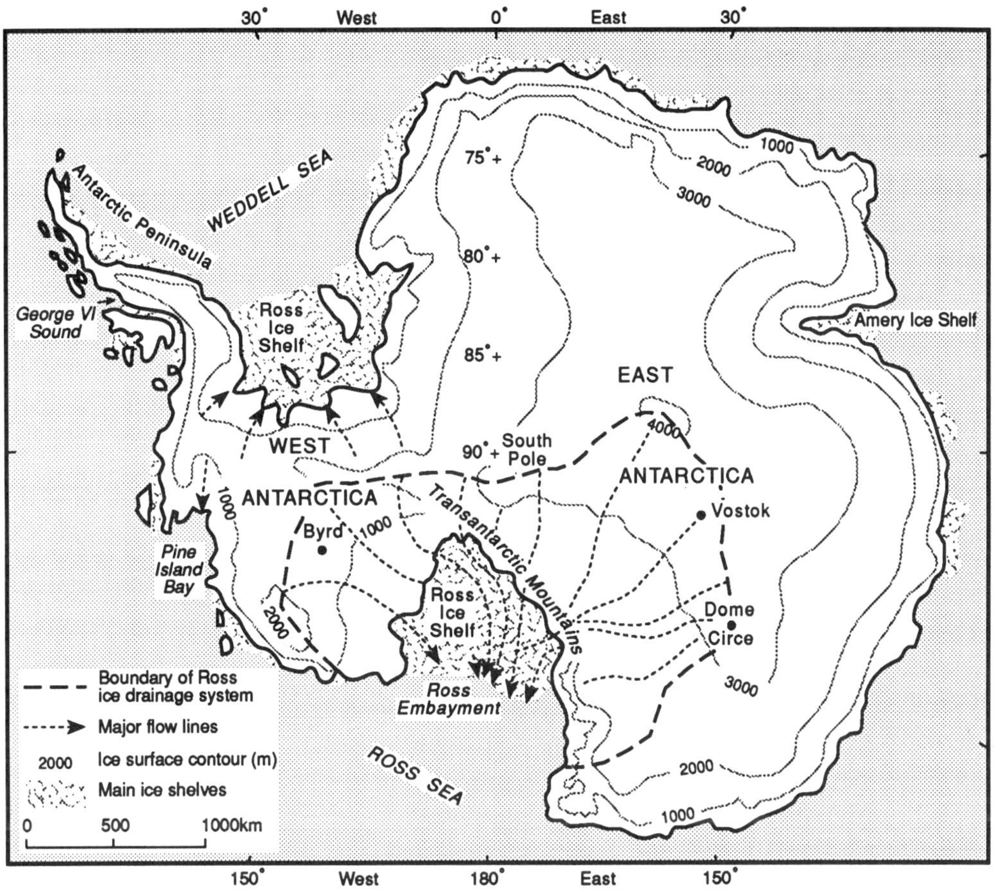

Fig. 5.2. Map of Antarctica, showing ice cover.

Since the smaller West Antarctic Ice Sheet is built over an archipelago, its base is largely grounded below sea level, making it sensitive to any global change that causes sea level to rise. This could lead to extensive breakup through iceberg calving around the marine margins, particularly in the Ross Embayment and Weddell Sea areas, where the ice sheet is drained by major ice streams feeding into the floating ice shelves. However, while field evidence shows that some ice shelves expanded and contracted during the Holocene (Clapperton and Sugden, 1982), there is no reason to believe the West Antarctic Ice Sheet collapsed catastrophically during the last glacial-interglacial cycle, although some scientists believe that it may have done so earlier (Barret et al., 1992). The narrow limb of the Antarctic Peninsula, as well as adjacent sub-Antarctic islands like the South Shetlands and South Orkneys, have a more maritime climate and lie in lower latitudes (lat.60–65°S). Although they are extensively ice-covered today, with glaciers terminating in fjords or on the modern beach, they also have ice-free areas that contain evidence that the ice fields were once much larger. Thus, this cold maritime region is sensitive to fluctuations in global climate (Clapperton, 1990).

South America

Despite the popular belief that the Andes are perennially ice-capped, large glaciers and ice fields are rather restricted (Clapperton, 1983). For example, in the tropical Andes, glaciers more than a few hundred meters in length are confined to Peru (Fig. 5.3, Table 5.1), where the biggest ice fields lie in the Cordillera Blanca, a 200km-

Fig. 5.3. Map of glacier-bearing and formerly glacier-covered ranges in the Andes.

TABLE 5.1. *List of the main glacier-bearing massifs in the Andes*[a] *(from Clapperton, 1993b)*

Country	Glacier massif	Coordinates	Highest summit(m) (approx)	Longest glacier (km)
Venezuela	Sierra Nevada de Mérida	8°30'N:71°W	5,000	<1
Colombia	Sierra Nev. de Santa Marta	11°N:73°'W	5,775	<1
	Cordillera Central (Nev. Ruiz-Tolima)	5°30'N:75°30'W	5,215	3
	Cordillera Oriental (Sierra Nev. del Cocuy)	6°30'N:72°W	5,493	2-3
Ecuador	Cordillera Occidental (Nev. Chimborazo)	1°25'S:78°45'W	6,310	3
	Cordillera Oriental (Nev. Antisana)	0°30'S:78°10'W	5,790	2-3
Peru	Cordillera Blanca	9°S:78°W	6,768	4
	Cordillera Huayhuash	10°30'S:77°W	6,632	2
	Cordillera Carabaya	13°30'S:71°50'W	6,384	1-2
	Cordillera Apolobamba	14°50'S:69°W	5,816	2
	Cordillera Central (Nev. Coropuna)	12°-17°S:72-77°W	6,425	1-2
Bolivia	Cordillera Real	16°S:68°W	6,500	1-2
	Cordillera Quimsa Cruz	17°S:67°W	6,105	<1
	Cordillera Occidnetal (Nev. Sajama)	18°S:68°W	6,520	<1
Argentina and Chile	Cordillera de los Andes (Nev. Ojos del Salado)	25°50'S:68°W	6,908	<1
	Cordillera Principal (Nev. Aconcagua)	30°-36°S:7°W	6,960	>5
	Cordillera de los Andes (Co. Tronador)	36°-50°S:71°50'W	3,410	3-5
	Hielo Patagonico Norte	47°S:73°30'W	4,058	45
	Hielo Patagonico Sur	48°-52°S:73°30'W	3,000	60
	Peninsula Muñoz-Gamero	53°S:73°W	1,585	<10
	Isla Riesco y Santa Inés	53°30'S:72°-54°W	1,665	>10
	Cordillera Darwin	54°30'S:69°-70°W	2,300	>25

[a] Many small glaciers exist also on individual mountains not named here.

long massif of granitic rocks with many peaks exceeding 6,000 m in altitude. Humid trade winds from the east bring abundant precipitation (>2,000 mm yr^{-1}) to nourish extensive ice fields that discharge glaciers up to 2–3 km long in places. Hundreds of small glaciers exist where the land rises above 5,000 m elsewhere in the tropical Andes, on mountain crests and in semicircular basins carved out of the ranges by glacial and frost action, but the only true ice cap is the one at Quelccaya in southeast Peru. This feature is 70 km^2 in size and lies on a volcanic plateau at 5,000–5,300 m.

Glaciers are even smaller south of Peru, where the climate becomes drier, as in the ranges of Bolivia, northern Chile, and northern Argentina. Although glaciers and permanent snow are fairly widespread in the more humid eastern ranges rising above 5,000 m they are rare or absent from the drier western ranges, even on peaks that exceed 5,500 m, and where the mean annual temperature is below zero. While 500–800 mm of precipitation falls annually in this region, evaporation is excessive and leads to moisture deficit. As seasonally arid conditions extend to about latitude 41°S, only the very highest peaks retain a permanent snow cover and support glacier ice. Thus an east-west flight across the mountains from Santiago, Chile, to Mendoza, Argentina, where the Andes are at their highest, will reveal only rather small glaciers, many of which look rather dirty because of an extensive cover of rock detritus shed by the surrounding peaks. In some valleys, glaciers apparently grade into tongue-shaped masses of boulders, known as rock glaciers and debris-covered glaciers, which are a feature of this dry region where intense freezing and thawing shatter the bedrock to produce frequent rockfalls.

The average altitude of the Andes drops from about 4,000 m to 2,000 m south of latitude 37°S, but seasonal precipitation is high enough to create sparkling ice caps on volcanic peaks rising above 2,500 m, such as Villarica, Tronador, and Osorno. South of latitude 45°S, a dramatic increase in snow and ice cover, the result of almost year-round precipitation from the maritime westerlies in the "Roaring Forties" and "Furious Fifties", culminates in the largest ice fields in the Southern Hemisphere outside Antarctica. These, the North and South Patagonian ice fields, cover areas of ca. 4,200 km^2 and >14,000 km^2 respectively and discharge outlet glaciers tens of kilometers long to fjords opening into the Pacific Ocean on the west and to large glacially deepened lake basins east of the Andes (Fig. 5.3). The largest glaciers descend into Argentina from the southern ice field, where the Upsala Glacier is ca. 60 km long and the scenically attractive Perito Moreno Glacier is ca. 25 km long (Fig. 5.4). Large ice fields and lengthy outlet glaciers also occur farther south, on Isla Santa Ines and in the Darwin mountains of Tierra del Fuego in southernmost South America.

The estimated altitude of the average snowline running north-south along the Andes is at about 5000 m in equatorial latitudes and rises to 5,500–6,000 m farther south in the drier subtropical Andes before descending steadily to ca. 250 m in the cold and wet climate of the southern Andes (Clapperton, 1993b). It is believed that this snowline was depressed on average by 500–1000 m, depending on moisture supply, during the most extensive stage of glaciation in the last glacial cycle (Broecker and Denton, 1990).

New Zealand, Australia, Papua-New Guinea, Africa

Situated between latitudes 42–45°S, the Southern Alps of New Zealand also lie across the path of precipitation-bearing westerlies and, like the southern Andes, contain large glaciers that in the west descend to evergreen beech forests. The largest are nourished by ice fields in the highest terrain of all (>3000 m) in the Mt. Cook National Park and include the Franz Josef and Fox glaciers flowing to the west and the Tasman and Murchison flowing to the east; some exceed 20 km in length (Suggate, 1990). Impressive arcs of morainic debris mark the positions of these glaciers during past intervals of colder climate, particularly east of the divide (many of those in the west terminated below present sea level).

There are no glaciers in any of the Australian mountains, but there is evidence that ice fields formerly existed in the Snowy Range of New South Wales and in western Tasmania when global climate was colder (Colhoun and Fitzsimons, 1990). Small ice fields are still present on the higher massifs of Papua-New Guinea just south of the equator; most have receded drastically this century, but moraines deposited during their last glacial extent suggest a snowline lowering of at least 1000 m (Brown, 1990). A similar situation prevails in Africa, where existing glaciers are confined to the high equatorial massifs of Mt. Kilimanjaro (5,800 m), Mt. Kenya (4800 m), and Mt. Ruwenzori (5000 m). The glaciers are all small, collectively amounting to ca. 10 km^2 of ice, compared to an estimated amount of ca. 640 km^2 for the last glacial cycle (Rosqvist, 1990); all of the glaciers have receded drastically during the present century. Past glaciation on the African massifs has been studied extensively (Hastenrath, 1984; Osmaston, 1989a, 1989b; Mahaney, 1990), revealing that the equatorial snowline was formerly more than 1,000 m lower.

Southern Ocean

South of latitude 55°S, the temperate water of the southern oceans gives way rapidly to the cold polar water surrounding Antarctica. This narrow zone of transition, above which dense freezing mists commonly hang, is known as the Antarctic Convergence. Most islands south of the Convergence support glaciers, particularly those with high mountains, such as South Georgia, Heard Island, and the South Orkney and South Shetland Islands (Fig. 5.1). Here the snowline lies at

Fig. 5.4. The Perito Moreno Glacier, a major eastern outlet of the South Patagonian Icefield. It terminates in Lago Argentino and is surrounded by beech forest.

200–500 m, depending on exposure to the westerlies, and many glaciers terminate either close to sea level or in the sea. Well over 50 percent of the area of these so-called sub-Antarctic islands is currently glacier-covered and during the last glacial cycle, most islands were almost completely buried by ice caps (Clapperton, 1990).

Evidence of Earlier Glaciers

In all of these regions, there is abundant glacial geomorphological and stratigraphical evidence that glaciers were formerly much larger and, in places, existed where there is no longer even a permanent snow cover. The detailed mapping of landforms of glacial erosion and deposition has helped delimit the former extent of ice-age glaciers in many parts of the Southern Hemisphere (Quaternary Science Reviews, vol. 9, 1990). It is clear from these studies that the glaciers fluctuated in size many times during the most recent glacial cycle and that the history of climatic change that drove the oscillations is extremely complex. While it is comparatively straightforward to delimit how far the glaciers once extended, it is much more difficult to establish when they did so. As a general rule, morainic loops across the valley floors are younger toward the valley head, and indicate a *relative* chronology of glacial stages as the glaciers became smaller. It is possible to derive an *absolute* chronology of events by the radiocarbon dating of organic material buried by, or contained within, or overlying the glacial sediments for stages up to ca. 40,000 yr ago, and by using techniques like lichenometry (based on the growth rate of lichens on boulders) and dendrochronology (based on the width of annual growth rings of trees) for dating stages within the last few hundreds to thousands of years. Although detailed studies of past glacial histories in mountain regions of the Southern Hemisphere are few, enough is known at reconnaissance level to establish a preliminary pattern of glacier fluctuations during the transition from the last glacial maximum to the present interglacial interval. This is outlined in this essay and in the preliminary time-distance diagram summarizing glacier fluctuations for the last ca. 70,000 yrs in the mountain ranges of the tropics and the Southern Hemisphere (Fig. 5.5).

Glacial Conditions at 30,000–21,000 yr BP

Analyses of the paleoenvironmental record in cores extracted from the Greenland Ice Sheet and from deep ocean sediments in the North Atlantic suggest that major cold periods during the last glacial cycle peaked at intervals of about 7,000–8,000 years (so-called Heinrich events) and minor cold periods at intervals of ca. 2,000–2,500 years (so-called Dansgaard-Oeschger events). Although equivalent high-resolution records are yet to emerge from other regions, it seems that these climatic events affected the atmosphere worldwide (Jouzel, 1994). The most intense cold peak of the early part of the last glacial cycle was at ca. 65,000 yr BP, an interval known from oxygen-isotope studies as marine isotope stage 4 (MIS 4). The four most recent significant climatic reversals occurred at ca. 27,000 yr BP, 21,000–18,000 yr BP, 14,000 yr BP and 11,000 yr BP (Bond et al., 1993), but smaller-scale events occurred in between, at intervals of 2,000–3,000 years (Bond and Lotti, 1995).

While we might assume that glaciers sensitive to global atmospheric change, like those in mountain regions with humid climates, would have expanded promptly during each cooling interval, the detailed record of such glacier fluctuations is not well known. This is to be expected in dynamic mountain environments where much erosion occurs, removing sedimentary evidence, and also if the deposits left by one glacial advance are buried or removed by a later one of larger extent. Nevertheless, preliminary studies in the Andes and in some other montane regions of the Southern Hemisphere suggest that during the last glacial cycle, glaciers advanced farthest sometime before ca. 40,000 yr BP, and attained slightly smaller dimensions at and after ca. 27,000 yr BP. This is suspected from the radiocarbon age of peat and organic soils overlying morainic deposits in the northern Andes, and of marine shells incorporated by glacier advances in the Magellan Strait area of the southernmost Andes (Clapperton, 1992). In parts of the humid tropical Andes, such as in the higher ranges of Venezuela and Colombia and Ecuador, glaciers expanded more than 10 km beyond the modern ice fields (Schubert, 1974; van der Hammen et al., 1981; Clapperton, 1986).

It is of growing interest that these glacial maxima in some tropical and Southern Hemisphere ranges occurred well before the recognised LGGM, when the Earth's volume of glacier ice was highest. Of course, in regions where glacier extent was influenced by sea level, as in areas such as Antarctica, sub-Antarctic islands and the west coasts of Chile and New Zealand, where the ice advanced offshore, glaciers generally extended furthest when global sea level was lowest, as at the LGGM. The reason that glaciers in some montane regions achieved their greatest advances well before the LGGM is not fully understood, but it is likely to be associated with atmospheric temperature and humidity. For example, cold air can hold less moisture than warm air. Thus, as the average global temperature fell in response to reduced solar radiation, and feedback effects associated with the growth of giant ice sheets in the Northern Hemisphere, the cooler atmosphere moving over cooler ocean surfaces was able to hold less moisture, thereby depriving the land masses of precipitation. Since this effect was at its maximum during the LGGM, it is understandable that mountain glaciers worldwide would advance less than during earlier cool intervals when the ice sheets did not achieve such large dimensions and when precipitation levels were presumably higher. It is also recorded that vast lakes filled the southern altiplano basin in Bolivia at ca. 31,000–27,000 yr BP, thus indicating much greater moisture availability in a region now characterized by salt flats.

In the southern Andes, however, the fall in global sea level and a northward shift in the belt of highest precipitation (in latitudes 50–40°S) meant that western outlet glaciers from the Patagonian Ice Cap reached their maximal limits coincident with the LGGM at ca. 21,000 yr BP (Porter, 1981). This also occurred in West Ant-

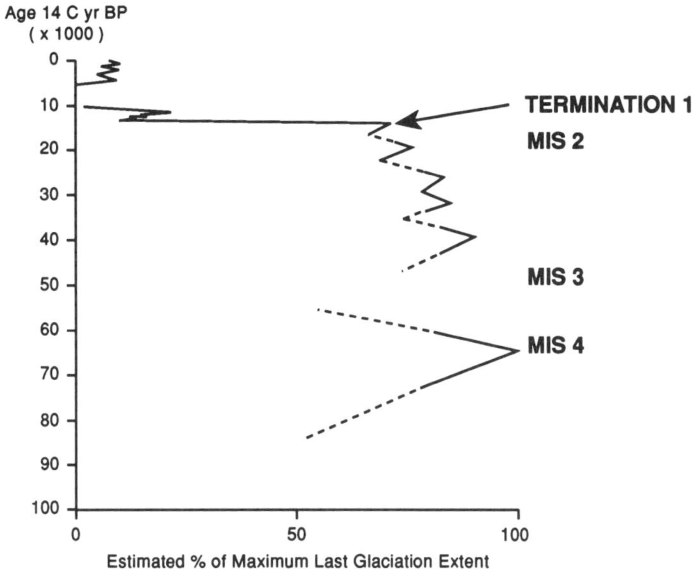

Fig. 5.5. Preliminary time-distance diagram showing glacier fluctuations in mountain ranges of the tropics and the Southern Hemisphere during the last glacial cycle and the Holocene.

arctica, where outlet glaciers flowing from this relatively maritime ice sheet moved across the Ross Embayment and penetrated the Dry Valleys area of the Transantarctic Mountains to deposit morainic sediment that has been dated to the LGGM (Denton et al., 1989). It is presumed that glaciers on islands in the Southern Ocean and in the sub-Antarctic reached their maximal extents also during the LGGM due to the influence of low sea level (Clapperton, 1990). Although the East Antarctic Ice Sheet probably extended farther offshore as global sea level fell, it may have actually shrunk in volume due to precipitation starvation. Certainly the colder temperatures were not significant in this already frigid environment where the snowline is currently at sea level. The main change in the Antarctic domain was the immense increase in the area of seasonal and permanent pack ice fringing the continent, more than doubling the area of white surface (Huybrechts, 1993) and thereby enhancing the amount of radiation loss through reflection (albedo effect); it also reduced moisture supply to the continent. While the chronology of last glaciation advances in New Zealand is poorly known, the more extensive "Early Otira" stage may have occurred during MIS 4 (Suggate, 1990), and the "Late Otira" stages possibly peaked during MIS 2 (ca. 30,000–14,000 yr BP).

Glacial Conditions at 21,000–14,000 yr BP

By the time of the global glaciation maximum at 21,000–18,000 yr BP, most glaciers and ice fields in southerly latitudes of the Southern Hemisphere lay at or not far inside their maximal limits. In lower latitudes, where climate had also reached its coldest, reduced precipitation caused glaciers and some paleolakes to become smaller, but they remained quite extensive compared to modern (interglacial) times. From ca. 18,000–15,000 yr BP, a slow but steady rise in global sea level indicated progressive melting of the large northern ice sheets. The presence of 16,000–15,000 yr old organic deposits in rock basins previously covered by glaciers on South Georgia (Rosqvist, 1994) and southern Patagonia (Porter et al., 1992), suggests relatively rapid ice recession in these areas due to iceberg calving as most glaciers terminated in the rising sea or in growing proglacial lakes. Analysis of fossil pollen in peat bogs of this age, however, indicates that the climate remained cold, dry, and windy in most latitudes of the Southern Hemisphere until well after ca. 15,000 yr BP (Clapperton, 1993a), implying that the recession of land-terminating glaciers was probably not as rapid as those breaking up into deep water.

A significant change in climatic conditions

culminated at about 14,000 yr BP, when many glaciers re-advanced, reversing the overall trend in recession (Clapperton, 1995). Simultaneously, large freshwater lakes began to fill the enclosed basins of the previously dry Altiplano plateau of Bolivia, north Chile and north Argentina, as well as western Patagonia (Clapperton, 1993a). This change in the "state of health" (mass balance) of glaciers suggests a significant increase in precipitation rather than a drastic climatic cooling, although thermal decline may also have been involved. That glacier expansion and paleolake development were coeval is demonstrated around the Salar de Uyuni in the Bolivian altiplano (Fig. 5.3), where glacial and lacustrine deposits are interbedded and date to ca. 14,000–13,000 yr BP (Clapperton, et al., 1995). Paleolake Tauca which filled the southern altiplano with more than 100 m of water, and cut prominent shorelines around the basin, endured until ca. 10,000 yr BP, possibly with fluctuations in level during the four millenia of its existence (Servant and Fontes, 1978).

In the Chilean Lakes Region at latitude 40°S and in the Southern Alps of New Zealand, outlet glaciers from the mountain ice fields re-advanced at least two thirds of the way to their previous limits, culminating their advance soon after ca. 14,000 yr BP (Porter, 1981; Suggate, 1990). In places, the glaciers advanced over organic soils, peat bogs, and even small forests that had developed since ice recession from the glacial maximum. That this climatic event was not confined to temperate southern latitudes is confirmed by similar evidence in the tropical Andes of southeast Peru, where glaciers in the Nevado Ausangate-Quelccaya Ice Cap area bulldozed peat beds at about 14,000 yr BP (Mercer and Palacios, 1977). The largest glacier on Chimborazo volcano in Ecuador re-advanced into a valley to block its drainage and form an ice-marginal lake soon after ca. 14,000 yr BP (Clapperton, 1993c), while morainic deposits studied in some of the Colombian ranges are believed to have formed sometime before ca. 13,000 yr BP (Helmens, 1988).

There is thus considerable evidence throughout most latitudinal belts of the Southern Hemisphere that a distinct climatic reversal culminated between ca. 15,000–13,000 yr BP. Climate might have cooled only slightly, reversing the earlier warming trend, but it certainly seems to have become much more humid, perhaps documenting a major change in the global atmosphere, possibly linked to sea surface temperature and atmospheric dynamics. It is evident from the ice core records that the atmospheric dust content dropped sharply after 14,000 yr BP (Broecket and Denton, 1990), indicating a reduction in atmospheric turbulence and greater moisture availability (thus vegetation cover) over the continents.

Glacial Conditions at 14,000–10,000 yr BP

It is believed that a rapid improvement in climate followed the glacial re-advance, peaking around 14,000 yr BP. For example, large outlet glaciers south of the Chilean Lakes Region had withdrawn from Isla de Chiloe across a marine embayment and into the mainland valleys before ca. 12,900 yr BP (Heusser, 1990). Moreover, interpretation of fossil pollens and fossil beetles contained in peat deposits suggest a rapid expansion of peat mires and pioneer shrubs and forests on the heels of the receding glaciers (Ashworth and Hoganson, 1984; Markgraf, 1991). Treeline rose in the tropical Andes as the grasslands receded to higher altitudes (Salgado-Labouriau, 1991), while peat bogs began to form even on isolated islands in the Southern Ocean before 11,000 yr BP (Smith, 1981). This improvement in climatic conditions, particularly the rise in temperature and the reduction in storminess, is possibly reflected by the migration of people into Tierra del Fuego by 12,000 yr BP (Martinic, 1992). It is believed that they crossed Magellan Strait by a land bridge formed by glacial deposits and breached only by the outflow stream from a proglacial lake; this was prior to the rise in global sea level that had converted Magellan Strait into a seaway, probably before the start of the Holocene (Clapperton, 1992).

A minor reversal of the warming trend in some mountain ranges of the tropics and the Southern Hemisphere appears to have occurred sometime between 12,000–10,000 yr BP, when small-scale fluctuations in climate caused glacier re-advances that have now been identified in a number of places. For example, distinctive moraines marking a re-advance of the Franz Josef Glacier in New Zealand buried tree fragments radiocarbon-dated to 11,100 yr BP (Denton and Hendy, 1994), while in the southernmost Andes an apparently coeval re-advance of a southern outlet of the South Patagonian Ice field bulldozed a volcanic ash (from Reclus volcano) that had been spread across the area at 11,877 yr BP (Marden, 1993; Marden and Clapperton, in press). Farther north in the Andes, outlet glaciers from the Quelccaya Ice Cap in southeast Peru re-advanced over peat beds with an age of 10,900 yr BP, while Reschreiter Glacier on Chimborazo volcano in Ecuador possibly re-advanced to form a second ice-marginal lake (following drainage of the first one) soon after 11,000 yr BP (Clapperton and McEwan, 1985). In addition, interpretation of fossil pollens at sites as far afield as southernmost Chile and Colombia suggested to some palynologists that a distinct climatic reversal occurred between ca. 11,000–10,000 yr BP, but others have disagreed with this view (reviewed in Clapperton, 1993b).

The precise nature and cause of this climatic oscillation in the Southern Hemisphere and adjacent tropical mountains is not fully understood, but some researchers have speculated (Clapperton, 1993c) that it was a muted, smaller-scale equivalent of the late-glacial climatic change that generated a strong response in northwest Europe (the Younger Dryas stade). Further work is necessary to establish whether late glacial climatic changes were synchronous between hemispheres, but preliminary evidence suggests that they were. The presence of two to three distinct moraine ridges in parts of the Andes, as in some parts of Europe, suggests sensitivity of the climate-glacier system to fluctuating global conditions during this critical changeover from late-glacial to interglacial climate.

Interglacial conditions: 10,000 yr BP to the Present

Paleoenvironmental records from the Southern Hemisphere document rapid warming around 10,000 yr BP (Grove, 1988). By this time ice fields had disappeared completely from plateaux and valleys that had nourished substantial thicknesses only a few millenia earlier, and only areas that were cold enough and wet enough to support permanent snow remained glacier-covered. Although full interglacial conditions appear to have spread throughout the Southern Hemisphere by the Holocene, this did not bring 10,000 years of climatic stability, but quite the reverse. The natural behavior of the Earth's interlinked atmosphere-ocean-glacier system is variable and periodically responds to mechanisms forcing changes, albeit on a small scale. For example, variations in solar output (as during sunspot cycles), atmospheric loading with volcanic aerosols (as from the Mt. Pinatubo eruption in 1992), and internal oscillations in the ocean-atmosphere system (as with El Niño events in the Pacific) all produce climatic variability worldwide. Many such changes are felt globally, although perhaps with different intensity in different regions.

The radiocarbon-dated record of Holocene glacier and environmental changes in the Southern Hemisphere is still at reconnaissance level due to the lack of detailed investigation. Nevertheless, broad trends have been identified in the Andes, South Georgia, and New Zealand and form the basis for the following discussion. The data also permit tenuous comparison with the Holocene record of the Northern Hemisphere.

Early Holocene: 10,000–5000 yr BP

Very little is known about the extent to which mountain glaciers receded as a consequence of warmer climate in the early Holocene, but it can be assumed that ice fields existed in high altitude regions receiving abundant precipitation. Around the glaciers, moraine and outwash deposits left during recession from late-glacial limits became covered with plants as skeletal soils developed and, in places, the first postglacial forests grew; large boulders became covered with lichens and thin weathering rinds began to form on rocks, like the skin of an onion. These land surfaces became potential dating horizons whenever they were buried by a renewed advance of the glaciers. This is known to have happened periodically during the Holocene. For example, in the high and tectonically active ranges of the Andes and the New Zealand Alps, vast amounts of rock waste are shed from the peaks and valley walls to form screes at the bases of cliffs, and any glaciers that lie there become covered with bouldery supraglacial moraine. This material is deposited at the margins of the glaciers in their lower (ablation) zones, and, if the glacier is expanding, the detritus will bury the previous land surface and its cover of soil, plants, and, possibly, trees. When the glacier recedes once more, the steep, unsupported walls of these lateral moraines often become gullied, exposing the internal structure of the deposit, including the buried former land surface, which can be sampled and dated to obtain an age for the glacier fluctuations. Moraines that are only a few hundred years old have been dated by measuring the size of lichens growing on surface boulders, using a lichen growth curve established from specimens on surfaces of known age, such as gravestones, and by counting the annual growth rings of trees knocked over by the glacier advance as well as those growing on moraines left by the advance. These techniques, together with radiocarbon dating, have provided the data shown in Table 5.2, which lists the ages of apparent cold and/or wet intervals that caused glacier re-advances during the Holocene.

It is believed by some that an early Holocene cooling caused some glaciers to advance following their retreat from late-glacial limits, but this remains uncertain for most parts of the tropics and the Southern Hemisphere. While glacial moraines document cold or wet intervals when glaciers expand, the glacial record does not indicate the warm climatic periods clearly. Thus, we must turn to the paleoecological evidence derived from the analysis of fossil pollens and insects. These remains show clearly that an interval of warmth culminated sometime during the interval ca. 8,000-6,000 yr BP, with temperatures probably a degree or two higher than those in the present century (Grove, 1988). Unless precipitation increased in the mountain regions, it can be assumed that most glaciers receded to their minimal Holocene extent during this so-called "thermal optimum". Similar warm conditions have been identified for this interval also in the Northern Hemisphere (Grove, 1988), indicating about two millenia of global warming quite unrelated to human activity.

The effect of this warming was most severe in the seasonally dry regions of South America and of other continents, where lake levels dropped drastically as evaporation and transpiration far exceeded precipitation. The surface of Lake Titicaca in Peru and Bolivia dropped by 55 m between 7,700–7,000 yr BP (Wirrman et al., 1988), while lakes in western Patagonia (Argentina) became very low and possibly dried up between 7,700–5,100 yr BP (Stine and Stine, 1990). Meanwhile, large areas of the pampas and adjacent areas east of the Andes became active sand deserts during the same interval (Servant et al., 1981). While optimum conditions for vegetation growth in marginal areas like South Georgia peaked around 6,000 yr BP (Clapperton et al., 1989), the warming caused substantial recession of some ice shelves in the Antarctic Peninsula area. For example, the narrow ice shelf in George VI Sound between Alexander Island and the Antarctic Peninsula (Fig. 5.2) is currently 500–200 m thick, but the presence of fossil barnacle shells 6500 yr old in moraines subsequently deposited at the edge of this ice shelf, show that open water must have existed here during the mid-Holocene (Clapperton and Sugden, 1982).

Coastal studies in parts of the Southern Hemisphere have shown that mean sea level was 2–6 m higher some 7,000–5,000 yr ago. The legacy is left in the form of relict cliffs, sometimes cut off from the modern shore by higher beaches of sand and gravel often rich in broken marine shells; abandoned shingle bars, spits and tombolos also exist as "fossilized" shorelines dating from the mid-Holocene in many places, as in the Falkland Islands, along Magellan Strait, and in southeast Brazil (Clapperton, 1993b). This high Holocene sea level appears to have occurred globally, and although its cause is not fully understood and may be complex, it may have been related in part to warmer conditions worldwide that resulted in some thermal expansion of the oceans.

Late Holocene Neoglaciation: 5000 yr BP to the Present

As indicated in Table 5.2, there is evidence from New Zealand, the sub-Antarctic and the southern Andes that glaciers expanded significantly again soon after ca. 5,000 yr BP, confirming the climatic deterioration recorded by paleoecological

TABLE 5.2. *Approximate ages of glacier advances in the South American Andes/sub Antarctica transect and in the New Zealand Alps during the Holocene; warmer intervals are also indicated for the Andes-Antarctica region. Data come from Grove (1988) and Clapperton and Sugden (1989).*

South America/sub-Antarctica		**New Zealand (S.Alps)**	
Cold	**Warm**	**West (Tasman Gl.)**	**East (Mt.Cook)**
(ages are in ^{14}C yr BP)		(Glacier advances: dates in ^{14}C yr BP)	
	9400–8600		
8400			
7500			
			7000
	>6500		
6300-6000			
5200			
	5000-4800	5000	
4700-4200		4500-4200	4200
		3700	
3800			
3200	3300-2230	3500-3000	3400-3000
2700-2000		2700-2200	2600-2100
		1800-1700	
1800			
		1500	
1500			
		1100	
1150			
700		900	
	700-100		
840- 400-100			

data. Even in the tropical Andes, pollen studies indicate the onset of cooler and more humid conditions by 5500 yr BP (Salgado-Labouriau, 1991), and this appears to have been the forerunner of a series of cooler-warmer-cooler climatic oscillations that have characterized the second half of the Holocene. Ecological changes included the rise and fall of water levels in lakes and bogs and the expansion and contraction of plant populations sensitive to temperature and humidity, such as some aquatic and moorland species. In temperate latitudes, cooler and wetter conditions led to expansion of evergreen beech forest at the expense of deciduous forest on the west side of the Andes, while steppe grassland was replaced by open beech woodland on the drier eastern side of the range due to increased precipitation and cloudiness (Markgraf, 1983).

Over the past 5,000 yr, glaciers in most temperate montane regions of the Southern Hemisphere appear to have re-advanced during each millenium (Grove, 1988). These advances clearly indicate intervals when the mass balance of the ice fields became positive due to increased accumulation (precipitation) or reduced ablation (melting) or a combination of both. In general, the larger glaciers advanced a few kilometers at most beyond their modern positions, a 20-km advance of the Upsala Glacier draining part of the South Patagonian Ice field being exceptional (Mercer, 1976). Smaller glaciers rarely expanded more than two or three hundred meters beyond their current termini. Estimated snowline lowering to achieve these so-called neoglacial advances ranges from ca. 150–250 m, implying a thermal decline of no more than 1–2°C, but in many places it may have been increased precipitation and only a smaller amount of cooling that were largely responsible. Some researchers have argued that climate has fluctuated cyclically in the Holocene, with ca. 500yr-long cool intervals every 2000 years (Denton and Karlen, 1973). This may well be the case, but the hypothesis could also be misleading since it depends on rather sketchy chronological data. For example, many of the radiocarbon dates assigned to moraines may not record the age of the glacial advance very accurately, as they are commonly based on vegetation growth that occurred some time before or after the moraine was deposited. Thus the temporal framework used to indicate Holocene glacier fluctuations in Table 5.2 should be viewed only as a first approximation of the pattern of climatic change. The evidence from New Zealand in particular implies that glaciers have expanded and contracted many times during the past two millenia, and it could be argued that this kind of behavior was typical throughout the late Holocene, and that the apparent 2,500-yr cyclicity actually reflects larger-scale advances superimposed on (and obliterating) smaller ones of greater frequency.

There is little doubt that the cool intervals of the last 500 years (ca. 14th–19th centuries), known as the "Little Ice Age," produced significant glacier advances in the Southern Hemisphere. Photographs and sketches showing 19th-century dimensions and positions of glaciers in the Andes and the New Zealand Alps are comparable to those well known from the European Alps and Scandinavia, and demonstrate the global nature of the Little Ice Age (Grove, 1988). In New Zealand, glaciers expanded farther during the last few centuries than they did earlier in the late Holocene, but in the Andes, the greatest advances may have occurred at 2,700–2,000 yr. BP (Clapperton and Sugden, 1989). The sparsely vegetated fresh moraines of the past few centuries lie a few tens to hundreds of meters inside the limits of those formed a millenium or more earlier. Careful dating of moraines at the Mueller Glacier on the eastern side of the Southern Alps, New Zealand, show that fifteen moraines were formed between 7,200 and 100 yr ago; at ca. 7,200, 4,200, 3,790, 3,350, 2,940, 2,540, 2,160, 1,830, 1,490, 1,150, 840, 580, 340, 135, and 100 yr BP (Gellatly et al., 1985). They were all deposited within a distance of 500–600 m, the outermost lying just over a kilometer from the modern terminus. Assuming that this type of behavior may be typical of many glaciers in maritime temperate mountains, it confirms the fluctuational nature of climate in the late Holocene.

Records of equivalent detail are lacking for Andean glaciers, but in Ecuador, historical documents from the early Spanish colony in Quito and sketches made by early travelers confirm that snow cover and glacier extent were greater from the 15th to the late 19th centuries (Hastenrath, 1981). As in New Zealand, recession of glaciers from their latest Little Ice Age termini was most rapid during the 1930s and 1940s, but even so, small-scale advances were superimposed on the overall trend of recession, as at Franz Josef Glacier, which advanced in 1900–09, 1922–34, 1947–50, 1965–68 and in 1984 (Gellatly et al., 1985).

Glaciers and Global Warming

There is little doubt that the present century has been dominated by a recession of glaciers worldwide, punctuated by minor advances that have not halted the overall trend. Apart from anomalies related to internal dynamics or local climatic conditions, most glaciers in the Southern Hemisphere, from the equatorial Andes to West Antarctica, are smaller now than they were during cold peaks of the Little Ice Age and other neoglacial cold intervals. This does not mean they were never smaller than now during the warmer conditions that sometimes separated the cold neoglacial intervals. However, it is clear that a warming trend has characterized the 20th century, as it did the 11th–13th centuries, and may well reflect the natural swings in the Earth's climate system in response to whatever mechanisms are forcing the change. It is generally accepted that variations in output of energy from the sun influences global climate on the decadal and millenial scales, as may series of volcanic eruptions that inject sulfur-based aerosols to the atmosphere. Oceanic circulation changes, with concomitant fluctuations in sea surface temperature may influence cloudiness, monsoonal activity, and other circulatory behavior of the atmosphere that could induce small-scale fluctuations in climate over a number of years. While pollution of the atmosphere during this century may enhance the natural warming trend, it is unlikely to be the only cause (Hulme, 1994). The natural atmosphere-ocean system is immense and complex, and the scale of energy supplied by solar radiation far exceeds the comparatively small levels generated by humans. Thus, it seems logical to believe that climatic changes induced by natural causes are likely to suppress or enhance whatever influence is generated by the current level of human activity.

Pre-Pleistocene Glaciation

Two fundamental factors have controlled the periodic growth and decay of ice sheets in the Northern Hemisphere during the present ice age. First, inputs to the atmosphere of solar radiation have cyclically changed due to the Earth's orbital geometry (the so-called Milankovich effect), and these appear to have modulated the primary periodicity of the major changes (Imbrie and Imbrie, 1979). Second, the large ice sheets "dug their own graves" by depressing the land under their weight and by extending with low-gradient surface profiles into low latitudes so that vast areas were exposed to melting when the global climate warmed again. They also extended out onto the continental shelves where their major ice streams were sea-terminating. Thus, as sea level rose and sea surface temperatures increased, accelerated melting of the ice streams caused rapid and irreversible deglaciation (Denton et al., 1986). While much less is known about the ice sheets of former ice ages, there is every reason to believe that some may have behaved in a similar fashion. This is to be expected on the assumption that the Earth's orbital geometry in relation to the sun has varied cyclically more or less in the same way throughout much of geological time, and that large continental ice sheets may often have extended off-shore.

The greatest ice age prior to the current one occurred some 310–270 M yr ago during the Permian and the Carboniferous geological periods, when the Gondwanaland supercontinent existed in the Southern Hemisphere. As plate-tectonic movements took this large landmass into southern temperate and polar latitudes, an immense ice sheet complex developed. Reconstruction of this sheet from paleoglacial evidence in the form of lithified glacial sediments, striations and other indicators of ice movement shows that the complex was centered for much of the time on southern Africa and spread onto adjoining continents that we now know as South America (the eastern part, including the Falkland Islands), India, Australia, and Antarctica (Fig. 5.1); the northern margins spread into temperate latitudes as far north as 35°S. In places the glacial deposits are 1600 m thick, and evidence from Brazil suggests there were at least seventeen glacial-interglacial cycles (John, 1979). Interpretation of rhythmically bedded glacial and nonglacial deposits indicates that sea level fluctuated (possibly 150–250 m lower) due to the periodic growth and decay of large ice sheets and that "catastrophic" deglaciation appears to have occurred, flooding the lush vegetation of continental lowlands to form some of the well-known coal measures of the Permian and the Carboniferous periods. Of course the centers of ice-sheet growth and dispersal changed due to plate-tectonic movements during the 30–40 million years of this ice age, and glacial conditions in various parts of the Gondwanaland continental complex presumably altered as time passed. Nevertheless, sufficient evidence has accrued to suggest that the pattern of ice-sheet buildup and rapid collapse during the Permo-Carboniferous ice age may have been broadly similar, at least on some occasions, to those evident in the last glacial-interglacial cycle of the present ice age. In this case, one may speculate that the style of Pleistocene glaciation serves as a reasonable model for glaciation during the ancient ice ages. Moreover, there is no reason to assume that the pattern of large-scale glacial-interglacial cycles that has affected Earth during the past 800,000 years will not continue into the geological future. The timing and magnitude of these cycles are controlled by variations in the Earth-Sun orbital geometry and by the response of the present arrangement of continents and oceans on Earth's surface to changing inputs of radiant energy. Thus, until plate tectonic movements alter the distribution of land and sea masses, changing the circulation of oceans and atmosphere, it seems inevitable that Earth will continue to experience glacial-interglacial cycles, as it has in the past three million years.

References

Ashworth, A.C. and Hoganson, J.W., 1983. Testing the Late Quaternary climate record of southern Chile with evidence from fossil coleoptera. In: J.C. Vögel (Editor), Late Cainozoic palaeoclimates of the southern Hemisphere. A.A. Balkema, Rotterdam, pp. 85–102.

Barret, P.J., Adams, C.J., McIntosh, W.C., Swisher, C.C. and Wilson, G.S., 1992. Radiometric ages support Antarctic deglaciaiion around three million years ago. Nature, 259: 816–818.

Bond, G. and Lotti, R., 1995. Iceberg discharges into the North Atlantic on millenial time scales during the last glaciation. Science, 267: 1005–1010.

Bond, G., Broecker, W.S., Johnsen, S., McManus, J., Labeyrie, L., Jouzel, J. and Bonani, G., 1993. Correlations between climate records from North Atlantic sediments and Greenland ice. Nature, 365: 143–147.

Broecker, W.S. and Denton, G.H., 1990. The role of ocean-atmosphere reorganisations in glacial cycles. Quaternary Science Reviews, 9: 305–341.

Brown, I.M., 1990. Quaternary glaciations of New Guinea. Quaternary Science Reviews, 9: 273–280.

Clapperton, C.M., 1983. The glaciation of the Andes. Quaternary Science Reviews, 2: 83–155.

Clapperton, C.M., 1986. Glacial geomorphology, Quaternary glacial sequence and palaeoclimatic inferences in the Ecuadorian Andes. In: V. Gardiner (Editor), International Geomorphology 1986, Part II. Wiley, London, 843–870 pp.

Clapperton, C.M., 1990. Quaternary glaciations in the Southern Ocean and Antarctic Peninsula Area. Quaternary Science Reviews, 9: 229–252.

Clapperton, C.M., 1992. La ultima glaciacion y deglaciacion en el Estrecho de Magallanes: implicaciones para el poblamiento de Tierra del Fuego. Anales del Instituto de la Patagonia, 21: 113–128 pp.

Clapperton, C.M., 1993a. Nature of environmental changes in South America at the last glacial maximum. Palaeogeography, Palaeoclimatology, Palaeoecology, 101:189–208.

Clapperton, C.M., 1993b. Quaternary Geology and Geomorphology of South America. Elsevier Science, Amsterdam, 779 pp.

Clapperton, C.M., 1993c. Glacier readvances in the Andes at 12 500–10 000 yr BP: implications for mechanism of late-glacial climatic change. Journal of Quaternary Science, 8: 197–215.

Clapperton, C.M., 1995. Fluctuations of local glaciers at the termination of the Pleistocene. Quaternary International, 28: 41-50.

Clapperton, C.M. and McEwan, C., 1985. Late Quaternary moraines in the Chimborazo area, Ecuador. Arctic and Alpine Research, 17: 135–142.

Clapperton, C.M. and Sugden, D.E., 1982. Late Quaternary glacial history of George VI Sound area, West Antarctica. Quaternary Research, 18: 243–267.

Clapperton, C.M. and Sugden, D.E., 1989. Holocene glacier fluctuations in South America and Antarctica. Quaternary Science Reviews, 7: 185–198.

Clapperton, C.M., Sugden, D.E., Birnie, J. and Wilson, M.J., 1989. Late-glacial and Holocene glacier fluctuations and environmental change on South Georgia, Southern Ocean. Quaternary Research, 31: 210–228.

CLIMAP Project Members, 1981. Seasonal reconstruction of the Earth's surface at the glacial maximum. Geological Society of America, Map and Chart Series, 36.

Colhoun, E.A. and Fitzsimons, S.J., 1990. Late Cainozoic glaciation in western Tasmania. Quaternary Science Reviews, 9: 199–216.

Dawson, A.G., 1992. Ice Age Earth. Routledge, London, 293pp.

Denton, G.H. and Hendy, C.H., 1994. Youner Dryas age advance of Franz Josef Glacier in the Southern Alps of New Zealand. Science, 264: 1434–1437.

Denton, G.H. and Karlen, V., 1973. Holocene climatic variations—their pattern and possible causes. Quaternary Research, 3: 155–205.

Denton, G.H., Bockheim, J.G., Wilson, S.C. and Stuiver, M., 1989. Late Wisconsin and Early Holocene glacial history, Inner Ross Embayment, Antarctica. Quaternary Research, 31: 151–182.

Denton, G.H., Hughes, T.J. and Karlen, W., 1986. Global ice sheet system interlocked by sea level. Quaternary Research, 26: 3–26.

Gellatly, A.F., Röthlisgerger, F. and Geyh, M.A., 1985. Holocene glacier variations in New Zealand (South Island). Zeitschrift für Gletscherkunde und Glazialgeologie, 21: 265–273.

Goudie, A., 1992 Environmental Change. Third Edition. Claredon Press, Oxford 329 pp.

Grove. J.M., 1988. The Little Ice Age. Methuen, London, 498 pp.

Hastenrath, S., 1981. The Glaciation of the Ecuadorian Andes. A.A. Balkema, Rotterdam, 159 pp.

Hastenrath, S., 1984. The Glaciers of Equatorial and East Africa. Reidel, Dordrecht, 353 pp.

Helmens, K., 1988. Late Pleistocene glacial sequence in the area of the High Plain of Bogota (Eastern Cordillera, Colombia). Palaeogeography, Palaeoclimatology, Palaeoecology, 67: 263–283.

Heusser, C.J., 1990. Chilotan piedmont glacier in the Southern Andes during the last glacial maximum. Revista Geologica de Chile, 17: 3–18.

Hulme, M., 1994. Historic records and recent climatic change. In: N. Roberts (Editor), The Changing Global Environment. Blackwell, Oxford, pp. 69–98.

Huybrechts, P., 1993. Glaciological modelling of the Late Cainozoic East Antarctic ice sheet: stability or dynamism? Geografiska Annaler, 75A: 221–238.

Imbrie, J. and Imbrie, K.P., 1979. Ice ages. Solving the mystery. MacMillan, London. 224 pp.

John, B.S., 1979. The great ice age: Permo-Carboniferous. In: B.S.John (Editor), The winters of the world. David and Charles, Newton Abbot, pp. 154–172.

Kennett, J.P., 1977. Cenozoic evolution of Antarctic glaciation, the circum-Atlantic ocean and their impact on global palaeooceanography. Journal of Geophysical Research, 82: 3843–3860.

Mahaney, W.C., 1990. Ice on the Equator. Caxton Press, Wisconsin, 386 pp.

Marchant, D.R., Denton, G.H. and Swisher, C.C., 1993. Miocene-Pliocene-Pleistocene glacial history of Arena Valley, Quatermain Mountains, Antarctica. Geografiska Annaler, 75A: 269–302.

Marden, C.J., 1993. Late Quaternary glacial history of the South Patagonian Icefield at Torres del Paine, Chile. Ph.D. thesis, University of Aberdeen, 298 pp.

Marden, C.J. and Clapperton, C.M., 1995. Fluctuations of the South Patagonian Icefield during the last glaciation and the Holocene. Journal of Quaternary Research, 10: 197-210.

Markgraf, V., 1991. Late Pleistocene environmental and climatic evolution in southern South America Bamberger Geographisches Schriften, 11: 271-281.

Markgraf, V., 1983. Late and postglacial vegetational and palaeoclimatic changes in subantarctic, temperate, and arid environments in Argentina. Palynology, 7: 43–63.

Martinic, M., 1992. Historia de la Region Magellanica. Santiago, Chile, 2 vols.

Mercer, J.H. and Palacios, O., 1977. Radiocarbon dating of the last glaciation in Peru. Geology, 5: 600–604.

Osmaston, H.A, 1989a. Glaciers, glaciations and equilibrium line altitudes on Kilimanjaro. In: W.C. Mahaney (Editor), Quaternary and Environmental Research on East African Mountains. A.A. Balkema, Rotterdam, pp. 7–30.

Osmaston, H.A., 1989b. Glaciers, glaciations and equilibrium line altitudes on the Ruwenzori. In: W.C. Mahaney (Editor), Quaternary and Environmental Research on the East African Mountains. A.A. Balkema, Rotterdam, pp. 31–105.

Porter, S.C., 1981. Pleistocene glaciation in the southern Lake District of Chile. Quaternary Research, 16: 263–292.

Porter, S.C., Clapperton, C.M. and Sugden, D.E., 1992. Chronology and dynamics of deglaciation along and near the Strait of Magellan, southernmost South America. Sveriges Geologiska Undersökning, Ser. Ca 81: 233–239.

Rosqvist, G., 1994. A continuous record of deglaciation and climatic change since 15,715±150 BP from South Georgia, sub-Antarctica. In: C.M.Clapperton (Editor), IGCP Project 253: Termination of the Pleistocene. Aberdeen Workshop 22nd May 1994, Programme and Abstracts, p. 9.

Salagado-Labouriau, M.L., 1991. Vegetation and climatic changes in the Merida Andes during the last 13,000 years. Boletim IG-USP, São Paulo, Brazil, Publ. Especial, 8: 157–170.

Schubert, C., 1974. Late Pleistocene Merida glaciation, Venezuelan Andes. Boreas, 3: 147–152.

Servant, M. and Fontes, J.C., 1978. Les lacs Quaternaires des hautes plateaux des Andes Bolivien: premiere interpretations palaeoclimatiques. Cahiers de l'ORSTOM, Série Geologique, 10: 9–23.

Servant, M., Fontes, J. Chi., Rieu, M. and Saliege, J-F., 1981. Phases climatiques arides holocenes dans le sud-ouest de l'Amazonie (Bolivie). Comptes Rendues Acad. Sci. Paris, 29, ser. II: 1295–1297.

Smith, R.I.L., 1981. Types of peat and peat-forming vegetation on South Georgia. British Antarctic Survey Bulletin, 53: 119–139.

Stine, S. and Stine, M., 1990. A record from Lake Cardiel of climatic change in southern South America. Nature, 345: 705–708.

Suggate, R.P., 1990. Late Pliocene and Quaternary glaciations of New Zealand. Quaternary Science Reviews, 9: 175–198.

Van der Hammen, T., Barelds, J., de Jong, H. and de Veer, A.A., 1981. Glacial sequence and environmental history in the Sierra Nevada del Cocuy (Colombia). Palaeogeography, Palaeoclimatology, Palaeoecology, 32: 247–340.

Williams, M.A.J., Dunkerley, D.L., Deckker, P.D.E, Kershaw, A.P. and Stokes, T., 1993. Quaternary Environments. Arnold, London, 329 pp.

Wirrman, D., Mourguirat, P. and de Oliveira Almeida, L.F., 1988. Holocene sedimentology and ostracods distribution in Lake Titicaca—palaeohydrological interpretations. Quaternary of South America and Antarctic Peninsula, 6: 89–128.

PART II

THE QUATERNARY: Macro-processes

6. SEDIMENT DEFORMATION BENEATH THE LAURENTIDE ICE SHEET

Peter U. Clark

Identifying evidence of subglacial sediment deformation has important implications for interpreting the geologic record left by glaciers and ice sheets as well as for understanding their present and past behavior. Sediments deform beneath glaciers and ice sheets when high pore water pressures are sustained in the sediment (Boulton and Jones, 1979; Alley et al., 1986). Under these conditions, the sediment becomes "softer" than the overlying ice, deforming to large strains and contributing to high ice velocities under the shear stress applied by the ice. Because the physics governing subglacial processes differ for glaciers resting on a deforming (or "soft") bed for those resting on a nondeforming (or "hard") bed, sediments and landforms associated with a soft-bedded glacier are likely to be distinct from those associated with a hard-bedded glacier. In addition, glacier behavior governed by soft-bedded processes may be significantly different from the behavior of hard-bedded glaciers and ice sheets (Alley et al., 1987a).

Because of the importance of sediment deformation to glacial geology and glaciology, one of the greatest challenges facing glacial geologic research is identifying sedimentological and geomorphological evidence of this process from areas formerly covered by glaciers and ice sheets. This research should be driven by our understanding of the physics of subglacial sediment deformation processes and associated aspects of subglacial hydrology and how they differ from hard-bedded processes. Based on the premise that these distinct processes should each leave a characteristic imprint on the glacial record, glacial geologists have developed several hypotheses that distinguish glacial sediments and landforms and ice-sheet behavior characteristic of soft bedded regions from those characteristic of hard-bedded regions (Alley, 1991; Hart and Boulton, 1991; Hicock, 1992; Hicock and Dreimanis, 1992a; Clark and Walder, 1994). Among these differences are the following:

1. Sediment deposited from a deforming bed will show evidence of having deformed to large strains and at strain rates much higher than that for ice. Because the total strain experienced by most deforming sediment will be large, many structures will be destroyed and the sediment will be homogenized (Boulton, 1987; Hart and Boulton, 1991). Some sedimentological structures, however, should preserve a unique signature of the deformation process.

2. Subglacial sediment flux is likely to be several times greater for deforming sediment than for debris transported in basal ice and deposited by melt-out or lodgement (Alley, 1991).

3. The subglacial drainage system that develops at the ice-bed interface over a rigid bed should be distinctly different from that over deforming sediment (Alley, 1989a; Walder and Fowler, 1994). These differences should be reflected in the geomorphic and sedimentological record of subglacial drainage (Clark and Walder, 1994).

4. During subglacial sediment deformation, the shear stress applied by the overlying ice is regulated by material properties of the sediment. Because the geometry of the ice surface is determined by the shear stress over long distances (Paterson, 1994), low shear stresses regulated by deforming sediment will result in thinner ice with a lower ice-surface gradient than will shear stresses controlled by the internal deformation of ice over a nonsliding rigid bed. Furthermore, because the velocity of the ice coupled to deforming sediment is controlled by the strain rate of the sediment, steady-state glaciers advancing over soft beds will experience high horizontal ice velocities in association with low driving stresses (Alley et al., 1987a; Alley, 1990), unlike steady-state glaciers advancing over rigid beds (unless basal sliding rates are high).

5. A number of variables can lead to unstable ice-sheet behavior over soft beds, indicating that ice sheets resting on deforming sediment will more likely (although not necessarily) behave in an unstable manner than will ice sheets resting on a rigid bed.

In this chapter I review our current understanding of the geologic evidence for subglacial sediment deformation as it has been derived from studies of glacial sediments and landforms associated with the Laurentide Ice Sheet. The Laurentide Ice Sheet was the largest of the former Northern Hemisphere ice sheets (Fig. 6.1), rivaling the modern East Antarctic Ice Sheet in size and volume. Because of its large size, the Laurentide Ice Sheet played a significant role in controlling sea level history (Peltier, 1994) as well as in affecting Ice Age climates (Manabe and Broccoli, 1985; Broecker, 1994). Identifying the extent of sediment deformation beneath the Laurentide Ice Sheet is thus important to understanding its potential effect on ice-sheet dynamics.

The mechanics of subglacial sediment deformation identifies several sediment properties that exert a fundamental control on whether sediment will deform, how it will deform, and the extent to which sediment deformation controls ice-sheet behavior. I evaluate characteristics of the substrate beneath the Laurentide Ice Sheet in order to identify those regions where sediment deformation would have been most likely. I then test the hypotheses outlined earlier against the geologic record of the Laurentide Ice Sheet with respect to the distribution of easily deformable sediment inferred from characteristics of the substrate. These criteria should be useful in examining and interpreting deposits from older glaciations.

Initiation of Subglacial Sediment Deformation

Ice dynamics and subglacial processes are controlled by three potential physical mechanisms: internal deformation of ice by creep and, if the basal temperature of ice is at the pressure melting point, sliding and sediment deformation (Fig. 6.2;

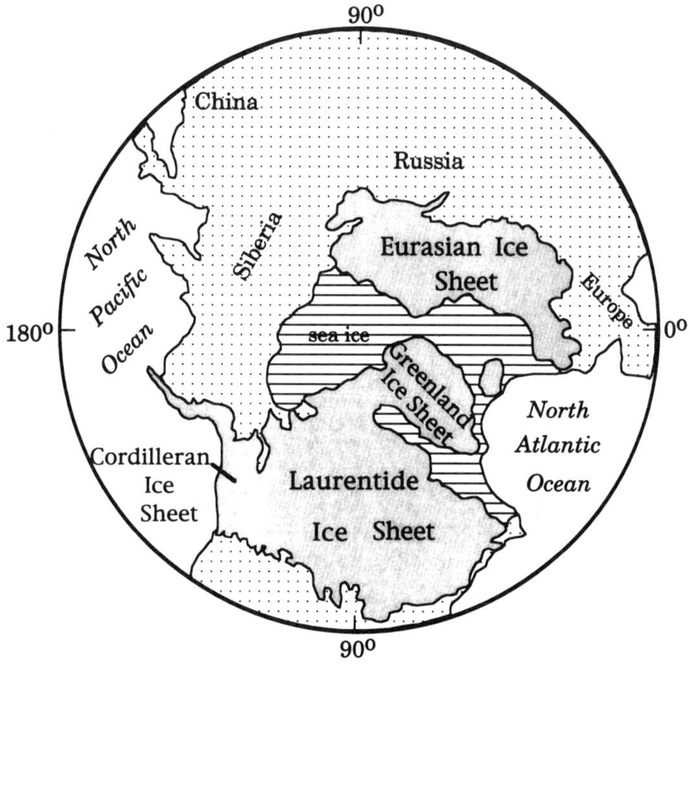

Fig. 6.1. Distribution of Northern Hemisphere ice sheets during the last glacial maximum ~21,000 years ago. Also shown are the distribution of land associated with lowered sea level and the distribution of sea ice (after CLIMAP Project Members, 1981).

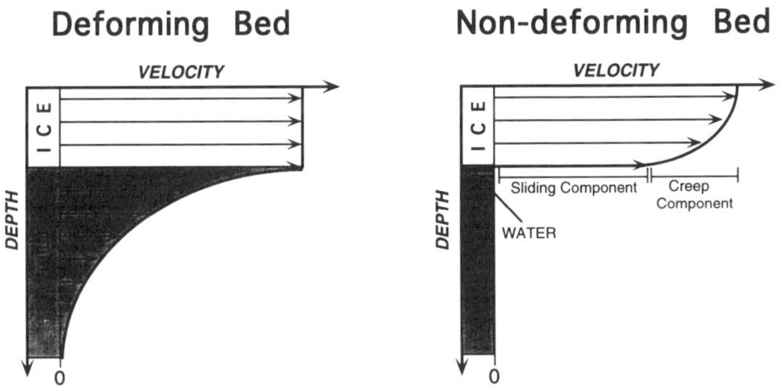

Fig. 6.2. Mechanisms for ice movement (after Alley, 1991). A. In the deforming bed mechanism, forward motion of ice occurs by deformation within the underlying sediment. B. Ice also moves by sliding at the ice-bed interface and by internal deformation of the ice (creep). Each of these three mechanisms represents an end member, and ice motion may occur by any combination of them.

Paterson, 1994). The influence of each of the three mechanisms on horizontal ice movement at any one time is determined first by the basal thermal regime; if the glacier base is frozen to the substrate, all forward ice motion is by internal deformation. If the base is at the pressure melting point, the relative contribution of sliding and sediment deformation to ice motion is dictated largely by subglacial hydrology, bed roughness, the availability, continuity, and hydraulic permeability of subglacial sediment, and the physical properties (such as grain size and mineralogy) that govern the physics of sediment deformation.

Water-saturated sediment first begins to deform beneath a glacier when the shear stress applied by the glacier on the sediment equals or exceeds the properties of the sediment that resist shear stress, collectively known as the *shear strength* of the sediment (Boulton and Jones, 1979). Stress applied to the sediment is distributed between the grain contacts and the water in the pores. When subglacial sediment is water saturated, the water pressure in the pore spaces will increase if the rate of water supplied by basal-ice melting exceeds the rate at which water drains from the sediment. This water pressure counteracts the pressure applied by the weight of the overlying ice on the sediment. The total pressure on the sediment, called the effective *pressure*, thus decreases, leading directly to a decrease in the sediment's shear strength and thus in its ability to withstand the stress applied by the overlying glacier without failing.

Following initial failure, sediment continues to deform, or "flow," as long as the applied stress remains greater than or equal to the sediment's shear strength. In general, flow behavior may be described by one of several relations; identifying the relation that best describes subglacial sediment deformation, however, remains a topic of debate (Kamb, 1991; Murray, 1994; Paterson, 1994). Specifically, this issue centers on whether deformation is distributed with depth (decimeters-to-meters scale) through the sediment beneath the ice-bed interface or whether it is concentrated in a very narrow zone (millimeters to centimeters scale) at the ice-bed interface (see Murray, 1994, and Paterson, 1994, for further discussion). Another important issue concerns the distribution of resistance to the shear stress applied by the ice between soft sediment and more rigid areas, such as bedrock knobs projecting through the sediment cover, commonly called "sticky spots" (Kamb, 1991; Alley, 1993). These questions are important with respect to interpreting, for example, sedimentological features of glacial diamictons (the depth of deformation) and controls on ice-sheet behavior (presence of sticky spots).

Subglacial Hydrology over Hard versus Soft Beds

Water may be supplied to the base of a glacier from surface and basal sources. For glaciers advancing into marine and lake basins, water already present in the overridden sediment is also important (Hicock and Dreimanis, 1992a). Most large ice masses, however, are polar except at their base, where melting may occur from frictional heating due to glacial movement and from the geothermal heat flux. If the underlying substrate can not drain all of this water, a drainage system will develop at the ice-bed interface. This system plays a fundamental role in controlling the dynamics of ice sheets resting on either a deforming or a rigid substrate. In both cases, an ineffective drainage system may result in high water pressure at the ice-bed interface, leading to rapid ice motion and glacier instability (Boulton, 1979; Clarke et al., 1984; Boulton and Hindmarsh, 1987; Kamb, 1987, 1991; Boulton and Dobbie, 1993; Iverson et al., 1995a).

When the substrate is impermeable bedrock, subglacial drainage will flow at the ice-rock interface as a continuous thin sheet, in channels, or as a series of linked cavities (Rothlisberger, 1972; Weertman, 1972; Walder, 1982; Kamb, 1987). In general, channels will only develop when surface meltwater is supplied to the bed; drainage is then predicted to occur in an arborescent network of channels cut into the basal ice (Rothlisberger or R-channels; Fig. 6.3).

The subglacial drainage system for a glacier resting on unconsolidated subglacial sediments differs from a drainage system over a rigid bed because subglacial sediments typically have higher hydraulic permeabilities than bedrock, and thus act as better aquifers, and because the bed is more easily erodible than bedrock (Boulton and Jones, 1979; Boulton and Hindmarsh, 1987; Clarke, 1987; Alley, 1989b; Walder and Fowler, 1994). The meltwater flux will thus be discharged by several mechanisms that differ from those involved in draining subglacial water over impermeable bedrock. In particular, when surface meltwater reaches the bed, the subglacial drainage system that develops over deforming sediment will be fundamentally different from that which develops over a rigid bed (Walder and Fowler, 1994). As discussed above, meltwater will flow over a rigid bed through an arborescent network of R-channels incised up into the ice. Conversely, where an ice sheet rests on deforming sediment, basal meltwater will drain through an anastomosing network of wide and shallow streams ("canals") incised into the sediment with a more or less flat roof of ice (Fig. 6.3; Walder and Fowler, 1994).

Implications of Sediment Failure to Ice Dynamics

Following the failure of subglacial sediment, the basal shear stress applied by the ice on the underlying sediment will be regulated by properties of the sediment and, if present, nondeforming areas such as bedrock knobs. Over long distances (>10 x ice thickness), the geometry of the ice surface is determined by the basal shear stress. A basal shear stress that is regulated by weak, deforming sediment, therefore, will result in thinner ice with a lower ice-surface gradient than will a basal shear stress controlled by stronger ice deforming over a rigid bed or a bed composed of many nondeforming areas. Furthermore, because the horizontal velocity of the ice coupled to deforming sediment is controlled by the strain rate of the sediment, high ice velocities (10^2–10^3 m/yr) may be reached at low driving stresses for glaciers advancing over soft beds, such as for currently active fast-flowing ice streams draining the West Antarctic Ice Sheet (Bentley, 1987).

An ice sheet resting on deforming sediment may respond in a stable manner to climate forcing if the rate of sediment deformation is maintained so as to ensure mass balance of the ice flux (Alley et al., 1987a,b). An ice sheet may

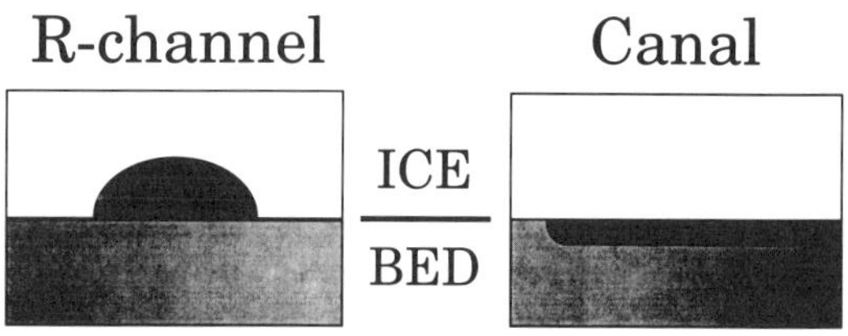

Fig. 6.3. Nature of channelized water drainage at the ice-bed interface over A. nondeforming bed (drainage in R-channels) and B. deforming bed (drainage in canals) (after Clark and Walder, 1994).

experience entirely dynamic fluctuations of its margin, however, in response to some internal pertubation of the deforming sediment system. These perturbations lead to a change in ice velocity that is independent of climate forcing, and resulting fluctuations of the ice margin represent internal instabilities.

Mechanisms that result in internal oscillations of ice sheets resting on deforming sediment may be generalized into three categories: parameters intrinsic to the sediment, subglacial hydrology, and basal temperature. Of these, only sediment parameters are unique to developing unstable behavior of ice sheets resting on soft beds; unstable behavior involving subglacial hydrology and basal temperature has previously been evaluated for ice sheets resting on rigid beds (Clarke et al., 1977; Kamb et al., 1985), but they clearly may play a role in unstable ice-sheet behavior over soft beds as well.

Sediment Deformation beneath the Laurentide Ice Sheet

Nature of the Laurentide Ice Sheet Substrate

Several sediment properties exert a fundamental control on when sediment will deform and how extensive that deformation will be. Among these controls are the grain size distribution of the sediment and its effect on material properties controlling permeability, effective pressure, and flow behavior. It is likely, however, that any sediment may deform under a wet-based glacier, given the importance of effective pressure in reducing sediment strength.

Continuity of sediment cover may play a more fundamental role in determining the effect of sediment deformation on ice-sheet behavior, however, as this determines the fraction of the bed that is composed of deformable sediment versus rigid bedrock, and thus the extent that sticky spots support basal shear stress and control ice-sheet dynamics. The shear strength measured on till sampled beneath several modern glaciers is unable to support the shear stress applied by those glaciers, suggesting that till cover is discontinuous and that bedrock protuberances contribute most of the basal resistance to ice motion (sticky spots) (Kamb, 1991; Humphrey et al., 1993; Iverson et al., 1995b). How to model such a "mixed bed" remains uncertain (Alley and MacAyeal, 1994). If till cover is continuous, however, so that bedrock is not exposed at the sole of a glacier, then basal shear stress will be supported by the properties of the sediment rather than concentrated at bedrock sticky spots.

The Laurentide Ice Sheet covered a variety of bedrock types, and physical properties (grain size, composition) of the till deposited by the ice sheet reflect the physical properties of the various bedrock types. Till that is derived from metamorphic, igneous, or sandstone bedrock generally has a coarser matrix (<2 mm) and greater gravel fraction than till derived from carbonate and shale (Scott, 1976; Dredge and Cowan, 1989). Exceptions to this general rule occur as a result of long-distance subglacial sediment transport (Clark, 1987; Hicock et al., 1989), glacial reworking of preexisting sediments (such as lake clays) (Dreimanis, 1976; Hicock and Dreimanis, 1992a), and presence of some lithologies (slates) in areas of crystalline bedrock that may produce fine-grained tills.

Thickness and continuity of till cover also reflect underlying bedrock, being generally thin and discontinuous over crystalline bedrock versus relatively thick and continuous over sedimentary bedrock (Dredge and Cowan, 1989; Dyke et al., 1989). Where fine-grained till that was derived from long-distance transport occurs over crystalline bedrock, it is generally discontinuous, and a large fraction of the substrate is exposed bedrock (Dyke et al., 1989).

Given the influence of source bedrock on till grain size and continuity, we can simplify the varied bedrock covered by the Laurentide Ice Sheet into two types: sedimentary and metamorphic/igneous (crystalline) (Fig. 6.4). Throughout glaciated North America, this classification distinguishes areas of discontinuous, usually sandy tills derived from underlying coarse-grained bedrock such as Precambrian shield areas from areas with continuous clayey-silty tills derived from underlying fine-grained sedimentary bedrock (Fig. 6.5).

These characteristics of the physical properties of tills deposited by the Laurentide Ice Sheet and their lateral continuity suggest that subglacial till

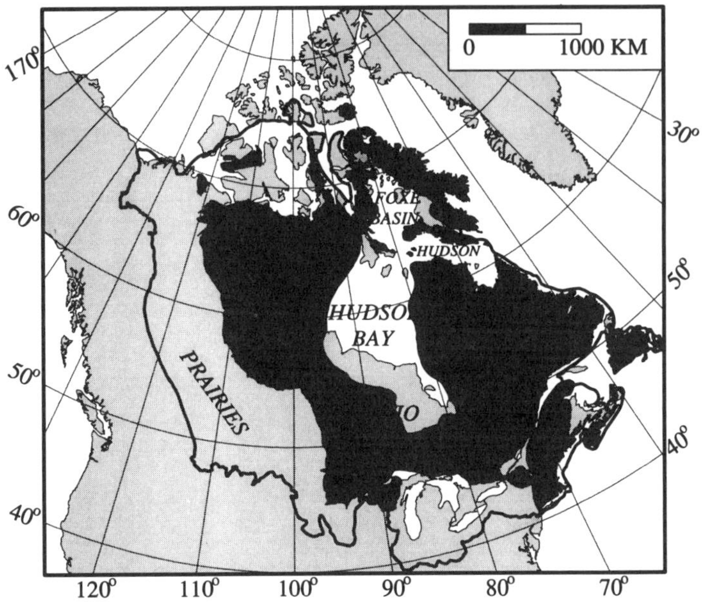

Fig. 6.4. Outline of the Laurentide Ice Sheet at 18,000 ^{14}C yr BP (shown by heavy dark line) (after Dyke and Prest, 1987). Also shown are the distribution of nearly continuous, fine-grained tills over sedimentary bedrock (shown in light gray within ice-sheet margin) and the distribution of discontinuous, sandy till over crystalline bedrock (shown in dark gray). Fine-grained, continuous tills would be the most likely sediments to deform subglacially and affect ice-sheet dynamics.

deformation is likely in regions of relatively continuous, fine-grained till with low hydraulic conductivity, which occur predominantly over sedimentary bedrock (exclusive of sandstone), but much less likely where the till is discontinuous and coarse-grained, and has relatively high hydraulic conductivity, or predominantly over crystalline bedrock and sandstone (Figs. 6.4, 6.5). Although coarser till over impermeable crystalline bedrock may have developed sufficiently high water pressures so that it deformed, the large fraction of exposed bedrock associated with these tills would have played the dominant role in providing basal resistance to ice flow.

In the following section, I evaluate the geologic record of the Laurentide Ice Sheet for evidence of subglacial sediment deformation from those areas where material properties suggest it would have occurred (Fig. 6.3). This discussion follows on the premise described in the Introduction that the geologic record should bear a characteristic imprint of the subglacial sediment deformation process that is distinct from subglacial processes operating on hard beds.

Sedimentological Evidence for Subglacial Sediment Deformation

One of the most active research areas of glacial sedimentology in the past twenty years has involved identifying criteria that distinguish the genesis of glacial diamictons. Prior to the recognition of sediment deformation as an important subglacial process, much of this research focused on distinguishing diamictons deposited from the base of actively sliding glaciers (lodgement till) from glacial diamictons formed by other processes (melt-out, sediment gravity flow; Dreimanis, 1976, 1988). Now that subglacial sediment deformation is recognized as a critical factor in the dynamics of glaciers and ice sheets, identifying sediments deposited by this process has become a subject of great interest.

Such genetic distinctions, however, remain uncertain. In particular, many characteristics of massive glacial diamictons that have been attributed to deposition from the base of an ice sheet on a rigid bed can be equally well explained by subglacial sediment deformation (Boulton, 1987; Alley, 1991; Hicock and Dreimanis, 1992b). Although there are no widely-accepted criteria for correctly interpreting the origin of these diamictons, I review several sedimentological and structural features of glacial diamictons that I believe are most consistent with the deforming bed process. Hicock et al. (1989) and Alley (1991) discuss several additional characteristics of glacial diamictons that may record the deforming bed process.

Fig. 6.5. A. A landscape in northern Labrador characteristic of areas of crystalline bedrock, showing large fraction of exposed bedrock that would provide strong resistance to ice flow. B. A landscape and sediment types in Illinois characteristic of areas of sedimentary bedrock, showing continuous cover of fine-grained till.

Clast Fabrics

The theory of particle motion in fluids deforming at a given strain rate (Jeffrey, 1922; Glen et al., 1957) suggests that clast fabrics that develop in deforming sediment may be distinctly different from those developed in ice (Lindsay, 1968; DesMarais, 1994; Hart, 1994). Particle motion in a fluid undergoing simple shear follows periodic orbits. Beginning early in their transport history, particles become aligned parallel to flow, where they spend most of their time. Following prolonged shear, however, particle orbits break down and develop a transverse orientation. When this transition from parallel to transverse orientation occurs is largely a function of strain rate. Strain rates in deforming sediment are several orders of magnitude higher than those in ice, so it is unlikely that this transition will occur for particles in ice within the timeframe of a glaciation, but it can occur relatively quickly (10^2–10^3 yr) in sediment deforming beneath a glacier (DesMarais, 1994).

Clast interactions will also affect fabric orientation and strength, redirecting particles into a transverse direction and weakening fabric strength (Glen et al., 1957). Their effects should be more pronounced in a rapidly straining sediment than in ice.

Fabrics measured on debris in basal ice have strong orientations parallel to ice flow (Lawson, 1979), or consistent with the theory. The few comparable data for sediments known to have deformed subglacially, from Iceland, suggest weakly developed fabrics that have significant variation in particle dip and in the mean direction (principal eigenvector) with respect to known ice flow (Dowdeswell and Sharp, 1986), or also consistent with the theory.

Fabrics measured from massive fine-grained tills associated with the Laurentide Ice Sheet are consistent with formation in a sediment that has experienced high strain rates. These fabrics are often weakly developed with a transverse component (Hicock, 1992; Hicock and Dreimanis, 1992a; DesMarais, 1994; Hart, 1994), although exceptions occur (Johnson and Hansel, 1990), suggesting either a different process (lodgement, melt-out) or sediment that experienced low strain rates. A range of fabric orientations and strengths from one till unit (Menzies, 1990), however, may be characteristic of sediment deformation, since strain rates vary both spatially (vertically) and through time.

Although fabric data may provide an important record for distinguishing strain rates, these data do not constrain thickness of the deforming sediment layer such that the thickness may be only on the order of the length of the pebbles being measured. Further work is required on this important problem.

Clast Pavements

Fine-grained debris flows, which are mechanically similar to subglacial deforming sediment, are able to support the weight of larger grains through their matrix strength and buoyancy (Johnson, 1970; Hampton, 1975). Deposits of debris flows commonly have clasts at their base or show evidence of grading, however, indicating that the clasts exceeded the competency of the fluid matrix and settled either to the bottom or to the level where the sediment could support them (Naylor, 1980; Lawson, 1981).

Settling of large clasts through a deforming viscous fluid explains characteristics of clast pavements that are commonly described from the base of fine-grained till deposited by the Laurentide Ice Sheet (Fig. 6.6; Clark, 1991; Hicock, 1991; Voorhees et al., 1994). These pavements are generally one clast thick, can extend over areas up to hundreds of square kilometers, usually occur at the lower contact of massive diamictons (although some occur as intra-till pavements), have lithologies reflecting the overlying diamicton, and rest on various types of sediment (Clark, 1991; Hicock, 1991). Other proposed mechanisms for the formation of clast pavements described from deposits of the Laurentide Ice Sheet have thus far failed to explain these characteristics (Wright et al., 1973; Mickelson et al., 1992).

Clast Striae

Hicock (1991, 1992; Hicock and Dreimanis, 1992a, 1992b) measured weakly developed striae, fine striae following curved surfaces on stones, and multiple striae orientations on clasts in fine-grained till deposited beneath the southern sector of the Laurentide Ice Sheet. Hicock argued that these characteristics are inconsistent with deposition directly from ice by lodgement or melt-out, where striae sets should be well developed with a single orientation parallel to ice flow, and instead indicate rotation of the clasts in a subglacial deforming sediment prior to deposition.

Sand Intraclasts

Sorted and stratified masses or blocks of sand are commonly described in fine-grained till (Fig. 6.7; Eyles et al., 1982; Hicock et al., 1989; Menzies, 1990). Their presence has been used as evidence of deposition by melt-out since, it has been argued, they would not have survived deposition by lodgement (Shaw, 1987) nor would they be preserved in sediment that has experienced large strain (Brown et al., 1987; Clayton et al., 1989).

Several mechanisms may result in the formation and preservation of sand intraclasts in deforming sediment. In some cases, sand intraclasts are undisturbed, and their geometry suggests deposition in a distributed drainage system at the ice-bed interface (canals) (Clark and Walder, 1994). If enough of a density contrast is established, sand deposited in channels at the ice-bed interface may founder into underlying water-saturated fine-grained till. Other sand intraclasts may be deposited in channels eroded through a deforming layer by summer meltwaters and then isolated from the ice-bed interface by channel

Fig. 6.6. A striated clast pavement near Tisdale, Saskatchewan. Pavement is at contact between grayish-brown clayey till above and buff-colored, compact till below (photograph courtesy of John Elson).

closure caused by sediment creep during the winter season (Alley, 1991). Many sand intraclasts have been rotated and deformed and their edges have been rounded (Fig. 6.7). These may have been entrained into the diamicton from underlying sediments (Johnson and Hansel, 1990; Menzies, 1990; Hart and Boulton, 1991) or formed at the ice-bed interface and subsequently transported within the deforming layer. More work is required to evaluate the ability of these features to survive as competent units within fine-grained deforming sediment.

Fig. 6.7. A deformed sand intraclast within a fine-grained till, exposed along bluffs of Lake Michigan near Milwaukee, Wisconsin. Note that primary sedimentary structures are present with the intraclast and that deformation involves both brittle (faulting) and ductile (folding) deformation.

Structural Data

Total strain associated with subglacial sediment deformation would generally be high enough that structural records of the strain (for example, folds) would be completely destroyed. Strain rates in a deforming bed vary both spatially and temporally, however, so that features recording low strain may be preserved. Strain rates are greatest at the top of a deforming layer and decrease to the base (Fig. 6.2A). Geological structures formed by simple shear may thus be preserved near the base, where the least strain has occurred, but be completely obliterated and homogenized at the top of the sequence (Hart and Boulton, 1991). Strain rates may also vary at any one site over the course of a glacial event, leading to a complex strain history that may be preserved as superimposed geological structures (Hicock, 1992; Hicock and Dreimanis, 1992a; Menzies and Maltman, 1992).

Sediment Flux

Sediment transport mechanisms by ice sheets include a relatively thick debris-rich basal ice layer (debris entrained by net basal freeze-on or tectonic processes), a thin, basal regelation layer (debris that may be repeatedly entrained and released by pressure melting and refreezing of basal ice), a subglacial deforming sediment layer, and sediment transport by subglacial water. Under steady-state conditions, sediment fluxes associated with subglacial sediment deformation will be significantly higher than fluxes by other basal processes (Alley, 1991).

Alley (1991) and Johnson et al. (1991) argued that the high sediment fluxes required for the formation of continuous, fine-grained till deposited along the southern margin of the Laurentide Ice Sheet are best accounted for by subglacial sediment transport in a deforming layer. Jenson et al. (1995) evaluated sediment flux by a deforming layer in a numerical model and found good agreement with sediment fluxes suggested from the field data (Alley, 1991; Johnson et al., 1991).

Transport of debris-rich ice may be a viable mechanism, however, for ice sheets that are not in steady state (Alley and MacAyeal, 1994). In his evaluation of the basal freeze-on mechanism, Alley (1991) considered only a steady-state ice flow regime, in which the dominant process for fast-moving ice, such as characterized the southern margin of the Laurentide Ice Sheet, would be net melting and thus the release, rather than the entrainment and transport, of basal debris. Ice flow along the southern margin during much of the last glaciation, however, was probably not in steady state, but instead underwent repeated advances and retreats (Clayton et al., 1985; Hansel and Johnson, 1992).

One possible mechanism for this unsteady behavior is the binge/purge model (MacAyeal, 1993). Alley and MacAyeal (1994) found that basal freeze-on can entrain a significant amount of sediment during a purge episode (ice advance), which is subsequently released from the ice by basal meltout following the advance. If such a mechanism was operating along the southern Laurentide margin, it may explain the high sediment fluxes in those regions.

Geomorphic and sedimentological records of subglacial hydrology

Clark and Walder (1994) evaluated the hypothesis that differences in drainage systems developed at the ice-bed interface for ice sheets resting on rigid versus deforming substrates should be expressed in the geomorphic and sedimentological record of these drainage systems. They identified the primary difference as the distribution of eskers, which are long and sinuous ridges of sand and gravel deposited in an arborescent network of subglacial tunnels, or R-channels. R-channels form only where the glacier bed is rigid, either because of the discontinuous sediment cover or because subglacial sediment did not deform pervasively under the shear stress applied by overlying ice. Eskers should thus be present over rigid beds (although other factors such as sediment supply also dictate their presence) and largely absent where the subglacial bed was deforming.

Clark and Walder (1994) used the characteristics of the substrate beneath the Laurentide Ice Sheet as a guide to distinguishing areas of rigid substrate from those of deforming beds, and thus as a basis for predicting the distribution of eskers based on the glacier hydrology model. They found that eskers deposited beneath the Laurentide Ice Sheet were largely confined to those areas argued to be rigid beds on the basis of discontinuous sediment cover over crystalline bedrock (Fig. 6.8), or consistent with the theory (Fig. 6.3). In contrast, eskers are rare or absent over areas where continuous, fine-grained till overlies sedimentary bedrock, or areas likely to have been subglacial deforming beds, also consistent with the theory. Clark and Walder (1994) proposed that well-sorted sand intrabeds commonly described from fine-grained till (discussed earlier) have the same dimensions as, and are thus likely records of, the subglacial drainage system that would have developed over deforming sediment.

Surface Geometry of the Laurentide Ice Sheet

The reconstructed surface geometry of former ice sheets provides important information on basal processes. Reconstructions of ice-surface geometry from glacial geologic evidence have largely used information on former ice flow direction and elevation of the ice margin (Mathews, 1974; Thorson, 1980; Clark, 1992); Shreve (1985) used a glacier hydrology model and esker characteristics. Indicators of former ice flow direction of the Laurentide Ice Sheet are common (Prest et al., 1968). Ice-marginal elevations are derived from lateral and terminal moraines, but significant information on ice-surface elevation at any distance from the terminus is possible only if the margin is lobate. For this reason, glacial geologic reconstructions of the Laurentide Ice Sheet surface elevation are limited with respect both to areas of the margin that are lobate and then to the length of flowline for which ice-marginal elevations are available.

Despite these limitations, reconstructions that exist identify significant differences in ice-surface geometry that are consistent with the inferred distribution of rigid and deforming beds (Fig. 6.9). Reconstructed longitudinal ice-surface profiles from the southern, southwestern, and northwestern sectors of the Laurentide Ice Sheet, or areas of fine-grained and continuous till, identify low-sloping and thin ice and corresponding low driving stresses (<5 kPa) (Fig. 6.9B, profiles 1, 2-6; Mathews, 1974; Beget, 1987; Clark, 1992). The reconstructed profile from the Green Bay Lobe (profile 2) exhibits a higher surface gradient than other lobes of the southern margin of the Laurentide Ice Sheet. Landforms and deposits of the Green Bay Lobe are fundamentally different from those of other lobes, however, including sandier tills and many eskers (Fig. 6.8; Mickelson et al., 1983). These features thus suggest that bed deformation may have been limited beneath the Green Bay Lobe.

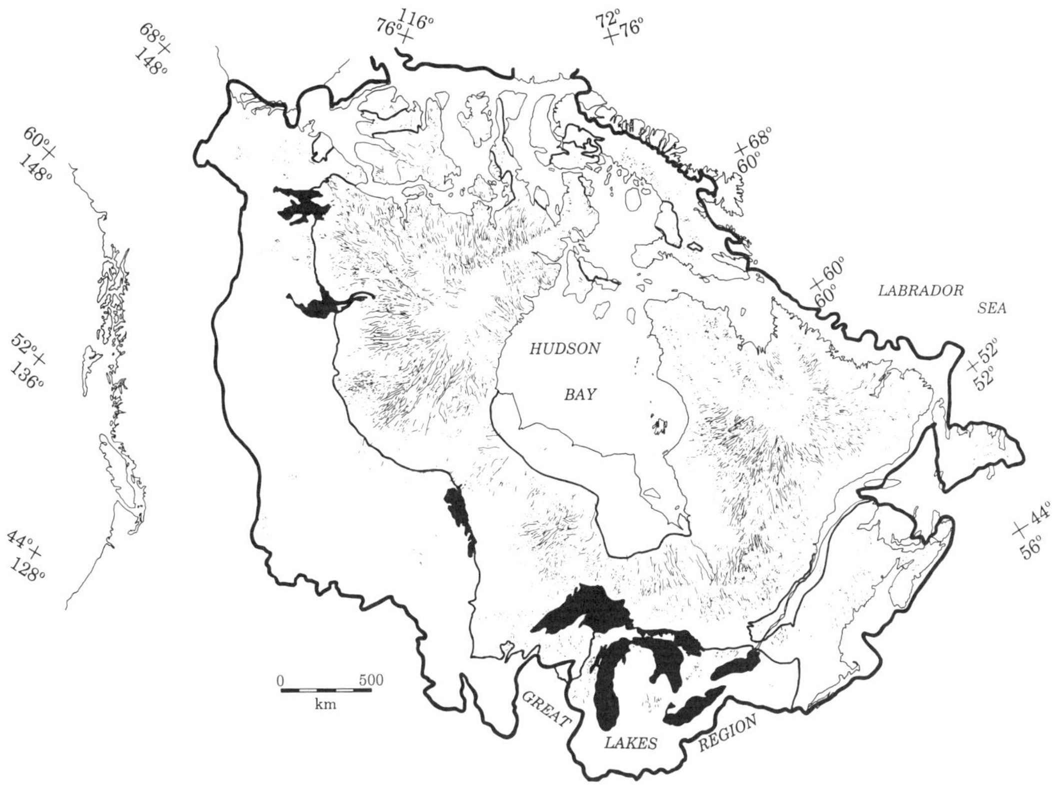

Fig. 6.8. Distribution of eskers deposited by the Laurentide Ice Sheet (from Clark and Walder, 1994). Margin of the ice sheet is shown by a heavy line, and contact between areas of crystalline and sedimentary bedrock is shown by an intermediate line. Note that most eskers are found on crystalline bedrock.

Longitudinal profiles from the eastern margin of the ice sheet, or areas underlain by crystalline bedrock, identify thicker ice with steeper gradients and higher driving stresses (30-125 kPa) (Fig. 6.9B, profiles 7 – 9; Buckley, 1969; Shreve, 1985; Clark, 1988; Klassen and Fisher, 1988). Two profiles are from northern Labrador (profiles 8A, 8B; Clark, 1988; Clark and Josenhans, 1986). These both show steep segments of former outlet glacier of the Laurentide Ice Sheet that flowed through the Torngat Mountains, northern Labrador, and profile 8B shows a flat segment where ice flowed out onto fine-grained sediments on the adjacent continental shelf.

Information on ice thickness is also suggested by the form of the ice margin. The position of an ice margin is controlled by topography only where the ice is too thin to override the topography completely. Thus, if the topographic relief is low in comparison to the ice thickness, the ice margin will be only perpendicular to the regional ice-surface gradient. As the ice thickness decreases in relation to the relief, however, the position of the margin will be increasingly controlled by the relief.

The margin of the Laurentide Ice Sheet at any one time was lobate over areas of fine-grained till and relatively straight or curvilinear over crystalline bedrock (Fig. 6.10; Prest, 1969; Dyke and Prest, 1987). With the exception of the mountainous regions of the eastern seaboard, topographic relief in both areas is generally low (Bostock, 1970), so that the differences in the ice margins must reflect the differences in ice thickness indicated by reconstructed ice surfaces (Fig. 6.9).

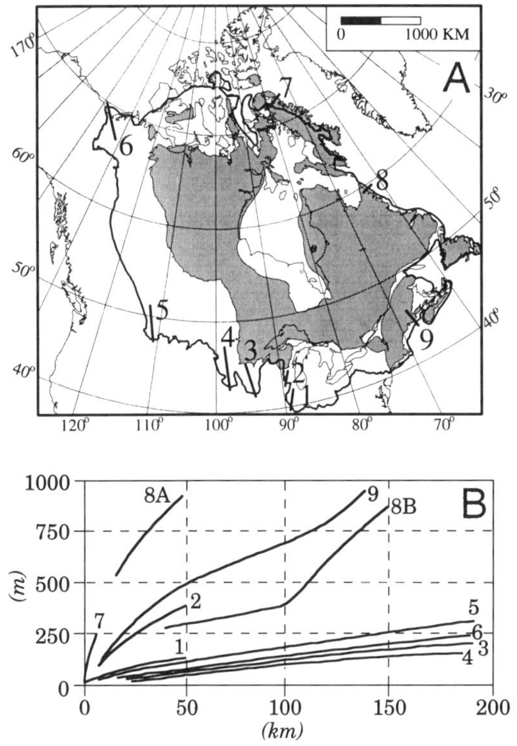

Fig. 6.9. A. Location map of reconstructed surface profiles of the Laurentide Ice Sheet. B. Reconstructed surface profiles for the Laurentide Ice Sheet. Profile 1 is for the Lake Michigan Lobe (Clark, 1992), profile 2 is for the Green Bay Lobe (Clark, 1992), profile 3 is for the Des Moines Lobe (Clark, 1992), profile 4 is for the James Lobe (Clark, 1992), profile 5 is from the southwestern sector of the ice sheet (Mathews, 1974), profile 6 is for the northwestern ice-sheet sector (Beget, 1987), profile 7 is from Bylot Island (Klassen and Fisher, 1988), profiles 8A and 8B are from northern Labrador (Clark and Josenhans, 1986; Clark, 1988), and profile 9 is from Maine (Shreve, 1985).

Ice-Sheet Behavior

Geologic records of ice-margin fluctuations are fairly well established by radiocarbon dating for several sectors of the Laurentide Ice Sheet during the most recent glaciation. Because of several potential problems in radiocarbon dating of samples (Clayton and Moran, 1982; Lowell, 1995), however, the radiocarbon chronology for some events may be in error. This problem can be resolved only by further dating.

Following the last glacial maximum (about 20,000 ^{14}C yr BP), several ice lobes along the southern ice-sheet margin (Erie, Huron, Lake Michigan, Des Moines, James) advancing over continuous, fine-grained sediment experienced large and rapid oscillations, with margins advancing at hundreds to thousands of meters per year over periods of hundreds of years (Fig. 6.11; Dreimanis, 1977; Clayton et al., 1985; Hansel and Johnson, 1992). Ice-surface gradients and driving stresses were extremely low over these areas (Fig. 6.9), indicating that rapid velocities during ice-margin advance resulted from elevated subglacial water pressures and/or subglacial sediment deformation (Clayton et al., 1985; Clark, 1994). In marked contrast to this behavior, after this glacial maximum, the southern margin retreated steadily where it rested on crystalline bedrock (Fig. 6.11), despite having steeper ice-surface gradients and driving stresses (Fig. 6.9; Shreve, 1985).

Fast ice-sheet flow is also identified from other areas where ice advanced over fine-grained sediment. Terrestrial and deep-sea records identify episodic fast ice-sheet flow through Hudson Strait, discharging massive numbers of icebergs into the Labrador Sea and North Atlantic (Heinrich Events) (Andrews and Tedesco, 1992; Bond et al., 1992; Kaufman et al., 1993; Bond and Lotti, 1995). Deep-sea records also identify episodic fluxes of icebergs into the North Atlantic from advances of the Laurentide Ice Sheet through the soft-bedded Gulf of St. Lawrence (Bond and Lotti, 1995).

Records of fast ice-sheet flow are thus only found in association with areas of fine-grained sediment; on crystalline bedrock, the margin did not experience such behavior (Fig. 6.11). These records of fast ice-sheet flow suggest two frequencies of ice-sheet variability superimposed on the lower-frequency (10^4–10^5 yr) ice-volume changes driven by orbital variations: Heinrich events at suborbital time scales of variablity (10^3–10^4 yr) and higher-frequency ice-margin fluctuations at millennial timescales (10^3 yr) that occur every 1–3 ka (Bond and Lotti, 1995; Clark, 1995). Whether this variability represents some internal ice-sheet instability associated with areas of soft beds (Clayton et al., 1985; MacAyeal, 1993; Clark, 1994) or simply a large and rapid

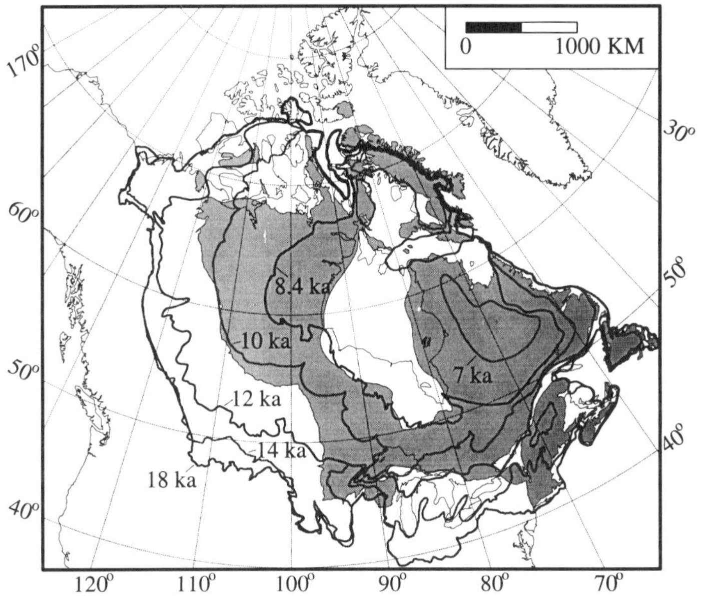

Fig. 6.10. Isochrones (in thousands of ^{14}C years BP) of recessional positions of the Laurentide Ice Sheet (after Dyke and Prest, 1987). Note the different form of isochrones where the ice margin rested on crystalline bedrock (dark gray area) and where the ice margin rested on fine-grained till (light gray area). These differences are interpreted to reflect differences in ice thickness.

response to some climate forcing (Bond et al., 1993; Bond and Lotti, 1995) remains an important, but unresolved, issue (Broecker, 1994).

Summary

Studies of modern glaciers and ice sheets resting on unconsolidated water-saturated sediment (soft beds) suggest that sediment deformation is an important process in affecting glacier dynamics. The specific description of the flow behavior of the sediment, however, remains uncertain (Boulton and Hindmarsh, 1987; Kamb, 1991). Furthermore, all modern soft-bedded glaciers studied thus far also identify bedrock knobs projecting through the sediment and contributing to basal resistance to ice flow (Alley, 1993).

Geologic studies of former Northern Hemisphere ice sheets suggest that deforming sediment may have been responsible for the origin of some glacial sediments and landforms. Continuous, fine-grained till deposited by the Laurentide Ice Sheet, found largely in association with sedimentary bedrock (Figs. 6.4, 6.5), will likely deform when water-saturated and subjected to a shear stress applied by overlying ice. Sedimentological and structural evidence from this fine-grained till that indicates subglacial sediment deformation includes massive diamictons, clast pavements, lenses of sorted sediments, and inclusions of boudins and attenuated laminae and fold noses. Preliminary estimates of till flux for the soft-bedded Lake Michigan Lobe in Illinois identify subglacial sediment deformation as the most likely mechanism of sediment transport. Differences between subglacial hydrology over rigid beds versus deforming sediment predicted by glaciological theory are recorded by the distribution of eskers, which form only in R-channels associated with a rigid bed. Reconstructed ice-surface gradients from areas of crystalline and sedimentary bedrock also identify fundamental differences in subglacial processes. Finally, geologic records of fluctuations of the Laurentide Ice Sheet identify rapid and irregular ice-margin fluctuations that occurred only over areas of fine-grained till, whereas largely uninterrupted ice-margin retreat with no evidence of significant re-advance occurred over rigid-bed areas.

Important problems that remain in fully describing subglacial sediment deformation include:

1. Identifying sedimentological and structural features that identify the high strain rates, depth of deformation, and rheological behavior of the sediment.
2. Developing an appropriate flow law for deforming till.

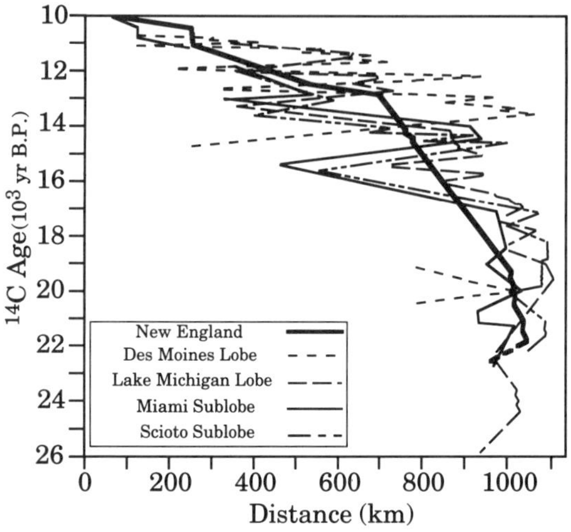

Fig. 6.11. Time-distance diagrams of the southern margin of the Laurentide Ice Sheet (see Clark (1994) for original references to sources of data). Note the different behavior during ice retreat for the margin in New England, where it rested on crystalline bedrock, compared to that of the margin further west, where it rested on fine-grained till (various lobes).

3. Modeling the relative contributions of sticky spots and properties of the deforming sediment that provide basal drag to ice flow (mixed-bed model).
4. Characterizing the subglacial hydrology with respect to its control on subglacial water pressure and possible ice-bed decoupling.
5. Further identifying mechanisms that lead to ice-sheet instability.

Although a full description of the physics of subglacial sediment deformation has yet to be developed, significant differences in bed processes are reflected by the distribution of glacial deposits and landforms. Our current understanding of the process has yielded important insights into the geologic record and the dynamic behavior of former Northern Hemisphere ice sheets. A fundamental result of these studies is that subglacial sediment deformation as a widespread process affecting ice-sheet dynamics resulted largely from the distribution of the bedrock geology (Figs. 6.4, 6.5). Whereas forward ice motion by sliding and internal ice deformation was not restricted to any particular substrate, the development of continuous, fine-grained deforming sediment required easily erodable sedimentary bedrock. The importance of this process to glacial geology and ice-sheet dynamics is largely a reflection of the widespread presence of sedimentary bedrock beneath the early Northern Hemisphere ice sheet. An interesting outgrowth to this conclusion is a consideration of the deposits and dynamics of an ice sheet resting only on crystalline bedrock or only on sedimentary bedrock.

References

Alley, R.B., 1989a. Water pressure coupling of sliding and bed deformation: I. Water system. Journal of Glaciology, 35: 107–118.

Alley, R.B., 1989b. Water pressure coupling of sliding and bed deformation: II. Velocity-depth profiles. Journal of Glaciology, 35: 119–129.

Alley, R.B., 1990. Multiple steady states in ice-water-till systems. Annals of Glaciology, 14: 1–5.

Alley, R.B., 1991. Deforming-bed origin for southern Laurentide till sheets? Journal of Glaciology, 37: 67–76.

Alley, R.B., 1993. In search of ice-stream sticky spots. Journal of Glaciology, 39: 447–454.

Alley, R.B., Blankenship, D.D., Bentley, C.R. and Rooney, S.T., 1986. Deformation of till beneath ice stream B, West Antarctica. Nature, 322: 57–59.

Alley, R.B., Blankenship, D.D., Rooney, S.T. and Bentley, C.R., 1987a. Till beneath ice stream B. 4. A coupled ice-till flow model. Journal of Geophysical Research, 92: 8931–8940.

Alley, R.B., Blankenship, D.D., Bentley, C.R. and Rooney, S.T. 1987b. Till beneath ice stream B. 3. Till deformation: evidence and implications. Journal of Geophysical Research, 92, 8921–8929.

Alley, R.B. and MacAyeal, D.R., 1994. Ice-rafted debris associated with binge/purge oscillations of the Laurentide Ice Sheet: Paleoceanography, 9: 503–511.

Andrews, J.T. and Tedesco, K., 1992. Detrital carbonate-rich sediments, northwestern Labrador Sea: implications for ice-sheet dynamics and iceberg rafting (Heinrich) events in the North Atlantic. Geology, 20: 1087–1090.

Beget, J., 1987. Low profile of the northwest Laurentide Ice Sheet. Arctic and Alpine Research, 19: 81–88.

Bentley, C.R., 1987. Antarctic ice streams: a review. Journal of Geophysical Research, 92: 8843–8858.

Bond, G., Heinrich, H., Broecker, W., Labeyrie, L., McManus, J., Andrews, J., Huon, S., Jantschik, R., Clasen, S., Simet, C., Tedesco, K., Klas, M., Bonani, G. and Ivy, S., 1992. Evidence for massive discharges of icebergs into the North Atlantic ocean during the last glacial period. Nature, 360: 245–249.

Bond, G., Broecker, W.S., Johnsen, S., McManus, J., Labeyrie, L., Jouzel, J. and Bonani, G., 1993. Correlations between climate records from North Atlantic sediments and Greenland ice. Nature, 365: 143–147.

Bond, G.C. and Lotti, R., 1995. Iceberg discharges into the North Atlantic on millennial time scales during the last glaciation. Science, 267: 1005–1010.

Bostock, H.S., 1970, Physiographic subdivisions of Canada. Geological Survey of Canada, Economic Geology Report No. 1, pp. 9–30.

Boulton, G.S., 1979, Processes of glacier erosion on different substrata. Journal of Glaciology, 23: 15–38.

Boulton, G.S., 1982. Subglacial processes and the development of glacial bedforms. In: R. Davidson-Arnott, W. Nickling and B. Fahey (Editors). Research in glacial, glaciofluvial and glaciolacustrine systems. Geo Books, Norwich, pp. 1–31.

Boulton, G.S., 1987. A theory of drumlin formation by subglacial sediment deformation. In: J. Menzies and J. Rose (Editors), Drumlin Symposium. A.A. Balkema, Rotterdam, pp. 25–80.

Boulton, G.S. and Jones, A.S., 1979. Stability of temperate ice caps and ice sheets resting on beds of deformable sediment. Journal of Glaciology, 24: 29–43.

Boulton, G.S. and Hindmarsh, R.C.A., 1987. Sediment deformation beneath glaciers: rheology and geological consequences. Journal of Geophysical Research, 92: 9059–9082.

Boulton, G.S. and Dobbie, K.E., 1993. Consolidation of sediments by glaciers: relations between sediment geotechnics, soft-bed glacier dynamics and subglacial groundwater flow. Journal of Glaciology, 39: 26–44.

Broecker, W.S., 1994. Massive iceberg discharges as triggers for global climate change. Nature, 372: 421–424.

Brown, N.E., Hallet, B. and Booth, D.B., 1987. Rapid soft bed sliding of the Puget glacial lobe. Journal of Geophysical Research, 92: 8985–8997.

Buckley, J.T., 1969. Gradients of past and present outlet glaciers. Geological Survey of Canada Paper 69-29, 13 pp.

Clark, P.U., 1987. Subglacial sediment dispersal and till composition. Journal of Geology, 95: 527–541.

Clark, P.U., 1988. Glacial geology of the Torngat Mountains, Labrador. Canadian Journal of Earth Sciences, 25: 1184–1198.

Clark, P.U., 1991. Striated clast pavements: products of deforming subglacial sediment? Geology, 19: 530–533.

Clark, P.U., 1992. Surface form of the southern Laurentide Ice Sheet and its implications to ice-sheet dynamics. Geological Society of America Bulletin, 104: 595–605.

Clark, P.U., 1994. Unstable behavior of the Laurentide Ice Sheet over deforming sediment and its implications for climate change. Quaternary Research, 41: 19–25.

Clark, P.U., 1995. Fast glacier flow over soft beds. Science, 267: 43–44.

Clark, P.U. and Josenhans, H.W., 1986. Late Quaternary land-sea correlations, northern Labrador, Canada. Geological Survey of Canada Paper 86-1B, pp. 171–178.

Clark, P.U. and Walder, J.S., 1994. Subglacial drainage, eskers, and deforming beds beneath the Laurentide and Eurasian ice sheets. Geological Society of America Bulletin, 106: 304–314.

Clarke, G.K.C., 1987. Subglacial till: a physical framework for its properties and processes. Journal of Geophysical Research, 92: 9023–9036.

Clarke, G.K.C., Nitsan, U. and Paterson, W.S.B., 1977. Strain heating and creep instability in glaciers and ice sheets. Reviews of Geophysics and Space Physics, 15: 235–247.

Clarke, G.K.C., Collins, S.G. and Thompson, D.E., 1984. Flow, thermal structure, and subglacial conditions of a surge-type glacier. Canadian Journal of Earth Sciences, 21: 232–240.

Clayton, L. and Moran, S.R., 1982. Chronology of late-Wisconsin glaciation in middle North America. Quaternary Science Reviews, 1: 55–82.

Clayton, L., Teller, J.T. and Attig, J.W., 1985. Surging of the southwestern part of the Laurentide Ice Sheet. Boreas, 14: 235–242.

Clayton, L., Mickelson, D.M. and Attig, J.W., 1989. Evidence against pervasively deformed bed material beneath rapidly moving lobes of the southern Laurentide Ice Sheet. Sedimentary Geology, 62: 203–208.

CLIMAP Project Members, 1981. Seasonal reconstruction of the earth's surface at the last glacial maximum. Geological Society of America Map and Chart Series, MC-36.

DesMarais, B.D., 1994. Development of clast fabric in subglacial tills: implications for identifying former deforming beds. Geological Society of America Abstracts with Programs, 26: A-125.

Dowdeswell, J.A. and Sharp, M.J., 1986. Characterization of pebble fabrics in modern terrestrial glacigenic sediments. Sedimentology, 33: 699–710.

Dredge, L.A. and Cowan, W.R., 1989. Quaternary geology of the southwestern Canadian Shield. In: R.J. Fulton (Editor), Quaternary geology of Canada and Greenland, Geological Survey of Canada, Geology of Canada, No. 1, Ottawa, pp. 214–235.

Dreimanis, A., 1976. Tills: their origin and properties: In: R.F. Legget (Editor), Glacial till. Royal Society of Canada, Special Publication 12, pp. 11–49.

Dreimanis, A., 1977. Late Wisconsin glacial retreat in the Great Lakes region, North America. New York Academy of Science Annals, 288: 70–89.

Dreimanis, A., 1988, Tills. Their genetic terminology and classification. In: R.P. Goldthwait and C.L. Matsch (Editor), Genetic classification of glacigenic deposits. A.A. Balkema, Rotterdam, pp. 17–83.

Dyke, A.S. and Prest, V.K., 1987. Late Wisconsinan and Holocene retreat of the Laurentide Ice Sheet. Geological Survey of Canada, Map 1702A, scale 1:5,000,000.

Dyke, A.S., Vincent, J.-S., Andrews, J.T., Dredge, L.A. and Cowan, W.R., 1989. The Laurentide Ice Sheet and an introduction to the Quaternary geology of the Canadian Shield. In: R.J. Fulton (Editor), Quaternary geology of Canada and Greenland. Geological Survey of Canada, Geology of Canada, No. 1, Ottawa, pp. 178–189.

Eyles, N., Sladen, H.A. and Gilroy, S., 1982. A depositional model for stratigraphic complexes and facies superimposition in lodgement till. Boreas, 11: 317–333.

Glen, J.W., Donner, J.J. and West, R.G., 1957. On the mechanism by which stones in till become oriented. American Journal of Science, 255: 194–205.

Hampton, M.A., 1975. Competence of fine-grained debris flows. Journal of Sedimentary Petrology, 45: 834–844.

Hansel, A.K. and Johnson, W.H., 1992. Fluctuations of the Lake Michigan lobe during the late Wisconsin subepisode. Sveriges Geologiska Undersokning, 81: 133–144.

Hart, J.K., 1994. Till fabric associated with deformable beds. Earth Surface Processes and Landforms, 19: 15–32.

Hart, J.K. and Boulton, G.S., 1991. The interrelation of glaciotectonic and glaciodepositional processes within the glacial environment. Quaternary Science Reviews, 10: 335–350.

Hicock, S.R., 1991. On subglacial stone pavements in till. Journal of Geology, 99:607–619.

Hicock, S.R., 1992. Lobal interactions and rheologic superposition in subglacial till near Bradtville, Ontario, Canada. Boreas, 21: 73–88.

Hicock, S.R., Kristjansson, F.J. and Sharpe, D.R., 1989. Carbonate till as a soft bed for Pleistocene ice streams on the Canadian Shield north of Lake Superior. Canadian Journal of Earth Sciences, 26: 2249–2254.

Hicock, S.R. and Dreimanis, A., 1992a. Deformation till in the Great Lakes region: implications for rapid flow along the south-central margin of the Laurentide Ice Sheet. Canadian Journal of Earth Sciences, 29: 1565–1579.

Hicock, S. and Dreimanis, A., 1992b. Sunnybrook drift in the Toronto area, Canada: reinvestigation and reinterpretation. In: P.U. Clark and P.D. Lea (Editors), The last interglacial-glacial transition in North America. Geological Society of America, Special Paper 270, pp. 139–162.

Humphrey, N.F., Kamb, B., Fahnestock, M. and Engelhardt, H., 1993. Characteristics of the bed of the lower Columbia Glacier, Alaska. Journal of Geophysical Research, 98: 837–846.

Iverson, N.R., Hanson, B., Hooke, R.LeB. and Jansson, P., 1995a. Flow mechanism of glaciers on soft beds. Science, 267: 80–81.

Iverson, N.R., Jansson, P. and Hooke, R. LeB., 1995b. In-situ measurement of the strength of deforming subglacial till. Journal of Glaciology, 40: 497–503.

Jeffrey, G.B., 1922. The motion of ellipsoidal particles immersed in a viscous fluid. Royal Society of London, Proceedings, Series A, 102: 161–179.

Jenson, J.W., Clark, P.U., MacAyeal, D.R., Ho, H. and Vela, J.C., in press. Numerical modeling of advective transport of saturated deforming sediment beneath the Lake Michigan lobe, Laurentide Ice Sheet. Geomorphology.

Johnson, A.M., 1970. Physical processes in geology. Freeman, Cooper and Co., San Francisco, 577 pp.

Johnson, W.H. and Hansel, A.K., 1990. Upper Wisconsinan glacigenic sequences at Wedron, Illinois. Journal of Sedimentary Petrology, 60: 26–41.

Johnson, W.H., Hansel, A.K. and Stiff, B.J., 1991. Glacial transport rates, late Wisconsinan Lake Michigan lobe in central Illinois: implications for transport mechanisms and ice dynamics. Geological Society of America Abstracts with Program, 23 (5): A-61.

Kamb, B. 1987. Glacier surge mechanism based on linked cavity configuration of the basal water conduit system. Journal of Geophysical Research, 92: 9083–9100.

Kamb, B., 1991. Rheological nonlinearity and flow instability in the deforming bed mechanism of ice stream motion. Journal of Geophysical Research, 96: 16,585–16,595.

Kamb, B., Raymond, C.F., Harrison, W.D., Engelhardt, H., Echelmeyer, K.A., Humphrey, N., Brugman, M.M. and Pfeffer, T., 1985. Glacier surge mechanism: 1982–1983 surge of Variegated Glacier, Alaska. Science, 227: 469–479.

Kaufman, D.S., Miller, G.H., Stravers, J.A. and Andrews, J.T., 1993. Abrupt early Holocene (9.9–9.6 ka) ice-stream advance at the mouth of Hudson Strait, Arctic Canada. Geology, 21: 1063–1066.

Klassen, R.A. and Fisher, D.A., 1988. Basal-flow conditions at the northeastern margin of the Laurentide Ice Sheet, Lancaster Sound. Canadian Journal of Earth Sciences, 25: 1740–1750.

Lawson, D.E., 1979. A comparison of the pebble orientations in ice and deposits of the Matanuska Glacier, Alaska. Journal of Geology, 87: 629–645.

Lawson, D.E., 1981. Distinguishing characteristics of diamictons at the margin of the Matanuska Glacier, Alaska. Annals of Glaciology, 2: 78–84.

Lindsay, J.F., 1968. The development of clast fabric in mudflows. Journal of Sedimentary Petrology, 38: 1242–1253.

Lowell, T.V., 1995. The application of radiocarbon age estimates to the dating of glacial sequences: an example from the Miami sublobe, Ohio, U.S.A. Quaternary Science Reviews, 14: 85–99.

MacAyeal, D.R., 1993. Growth/purge oscillations of the Laurentide Ice Sheet as a cause of the North Atlantic's Heinrich events. Paleoceanography, 8: 775–784.

Manabe, S. and Broccoli, A.J., 1985. The influence of continental ice sheets on the climate of an ice age. Journal of Geophysical Research, 90: 2167–2190.

Mathews, W.H., 1974. Surface profiles of the Laurentide Ice Sheet in its marginal areas. Journal of Glaciology, 13: 37–43.

Menzies, J., 1990. Sand intraclasts within a diamicton melange, southern Niagara Peninsula, Ontario, Canada. Journal of Quaternary Science, 5: 189–206.

Menzies, J. and Maltman, A.J., 1992. Microstructures in diamictons-evidence of subglacial bed conditions. Geomorphology, 6: 27–40.

Mickelson, D.M., Clayton, L., Fullerton, D.S. and Borns, H.W., Jr., 1983. The late Wisconsin glacial record of the Laurentide Ice Sheet in the United States. In: H.E. Wright, Jr. and S.C. Porter (Editors), Late-Quaternary environments of the United States. University of Minnesota Press, Minneapolis, pp. 3–37.

Mickelson, D.M., Ham, N.R., Jr. and Ronnert, L., 1992. Striated clast pavements: products of deforming subglacial sediment?: Comment. Geology, 20: 285.

Murray, T., 1994, Glacial deformation. In: A. Maltman (Editor), The geological deformation of sediments. Chapman & Hall, London, pp. 73–94.

Naylor, M.A., 1980. The origin of inverse grading in muddy debris flow deposits-A review. Journal of Sedimentary Petrology, 50: 1111–1116.

Paterson, W.S.B., 1994. The physics of glaciers. Pergamon Press, London, 480 pp.

Peltier, W.R., 1994. Ice Age paleotopography. Science, 265: 195–201.

Prest, V.K., 1969. Retreat of Wisconsin and Recent ice in North America. Geological Survey of Canada, Map 1257A, scale 1:5,000,000.

Prest, V.K., Grant, D.R. and Rampton, V.N., 1968. Glacial Map of Canada. Geological Survey of Canada, Map 1253A, scale 1:5,000,000.

Rothlisberger, H., 1972. Water pressure in intra- and subglacial channels. Journal of Glaciology, 11: 177–203.

Scott, J.S., 1976. Geology of Canadian tills. In: R.F. Legget (Editor), Glacial till: an interdisciplinary study. Royal Society of Canada, Special Publication No. 12, Ottawa, pp. 50–66.

Shaw, J., 1987. Glacial sedimentary processes and environmental reconstructions based on lithofacies. Sedimentology, 34: 103–116.

Shreve, R.L., 1985. Late Wisconsin ice-surface profile calculated from esker paths and types, Katahdin esker system, Maine. Quaternary Research, 23: 27–37.

Thorson, R.M., 1980. Ice sheet glaciation of the Puget Lowland, Washington. Quaternary Research, 13: 303–321.

Voorhees, D.H., Johnson, W.H. and Hansel, A.K., 1994. Glacigenic stone horizons as indicators of subglacial bed conditions in north-central Illinois. Geological Society of America Abstracts with Programs, 26: A-125.

Walder, J.S., 1982. Stability of sheet flow beneath temperate glaciers and implications for glacier surging. Journal of Glaciology, 28: 273–293.

Walder, J.S. and Fowler, A., 1994. Channelized subglacial drainage over a deformable bed. Journal of Glaciology, 40: 3–15.

Weertman, J., 1972. General theory of water flow at the base of a glacier or icesheet. Reviews of Geophysics, 10: 287–333.

Wright, H.E., Jr., Matsch, C.L. and Cushing, E.J., 1973. Superior and Des Moines lobes. In: R.F. Black, R.P. Goldthwait and H.B. Willman (Editors), The Wisconsinan Stage. Geological Society of America, Memoir 136, pp. 153–185.

7. MEGAFLOODS AND GLACIATION

Victor R. Baker

Most glacial deposits in the geological record, from Precambrian to late Paleozoic, consist of massive diamictites. From contextual evidence we can infer that many of these were deposited subaqueously, accumulating in glacially influenced marine settings. Rainout and debris-flow processes are commonly invoked for their origin (Visser, 1989). Although the properties of these diamictites have also been ascribed to ejecta emplacement by bolide impact processes (Oberbeck et al., 1993), numerous associations with cold-climate evidence seem to favor the glaciogenic hypothesis. Could cataclysmic flooding processes also have played a role in glaciogenic sedimentation? This chapter explores this question by reviewing geological relationships for the latest large-scale deglaciation on the planet. It is interesting that megaflood/glacial associations have recently been hypothesized for Mars.

Deglaciation and Late Pleistocene Continental Ice Sheets

It is commonly assumed that it is best to infer the nature of ancient processes from processes acting now. This philosophy of actualism is commonly equated to uniformitarianism. When dealing with large, rare processes, however, it is obvious that the present affords us a very limited sample in time that is unlikely to capture such cataclysms. It follows that the best guide to the possible ancient cataclysms of late Paleozoic or older glaciations may not be found in the present. Fortunately, the last major glaciation of the planet has left detailed evidence, both landforms and sediments, that can be interpreted as to genesis with considerable accuracy.

In both Eurasia and North America, Pleistocene glaciations are accompanied by the inception and growth of gigantic ice-dammed lakes. Recent work at the margins of the former North American ice sheets (Baker and Bunker, 1985; Kehew and Teller, 1994) has documented pervasive evidence of Pleistocene cataclysmic outburst floods (jökulhlaups). Such floods, which are a direct consequence of deglaciation, may have exerted major short-term influences on global fluxes of water and sediment (Baker, 1994). Shaw (1989) proposed that subglacial meltwater may be a major contributor to proglacial cataclysmic flooding. Although the details of flood causation are the subject of considerable current debate, the evidence of megaflood/ice-sheet association is extensive and can guide our inferences about ancient glacial sediments.

North America

The usual model of proglacial drainage is one in which outwash streams gradually aggrade their beds, producing a train of sediment that coarsens and sometimes thickens proximally toward the glacial front. The meltwater from North America's late Pleistocene Laurentide Ice Sheet initially produced such patterns in the marginal valley systems. After the ice sheet retreated behind its major morainal features and landform divides, however, meltwater mainly ponded as proglacial lakes. As lake levels rose, the water was released as glacial-lake outburst floods (Kehew and Teller, 1994). These floods carved complex glacial-lake spillways (Fig. 7.1), thereby redistributing great volumes of sediment. More than a dozen spillways rapidly conveyed the floods to the ocean, particularly through the Mississippi River system to the Gulf of Mexico. As summarized by Teller (1990), much of this proglacial discharge passed through the Mississippi River to the Gulf of Mexico until about 11 ka. The immense Lake Agassiz Basin then was integrated into the St. Lawrence drainage to the Atlantic Ocean. At about 10 ka the Lake Agassiz discharge briefly returned to the Mississippi system.

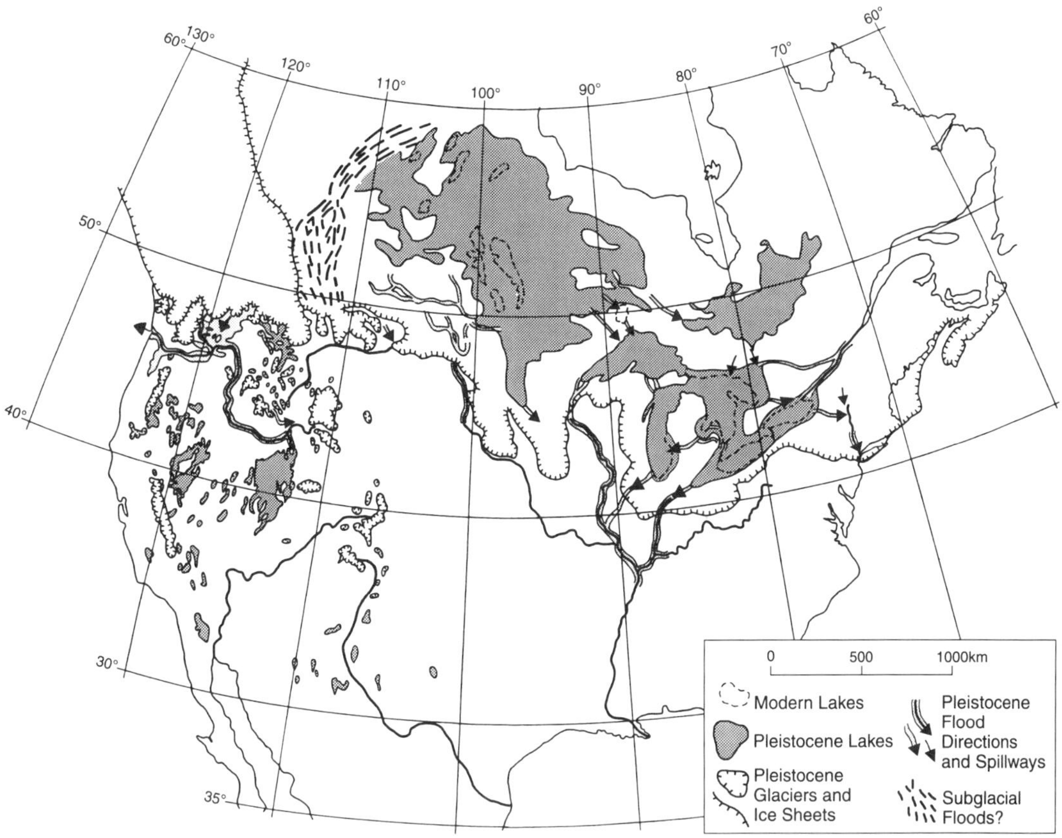

Fig. 7.1. Cataclysmic flood spillways and hypothesized subglacial flood routes (Rains et al., 1993) for North America during the Late Pleistocene.

Eurasia

In a reconstruction of the late Pleistocene Eurasian ice sheets, Grosswald (1993) envisions damming of the large, north-flowing Siberian rivers with associated spillovers and drainage diversions (Fig. 7.2). Alternative reconstructions involve much less ponding and diversion (Velichko et al., 1984), but incomplete field mapping and geochronology preclude scientific preference for either of these hypothetical alternatives. Nevertheless, it is clear that the phenomenon of massive, regionally extensive, high-energy flooding was an important feature of glaciation in the area. Terminal glacial features, extending west-to-east across the mouths of the Ob'-Irtysh and the Yeneisei rivers, suggest that a blockage prevented them from reaching the Arctic Ocean. The causative glaciation has been hypothesized to originate from an ice dome in what is now the Kara Sea (Grosswald, 1993). The great Mansi paleolake was generated south of the Kara dome, inundating much of the West Siberian Plain (Grosswald, 1993).

The Turgai Hollow leads southward from the Mansi paleolake basin to the Aral Sea basin. Its threshold elevation of 126 m (Arkhipov, 1984) determines the level of Lake Mansi. Fluvial and lake sediments in the Turgai spillway document the conveyance of flows southward during the latest Pleistocene. The paleodischarges then entered the Aral Sea basin, which at its late-glacial highstand of 70 m had two spillways to the Caspian Basin. The latter had a late-glacial highstand of 48 m (the Khvalyn Sea) which may have been generated by the floodwater influx (Grosswald, 1993). The Manych spillway, which connected the Khvalyn Sea to the Black Sea,

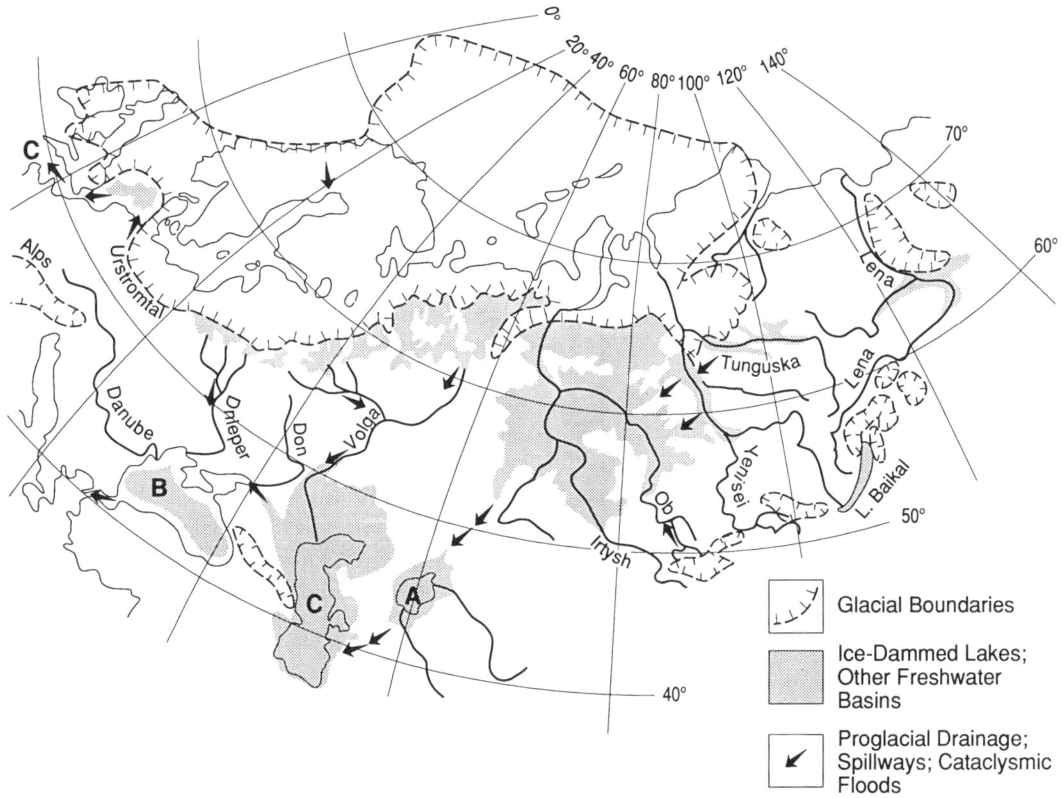

Fig. 7.2. Late Pleistocene Eurasian Ice Sheet and related drainage features, showing location of Altay Mountain flood area described by Baker et al. (1993). Glacial spillways carried outflow floods to the Pleistocene predecessors of the Aral Sea (A), Caspian Sea (C), and Black Sea (B) (after Rudoy and Baker, 1993).

displays a suite of landforms characteristic of cataclysmic flooding: streamlined hills, elongate depressions (now lakes), and scabland erosion. The immense channel width (35 km) further indicates the great scale of the paleoflow. The paleochannel had a cross-sectional area of approximately 10^6 m^2 at the 48-m Khvalyn level (28-m flow depth). At a flow velocity of 10 ms^{-1}, required for bedrock scabland erosion of this type (Baker and Costa, 1987), the estimated flood discharge is 10^7 m^3s^{-1}, which is comparable to that of ice-dammed lake outbursts like those of glacial Lake Missoula (Baker et al., 1993).

Cataclysmic Flood Landforms and Deposits

The most detailed studies of cataclysmic flooding features are from intermontane paleolakes. Although the best-known examples are associated with the ice-dammed glacial Lake Missoula (Baker, 1973), more examples have been documented in recent years. In Kygyzstan, Lake Issyk-Kul, the second largest mountain lake in the world, was transformed into an ice-dammed lake during the late Pleistocene (Grosswald et al., 1994). Evidence for this includes iceberg-emplaced landforms rising to 300 m above the present lake level. Spectacular erosion at the full-glacial lake outlet (Boam Canyon) indicates lake burst flooding, which also emplaced an enormous outwash fan extending more than 85 km from the canyon mouth (Grosswald et al., 1994). It is estimated that the maximum flooding released a water volume of 1,300 km^3. Rudoy and Baker (1993) described the evidence for late Pleistocene cataclysmic flooding in the Altay Mountain region of central Asia. The Pleistocene glacial lake bursts of that region are estimated to have had discharges of $18x10^6$ m^3s^{-1}, equaling or even exceeding those of the Channeled Scabland region (Baker et al, 1993). As in the Channeled Scabland of western North America (Baker, 1973), the cataclysmic flood sediments of the Altay region can be divided into two general facies: coarse-grained channel deposits associated with the high-velocity (high-energy) flood waters of the main channelways, and finer-grained slackwater deposits, which accumulated in low-velocity (low-energy) regions marginal to the main flood channelways.

High-Energy Facies

In the Channeled Scabland and the Altay areas, coarse-grained flood deposits form a series of longitudinal bars, which rise to about 200 m above the modern river level. These bars are as much as 2 to 5 km in length and occupy alternating positions along the flood channelway, conforming to valley meander bends. The bars represent depositional macroforms, with their magnitudes scaled to channel width. The internal structures of the bars are dominated by huge foresets, sometimes 10 or 20 m thick, composed of gravel and coarse sand. Plane bedding may occur locally but does not dominate as a stratigraphic type. Bar surfaces may be armored with large boulders.

A complete sequence of high-energy cataclysmic flood landforms is recognized in various lake-outburst channelways (Table 7.1). The depositional forms are scaled to flow width (macroforms), flow depth (mesoforms), and the flow boundary layer (microforms). The mesoforms are giant current ripples (Baker, 1973), also termed "giant current dunes" (Waitt, 1985). These flood landforms (Fig. 7.3) range in spacing from 30 to 200 m and are composed of pebble and cobble gravel (Baker, 1973; Rudoy and Baker, 1993). Internally the giant current ripples consist of foreset-bedded gravels at angles of about 20–30° (Baker, 1973). Layers of cobbles may alternate with discrete layers of pebbles or granules. The sorting gives the gravel a distinctive open-work texture.

Kehew and Lord (1987) describe the erosional and depositional characteristics of flood spillways in central North America. The main channels are on the order of 10 km in width and up to several hundred kilometers in length. Unlike river channels and valleys, which increase in width downstream, flood spillways maintain a constant width along their length. Often there are prominent inner channels, which are 1–3 km in width and 25–100 m deep. Streamlined erosional hills and longitudinal grooves occur on the terrace-like outer zone adjacent to the inner channel. Bars of coarse boulder gravel occur locally in the inner channel. The bars average 2 km in length and 20 m in thickness. The gravel is massive, matrix-supported, and very poorly sorted. These characteristics probably reflect much lower intensities of flow processes (power per unit area) than are achieved in the narrow-deep valleys impacted by ice-dammed lake releases.

Low-Energy Facies

The low-energy facies of cataclysmic proglacial flooding occurs in areas marginal to the main channelways, such as the back-flooded mouths of tributary valleys. This facies also occurs in basins into which the main channelways debouch their floodwaters. In the Missoula Flood slackwater areas (Baker and Bunker, 1985), the low-energy facies consists of rhythmite sections (Waitt, 1985; Smith, 1993). Individual beds are graded, fining upward from fine gravel or sand to silt. The beds may contain ice-rafted glacial erratics, and they form laterally extensive drapes over the pre-existing valley floor topography. Scores of such units, each 0.5 to 1.5 m thick, may make up sections of the flood slackwater facies (Waitt, 1985). The sequences coarsen markedly in a proximal direction toward the high-energy main flood channelways.

Sedimentary Basins

The geological record favors the long-term preservation of relatively thick sedimentary sequences emplaced in sedimentary basins. The high-energy flood facies, described earlier, are largely restricted to erosional terrains, cut into

TABLE 7.1. *High-energy cataclysmic flood bedforms observed in Pleistocene glacial lake outburst channels*

	Scoured in rock	Scoured in sediment	Depositional
Macroforms (scale controlled by width)	Undulatory bed profile Anastomosis	Streamlined hills	Alternate bars Expansion bars Pendant bars Eddy bars
Mesoforms (scale controlled by depth)	Longitudinal grooves Pothole complexes Inner channels Cataracts	Scour marks	Giant current ripples (dunes, gravel waves)
Microforms	Scallop pits Striations Flute marks	Not preserved	Percussion marks

continental cratons, as in Eurasia and central North America. Local cratonic basins may preserve high-energy fan complexes and distal slackwater facies, as in the Missoula Flood deposits of the Quincy Basin in the Channeled Scabland (Baker and Bunker, 1985). The Ephrata Fan of the Quincy Basin covers an area of 1,000 km^2. It is characterized by a very pronounced proximal-to-distal decrease in maximum grain size from large boulders to coarse sand in a distance of 20 km (Baker, 1973).

Cataclysmic floodwater reaching the sea, if charged with suspended sediment, is highly likely to produce hyperpycnal plumes when floodwater density exceeds that of sea water (Mulder and Syvitski, 1995). Griggs et al. (1970) traced gravel lithologies of Missoula Flood provenance across the eastern Pacific floor and on to the Juan de Fuca Rise. Transport of the gravel was presumably by flood-induced turbidity flows at the mouth of the Columbia River, which conveyed the Missoula flows to the ocean.

Fig. 7.3. Giant current ripples (gravel waves) at Marlin, Washington (Baker, 1973). These ripples are spaced about 80 m apart and rise about 4 m in height.

Flow Processes

The understanding of processes in cataclysmic floods has been made possible through computer flow modeling (O'Connor and Baker, 1992; O'Connor, 1993). These models compute energy-balanced water surface profiles for steady, gradually varied flow on the basis of the one-dimensional energy equation (a form of the Bernoulli Equation) in conjunction with the step-backwater method of profile computation. For flow conditions (discharge, stage, and channel geometry) specified at one cross section, stage is predicted at an adjacent cross section of known geometry and reach length by conserving mechanical energy and accounting for estimated flow-energy losses associated with roughness and channel expansions and contractions. Calculated conditions at the second cross section are then taken as "known" for the next incremental reach. In this manner, the stage established for each cross section and the water-surface profile can be determined for an indefinite length of channel. For application to cataclysmic flood channel geometries, various water-surface profiles are computed with different combinations of discharge and energy loss coefficients until a profile is achieved that most closely matches the water-surface profile suggested by channel geometry (O'Connor and Baker, 1992).

Hydraulic parameters for some cataclysmic flood channels are reported in Table 7.2. Note that slopes (S) and flow depths (D) are much larger than those for the Amazon River. Despite a large discharge (Q), the Amazon has flow velocities (V) and bed shear stresses (t) comparable to those of much smaller alluvial rivers (Baker and Kochel, 1988). Particularly important is the rate of energy dissipation, or power, per unit area of streambed, which can be expressed

$$\omega = \frac{\gamma QS}{W} = \tau\bar{V}$$

where W is bed width, Q is discharge, S is energy slope, γ is the specific weight of the transporting fluid, $\bar{V}$ is the mean flow velocity, and τ is the bed shear stress. The cataclysmic flood features of very deep high-gradient flows are generated by power (ω) values many orders of magnitude larger than those produced in alluvial rivers (Baker and Costa, 1987).

Subglacial Flooding

A variety of landforms, involving both erosion and deposition, form beneath thick continental glaciers. These landforms have generally been explained by various subglacial deformation processes. However, modern glaciers may not provide satisfactory analogs to the ancient ice sheets that produced these landforms. In recent years an hypothesis involving catastrophic subglacial flooding has been posed to explain the spatial and temporal associations of these features. Large-scale, subglacial meltwater floods are inferred to have formed certain drumlins (Shaw and Sharp, 1987), tunnel channels (Brennand and Shaw, 1994), bedrock erosional marks (Shaw, 1994a) and Rogen moraine (Fisher and Shaw, 1992). Shaw (1994b) envisions a whole sequence of erosional/depositional processes beneath giant continental ice sheets, involving four major phases: (1) ice-sheet advance, (2) surging, stagnation, and melt-out, (3) cataclysmic events, including meltwater outburst floods, and (4) regional stagnation with esker formation. The subglacial landforms are superimposed in a sequence that reflects the temporal succession of these phases (Table 7.3). Thus, the cataclysmic flooding is part of a connected sequence of glacial processes that are inferred to be physically reasonable for large continental warm-based ice sheets (Shaw, 1994b).

Parts of the sequence shown in Table 7.3 may be recognized in ancient glacial settings. Fluting in Permo-Carboniferous glacial pavements is attributed to ice erosion (Visser et al., 1986), but subglacial meltwater may offer an alternative hypothesis for their generation (Shaw, 1994a). However, no one landform is likely to be diagnostic. The entire assemblage of genetically related landforms needs to be documented in order for us to understand these enigmatic features.

Global Consequences of Glacial Megafloods

The late-glacial oceanographic record shows pulsed inputs of glacial meltwater (Fairbanks et al., 1992). This is generally attributed to irregular melting of the Laurentide Ice Sheet. However, the hypothesized subglacial megaflooding would also have had immense consequences for oceanic response (Shaw, 1989). This hypothesis also seems consistent with the late-glacial flattening of the Laurentide Ice Sheet profile at its outer periphery, a phenomenon attributed to late glacial surging (Clark, 1994). Shoemaker (1992a) presents a physical argument explaining the remarkable low gradients of Laurentide Ice Sheet lobes as the product of meters-thick floodwater moving along the ice-bed interface. This argument is

TABLE 7.2. *Paleohydraulic parameters for cataclysmic flood channelways compared to those for the Amazon River and a Martian outflow channel (Ares Vallis)*

Flood Channel	Q m^3s^{-1}	W km	D m	S	V ms^{-1}	t Mn^{-2}	w Wm^{-2}
Chuja (Altay)	$2x10^7$	3	400	0.02	20–45	$5x10^3$–$2x10^4$	10^5–10^6
Rathdrum (Missoula)	$2x10^7$	6	175	0.01	25	$1x10^4$	$2x10^5$
Grand Coulee(Missoula)	$5x10^6$	1.7	100	0.01	30	$1x10^4$	$3x10^5$
Amazon	$3x10^5$	2	60	$1x10^{-5}$	2	6	12
Ares Vallis (Mars)	10^8–10^9	25	200	0.001	15	$2x10^4$	$3x10^5$

extended to include large subglacial lake source areas and regional-scale outburst flooding (Shoemaker, 1992b).

From evidence of late Pleistocene reef-drowning in the Caribbean-Atlantic region, Blanchard and Shaw (1995) infer three catastrophic, meter-scale sea-level-rise events. They relate these to ice-sheet collapse, reorganization of ocean-atmosphere circulation, and release of subglacial and proglacial meltwater. Catastrophic release of meltwater megafloods is hypothesized to have triggered ice-sheet instability in the Laurentide Ice Sheet. Independent evidence of such instability is provided by ice-rafted debris in North Atlantic cores (Bond et al., 1992). The responsible debris-laden icebergs are inferred to have been released episodically and catastrophically through Hudson Strait into the Labrador Sea. Named "Heinrich events," the iceberg releases immediately preceded abrupt cooling events that are recognized in the Greenland ice-core record of the most recent glacial period (Dansgaard et

TABLE 7.3. *Subglacial processes and landforms associated with an evolutionary sequence of large, warm-based ice sheets (Shaw, 1994b).*

Phase	Glacial Processes	Landforms	Time
4	Ice stagnation Multiple floods	Eskers	Younger
3	Cataclysmic outburst flooding	Tunnel valleys Rogen moraine Drumlins	
2	Stagnation/surge episodes	Meltwater tills Meltwater scours/plucking	
1	Sliding and subglacial deformation	Striation and deposition	Older

al., 1993). The climatic connection may have been achieved through the impact of a flood-induced, low-density cap of cold-fresh water over the zone of deep-water formation in the North Atlantic, a site that is critical to the thermohaline circulation of the global ocean that controls world climate (Broecker et al., 1992). Thus, in this highly conjectural scenario, glacial megafloods may correlate to planet-wide patterns of ice-sheet dynamics, sea-level rise, glaciomarine sedimentation, and climate change.

Megafloods and Glaciation on Mars

While cataclysmic megafloods on Earth are inferred to have glacial causes, on Mars the cataclysmic megafloods seem to be the cause, and glaciation is the response. The Mars floods are inferred to have been much larger than any that occurred on Earth (Baker, 1982). The flooding evidence includes an assemblage of bedforms similar to that shown in Table 7.1, developed in huge outflow channels tens of kilometers wide, hundreds of meters deep, and up to 2,000 kilometers long (Baker, 1982). These channels are ancient by Earth standards, with ages approximately 10^9 years. In addition to the channels, numerous relict valley network systems (Baker, 1982), concentrated in the Martian highlands, imply formation by dynamical cycling of water.

The discovery of evidence for extensive inundation of the northern-hemisphere plains of Mars (Parker et al., 1993) and the discovery of evidence for widespread apparently contemporaneous glaciation (Kargel and Strom, 1992; Kargel et al., 1995), led to a proposal reagarding episodic ocean formation and related climatological change during Martian geological history (Baker et al., 1991). This hypothesis is consistent with a range of otherwise enigmatic observations, including the presence of anomalously young valley networks and local occurrences of layered deposits in probable paleolake basins. The trigger for each ocean/glacial episode was probably a massive, short-duration phase of volcanism that could have resulted from an immense mantle plume heating a huge area of lithosphere. The resulting thermal bulging, melting of ground ice, and hydrothermal circulation to fracture systems led to immense water outbursts from the subsurface. The latter are evidenced by the cataclysmic flood landforms of the outflow channels that drained the water from Tharsis to the northern plains, where it created a vast, short-lived ocean, persisting perhaps 10^4 to 10^5 years (Baker et al., 1991). The warm, wet climate associated with the ocean and with released greenhouse gases led to enhanced precipitation over the Martian highlands and to extensive glaciation.

The evidence for glaciation includes an assemblage of landforms resembling erosional grooves, streamlined/sculptured hills, drumlins, arêtes, horns, cirques, tunnel channels, eskers, moraines, kame terraces, outwash plains, kettles, glaciolacustrine plains, and wave-cut cliffs. While individual landforms in this assemblage have been explained by various alternative hypotheses, the entire sequence of forms occurs in a spatial association (Table 7.4) analogous to what would be expected for Earth glaciation. To have a nonglacial process produce this remarkable association in time and space would be highly unlikely. By analogy to Earth, it is hypothesized that prolonged snow accumulation occurred during the temporary planetwide warm, wet climate epochs associated with the hypothesized short-duration glacial-ocean phases.

The warm conditions that allowed ponded water to occur temporarily at the north pole of Mars may possibly have occurred at times of high orbital obliquity. It is interesting that high obliquity may also have contributed to Earth's late Proterozoic glaciation, during the period 800–600 Ma. Williams (1994, p. 146) argues that high obliquity offers the best explanation for the "enigma of frigid, strongly seasonable climates, with permafrost and grounded ice-sheets near sea-level, in low latitudes." The possibility of parallels between the newly discovered Mars glaciation and Precambrian Earth glaciations opens new opportunities for understanding this phenomenon.

Discussion

Confusion concerning the role of cataclysmic processes in the sedimentary record largely derives from the imposition of philosophy on the

TABLE 7.4. *Spatial asociation of glacial features in the Hellas region of Mars (after Kargel and Strom, 1992).*

Erosional grooves	Proximal
Eskers and drumlins	
Kames and kettles	
Morainal ridges	
Glaciolacustrine plain	
Shorelines	Distal

natural reasoning processes of geologists. Clifton (1988, p. 3, 4) exemplifies this confusion by stating, "Parsimony demands that we attribute phenomena in the sedimentary record to the most probable explanation, and convulsive geologic events are, by nature, improbable." In a natural science, however, the only valid limitations on our explanations are those made by nature. If nature contradicts our philosophy, parsimony included, it is nature that our explanations must follow, no matter how seductive the philosophy. Despite confusion on this point, it has become clear that megafloods were extensively associated with deglaciation of Earth's large continental ice sheets in the late Quaternary. Megafloods and glaciation also seem to have been intimately associated in the geological history of Mars, although relationships are quite unusual in comparison to Earth. The prospects are good for discovering cataclysmic flood indicators in late Paleozoic and other ancient glaciogenic sequences. The only philosophical principle necessary for such discovery is an open mind, unfettered by confused notions of actualism, parsimony, or uniformitarianism.

In local context it is rare to find sedimentary signatures uniquely indicative of glacial megaflooding. Proximal flood channelways may contain high-energy facies, including bouldery gravels, large-scale foreset bedding, and openwork textures of giant ripple cross-strata. However, poorly sorted bouldery gravels with massive textures also occur in the bars of flood spillways (Kehew and Lord, 1987). The low-energy facies, which occur in basin settings, are most likely to be preserved in ancient rocks, but these are even more problematic for unequivocally indicating cataclysmic flooding origins. Rhythmites may be produced by a variety of local sedimentary processes. Megafloods are only one of many possible hypotheses, and one that historically has been prematurely rejected on erroneous philosophical grounds.

As classically illustrated in the geological controversies surrounding the origin of Channeled Scabland (Baker and Bunker, 1985), megaflood origins can be established through the painstaking identification of regional patterns of mutually consistent indicators. Locally, individual sedimentary or landform units may be explicable in terms of local causes that may have noncataclysmic sources. It is only when a larger, genetically related pattern emerges (Figs. 7.1, 7.2) that a catastrophic flooding hypothesis becomes viable. Recent discoveries, summarized in this brief review, suggest that glacial megaflooding origins may well be seriously considered for appropriate sedimentary sequences. The regional patterns necessary for corrobation will not be identified unless one is open to their possibility and acts accordingly to explore their probable consequences.

References

Arkhipov, S.A., 1984. Late Pleistocene glaciation of western Siberia. In: A.A. Velichko (Editor), Late

Quaternary environments of the Soviet Union. University of Minnesota Press, Minneapolis, pp. 13–19.
Baker, V.R., 1973. Paleohydrology and sedimentology of Lake Missoula flooding in eastern Washington. Geological Society of America, Special Paper 144, 79 pp.
Baker, V.R., 1982. The channels of Mars. University of Texas Press, Austin, Texas, 198 pp.
Baker, V.R., 1994. Glacial to modern changes in global river fluxes. In: National Reserach Coundil (Editor), Material fluxes on the surface of the Earth. National Academy Press, Washington, DC, pp. 86–98.
Baker, V.R. and Bunker, R.C., 1985. Cataclysmic Late Pleistocene flooding from glacial Lake Missoula: a review. Quaternary Science Reviews, 4: 1–41.
Baker, V.R. and Costa, J.E., 1987. Flood power. In: L. Mayer and D. Nash (Editors), Catastrophic flooding. Allen and Unwin, London, pp. 1–21.
Baker, V.R. and Kochel, R.C., 1988. Flood sedimentation in bedrock fluvial systems. In: V.R. Baker, R.C. Kochel and P.C. Patton (Editors), Flood geomorphology. Wiley, N.Y., pp. 123–137.
Baker, V.R., Strom, R.G., Gulick, V.C., Kargel, J.S., Komatsu, G. and Kale, V.S., 1991. Ancient oceans, ice sheets and the hydrological cycle on Mars. Nature, 352: 589–594.
Baker, V.R., Benito, G. and Rudoy, A.N., 1993. Paleohydrology of Late Pleistocene superflooding, Altay Mountains, Siberia. Science, 259: 348–350.
Blanchard, P. and Shaw, J., 1995. Reef drowning during the last deglaciation: evidence for catastrophic sea-level rise and ice-sheet collapse. Geology, 23: 4–8.
Bond, G., Heinrich, H., Broecker, W., Labeyrie, L., McManus, J., Andrews, J., Huon, S., Jantschik, R., Clasen, S., Simet, C., Tedesco, K., Mieczyslawa, K., Bonani, G. and Ivy, S., 1992. Evidence for massive discharges of icebergs into the North Atlantic during the last glacial period. Nature, 360: 245–249.
Brennand, T.A. and Shaw, J., 1994. Tunnel channels and associated landforms, south-central Ontario: their implications for ice-sheet hydrology. Canadian Journal of Earth Science, 31: 505–522.
Broecker, W.S., Bond, G., Klas, M., Clark, E. and McManus, J., 1992. Origin of the North Atlantic's Heinrich Events. Climate Dynamics, 6: 91–109.
Clark, P.U., 1994. Unstable behavior of the Laurentide Ice Sheet over deforming sediment and its implications for climate change. Quaternary Research, 41: 19–25.
Clifton, H.E., 1988. Sedimentologic relevance of convulsive geologic events. Geological Society of America, Special Paper 229, pp. 1–5.
Dansgaard, W., Johnson, S.J., Clausen, H.B., Dahl-Jensen, D., Gundestrup, N.S., Hammer, C.U., Hvidberg, C.S., Steffensen, J.P., Sveinbjörnsdottir, A.E., Jouzel, J. and Bond, G., 1993. Evidence for general instability of past climate from a 250-kyr ice-core record. Nature, 364: 218–220.
Fairbanks, R.G., Charles, C.D. and Wright, J.D., 1992. Origin of glacial meltwater pulses. In: R.E. Taylor, A. Long and R. Kra (Editors), Radiocarbon after four decades. Springer-Verlag, Berlin, pp. 473–500.
Fisher, T.G. and Shaw, J., 1992. A depositional model for Rogen moraine, with examples from the Avalon Peninsula, Newfoundland. Canadian Journal of Earth Science, 29: 669–686.
Griggs, G.B., Kulm, L.D., Waters, A.C. and Fowler, G.A., 1970. Deep-sea gravel from Cascadia Channel. Journal of Geology, 78: 611–619.
Grosswald, M.G., 1993. Extent and melting history of the late Weichselian ice sheet, the Barents-Kara continental margin. In: W.R. Peltier (Editor), Ice in the climate system. Springer-Verlag, Heidelberg, pp. 1–20.
Grosswald, M.G., Kuhle, M. and Fastook, J.L., 1994. Wurm glaciation of Lake Issyk-Kul area, Tian Shan Mts: A case study in glacial history of central Asia. Geojournal, 33: 273–310.
Kargel, J.S. and Strom, R.G., 1992. Ancient glaciation on Mars. Geology, 20: 3–7.
Kargel, J.S., Baker, V.R., Begét, J.E., Lockwood, J.F., Péwé, T.L., Shaw, J.S. and Strom, R.G., 1995. The collapse of a Martian continental glacier. Journal of Geophysical Research, 100: 5351–5368.
Kehew, A.E. and Lord, M.L., 1987. Glacier-lake outbursts along the mid-continent margins of the Laurentide ice-sheet. In: L. Mayer and D. Nash (Editors), Catastrophic Flooding. Allen and Unwin, Boston, pp. 95–120.
Kehew, A.E. and Teller, J.T., 1994. History of late glacial runoff along the southwestern margin of the Laurentide Ice Sheet. Quaternary Science Reviews, 13: 859–877.
Mulder, T. and Syvitski, J.P.M., 1995. Turbidity currents generated at river mouths during exceptional discharges to the world oceans. Journal of Geology, 103: 285–299.
Oberbeck, V.R., Marshall, J.R. and Aggarvad, H., 1993. Impacts, tillites, and the breakup of Gondwanaland. Journal of Geology, 101: 1–19.
O'Connor, J.E., 1993. Hydrology, hydraulics, and sediment transport of Pleistocene Lake Bonneville flooding on the Snake River, Idaho. Geological Society of America, Special Paper 274, 83 pp.
O'Connor, J.E. and Baker, V.R., 1992. Magnitudes and

implications of peak discharges from glacial Lake Missoula. Geological Society of America Bulletin, 104: 267–279.

Parker, T.J., Gorsline, D.S., Saunders, R.S., Pieri, D.C. and Schneeberger, D.M., 1993. Coastal geomorphology of the Martian northern plains. Journal of Geophysical Research, 98: 11,061–11,078.

Rains, R.B., Shaw, J., Skoye, R., Sjogren, D. and Kvill, D., 1993. Late Wisconsin subglacial megaflood paths in Alberta. Geology, 21: 323–326.

Rudoy, A.N. and Baker, V.R., 1993. Sedimentary effects of cataclysmic Late Pleistocene glacial outburst flooding, Altay Mountains, Siberia. Sedimentary Geology, 85: 53–62.

Shaw, J., 1989. Drumlins, subglacial meltwater floods, and ocean responses. Geology, 17: 853–856.

Shaw, J., 1994a. Hairpin erosional marks, horseshoe vortices and subglacial erosion. Sedimentary Geology, 91: 269–283.

Shaw, J., 1994b. A qualitative view of sub-ice-sheet landscape evolution. Progress in Physical Geography, 18: 159–184.

Shaw, J. and Sharpe, D.R., 1987. Drumlin formation by subglacial meltwater erosion. Canadian Journal of Earth Sciences, 24: 2316–2322.

Shoemaker, E.M., 1992a. Subglacial floods and the origin of low-relief ice-sheet lobes. Journal of Glaciology, 38: 105–112.

Shoemaker, E.M., 1992b. Water sheet outburst floods from the Laurentide Ice Sheet. Canadian Journal of Earth Sciences, 29: 1250–1264.

Smith, G.A., 1993. Missoula flood dynamics and magnitudes inferred from sedimentology of slack-water deposits on the Columbia Plateau, Washington. Geological Society of America Bull etin, 105: 77–100.

Teller, J.T., 1990. Meltwater and precipitation runoff to the North Atlantic, Arctic, and Gulf of Mexico from the Laurentide Ice Sheet and adjacent regions during the Younger Dryas. Paleoceanography, 5: 897–905.

Velichko, A.A., Isayeva, L.L., Makeyev, V.M., Matishov, G.G., and Faustova, G.G., 1984. Late Pleistocene glaciation of the Arctic shelf, and the reconstruction of Eurasian ice sheets. In: A.A. Velichko (Editor), Late Quaternary environments of the Soviet Union. University of Minnesota Press, Minneapolis, pp. 35–44.

Visser, J.N.J., 1989. The Permo-Carboniferous Dwyka Formation of southern Africa: deposition by a predominantly subpolar marine ice sheet. Palaeogeography Palaeoclimatology Palaeoecology, 70: 377–391.

Visser, J.N.J., Lowe, D.R. and Beukes, N.J., 1986. A Permo-Carboniferous glacial pavement and related Karoo rocks along the Ghaap escarpment near Boetsap, Northern Cape Province. Transactions of the Geological Society of South Africa, 89: 385–388.

Waitt, R.B., 1985. Case for periodic, colossal jökulhlaups from Pleistocene glacial Lake Missoula. Geological Society of America Bulletin, 96: 1271–1286.

Williams, G.E., 1994. The enigmatic Late Proterozoic glacial climate: an Australian perspective. In: M. Deynoux, J.M.G. Miller, E.W. Domack, N. Eyles, J.J. Fairchild and G.M. Young (Editors), Earth's Glacial Record. Cambridge University Press, Cambridge, pp. 146–164.

PART III

THE CARBONIFEROUS-PERMIAN: A Paleogeographic Reconstruction

8. PERMIAN WORLD TOPOGRAPHY AND CLIMATE

Alfred M. Ziegler, Michael L. Hulver and David B. Rowley

Current paleogeographic maps of the Permian world give a cursory view of mountain ranges and other features of the land surface (Ziegler et al., 1979; Ziegler, 1990; Scotese and Langford, 1995). Paleotopography is much more difficult to reconstruct than paleobathymetry because of the relative lack of direct evidence, so features portrayed usually are limited to "active" mountain chains and rift shoulders of the interval of interest. Yet a knowledge of topography is critical for a variety of reasons in addition to our wish to draw the complete map. First, modern concepts of basin evolution have demonstrated the regional tectonic controls that operate in so many cases, so that mountain building is now regarded as "in phase" and in control of basin development, rather than as terminating or punctuating the stratigraphic sequences. These interrelationships may be obvious locally, but the paleogeographer is faced with extending the features across broad areas, and a knowledge of tectonic stress fields is crucial in projecting trends in places where direct evidence may have been destroyed by erosion. Second, mountain chains and inland seaways have an important influence on atmospheric circulation and precipitation patterns, so climate modelers use the paleotopography as direct input in their studies (Kutzbach and Ziegler, 1993). Climate modelers often assume that the current reconstructions are correct (Crowley et al., 1989) and derive climate extremes that could not have been tolerated by the plants and animals known to have lived there (Yemane, 1993; Ziegler, 1993). Third, and important for the present volume, the existence of mountains in high latitudes in the Late Paleozoic was probably crucial in the development of the ice sheets (Visser, this volume). By the Permian most of the ice caps were centered on upland regions, as will be seen.

A number of the mountain ranges shown on the Permian maps herein were not tectonically active at the time and indeed had their origins as far back as the Precambrian. It is sometimes claimed in the geomorphological literature that mountain ranges have a limited duration. Ahnert (1970), in denudation studies of mid-latitude drainage basins, concluded that relief would be reduced to 10 percent in 18.5 m.y., and to 1 percent in 37 m.y., even allowing for isostatic rebound. Others have assumed from this that seemingly ancient features, like the Appalachian Mountains, must have been rejuvenated (Hack, 1982), and, indeed, that the Neogene was a tectonically active period worldwide, judging by the numerous existing mountain ranges.

Pitman and Golovchenko (1991) questioned Ahnert's assumptions and produced their own erosional constant on the basis of the amount of sediment in the Atlantic Basin. They concluded that present-day denudation rates are 2 to 4 times the long-term average rate because of the dual effects of human and glacial influence, and calculated that it would take 450 m.y. to degrade a Himalayan scale mountain chain, while a long-term eustatic drop would prolong the process. Their calculations do seem to support the old notion of a Cretaceous transgression across an Appalachian peneplain. It can be shown from abundant subsurface information (Brown et al., 1972) that Late Cretaceous shorelines did not extend inland from Chesapeake Bay, and sediment isopach data (Poag and Sevon, 1989) assure us that the Appalachians continued as a sediment source during this time. In fact the Cretaceous sea level with respect to the Atlantic coastal plain was about the same as today, implying that Pitman's (1978) "hinge line" was inboard of the present coast and that the thermal subsidence has affected the entire system. It is certainly true that relief has been rejuvenated by the generally dropping eustatic sea levels through the Tertiary, and especially by the rapidly fluctuating sea levels in the Pleistocene. The increase in relief would naturally stimulate isostatic uplift which would in turn add to the total relief (England and

Molnar, 1990). We therefore conclude that both relative and total relief have increased on a worldwide basis during the Tertiary, but independently of tectonics.

In our treatment of ancient mountain ranges we follow the lead of Veevers (1984) who provided a detailed analysis of the "morphotectonics" of Australia through geologic time. Veevers attributed Precambrian origins to a number of present upland regions, and more recent analyses of the geomorphology of this continent provide additional support for the antiquity of the landforms (Wellman, 1987; Young and McDougall, 1993; Nott, 1995). We can assume that average denudation rates derived by Pitman and Golovchenko (1991) are of the right order, and emphasize that the mountains that survive long periods are not composed of average rocks, being held up by such chemically and physically resistant lithologies as quartzite and chert. Moreover, the precipitation needed to erode the features is not consistent over geologic time, particularly well within supercontinents like Gondwana where foldbelts could have been geographically isolated from moisture sources for long intervals.

Paleogeographic Map Reconstruction

The main focus of this chapter is orography, but coastlines, shelf margins and deeper basins are also included on the maps. These marine features are discussed only with reference to the mountain ranges with which they were tectonically linked. We have relied most heavily on the literature that contains the basic lithofacies control points as well as their paleogeographic interpretation (Vinogradov, 1969). Reinterpretation is then possible where control is weak, although we have generally followed published paleogeographic maps unless they are at variance with the current plate tectonic paradigm.

The lines depicted on our maps (in pocket) represent the shoreline, the 200 m, 1,000 m and 2,000 m contours above sea level, and the 200 m contour below sea level. Only the shoreline can be known with any certainty, and even this would fluctuate considerably during the course of a stage length interval, so its position on the maps is approximate. The other lines are to be regarded as even less precise, and in fact most published paleogeographic maps depict features such as deltas and rift valleys and avoid the use of contours altogether. We use contours because of the simplicity of portrayal, and we believe that the nature of the feature should be obvious from its map pattern. Further description of this approach to paleogeography is detailed in Ziegler et al. (1985).

The basemaps upon which we have mapped the Permian paleogeography are updates of our previous work (Lottes and Rowley, 1990; Nie et al. 1990; Nie et al. 1993). In Ziegler et al. (in press), we discussed in some detail the approach that we have adopted regarding the incorporation of paleomagnetic data. Specifically, we have taken paleomagnetic data from each of the major continents whose relative positions are independently determined by plate kinematic data from the ocean basins, then computed 20 m.y. sliding-window averages. Ziegler et al. (in press) used the weighted global mean apparent polar wander path (APWP) rather than any other clustering scheme, and we adopt this here as well. Also, we computed our global APWP back to 260 Ma, and have simply extended the interval to the base of the Permian. For consistency we have again employed the data summarized by Van der Voo (1993) and the relative rotation parameters for each of the major Pangean continents, derivable from table 3 of Ziegler et al. (in press). Van der Voo (1993) qualified each pole based on seven reliability criteria (Q) as a proxy for quality, and we have limited our analysis to those poles that have $Q \geq 3$, an age of magnetization adequately determined (Q1 checked), and that have been assessed to be useful for APWP determination (Van der Voo, 1993). Table 8.1 lists the continental mean poles and the global weighted mean based on these poles. The reconstructions (Figs. 8.1–8.4; in pocket) are based on these data and on the relative rotation parameters appropriate to 260 Ma listed in Ziegler et al. (in press) for the entire Permian for the major Pangean continents. Clearly the smaller continental blocks within Paleo-Tethys moved independently throughout the Permian and so their rotations are not constant.

The map intervals selected for this chapter are the Sakmarian (Fig. 8.1) and Artinskian (Fig. 8.2)

TABLE 8.1. *Paleomagnetic Summary*

	Interval age	Average pole age	N	kappa	A95	Mean lat.	Mean long.	Rotated lat. based on global mean	Rotated long. based on global mean	Angular Distance from Global Mean
Weighted Global Mean	240	239	9	73	6	57	152	90		0
N. America	240	239.2	12	153	4	53	146	85	110	5
Eurasia	240	241.4	14	99	4	52	151	85	149	5
Africa	240	235	1	20	8	71	151	76	-28	14
S. America	240	234.5	2	17	14	57	148	88	68	2
India	240	242.3	3	321	7	67	-172	71	-74	19
Weighted Global Mean	250	246.8	8	61	7	57	155	90	-	0
N. America	250	245.5	10	181	4	54	154	87	145	3
Eurasia	250	245.5	17	82	4	52	154	85	147	5
S. America	250	256	1	0	14	72	120	69	4	21
India	250	248	3	36	21	60	-177	75	-90	15
Weighted Global Mean	260	255.8	6	69	8	52	163	90	-	0
N. America	260	253	3	1294	3	52	162	89	73	0
Eurasia	260	256.3	7	105	6	49	157	85	108	5
S. America	260	262.8	4	57	12	65	146	74	10	16
India	260	254	1	411	2	46	175	81	-138	9

stages of the Lower Permian, and the Kazanian (Fig. 8.3) and Tatarian (Fig. 8.4) Stages of the Upper Permian. These terms are derived from the standard southern Ural succession, and it should be mentioned that the International Subcommission on Permian Stratigraphy has proposed an "operational scheme" that interleaves terms from various sections around the world (Jin et al., 1994). The Upper Permian stage scheme is thus now more refined, and the terms that apply to our maps are Wordian and Dzulfian. Ideally, the information included on a paleogeographic map should represent as narrow a time interval as possible, but it must be admitted that it is not always possible, especially in continental sequences, to correlate at this level. In fact, most paleogeographers represent the Permian by four maps while the new correlation scheme contains ten stages. The main contribution of the Subcommission has been to improve the correlation of reference sequences around the world, and especially to bring Chinese usage into phase with a common standard. Until recently, the Chinese applied the term Lower Permian to rocks ranging through the late Lower and early Upper Permian, and this has caused confusion in the tectonic and paleomagnetic literature. For the Southern Hemisphere, an extremely useful compilation of 135 Permian stratigraphic sequences now exists (Langford, 1992) and we have accepted these correlations although the extent of the glacial advances through the Lower Permian in the Paraná (51), Karoo (67) and Kalahari basins seems to have been underestimated (Visser, 1990; Visser, this volume). Recognizing that differences exist between absolute time scales, we have chosen the DNAG time scale (Palmer, 1983) for these map reconstructions.

The mountain chains that we reconstruct for the Permian (Figs. 8.1–8.4) are grouped loosely by origin in the text of this chapter. Comments specific to the recognition of each tectonic type are included at the beginning of each section, but first, some general remarks on establishing the geologic age range of orographic features follow:

1. The most direct information comes from volcanic edifices, the products of which can be dated radiometrically and, when interbedded with sediments, paleontologically.

2. Deeper-seated processes, mainly metamorphism and intrusion, also yield radiometric dates, but the ages generally apply not to original events but to subsequent unroofing and cooling. Depending on the technique and the minerals dated, a variety of ages can be derived from a single rock, representing the protracted uplift and erosion of the orogen, and thus are extremely useful to the paleotopographer (Dallmeyer, 1989).

3. In limited cases, tectonics and sedimentation are so intimately linked that direct dating of the deformation can be extracted from the stratigraphic patterns. For instance, synsedimentary folding of the Cape fold/thrust belt has been demonstrated (DeBeer, 1992), while extensional rift valley fills typically show the progressive tilting of the footwall sequence (Cox, 1970).

4. Regional unconformities sometimes reveal buried topography; Visser (1987) was able to reconstruct the 2,000 km long Cargonian Highlands of South Africa based on 2,000 borehole records through the Permian and Triassic sequences that eventually smothered this 2,500 m high mountain range.

5. Mountains are often dated by their erosion products in surrounding basins as revealed by combining sedimentary provenance and transport direction studies. A special case of great use in restoring Permian topography is the existence of tillites in glacially carved valleys in Africa (Cahen and LePersonne, 1981; Martin, 1981; Visser, this volume).

6. The mere existence today of a significant topographic feature that is known to have originated prior to the Permian must be examined through its possible sedimentation effects on surrounding basins through time. A number of Panafrican fold/thrust belts in Gondwana still project 1,000 m above their foreland basins, and even though the possibility of reactivation always exists, a very long history is a real possibility.

In the following sections, the tectonically active mountains are treated first and are grouped by type. Then the mountains generated prior to the Permian are discussed in age categories. Finally, comments on climate belts in relation to the changing geography are offered. The categories used to group the various types of mountains are meant to be informal, because of uncertainties regarding the origin of a number of them and also because transitions can occur from one type to another within the Permian. Permian mountains have been named for their modern counterparts (Urals), for their discoverers (Karpinsky

Swell), and for their age (Cargonian Highlands, from Carboniferous and Gondwanan). Where choices exist and where paleo-features remain unnamed, we use modern counterparts or well known tectonic names when counterparts do not exist. Brevity and recognizability are paramount in this exercise, which we hope will lead eventually to a general compendium of paleotopographic features. A complete list of the Permian geographic names, as well as their present day coordinates and country, is provided in Appendix 8.1. The numbers in the subsequent location titles refer to this appendix.

Collisional Mountains

The assembly of Pangea began in the Late Paleozoic, so a number of continental collisions were in progress during the Permian. These involved Gondwana, Laurussia, Kazakhstania, Siberia, Kara (north of Siberia), Tarim (36) (western China), and the North China Block. Extremely high mountains must have been generated locally, to judge by metamorphic rocks now exposed at the surface, but these are mainly limited to zones where salients on opposing margins came in contact. Typically associated with these systems are foreland basins generated by the tectonic and sediment loading of the crust adjacent to the orogen. In fact, these thrust-loaded basins provide some of the best information on the initiation of contraction in the region, which can be detected by increases in the rate of subsidence, the supply of coarse sediment, and the transition from marine to terrestrial conditions. No attempt is made to describe in detail these extremely complicated collision zones, except to comment on their effects on Permian paleogeographic developments and to supply a few general references. It is interesting to note that remnants of many of these late Paleozoic collisions are still impressive in today's landscape.

Alleghanian (7), Atlas (56), and Hercynian (15) Mountains

These mountains represent the American, African, and European segments, respectively, of the collision zone between Laurussia and Gondwana. Well-developed foreland basins are represented by the Appalachian (Hatcher et al., 1989), Variscan (P.A. Ziegler, 1988) and Tindouf thrust-loaded basins (Cavaroc et al., 1976). These all contain marine sequences of Carboniferous age, but by the Permian the basins had filled and only terrestrial sequences are observed. The deformation in the Appalachians is bracketed by radiometric studies as occurring between 330 and 265 Ma (Hatcher et al., 1989), so the waning phases of the orogeny occurred about the middle of the Permian. Most of the Permian cover has been removed from the foreland basins, and a summary of the information on maturation indices and radiometric dates in the Appalachian Basin (Hulver, 1995) shows that from 1 to 6 km of Permian and older rocks have been removed from the region. Our portrayal of the Appalachian mountains and basin is based on a restoration of the material removed minus the appropriate corrections for isostasy. To the east in the Mediterranean and the Black Sea regions, the Proto-Tethys remained open and normal subduction processes apparently continued, although the orogen was subject to regional extension in the Permian, as indicated by widespread pull-apart basins in southern Europe (P.A. Ziegler, 1988) and in Africa north of the Atlas Fault (El Wartiti et al., 1990). At the western end of the orogen, thick clastic sequences in Venezuela (Benedetto, 1980) and Mexico (López-Ramos 1981) seem to define the southern and northern margins of the collision zone in the Permian, and in both basins marine conditions continued well into the Late Permian.

Ural Mountains (22)

These mountains resulted from the collision of the Russian Platform portion of Laurussia with the microcontinent, Kazakhstania. Thrusting was well under way by the end of the Carboniferous and continued into the Early Triassic (Zonenshain et al., 1990). The seaway on the platform to the west remained through the Permian but was filled progressively from the east by the erosional products of the rising foldbelt. Early in the Permian a well-defined foreland trough occurred between the Urals and a line of reefs to the west, but by the Late Permian the entire seaway was nearly filled by the advancing clastic wedge and marine conditions were ephemeral (Vinogradov,

1969). The eastern margin of the foldbelt is much less well known as it underwent extension late in the Triassic and lies buried beneath the West Siberian Lowlands. The southern terminus of the Urals is reconstructed from subsurface fold belts, and the character of these is uncertain. It has been traditionally assumed that the Urals were peneplained in the Mesozoic, but "detailed geomorphological studies have shown that the ranges and mountains are not young uplifts, but outliers that rose above the surface of the peneplain in the Cretaceous" (Borisevich, 1992, p. 41).

Zaisan Mountains (24) and Yenesei Ridge (25)

The "Atlas of Lithological and Paleogeographical Maps of the U.S.S.R." (Vinogradov, 1969) shows a termination of marine sedimentation and andesitic volcanism between Kazakhstania and Siberia in the Late Carboniferous, and a mountain range (Ob′-Zaisan) on the site during the Permian. Zonenshain et al. (1990) indicate a collision of the continents at this time along the Irtysh-Zaisan zone, and because we are uncertain that the range extended as far as Ob′ in the West Siberian Lowlands, we use the name Zaisan on our maps. The Yenesei Ridge is a present-day feature along the western margin of the Siberian craton, which appears to be the limit of late Paleozoic deformation. Here parallel bedded sequences of late Precambrian through Early Permian strata were brittly deformed and subsequently intruded by Early Triassic basalt which remains undeformed (Goryainov, 1991).

Kyzylkum Upland (18)

This upland was included on Vinogradov's Permian maps (1969) and is thought to have resulted from Late Carboniferous through Permian thrusting due to the collision of the hypothetical Karakum-Tarim microcontinent with the southern portion of Kazakhstania (Zonenshain et al., 1990). If Karakum (centered in Turkmenia), and Tarim (36) (in western China) were indeed part of the same microcontinent, the junction has been disrupted by strike-slip motion and has been overthrust along the Pamir syntaxis. It must be admitted that practically nothing is known of the Paleozoic record of the Karakum area, so its linkage with Tarim is perhaps more of a convenience than scientific fact.

Byrranga Range (23)

This impressive range along the axis of the Taimyr Peninsula of northern Siberia is the remnant of a southward prograding wedge of Late Carboniferous and Permian age (Filippov, 1992). We interpret this fold-thrust belt to be the effect of a protracted collision between Siberia and the Kara microcontinent (which includes the Kara Sea, the Severnaya Zemlya islands, and northern Taimyr). Recently it has been argued that this collision occurred in the mid-Mesozoic (Zonenshain and Natapov, 1989; Ufyland et al., 1991), but no Mesozoic oceanic or island arc deposits occur in this zone, and the 6km-thick Late Paleozoic clastic wedge is definitely on the Siberian margin, linking Kara with Siberia well before the Permian. There is clearly an angular unconformity of later (Cretaceous) Mesozoic age but the thick Carboniferous to Permian marine-to-continental wedge is the herald of the orogenic activity and links these elements together by this time.

Mongolian (32) and Yin Shan (33) Arcs

These arcs collided at the end of the Permian to bring North China and Mongolia into contact, but since they had separate histories during most of the Permian, they are described later in the section on island arcs.

Andean-type arcs

We restrict this term to arc systems with a well developed fold-thrust belt and foreland basin on the continent side of the volcanic arc. To judge by the modern Andes, extremely high relief may be generated in such systems. However, this may be the result of the fact that western South America was subject to the same tectonic and erosive effects through much of the Mesozoic and the Cenozoic (Ziegler et al., 1982). The existing data are insufficient to determine whether such high relief was associated with any of the non-collisional systems in the Permian.

Andean (49), Cape (68) and New England (90) Arcs

This is the "Samfrau Orogenic Zone" of Du Toit (Powell and Veevers, 1994) and is a continuous system that developed through the Permian and extended along the margin of Gondwana from southern South America through South Africa and across Antarctica to eastern Australia. Recently, Veevers et al. (1994) plotted the incidence of volcanics, tuffaceous sediments and unconformities along each segment of the margin and proposed that this Andean-type margin developed diachronously from west to east, a distance originally exceeding 10,000 km. Thus the magmatic arc and the yoked basin initially developed in the Sierra de la Ventana of eastern Argentina in the Late Carboniferous and propagated to the Cape Foldbelt of South Africa and the Ellsworth Mountains of Antarctica during the Early Permian; eastern Australia was enveloped in the system by Late Permian time. These areas have been separated by rifting in the Mesozoic, and much of the original evidence lies inaccessible in the highly extended zones of the Falkland and Agulhas plateaus, the Weddell and Ross seas, and the Lord Howe Rise, so that many of the details of the foldbelt are unknown. The foreland basins, however, are largely intact and include the San Rafael and the Sauce Grande (Argentina), Karoo (S. Africa), Transantarctic (85), and Sydney Bowen basins (95, 97) (Australia). The first three formed a continuous "Inland Sea" together with the Paraná (Brazil) and the Kalahari (Botswana) basins and shared limited marine faunas in the Early Permian (Veevers et al., 1994). By the beginning of Late Permian times an endemic bivalve fauna was common to the Paraná (51) and Karoo (67) basins and is thought to indicate a "tenuous oceanic connection" that was "insufficient to allow the development of normal marine conditions" (Cooper and Kensley, 1984, p. 1362). The location of this connection is uncertain but is thought to be in Antarctica. By latest Permian times, huge lakes were all that remained of this inland seaway (Yemane and Kelts, 1990).

Island arcs

Included in this category are subduction generated volcanic arcs of an extensional or neutral regime, as opposed to the contractional regime of Andean-type arcs. These arcs are quite easy to trace in the geologic record because their products, whether intrusive, volcanic, or reworked sedimentary, are so characteristic and voluminous. Indeed, this is one of the main ways in which continental crust is generated. However, island arcs generally do not constitute major orographic systems, and even though individual volcanic edifices often reach 2,000 m, they are typically isolated from each other by tens of kilometers. We have deliberately overemphasized their size on our maps so that they can be seen at the scale we have used.

McCloud Arc (1) and Golconda Belt (3)

The McCloud Arc spans the northern Sierra and eastern Klamath terranes of California and the Quesnel terrane of British Columbia (Saleeby, 1992). The McCloud rocks contain a fauna that is distinct from those of both the North American platform and the Tethyan seaway, and recently it has been claimed, on the basis of "probabilistic estimates of diversity and similarity" (Belasky and Runnegar, 1994, p. 1095), that these terranes ranged from 3,000 to 6,700 km west of North America. These authors noted that there is a general eastward decline in these parameters as measured in present-day Pacific islands, and applied a calibration of these data to corals of the Permian terranes. Miller et al. (1992, p. 83) have pointed out that "geologic relations suggest that the McCloud belt terranes had an episodic relation to a continental mass that was most likely North America." For instance, the intervening deep water Havallah Basin of Nevada contains arc as well as continent-derived sediments. We follow Saleeby (1992) in mapping the McCloud Arc as an east dipping extensional system and feel that the nearby platform faunas may owe their differences to the effects of upwelling currents along the shelf margin (Sheldon, 1964), or to high or variable salinities in the epeiric seas. "By Late Permian time, the McCloud island arc moved closer to the continental margin by shortening or subducting the depositional basement of the Havallah sequence, resulting in imbricate thrust faulting of the sedimentary fill of the basin" (Miller et al., 1992, p. 95). We show this Golconda thrust belt as a low land ridge on Figure 8.4.

Chios (16) and Caucasian (17) Arcs

These arcs are thought to represent north dipping subduction zones along the northern margin of Tethys and are seen in two limited regions in the Aegaean and Caucasus, respectively. The concept of the Chios Arc is based on a Permian accretionary complex that is thought to have collided with the Pelagonian Block (45) about the end of the Permian while leaving the Vardar back-arc basin in its wake (Stampfli et al., 1991). The Caucasian Arc is indicted by calc-alkaline volcanics in the Caucasus and in areas along strike to the east (Zonenshain et al., 1990).

Mongolian Arc (32)

This arc rims the southern margin of Mongolia and is continuous into the Altaids of Russia and China to the west, and to the Amuria region of these two countries to the east. Our interpretation is based on lithofacies maps of Mongolia (Amantov et al., 1970), which imply a gradual accretion of deep water clastics and siliceous sediments from the south to an arcuate zone of intermediate and acidic lavas and pyroclasts through the late Paleozoic. The northern margin of this arc is the Mongolo-Okhotsk Seaway, which extends into northeastern Mongolia along the present Amur River and extends southwest through Ulan Bator. This seaway did not close until the Late Jurassic (Nie et al., 1990). The Mongolian Arc collided with the Yin Arc of the North China microcontinent in late Permian times (see next section).

Yin Shan, or Yin Mountains (33)

These mountains along the northern margin of the North China platform (Han and Yang, 1980) formed above a south-dipping subduction zone in the Carboniferous and the Permian (H.Z. Wang, 1985). They provided sediments to the broad coastal plain to the south (Liu, 1990), but a fold-thrust geometry does not seem to have developed. Collision with the Mongolian Arc to the north is usually stated to be Upper Permian, but it must be remembered that the usage of this term in the traditional Chinese literature is restricted to the upper half of the Late Permian. The termination of marine conditions did not occur until the final, Changxinain Stage of the period (to judge by a lithofacies database supplied by Jin Yugan, of the Nanjing Institute of Geology and Paleontology). This collision brought the southern Tethyan faunas and Cathaysian floras in contact with the northern Boreal faunas and Angaran floras at the very end of the Paleozoic. Many details of the paleontology, volcanics, and tectonics of this zone are included in Ishii et al. (1991). Unlike the collisions described earlier, this one does not seem to have generated much topography.

Yunkai Arc (39)

This arc refers to the coastal region of the South China Block that began the Permian as part of the southeast-facing passive continental margin but evolved into a neutral arc in the later Permian. The transition from passive- to active-type margins has been interpreted to involve collision with an oceanic arc and the obduction of the Qinzhou "trough" assemblage during the Kazanian, followed by a flip in arc polarity and the establishment of the east-facing active margin on South China into the Mesozoic (Rowley, 1993). This active margin is again observed farther west as an elongate region of mostly Upper Permian and Triassic, often marine andesites and basalts developed north of the Jinshiajiang suture and south of the Litang suture (Sengör et al., 1993). This arc system includes both the Shaluli Shan and the Chola Shan and connects to the southeast with South China in Yunnan where coeval volcanics and plutons, often lumped with the extensive Emei Shan basalts, are exposed on both sides of the Ailao Shan. The Yunkai Arc bears the same relationship to the South China Block as the Aleutian Arc does to southern Alaska today.

Sibumasu (41) and Indochinese (40) Arcs

These terms refer not to individual arcs but to microcontinents in Southeast Asia, and our portrayal of the volcanic chains associated with them follows Hutchison (1989). He points out that these elements collided in the mid- to Late Triassic so they could not have been far apart in the Permian. Moreover, the Sibumasu (or Sinoburmalayan) microcontinent is characterized by glaciomarine strata extending through Sumatra, Thailand, and Burma (Metcalfe, 1993) into the Beishan area of Yunnan and seems to have rifted from Gondwana in the Carboniferous as suggested by the widespread Carboniferous rifts (Hutchison, 1989). Indochina has Cathaysian floral affinities, pointing to a low latitude position in the Permian separate from both Sibumasu and South China. We believe that Indochina collided with both of these blocks in the Triassic.

Intraplate contractional zones

The mountain chains in this category range from inverted aulacogens to transpressional zones to reactivated thrust belts. All are well away from continental margins, but many are traceable to stresses generated by collisions or other marginal effects. Veevers (1994) included in this category the Gamburtsev Subglacial Mountains of Antarctica as "originating by Mid-Carboniferous shortening of an intracratonic basin" (p. 593). The Isostatically Adjusted Bedrock Surface Map of Antarctica (Drewry, 1983) shows an impressive 4km summit within a 500km diameter area above 2 km. Veevers proposed a late Paleozoic age for these mountains to account for glacial flow patterns, but we feel that the Wegener Mawson Belt was the source of the ice sheets. It seems unlikely that a 4km mountain range could have survived for 300 million years, and its geometry, as judged from Drewry (1983), compares more closely with an underplated region associated with perhaps the widespread Jurassic (Ferrar), or some more recent magmatism.

Ancestral Rocky Mountains (4)

These mountains include the transpressional Front Range and Uncompahgre Uplifts centered in Colorado, but associated tectonic activity has been recognized throughout many of the western United States. Kluth (1986) shows the evolution of the uplifts and associated basins in a detailed series of maps ranging from the Late Mississippian to the Early Permian and relates this activity to "intraplate response of the foreland to events that produced the Ouachita-Marathon Orogeny" (p. 361), that is, the collision of North America with Gondwana. Areas more than 1,000 km from the collision zone were affected. The tectonic activity decreased and stopped within the Early Permian, and the mountains were eroded flat and covered by the Cretaceous Western Interior Seaway, so the modern Rockies bear little relation to their forebears.

Karpinsky Swell (19)

This structure, named for A.P. Karpinskiy who first proposed it, resulted from the partial inversion of the eastern segment of the Donets Basin. It extends from remnant low hills along the Ukraina-Russia border to the subsurface of the North Caspian Depression (Botneva et al., 1991). The Donets Basin is thought to have had Late Proterozoic antecedents, but it was reactivated in the Middle Devonian through the Carboniferous, when it accumulated in excess of 20 km of fill. Some 15–18 km of strata are exposed at the surface, and this is taken as the total magnitude of uplift. The Late Carboniferous strata contain "cosedimentation structures" that have been attributed to "embryonic folds," while by the Artinskian, Sakmarian fossils were being redeposited, indicating substantial folding and uplift during the Early Permian (Nagornyy and Nagornyy, 1976). A more recent reactivation is shown by a 450m offset of Upper Cretaceous deposits.

Asunción Arch (50)

This nearly buried ridge of Late Proterozoic rocks defines the western margin of the Paraná Basin (51) and extends north from the capital city of Paraguay into Brazil. Some 4 km of motion of the arch relative to the basin is implied by isopachs of Paleozoic and Mesozoic sediments, and it clearly was the source of the Mato Grosso and Santa Catarina glacial lobes during the Early Permian (França and Potter, 1991). An origin as a peripheral bulge of the sub-Andean foreland basins in the Devonian (Zalan et al., 1991) seems unlikely because of its scale, but we have no insight as to as its true origin.

Mauritanide Belt (55)

This east-verging thrust belt in west Africa was at one time regarded as the eastern margin of the Appalachian foldbelt, but an intervening "Senegal Microplate" recently has been proposed on he basis of geophysical and geometrical considerations (Venkatakrishnan and Culver, 1988). The main foldbelt is therefore restricted to the coastal zone, but the original Panafrican Mauritanide zone was reactivated during the Appalachian Orogeny. Late Carboniferous and Early Permian $^{40}Ar/^{39}Ar$ plateau ages on mica and slate indicate tectonothermal activity, and early to middle Paleozoic rocks of the foreland have been overthrust (Dallmeyer and Lecorche, 1990).

Ougarta (57) and Iforas (59) Mountains

These Paleozoic foldbelts follow along the east margin of the West African Shield. The Ougarta Chain of northeastern Algeria is an inverted aulocogen that received thick but shallow water sediments from the Ordovician through at least Devonian (Donzeau et al., 1981). We propose that the inversion probably began during the Upper Carboniferous (Moscovian Stage) when bioclastic limestones of the Bechar Basin to the northeast gave way to terrigenous coal-bearing clastics that extended into the Permian (Lemosquet and Pareyn, 1983). There is more than 5 km of structural relief on this feature which today consists of low hills (Choubert and Faure-Muret, 1971). Along strike to the south, the foldbelt reappears along the western margin of the Hoggar Massif at Adrar (=hill) d'Ahnet and continues to the Adrar des Iforas region of Mali. Here the Cretaceous rocks rest on the Lower Moscovian (Conrad, 1983) with angular unconformity, but little is known of the timing of the folding, and we assume that this structure represents a reactivation of the margin of the West African Shield during the collision of Africa and North America.

Ross High (84)

A linear basement high, subparallel with the Transantarctic Mountains (85), has been proposed to account for stratigraphic and sediment derivation patterns (Collinson et al., 1994). Early Triassic rocks buried this structure, which preserves 2–3 km of late Paleozoic sediments on either side. These authors state that "the Ross High appears to have been episodically active as early as the Devonian and as late as the Early Triassic" (p. 177) and label this structure a "Foreswell".

Intraplate extensional zones

Many of the rift zones active in the Permian are concentrated in east Africa and adjacent areas of Gondwana to the east; good general reviews on the subject include Cox (1970) and Daly et al. (1989). Most of these rifts can be shown to have developed along Precambrian foldbelts and show significant extensional activity even today. Volcanics are generally absent until the Mesozoic, and fault activity during the Permian is inferred from arkoses derived from the foot wall side, from syndepositional tilting of sequences on the hanging wall side, and by the channeling of the glacial lobes along the valley axes. From these observations we can conclude that most of the rifts were active throughout the Permian, although the evidence of the earlier phases tends to get obscured by the subsequent rift fill. Extension also affected the Tethyan and the North Atlantic regions in the Permian. In the following sections we focus on the rift shoulders, since this chapter concerns paleotopography, but the respective rifts are mentioned as well.

Lomagundi Highlands (69)

This modern day upland area is centered on the country of Zimbabwe and underlain by the Archaean Zimbabwe Province (Tankard et al., 1982). It is bounded to the north and south by the valleys of the Zambezi and the Limpopo, respectively, which were occupied by west flowing glacial lobes in the Early Permian (Visser, this volume). The concept of a Permian "Lomagundi Higlands" as a "local mountain gathering ground" for small ice lobes derives from the fact that "Lower Karoo rocks are banked against a buried scarp, which present erosion has partly exhumed" along its western margin (Bond, 1970, p. 467–8). Current studies of the geology of the Zambezi Valley are available (Orpen et al., 1989; Oesterlen and Blenkinsop, 1994), although there is disagreement as to the timing and regional stresses involved.

Muchinga Scarp (70) and Makutu (71) Mountains

These presently reactivated rift shoulders are associated with the 800km long Luangwa rift valley of Zambia. Early Permian rocks are apparently unknown, but the Late Permian lacustrine Madumabisa Mudstone is at least 1700 m thick and contains arkosic grits indicating periodic rejuvenation of the source-areas (Drysdall and Weller, 1966).

Tanzanian (74) and Uluguru (75) Uplands

The Rukwa rift valley bounds the southwestern margin of the Tanzanian Craton (Dypvik et al., 1990) and the Selous (=Luwegu) and the Mikumi rift valleys and the Uluguru horst bound the eastern margin (Wopfner and Kaaya, 1991). Permian and later extension is documented for these basins, and Early Permian tillites are known from the Ruhuhu Basin (72) at the junction of the major rift trends (Wopfner and Kreuser, 1986). Marine conditions extended along the conjugate margin in Madagascar by the Early Permian (Langford, 1992) and into the Mikumi Basin by Late Permian times (Nyblade et al., 1993) testifying to the importance of this broad zone of crustral extension between east and west Gondwana in the Permian.

Huqf Axis (78)

The southeastern rim of the Arabian Peninsula was tilted northwest prior to the Carboniferous-Permian glacials, which rest on Precambrian to Devonian rocks (Levell et al., 1988). The Al Khlata diamictites were deposited in a basin 800 km long and 200 km wide on what must have been the lower dip slope of the hanging wall. This area was in relatively low latitudes during the glaciation, leading to speculation that this uplift must have been substantial to reduce the temperatures enough to allow for glacial development.

Lambert Rift Shoulders (82)

The modern Lambert Glacier has reoccupied the 650km long Lambert Rift complex, which still has rift shoulders more than 2 km high (Drewry, 1983), with isostatic corrections for ice load. The depth to magnetic basement is more than 5 km, and the rift complex is north-south trending, with many side branches and horsts (Fedorov et al., 1982). The stratigraphic record is fragmentary but 2500 m of the Permian Amery Group that were deposited by north-flowing axial braided river systems are exposed (Webb and Fielding, 1993). K-Ar studies indicate crustal fracturing and emplacement of mafic dikes in the Late Carboniferous as well as reactivation at a number of times in the Mesozoic and the Cenozoic (Hofmann, 1987). It is commonly assumed that the Lambert Rift found its continuation in the Mahanadi Valley (80) of India, but recent work has shown that, while some faulting is observed locally, the structure is not a rift (Casshyap and Srivastava, 1987). Indeed, the gravity map shows no steep gradients in this region (Narain, 1975). We assume that the Lambert Rift continued northeast along the margin of India and Antarctica and thence along the west coast of Australia. This idea was stimulated by the fact that the

Lambert Rift curves northeast at its intersection with the Indian Ocean in Prydz Bay (Stagg, 1985). We suggest that the recently recognized Gondwanan graben in Bangladesh and its counterparts in the Perth Basin (86) and offshore western Australia would thus be the along-strike continuations of the Lambert rift. Little is known about the existence of syn-rift topography in these areas, and thus we have limited the orography to the area adjacent to the Lambert.

Panjal (79), Akhdar (77), and other Tethyan Rift Shoulders

Our reconstruction of the Tethyan region follows the imaginative scenario of Stampfli et al. (1991). This region has been overprinted by the Mesozoic and Cenozoic contractional phases and has defied a consensus on its evolution. The Stampfli et al. (1991) view makes some basic assumptions and applies some ideal models of rift margin development, but the reconstructions are detailed and plausible. The assumptions are that the Gondwanan margin has always been passive or extensional, while the Eurasian margin has been active, albeit with some back-arc extension. Their "simple shear model of extension leads to an asymmetric model of passive margins: a lower plate tilted block margin and an upper plate flexural, ramp-like margin" (p. 373). They have applied this model in some detail to the stratigraphic and volcanic patterns of the Panjal area of northern India and of the Jebel Akhdar area of Oman, while sketching some details of less well exposed areas to the west. To a large extent, the uplifted margins as shown on the maps are diagrammatic. Stampfli et al. (1991) show two ocean basins opening in the Late Carboniferous and Permian; one is termed Permotethys and separated the Pelagonian Block (45) (northern Turkey and the Balkans), Iranian Block (44), Afganistan, and the Lhasa Block (43) from Arabia and India, and the other is termed the Eastern Mediterranean and separated the Adria Turkish Plate (46) from northern Africa.

Eskimonia High (10)

Eskimonia is a term applied to the east coast of Greenland in the Permian (Haller, 1971). Active extension is inferred because "rotational block faulting took place in the Late Carboniferous to Early Permian along the basin margin over shoulders of the underlying Devonian basin" (Surlyk, 1990, p. 110). By the middle of the Permian, "sedimentation took place in a bajada formed by coalescent, low gradient alluvial fans passing into a coarse-grained braidplain and extensive intermontane lakes" (p. 110). Surlyk concludes that the Late Permian seas advanced over the eroded marginal rift blocks and that this defines the initial phase of subsidence by thermal contraction.

Intraplate volcanism

Included in this category are several diverse areas of extensive volcanism that formed well away from subduction zones. The origin of each is a matter of debate, with the possibilities ranging from mantle plumes to broad zones of crustal extension. The issue to be discussed is the extent to which topography was generated, either through volcanic edifices or through the process of underplating. Not included is the Siberian Flood Basalt Province, because it was formed about the Permo-Triassic boundary (Renne and Basu, 1991) and therefore could not have had a topographic effect in the Permian. It is one of the world's most voluminous flood basalt provinces and is thought to have been the result of a mantle plume (Sharma et al., 1992).

Oslo Dome (13)

Northern Europe was the site of extensive volcanism at about the Carboniferous-Permian transition. Areas affected include a variety of basement provinces spanning the margin of the Fenno-Scandian Shield, to the Scottish and North German Caledonides, and south to areas of Germany and Poland involved in Variscan crustal thickening. These are "characterized by a distinct alkaline bimodality," "derived by partial melting from an asthenospheric MORB-type source," and "generated during the Autunian phase of wrench faulting" (P.A. Ziegler, 1990, p. 61 and Enclosure 17). The area of the North Sea basin (14) was apparently thermally uplifted and subjected to regional erosion but after the volcanism terminated, the area began to subside regionally in response to thermal relaxation of the lithosphere (P.A. Ziegler, 1990). Adjacent areas of the Fenno-Scandian Shield, centered on Oslo, seem to have remained high even to the present, and we assume that underplating is responsible, simply because this was evidently the last major tectonothermal event in the region. It must be admitted that there is good evidence that the region is being uplifted today and that the reasons are not at all clear (Jensen et al., 1992; Larsen et al., 1992). Late Triassic fission-track ages on apatite seem to indicate that 3–4 km of Paleozoic cover were in the process of being removed from southern Sweden (Zeck et al., 1988), and this would imply substantial late Paleozoic or early Mesozoic denudation. Finally, we believe that much of Norway has a remnant Caledonian root, described later as the Scandes Mountains, and this zone partly overlaps the Oslo Dome.

Transbaikalian Uplands (29)

This extensive area of Permian alkaline granitoids and related volcanics along the southern margin of the Siberian Craton in Russia and Mongolia has been described as a hot spot (Zonenshain et al., 1985) and as a tensional rift regime (Yarmoluk et al., 1990). The numerous northeast-trending graben suggest a broad zone of crustal extension at least 2,000 km long that was last penetratively deformed about the middle of the Paleozoic (Wickham et al., 1995). "The repeated cycles of magmatism involved significant anorogenic crustal growth and a gradual cratonization of the continental crust of East Asia, caused by multiple melting and remelting events and associated large influxes of mantle-derived alkalic mafic magma" (Wickham et al., 1995, p. 15641). Substantial thicknesses of clastic sediments in the Verkhoyansk Complex of eastern Siberia are thought to have been derived from Transbaykalia, implying considerable uplift of the region in association with the volcanism and extension (Zonenshain et al., 1990). The area has been topographically rejuvenated in the late Tertiary in association with the Baikalian rift system (Ufimtsev, 1986).

Omeishan (=Emei Mountain) Basalt Plateau (37)

This middle Late Permian lava outpouring covered a 330,000 km^2 area of the South China platform, extending inland about 700 km from the western, active margin and spanning 900 km from north to south with an average thickness of 700 m (Sheng and Jin, 1994). Initial flows were deposited on a carbonate platform, but the unit is dominantly continental, and we envisage a broad, low-elevation plateau, because the region continued as a coal basin at the end of the volcanism. The basalts are tholeiitic with high Fe, Ti, and low Mg (Zheng, 1985), and have been variously attributed to mantle plume activity (Huang, 1988) and to taphrogenesis of the Yangtze Platform (Luo et al., 1988).

Air Massif (60) and Zinder Hot Spot (61)

"In Niger and Nigeria over 80 anorogenic centers representing the eroded roots of volcanoes form an extensive chain of syenite-granite ring complexes extending for about 1,300 km" (Rahaman et al., 1984, p. 173). There is an age progression from Ordovician to Devonian in the Air Massif, south to Carboniferous and Permian around Zinder and thence to the Jos Plateau of Nigeria in the Mesozoic. If this is indeed a hot-spot track, there appears to have been some activity around Air, as late as the Upper Permian, to judge by the volcano-sedimentary series of the Upper Izegouanane Formation (Kogbe, 1991). Even today there is 1,300 m of relief associated with the mid-Paleozoic intrusions of Air which we interpret as evidence of underplating.

Remnants of Paleozoic Mountains in the Permian

Since remnants of Permian mountain building, like the Appalachians and Urals, exist today, it should come as no surprise that remnants of still older foldbelts were important features of the Permian landscape. Evidence for their existence is often subtle, but they can be detected through sedimentary provenance studies, "mega-unconformity" relationships, and radiometric cooling ages in the low temperature range. In addition, the roots of mountain chains can be detected with Bouguer gravity profiles, helping to confirm that even today the isostatic uplift of an ancient feature is incomplete and must be "amortized" over its entire history. This argument can be applied only where there is strict parallelism of the gravity and the tectonic patterns, since mountains can obtain new roots through underplating.

Antler Belt (2)

This belt is a northeast trending foldbelt centered in Nevada and formed in the Early Carboniferous through contraction of a back-arc basin along the continental shelf margin (Miller et al., 1992). By the Permian, the pattern was reversed and extension prevailed (see section on the McCloud Arc), but evidently some areas of "high relief" remained into the Early Permian and were progressively reduced during the period (McKee et al., 1967). Miller et al. (1992) show the belt extending into Idaho, and their detailed lithofacies maps show the rather fragmentary data on the clastic fringes and isopach patterns around these islands. A parallel situation may exist in the Caribou Terrane in central British Columbia, although there is no direct evidence from provenance patterns (Henderson et al. 1993).

Scandes Mountains (12)

By this term we refer to the mountains of western Scandinavia, mainly Norway, that were last penetratively deformed in the Caledonian Orogeny around the Siluro-Devonian transition (Roberts et al., 1985). There is some overlap of this long belt with the Oslo Dome discussed earlier. The high topography today in Scandinavia (Riis and Fjeldskaar, 1992) is associated with the hanging wall of the Caledonian thrust belt, except west of Oslo. A pronounced negative Bouguer anomaly parallels the axis of the mountains (Balling, 1980), suggesting that some relief remains from this mid-Paleozoic mountain range. We know of no direct evidence, however, that the Scandes were up in the Permian.

Patom Uplands (28)

This is a dissected upland area of Precambrian rocks northeast of Lake Baikal that today achieves elevations above 1500 m. These rocks were thrust north over the Siberian Platform, deforming rocks as young as Silurian, and this has been termed an "important Late Caledonian tectonic-magmatic reactivation of a large part of the craton" (Khain, 1985, p.111). Rocks of the Patom Uplands have been affected by zonal metamorphism of kyanite-sillimanite type, and cooling of the orogen through the biotite isotherm occurred during the Silurian and the Devonian (Manuilova, 1991). Although a well-defined foreland basin of late Paleozoic age does not exist, Russian authors suggest the Patom region as a source of terrigeneous material to the Siberian Platform on their late Paleozoic lithofacies maps (Vinogradov, 1969; Il'yukhina and Verbitskaya, 1977). Reactivation of this thrust zone occurred during the Jurassic, and the region, along with the adjacent Transbaikalian Uplands, was rejuvenated in the late Tertiary as part of the Baikal Rift extension (Ufimtsev, 1986).

Pan Xi Mountains (35)

This term comes from the Gansu Province Bureau of Geology and Mineral Resources (1989) paleogeographic maps for the Permian and is derived from Pan, the traditional name for the Gansu-Qinghai region of China, and Xi, for west. The Pan Xi Mountains parallel the southwest margin of the North China Platform and, like the active Yin Shan Arc, discussed earlier, along the northern margin, they are evidenced by a fringe of coarse clastic sediments (Han and Yang, 1980; Gansu Province Bureau of Geology and Mineral Resources, 1989). The last Paleozoic tectonic activity in the Pan Xi area, however, was evidently in the mid-Paleozoic after an island arc accreted to the North China Block, and Late Silurian to Early Devonian granites formed apparently above a short-lived, north-dipping subduction zone (Lerch et al., 1995).

MacDonnell Ranges (90)

This is the Devonian Alice Springs Orogen in central Australia that has survived to the present day in the form of 1,500m peaks. Thick-skinned thrusting due to an unusually high regional compressive stress field produced about 20 km of structural relief and brought granulite zone rocks to the surface along a mid-Proterozoic province boundary (Shaw et al., 1992). These authors conclude that "final uplift along the thrust front at 300–320 Ma (Late Carboniferous) was rapid, as the time of final closure of small domains in K-feldspar approximates (the time of) apatite fission track ages" (p. 937). Early Permian glacigene sediment transport directions in basins to the east and south suggest that the McDonnell and the Musgrave ranges were the sites of ice caps (Veevers, 1984). Veevers explains that the Musgrave Ranges had a Precambrian history as an upland sediment source.

Flinders Range (93)

The Adelaide Fold Belt formed here along the eastern margin of the Australian craton in the early Paleozoic. The direct evidence for its topographic manifestation in the Permian includes apatite fission track ages in the range of 260–175 Ma (Foster et al., 1994), and glacial transport directions determined from adjacent basins (Wopfner, 1981). Cretaceous sediments are preserved at "high" elevations and have been correlated with an erosion surface at 900 m, suggesting that there has been a rejuvenation of the Flinders Range (Foster et al., 1994).

Lachlan Highlands (94)

This area of eastern Australia had a long Paleozoic history as an active plate margin involving the accretion of a 1,000km-wide zone of arcs, microcontinents and flysch prisms (Scheibner, 1987). Carboniferous apatite fission track determinations indicate cooling and uplift (Foster and Gleadow, 1993) while Permian tillites occur both on top of the range and in the Murray Basin to the north (O'Brien, 1986). O'Brien shows the inferred limit of grounded ice in this marine embayment, confirming that by Permian time the depositional strike was east-west, and thus at right angles to the earlier tectonic strike.

Remnants of Precambrian Mountains in the Permian

Most of the Precambrian areas included in this section have remnant peaks in the 1–1.5km range today, and many represent foreland thrust zones, so their elevations relative to their own foreland basins may be determined. Excess topographic features of this type must be regarded as potentially positive for all Phanerozoic intervals, and so we have searched for direct evidence of their influence in surrounding Permian basins. A number of the Precambrian ranges in Gondwana have Permian tillites preserved within U-shaped valleys up to 1.5 km deep, allowing minimum elevations of late Paleozoic mountains to be directly measured. No other remote geologic period provides such a wealth of information on paleotopography.

Grenville Mountains (8)

The Grenville Orogen constitutes the southeast margin of the Canadian Shield and formed between 1.2 and 1.0 Ga as a northwest-verging, crustal-scale thrust belt (Hoffman, 1989). The northwestern marginal zone, of greenschist to granulite grade, is coaxial with a 750km-long gravity low (-100 mgal) across portions of Quebec and Labrador (Hanna et al., 1989). A broader zone of uplands in this area is in the 500–1,000m elevation range, and we assume that it is isostatically supported by a thick crustal root formed 1 billion years ago. Some Middle Paleozoic apatite fission track ages (317–431 Ma) have been obtained from this part of the Canadian Shield (Crowley, 1991) and suggest long continued denudation of this region.

Canastra (52) and Espinhaço (54) Ranges

These are long north-south-trending ranges in eastern Brazil with peaks up to 1,850 m embedded in a broad upland region. The Meso-Cenozoic cover lies at elevations ranging from 800 m to 1,300 m (Chemale et al., 1993), implying considerable recent uplift, due, we think, to underplating in association with Mesozoic and Tertiary igneous activity much in evidence to the north and south (Petri and Mendes, 1983). These ranges also had a Paleozoic history, to judge by lithofacies patterns along the southeast margin of the Maranhao basin, especially in the Devonian (Petri and Fulfaro, 1983; de Melo, 1988) but also in the Carboniferous and Permian (Mesner and Woolridge, 1964). The Canastra and the Espinhaço ranges originally formed in the late Precambrian as thrust belts along the west and east margins, respectively, of the narrow São Francisco craton (Chemale et al., 1993).

Atacora Mountains (62)

This is presently a 500 m high north-north-east trending ridge of quartzite that formed as a west-verging thrust belt over the Volta Basin of Ghana during the late Precambrian to the Cambrian (Affaton et al. 1980; Trompette, 1983). No direct evidence for its existence in the Permian is known to us.

Mayombe (64) and Mitumba (73) Ranges

These ranges define the western and the eastern margins of the Congo (Zaire) basin, respectively. The Mayombe Range is a low (500–1,000 m) coast-parallel ridge formed in the late Precambrian as a northeast-verging fold-thrust belt (Vellutini and Vicat, 1983). Circumstantial evidence for its existence as a topographic feature in the Permian comes from the occurrence of Permian tillites at its northern terminus in Gabon (Micholet et al., 1970). The paleolatitude of this site in the Early Permian would have been low, about 38°S, and the presence of ice implies the proximity of an upland source region. The origin of the 2–3km-high Mitumba Range is uncertain, but its existence in the Permian is abundantly confirmed by the Lukuga tillite, "which occupies stepped U-shaped and hanging paleovalleys, and cirques, and rests on striated surfaces and roches moutonnées " (Cahen and Lepersonne, 1981). It appears that the present drainage is relict from the Permian, and it may be that rejuvenation of the Mitumba range has occurred as a result of the late Tertiary phase of rifting along Lake Tanganyika to the east.

Windhoek Highlands (65)

Late Precambrian to Early Cambrian folding and amphibolite grade metamorphism resulted in the Damara orogen between the Congo and the Kaapvaal cratons and along their western margins in Namibia. A late Paleozoic westward-directed drainage was established across the belt, and this glacially modified system has been exhumed today (Martin, 1975). "In the Kunene Valley the difference in elevation between the valley floor with its beautifully striated roches moutonnées and glacial sediments, and the highest basement ridges in the area is about 1,500 m" (Martin, 1981, p. 487). Visser (1987) has updated the glacial valley map of southern Africa and has applied the term Windhoek Highlands to this area, which today lies above 2 km. An Early Cretaceous exhumation of the coastal ranges to the south has been documented by apatite fission track analysis and has been related to "an accelerated phase of erosion associated with the early stages of rifting and break-up" of the South Atlantic (Brown et al., 1990, p. 339). Some of the present topography is therefore more recent than the Permian.

Cargonian Highlands (66)

This 2–2.5km-high Permian mountain range lies nearly buried beneath Mesozoic rocks and has been reconstructed from 2,000 borehole records covering a region 1,000 km long 500 km wide

(Visser, 1987). The contoured reconstruction of this mountain range, showing the details of every valley and ridge, is both unique and extraordinary. The Dwyka tillite map, with isopachs at the 100m level and ranging up to 700 m, is useful in outlining the shapes of the bounding Karoo (67) and the Kalahari basins to the south and to the north, respectively, and paleoice-flow and clast composition studies have been used to show how major ice spreading centers sometimes flowed across interfluves that otherwise defined individual ice streams (Visser, 1987). This study also reconstructs the shape of the Cargonian Highlands during the postglacial flooding sequence using stratigraphic marker horizons such as the distinctive early Late Permian Whitehill shale that has been traced laterally to the Irati Shale of the Paraná Basin (51) of Brazil (Oelofsen, 1987). Stratigraphic profiles demonstrate that the Cargonian Highlands subsided 700–800 m with respect to sea level during the Early Permian (Visser, this volume), probably due to the combined effects of structural loading of the Cape Orogen to the south and eustatic sea level rises. Unlike many of the remnant Precambrian mountains, the Cargonian Highlands are underlain by unfolded Late Archaean and early Proterozoic "cratonic domains" of the Kaapvaal Province (Tankard et al., 1982). How such a broad region became uplifted by the late Paleozoic is a mystery to us.

Eastern Ghats (81) and Wegener-Mawson Belt (83)

These areas represent the conjugate marginal zones of southeast India and east Antarctica, respectively, and have been linked together in the Indo-Antarctic Metamorphic Terrain (Grew and Manton, 1986). The Eastern Ghats extend for hundreds of kilometers with peaks above 1,500 m, while the Wegener-Mawson Belt (Kamenev, 1993), is 2,000 km long with peaks in the 2–3 km range (Drewry, 1983). These structures could be Mesozoic rift shoulders, but they are not associated with major manifestations of basaltic volcanism, as indeed is the case in the Western Ghats, which seem to be associated with the Deccan Traps of Late Cretaceous age. The combined Eastern Ghats and Wegener-Mawson belt experienced granulite-facies metamorphism about 1 Ga ago and has been interpreted as a Himalayan-type collisional zone (Radhakrishna, 1989). Foreland basins are present on the Indian side (Kale, 1991) where present gravity data supports the persistence of a flexurally controlled depression (Subrahmanyam, 1978). Gravity studies on Antarctica are fragmentary, but a study of the Napier Mountains shows a positive free-air anomaly "interpreted as an area of relatively high-density upper crust with regional isostatic compensation" (Wellman and Tingey, 1982, p. 937). Recent geochronologic work in the region extending from southern Madagascar (Paquette et al., 1994), from southeastern Sri Lanka (Kroner and Williams, 1993), and from the Lützow-Holm complex of Antarctica (Shiraishi et al., 1994) provides support for latest Precambrian to Early Paleozoic deformation and metamorphism within this realm of Gondwanaland that is broadly cospatial with the uplands that we portray. Permian tillites occur immediately north of the Eastern Ghats and tillite fabrics indicate paleotransport north, away from the mountains (Casshyap and Srivastava, 1987). Permian fluvial transport directions in Queen Maud Land, at the western terminus of the Wegener-Mawson belt, indicate poleward transport (Barrett, 1991), again away from the inferred mountain range. In summary, the evidence for Permian topography along the Indo-Antarctic contact zone is circumstantial at best and is based on the idea that a major zone of crustal thickening was generated there perhaps recurrently between 1–0.5 Ga and has not been completely dissipated through erosion or crustal extension during the opening of the Indian Ocean.

Hammersley (87), Kimberley (89), Musgrave (91), and Gawlor (92) Ranges

Australia has been the most stable of continents, and "most of the present highland areas of the Great Western Plateau date back to the beginning of the Phanerozoic, and the Western Shield much earlier" Veevers, 1984, p. 167). Permian glacial transport directions radiate from these highland regions (Wopfner, 1981; Veevers, 1984) giving confidence that they were indeed positive features at that time. Our portrayal of these ancient uplands is based simply on the assumption that the areas containing peaks about 1 km high today must have been at least that high in the Permian.

Conclusion on paleotopography

The sixty three mountain systems described in this chapter cover the spectrum from convergent and collisional plate boundaries, to rift and drift extensional effects, to intraplate tectonic and volcanic events, and to systems of various types that were remnant from earlier times. From the point of view of variety, the Permian seems to be similar to the present. It is perhaps surprising that a number of Paleozoic and even Precambrian mountain masses have survived to the present. This is partly due to the progressively lowered base-level of the Tertiary, which has resulted in the exhumation of mountains preserved during the Mesozoic high-sea-level intervals. A number of Permian mountains have also been uplifted through renewed extension, underplating, or compression. Still others have survived because they formed in major collision belts that doubled the crust and drove high-rank metamorphics and other resistant rocks onto continental forelands where they escaped subsequent extension. The Appalachians are typical of many ranges in that the preserved portions of the orogen consist of the foreland thrust belt and even the foreland basin wedge, while the "internides" were reduced not just by erosion but also through crustal extension soon after the folding.

In this chapter we have done our best to give the essential evidence for each mountain range in the Permian, but we are aware that much remains to be done to test the form and the evolution of these systems. The tectonically active zones in the Permian can be reconstructed in some detail by linking the stratigraphic patterns of the yoked basin with the metamorphic effects in the uplift zone. The biggest problem comes with interpreting the topographic evolution after the contraction, extension, or volcanism has ceased. This research area has been neglected, but recent developments in the field of low pressure and temperature systems show great promise in paleotopographic studies. The application of inverse models to fission track studies can be used to elucidate cooling histories in the 120°C and lower temperature range (Corrigan, 1991; Lutz and Omar, 1991). Also, cooling histories can be deduced by $^{40}Ar/^{39}Ar$ analysis of individual K-feldspar samples that have passed through the 350–150° C range (Lovera et al., 1989; Richter et al., 1991). The most fruitful approach is to combine a variety of geochronological techniques appropriate for a spectrum of blocking temperatures to deduce the long-term cooling and erosional history of each region (Dallmeyer, 1989; Shaw et al., 1992).

Climate Zonation

The geographic zonation of climate has been described in terms of florally defined biomes for intervals of the Lower (Sakmarian) and Upper (Kazanian) Permian (Ziegler, 1990). Starting at the equator, the tropical biomes are comprised of the: (1) Everwet, (2) Summerwet, and (3) Subtropical Desert biomes, based on trends in decreasing floral diversity and physiognomic adaptations to increasing water stress. The temperate biomes are termed: (4) Winterwet, (5) Warm Temperate, (6) Cool Temperate, (7) Midlatitude Desert and (8) Cold Temperate, recognized by the effects of limitations in the growing season due to lowered temperature and precipitation. The polar zone contains the, (9) Arctic, and (10) Glacial biomes, with biome 9 representing the effective limit of biological tolerance. Climatically sensitive sediments, especially coals, evaporites, desert dune sands, and tillites are also useful in defining the biomes. Coals are best developed in biomes 1, 5, 6, and, to a certain extent, 8; evaporites and dune sands characterize biomes 3 and 7; tillites constitute biome 10. In general the biome numbers increase poleward, but their boundaries are not usually found to be parallel with latitude because of the modification of temperature and precipitation gradients imposed by the land masses. An important point is that while temperature parameters decrease poleward, the precipitation shows the greatest deficit in the subtropics, just as today. The biome approach was adapted from a climate study of the Recent (Walter, 1985) and the Permian biome patterns proved to be strikingly similar to today's.

Permian climate has also been the subject of a numerical simulation using the Community Climate Model and based on the Kazanian Stage topography mentioned earlier (Kutzbach and Ziegler, 1993). Maps were generated for opposite seasons showing surface temperature, sea level pressure, wind direction, precipitation, and snow depth. Summary maps were also included showing annual precipitation and soil moisture, as well as a biome classification map based on the computer output. The biome map compares quite well with the one generated from the floral data and has the distinct advantage that all parts of the earth's dry-land surface are characterized, including mountain tops and other regions that have not preserved sedimentary rocks and fossils. In this study, the Hadley Cell zonation is well shown, with the major desert belts centered about 25° north and south. The tropical rainy zone is centered over the nearly equatorial Appalachian Mountains in Pangea but rises in latitude to about 30° north following the northern coast of Tethys to North China. This asymmetry is interesting because it is confirmed by the trend of the Cathaysian tropical rain forest province. The effect of the Appalachian chain is instructive because it obviously "pins" the Intertropical Convergence Zone to the orography throughout the seasonal cycle, and may explain why so much clastic sediment was supplied to evaporite basins in western North America. Here the climate of the source area contrasted with that of the depositional basin where the geological evidence is preserved. Climates in higher latitudes are modeled as zonal but with a strong maritime amelioration. This is particularly seen in the northern polar region, where the pole is ocean-dominated, and both warm and cool temperate biomes are shown to the north of the cold temperate biome. The earth's thermal pole was at about 75°, centered over the landmass, a situation that may have prevailed during many geologic intervals.

We present in this chapter distributional data for Permian coals, evaporites, and desert dune sands compiled from the literature (Figs. 8.1–8.4; detailed reference list supplied on request). The distribution of modern peat bogs correlates well with areas of continuous rainfall because water is necessary for both vegetational productivity and preservation (Lottes and Ziegler, 1994). The exception is in higher latitudes, where productivity may be lower but evaporation is reduced and eliminated in the winter, keeping organic oxidation at a minimum. The coal distribution is therefore important in detecting the equatorial and temperate rainy zones. The evaporites and desert sand dunes correlate very well with biomes 3 and 7 at the present time, that is, with dry conditions through the annual cycle. Also shown in Figures 8.1 and 8.2 are the known Early Permian glacial extents as reconstructed, somewhat tentatively, from the papers in the present volume. They alternated with temperate zone conditions as shown by the interbedding of tillites and coals in many Gondwanan sequences, so there is a very good parallel with the Quaternary of the Northern Hemisphere today.

The paleolatitudinal paths of selected basins that contain a good record of Permian climatically sensitive sediments are shown in Figure 8.5. These basins are grouped by latitude below into polar, temperate, subtropical, and equatorial categories. The movement of Pangea was such that most basins show a northward progression of 5° to 15° during the period, except for eastern Australia and adjacent parts of Antarctica, which remained at the same latitude. Accordingly, many basins record this latitudinal shift with a change in sediments, floras, and faunas. If a global change occurred in Permian climate, then the latitudinal transitions of the climate zones should move independent of continental motions. This obviously happened in the Early Permian with the transition from the Late Carboniferous glacial maximum, but otherwise the sediments and faunas clearly track continental motions.

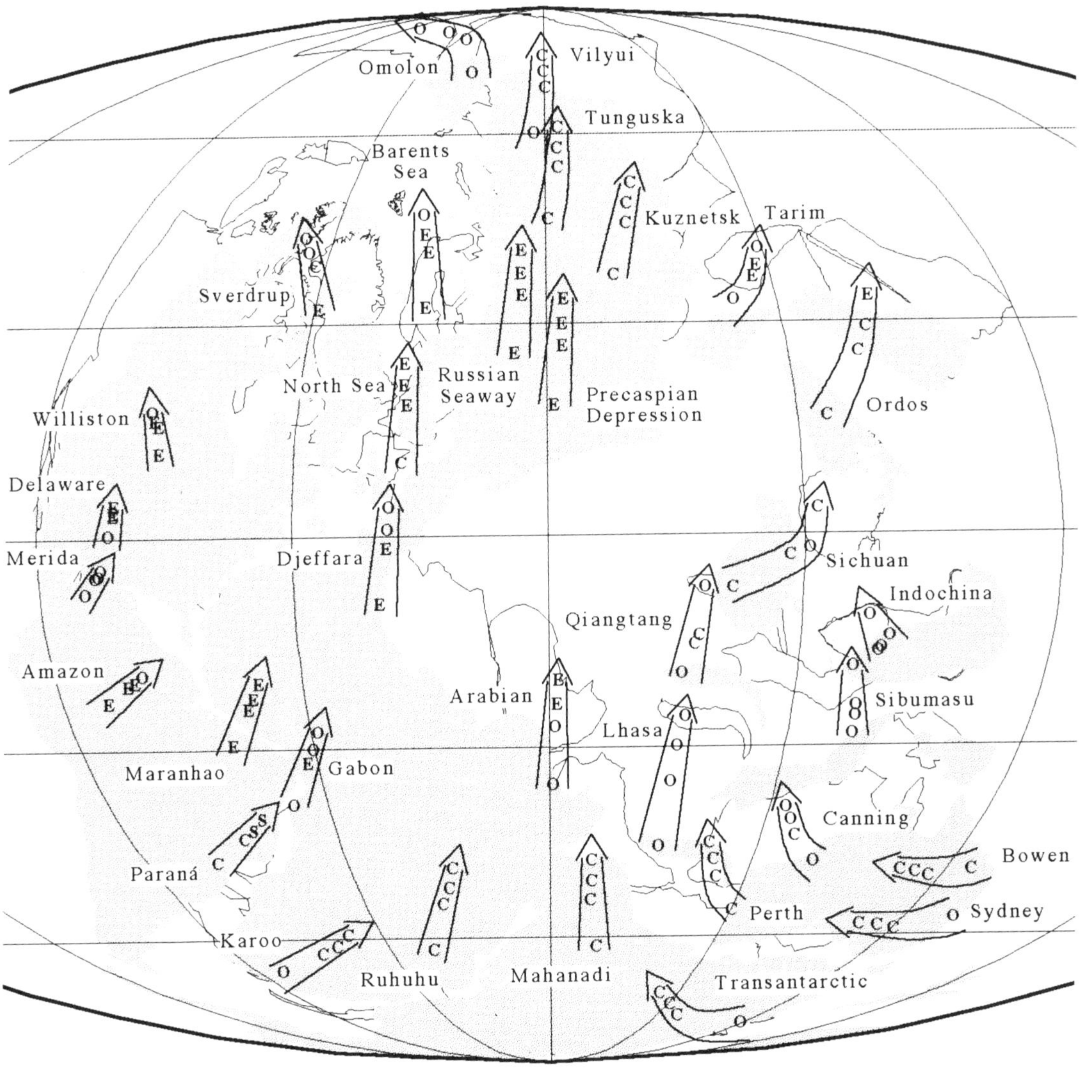

Fig. 8.5. Permian wander paths of selected basins and blocks. Arrows indicate motion of basin centres from the earliest Permian (Sakmarian, shaded outlines) to the latest Permian (Tatarian, unfilled outlines). Climate sensitive sediments for each of the four Permian intervals illustrated in Figs. 8.1-8.4 are shown by the symbols with arrows: **C** for coal, **E** for evaporites, **S** for sand dunes, and **O** for no recognized climate sensitive sediments.

North Polar Zone

The limited north polar areas of the Permian world include the Vilyui Basin (30) along the margin of the Siberian Platform, and the Omolon Block (31), which has been rotated with respect to Siberia but was also in high latitudes in the Permian, to judge by paleomagnetic data (Neustroev et al., 1993). The areas were linked by Carboniferous times by the temperate zone Angaran floral province (Meyen, 1976, and pers. com., 1986) and in the Late Permian by widespread ice-marine deposits in the thick outer-shelf Verkhoyansk Complex (Epshteyn, 1981). Glaciers are not implicated, rather, the rounded pebbles of the diamictites are thought to have been rafted to offshore sites by beach ice. Present correlations place the Omolon ice-marine deposits in the early Tatarian (Kashik, 1990). The source area for the pebbles of the Omolon Block was volcanic, while the Siberian margin material came from Precambrian and Paleozoic terrigenous and carbonate rocks. The Early Permian rocks of Omolon are dominantly marine carbonates, including a reef at one horizon (Kashik, 1990), implying a rapid northward transit or formation in a temperate latitude setting with no clastic dilution. Coals are present in the Vilyui Basin (30) through much of the Permian, but Meyen (1982) refers to the floras as impoverished in comparison with the temperate zone Angaran province.

North Temperate Zone

The Siberian Platform basins of Tunguska (26) and Kuznetsk (27) are representative of a broad area of coal deposition (Vinogradov, 1969; Yuzvitskii et al., 1991) and high diversity temperate-zone floras (Meyen, 1982) throughout the Permian Period. The southern transition of this zone is seen in Kazakhstan, where relatively impoverished floras have been linked with aridity. The northern margin of Canada moved into the temperate zone during the Permian. According to Beauchamp's (1994) study of marine assemblages, "highly diversified, tropical-like associations prevailed during the Asselian and Sakmarian, whereas poorly diversified, temperate-like associations, dominated by bryozoan, echinoderm, and brachiopod, characterized the Artinskian to Kazanian interval; polarlike, siliceous sponge-dominated biota prevailed during the latest Permian" (p. 229). Beauchamp noted that evaporites and reefs occurred initially and were followed by coaly seams and possible dropstones by the Late Permian; he thought that the cooling trend represented a "dramatic climatic deterioration in northern Pangea at the end of the Paleozoic" (p. 229). Beauchamp felt that the change was far greater than one might expect from a 10°–15° northward motion of the continent and considered but rejected upwelling as the mechanism for the cooling. We prefer his upwelling alternative.

North Subtropical Zone

Three major evaporite regions developed in this zone. In North America a broad (2,000 x 500 km) epeiric seaway is represented by the Delaware (5) and the Williston (6) basins and the intervening zone. Evaporites occur throughout (McKee et al., 1967) except in the southern portion during the earliest Permian, where coal deposition was continuous from the great equatorial swamps of the Carboniferous (Schoewe, 1951). Desert dune fields were extensive in the region to the west, and the winds were dominantly northeasterlies, as might be expected (Parrish and Peterson, 1988). In western Europe the area of evaporite formation began in the late Early Permian in a huge Rotliegend lake (1,200 x 200 km) (Glennie, 1984), which was suddenly flooded in the Late Permian by the (1700 x 600 km) Zechstein Sea (P.A. Ziegler, 1988). Like North America, Europe was moving north from the equator and limited coals persisted into the earliest Permian, but the dominant sedimentation was subtropical, with desert dunes being extensively developed. On the Russian Platform just to the west of the Urals occurred the largest area of Permian evaporite deposition (3,000 x 700 km), extending from the Precaspian depression (20) through the Russian Seaway (21) to the Barents Sea (11) (Vinogradov, 1969; Heafford, 1988). Coals do appear at the northern end of the Urals in the Pechora Basin by late in the Early Permian, but this portion of the seaway extended into higher latitudes. The numbers quoted for the size of the evaporite basins are maxima, and at any given time the evaporative area was probably much more limited.

Equatorial Zone

As Pangea drifted north beneath the equator, the equatorial floras of eastern North America and central Europe were displaced to areas like Venezuela, Spain, and China during the Early Permian (Ziegler, 1990). Coal swamps were particularly well developed in the Sichuan Basin (38) of South China and the Ordos Basin (34) of North China, and their Cathaysian floras were the evolutionary derivatives and ecological equivalents of the Euramerican floras. The Indochina block also contains the Cathaysian floras suggesting a biogeogeographic relationship with South China (Metcalfe, 1994). Liu (1990) concluded that the thick coals of China generally developed in a variety of continental settings, except during the Late Permian in North China. Here, thin and impersistent coals were replaced by evaporites as this microcontinent drifted north into the subtropical zone. Parallel transitions in the floras have been recognized, so a single arid climate Zechstein flora extended from Europe to North China by the end of the period (Z.Q. Wang, 1985).

South Subtropical Zone

Huge evaporite basins characterized the south as well as the north subtropical zone during the Permian. Extending 3,000 km from the Pacific, the 400km-wide Amazon Basin (48), a zone of Early Paleozoic crustal thinning (Szatmari et al., 1979; Mossman et al., 1984; da Silva, 1988), connected with the Maranhao Basin (53) (Mesner and Wooldridge, 1964) and possibly with the Gabon Basin (63) of coastal Africa. The Gabon Basin contains, in addition to gypsum and anhydrite, pyritic dark shale, chert, and phosphatic debris (Micholet et al., 1970), suggesting a marine connection. We show this connection through the Maranhao Basin because the intervening terrain was low in the Paleozoic, to judge by outliers of marine Devonian in eastern Brazil (DeMelo, 1988). On the Tethyan coast of Gondwana, the Djeffara Basin (58) of Tunisia and Libya contains Early Permian evaporites (Busson and Burollet, 1973), and the Arabian Basin (76) contains Late Permian evaporites (Szabo and Kheradpir, 1978; Sharief, 1982). The timing of evaporites in these two regions is consistent with the progressive movement of Pangea through the subtropical dry belt. What is surprising is that Early Permian glaciation affected both the Gabon Basin and the southern portion of the Arabian Basin (Micholet et al., 1970; Levell et al., 1988). The ice sheets therefore enveloped areas as low as 37° in the subtropical zone, as well as the temperate and the polar zones. As has been shown, the upland areas of Mayombe and Huqf were adjacent to these sites, so perhaps high elevations created cooler temperatures necessary to nucleate ice sheets. However, Pleistocene ice sheets extended to 39° N in North America, and were not near upland regions.

South Temperate Zone

This is the archetypal Gondwanan zone involving southern South America, southern Africa, India and Australia (see chapters by Lindsay, López-Gamundí, Visser, and Wopfner and Casshyap in this volume). Glaciers were ubiquitous in the earliest Permian and were limited to upland sites by the end of the Early Permian. Our representation of the ice sheet in Figure 8.1 is based on the Early Sakmarian reconstruction of Visser as well as other papers in this volume. The portrayal on Figure 8.2 is based on Visser (this volume, Fig. 10.12 and Brakel and Totterdell, 1993). The control is highly variable because of uncertainties in correlation, which in many areas is based solely on palynology. As far as we can tell, it is not known whether the zonation developed for these microfossils, or for any other fossils, was controlled by evolution or climate. If the latter proves to be the case, the entire exercise will be shown to have been circular, given the diachronous nature of climate change.

How realistic is the inferred ice cover? Rowley and Markwick (1992) and Markwick and Rowley (in press) showed that a simple relationship among sea level, grounded ice volume, and area of grounded ice can be constructed using combination of the present and Last Glacial Maximum data. In our Sakmarian reconstruction (Fig. 8.1), the area of ice coverage is 2.26 x 10^7 km^2. Using the relationships of Markwick and Rowley (in press), this areal coverage corresponds to the sequestration of 4.5 x 10^7 km^3 of water. This translates to about 100 m of sea-level change, well within the range observed during the Pleistocene.

Coal swamps were widespread throughout the period (Langford, 1992), except for sites that were moving into the subtropics by the end of the period, such as the Paraná Basin (51) and northwestern Argentina (Machado, 1975; Limarino and Spalletti, 1986). By latest Permian times, vast lakes occurred across southern Africa (Yemane and Kelts, 1990). An important point is that the coal deposits are interbedded with the tillites in many temperate zone sites, showing that climate fluctuated between the polar biome 10 and the cool temperate biome 6 (Ziegler, 1990) on time scales that were much too short to be accounted for by continental drift. In this respect, the Permian is similar to the Pleistocene. A number of excellent reviews of the climatic evolution of this zone through the entire Permian are available, based on macrofloras from eastern Australia (Retallack, 1980) and from India (Chandra and Chandra, 1987), microfloras from southern Africa (Falcon, 1986) and from India (Tiwari and Tripathi, 1987), and sediments from Tanzania (Kreuser et al., 1990). These widely spaced areas have remarkably similar records, with cool temperate humid conditions dominant in the Permian, except for the glacial effects at the beginning and warmer and possibly drier climates at the end. India moved the most, from 47° S to 62° S (1630 km), while eastern Australia stayed the same at about 58° S, given our interpretation of the paleomagnetic data.

A number of the eastern Tethyan microcontinents transited the cool temperate zone during the Permian and show significant changes in climate during the period. The Lhasa (43) and the Qiangtang (42) blocks have Early Permian glacial records followed by transitional and then warm-water faunas (Sun, 1993). Sun compared the stratigraphy of these areas with the Himalayan margin of India to the south and showed a diachroniety in the climate transitions in the three regions; Qiangtang entered the subtropics by the middle of the period and later acquired the tropical Cathaysian flora, while Lhasa followed in the mid-Late Permian and India only by the latest Permian. Sibumasu had a history similar to that of Lhasa (Metcalfe, 1994) and displays a carbonate rocks sequence interpreted to have been deposited in cool temperate conditions followed by temperate and then tropical conditions by the Triassic (Rao, 1988).

South Polar Zone

Antarctica alone occupied the polar zone and gradually drifted off the pole during the Permian. Glaciation seems to have been limited to the earliest Permian and was succeeded by coal swamp environments (Isbell et al., this volume). The shape of the ice sheet in Antarctica is very uncertain due to limited exposure. By the Late Permian, *Glossopteris* forests were common with trunks up to 18 cm and growth rings averaging 4.5 mm (Taylor et al., 1992). These authors pointed out that "Permian forests were present at latitudes where no vascular plants can survive today" and "a lack of latewood development in the tree rings and the absence of frost rings suggests that temperatures rarely went below freezing" (p. 1677). On the other hand, Antarctic Permian floras were low in diversity, with a lack of understorey plants such as ferns and horsetails that are typical elsewhere, so a cool temperate humid climate designation seems appropriate for all but the glacial beginning of the period (Cuneo et al., 1993).

Conclusion on paleoclimatology

Clearly, the major climatic event of the Permian was the deglaciation sequence, with its minor readvances in the Early Permian (Visser, this volume). The causes of the glaciation, which clearly reached its height in the Late Carboniferous, must relate to a combination of poleward drift and uplift along the many rift zones in Gondwana. The Early Permian glaciers extended across the polar and temperate zones and even to the margin of the subtropical zone. During glacial retreats, the temperate zone climates were reestablished, so it is clear that the cold polar climates were exported to very low latitudes during the advances. The parallels with the Pleistocene are obvious. By the Early Permian, the continent was moving off the pole, and the glaciers disappeared altogether. Apart from this, the Permian climate zones, as represented by the climate-sensitive sediments and the floras, seem to have been latitudinally stable. They were not strictly latitude-parallel, but their deviations are understandable in terms of circulation dynamics. In general, Pangea moved beneath the climate zones, and this is detectable in many sedimentary basins.

References

Affaton, P., Rahaman, M.A., Trompette, R. and Sougy, J., 1991. The Dahomeyide orogen: tectonothermal evolution and relationships with the Volta basin. In: R.D. Dallmeyer and J.P. Lécorché (Editors), The West African orogens and circum-Atlantic correlatives. Springer-Verlag, Berlin, pp. 107–122.

Ahnert, F., 1970. Functional relationships between denudation, relief and uplift in large mid-latitude drainage basins. American Jounal of Science, 268: 243–263.

Amantov, V.A., Blagonravov, V.A., Borzakovsky, Y.A., Durante, M.V., Zonenshain, L.P. and Luwsandarsan, B., et al., 1970. Stratigraphy and tectonics of the Mongolian People's Republic. Akademia Nauk SSSR Transactions, 1, 86 pp.

Balling, N., 1980. The land uplift in Fennoscandia, gravity field anomalies and isostacy. In: N.A. Mörner (Editor), Earth rheology, isostacy and eustacy. John Wiley and Sons, Chichester, pp. 297–321.

Barrett, P.J., 1991. The Devonian to Triassic Beacon supergroup of the Transantarctic Mountains and correlatives in other parts of Antarctica. In: R.J. Tingey (Editor), The Geology of Antarctica. Clarendon Press, Oxford, pp. 120–152.

Beauchamp, B., 1994. Permian climatic cooling in the Canadian Arctic. In: G.D. Klein (Editor), Pangea: paleoclimate, tectonics, and sedimentation during accretion, zenith, and breakup of a supercontinent. Geological Society of America Special Paper 288, pp. 229–246.

Belasky, P. and Runnegar, B., 1994. Permian longitudes of Wrangellia, Stikinia, and Eastern Klamath terranes based on coral biogeography. Geology, 22: 1095–1098.

Benedetto, G., 1980. Bioestratigrafía del Neopaleozoico de los Andes de Venezuela: una síntesis actualizada. Anais da Academia Brasileira de Ciencias, 52: 841–850.

Bond, G., 1970. The Dwyka Series in Rhodesia. Proceedings of the Geological Association, 81: 463–472.

Borisevich, D.V., 1992. Neotectonics of the Urals. Geotectonics, 26: 41–47.

Botneva, T.A., et al., 1991. Geological-geochemical basis for oil-gas exploration in the north Caspian depression. Petroleum Geology, 25: 122–126.

Brakel, A.T. and Totterdell, J.M., 1993. The Sakmarian-Kungarian palaeogeography of Australia. In: R.H. Finlay, R. Unrug, M.R. Banks and J.J. Veevers (Editors), Gondwana eight: assembly, evolution, dispersal. A.A. Balkema, Rotterdam, pp. 385–396.

Brown, P.M., Miller, J.A. and Swain, R.M., 1972. Structural and stratigraphic framework, and spatial distribution of permeability of the Atlantic Coastal Plain, North Carolina to New York. United States Geological Survey Paper 796, pp. 79.

Brown, R.W., Rust, D.J., Summerfield, M.A., Gleadow, A.J.W. and De Wit, M.J.C., 1990. An Early Cretaceous phase of accelerated erosion on the south-western margin of Africa: evidence from apatite fission track analysis and the offshore sedimentary record. Nuclear Tracks and Radiation Measurements, 17: 339–350.

Busson, G. and Burollet, P.F., 1973. La limite Permien-Trias sur la plate-forme Saharienne (Algerie, Tunisie, Libye). In: A. Logan and L.V. Hills (Editors), The Permian and Triassic systems and their mutual boundary. Canadian Society of Petroleum Geologists, Memoir 2, Calgary, pp. 74–88.

Cahen, L. and Lepersonne, J., 1981. Late Palaeozoic tillites of the Congo Basin in Zaire. In: M.J. Hambrey and W.B. Harland (Editors), Earth's pre-Pleistocene glacial record. Cambridge University Press, Cambridge, pp. 43–47.

Casshyap, S.M. and Srivastava,V.K., 1987. Glacial and proglacial Talchir sedimentation in Son-Mahanadi Gondwana Basin: paleogeographic reconstruction. In: G.D. McKenzie (Editor), Gondwana six: stratigraphy, sedimentology and paleontology. Geophysical Monograph 41. American Geophysical Union, Washington, DC, pp. 167–182.

Cavaroc, V.V., Padget, G., Stephens, G., Kanes, W.H., Boudda, A. and Woollen, I.D., 1976. Late Paleozoic of the Tindouf Basin-North Africa. Journal of Sedimentary Petrology, 46: 77–88.

Chandra, S. and Chandra, A., 1987. Vegetational changes and their climatic implications in coal-bearing Gondwana. Palaeobotanist, 36: 74–86.

Chemale, Jr., F., Alkmim, F.F. and Endo, I., 1993. Late Proterozoic tectonism in the interior of the São Francisco craton. In: R.H. Finlay, R. Unrug, M.R. Banks and J.J. Veevers (Editors), Gondwana eight: assembly, evolution, dispersal. A.A. Balkema, Rotterdam, pp. 29–41.

Choubert, G. and Faure-Muret, A., 1971. Carte tectonique de l'Afrique (1968). In: UNESCO (Editor), Tectonique de l'Afrique, Sciences de la Terre 6. 1: 15,000,000 map. UNESCO, Paris.

Collinson, J.W., Isbell, J.L., Elliot, D.H., Miller, M.F., Miller, J.M.G. and Veevers, J.J., 1994. Permian-Triassic Transantarctic basin. In: J.J. Veevers and C.McA. Powell (Editors), Permian-Triassic Pangean basins and foldbelts along the Panthalassan margin of Gondwanaland. Geological Society of America Memoir 184, pp. 173–222.

Conrad, J., 1983. Reggan formation. In: R.H. Wagner et al. (Editors), Australia, Indian subcontinent, South Africa, South America, and North Africa. IUGS Publication 20, pp. 322–323.

Cooper, M.R. and Kensley, B., 1984. Endemic South American Permian bivalve molluscs from the Ecca of South Africa. Journal of Paleontology, 58: 1360–1363.

Corrigan, J., 1991. Inversion of apatite fission-track data for thermal history information. Journal of Geophysical Research, 96: 10347–10360.

Cox, K.G., 1970. Tectonics and vulcanism of the Karroo Period and their bearing on the postulated fragmentation of Gondwanaland. In: T.N. Clifford and I.G. Gass (Editors), African magmatism and tectonics. Hafner Publishing Company, Darien, CT, pp. 211–235.

Crowley, K.D., 1991. Thermal history of Michigan Basin and southern Canadian Shield from apatite fission track analysis. Journal of Geophysical Research, 96: 697–711.

Crowley, T.J., Hyde, W.T. and Short, D.A., 1989. Seasonal cycle variations on the supercontinent of Pangea. Geology, 17: 457–460.

Cuneo, N.R., Isbell, J., Taylor, E.L. and Taylor, T.N., 1993. The Glossopteris flora from Antarctica: taphonomy and paleoecology. Comptes Rendus XII ICC-P, 2: 13–40.

Dallmeyer, R.D., 1989. Late Paleozoic thermal evolution of crystalline terranes within portions of the U.S. Appalachian orogen. In: R.D. Hatcher, Jr., T.A. Thomas, and G.W. Viele (Editors), The Appalachian-Ouachita orogen in the United States. Geology of North America series, vol. F-2 , Geological Society of America, pp. 417–444.

Dallmeyer, R.D. and Lécorché, J.P., 1990. $^{40}Ar/^{39}Ar$ polyorogenic mineral age record within the southern Maurintanide orogen (M'Bout-Bakel region) of west Africa. American Journal of Science, 290: 1136–1168.

Daly, M.C., Chorowicz, J. and Fairhead, J.D., 1989. Rift basin evolution in Africa: the influence of reactivated steep basement shear zones. In: M.A. Cooper and G.D. Williams (Editors), Inversion tectonics. Blackwell Scientific Publications, Boston, pp. 309–334.

da Silva, O.B., 1988. Revisao estratigrafica da bacia do solimoes. Anais do XXXV Congresso Brasileiro de Geologia, Belém, Pará, 6: 2428–2438.

de Beer, C.H., 1992. Structural evolution of the Cape Fold Belt syntaxis and its influence on syntectonic sedimentation in the SW Karoo Basin. In: M.J. de Witt and I.G.D. Ransome (Editors), Inversion tectonics of the Cape Fold Belt, Karoo and Cretaceous Basins of Southern Africa. A.A. Balkema, Rotterdam, pp. 197–206.

de Melo, J.H.G., 1988. The Malvinokaffric Realm in the Devonian of Brazil. In: N.J. McMillan, A.F. Embry, and D.J. Glass (Editors), Regional syntheses, Devonian of the world series. vol. I. Canadian Society of Petroleum Geologists, Memoir 4, Calgary, pp. 669–703.

Donzeau, M., Fabre, J. and Moussine-Pouchkine, A., 1981. Comportmement de la dalle Saharienne et orogenese Varisque. Essai d'interpretation. Bulletin de la Sociéte Histoire Naturelle Afrique du Nord, 69: 137–172.

Drewry, D.J. (Editor), 1983. Antarctica: Glaciological and geophysical folio. Scott Polar Research Institute, Cambridge, GB.

Drysdall, A.R. and Weller, R.K., 1966. Karoo sedimentation in northern Rhodesia. Geological Society of South Africa Transactions, 69: 39–69.

Dypvik, H., Nesteby, H., Ruden, F., Aagaord, P., Msindai, T. and Massay, C., 1990. Upper Paleozoic and Mesozoic sedimentation in the Rukwa-Tukuyu Region, Tanzania. Journal of African Earth Sciences, 11: 437–456.

El Wartiti, M., Broutin, J., Freytet, P., Larhrib, M. and Toutin-Morin, N., 1990. Continental deposits in Permian basins of the Mesetian Moroco, geodynamic history. Journal of African Earth Sciences, 10: 361–368.

England, P. and Molnar, P., 1990. Surface uplift, uplift of rocks, and exhumation of rocks. Geology, 18: 1173–1177.

Epshteyn, O.G., 1981. Late Permian ice-marine deposits of the Atkan Formation in the Kolyma river headwaters region, U.S.S.R. In: M.J. Hambrey and W. B. Harland (Editors), Earth's pre-Pleistocene glacial record. Cambridge University Press, Cambridge, pp. 270–273.

Falcon, R.M.S., 1986. A brief review of the origin, formation, and distribution of coal in southern Africa. In: C.R. Anhaeusser and S. Maske (Editors), Mineral deposits of southern Africa, vol. 2. The Geological Society of South Africa, Johannesburg, pp. 1879–1898.

Fedorov, L.V., Grikurov, G.E., Kurinin, R.G. and Masolov, V.N., 1982. Crustal structure of the Lambert Glacier area from geophysical data. In: C. Craddock (Editor), Antarctic Geoscience. IUGS Series B, 4: 931–936.

Filippov, Yu.F., 1992. Formational composition and structure of Taimyr Carboniferous-Triassic complex. Russian Geology and Geophysics, 33: 6–14.

Foster, D.A. and Gleadow, A.J.W., 1993. The architecture of Gondwana rifting in southeastern Australia: evidence from apatite fission track thermochronology. In: R.H. Finlay, R. Unrug, M.R. Banks and J.J. Veevers (Editors), Gondwana eight: assembly, evolution, dispersal. A.A. Balkema, Rotterdam, pp. 597–603.

Foster, D.A., Murphy, J.M. and Gleadow, A.J.W., 1994. Middle Tertiary hydrothermal activity and uplift of the northern Flinders Ranges, South Australia: insights from apatite fission-track thermochronology. Australian Journal of Earth Sciences, 41: 11–17.

França, A.B. and Potter, P.E., 1991. Stratigraphy and reservoir potential of glacial deposits of the Itararé Group (Carboniferous-Permian), Paraná Basin, Brazil. American Association of Petroleum Geologists Bulletin, 75: 62–85.

Gansu Province Bureau of Geology and Mineral Resources (Editor), 1989. Regional Geology of Gansu Province. People's Republic of China Ministry of Geology and Mineral Resources Geological Memoirs Series 1, No. 19. Geological Publishing House, Beijing, 690 pp. (in Chinese with English summary).

Glennie, K.W., 1984. Early Permian-Rotliegend. In: K.W. Glennie (Editor), Introduction to the geology of the North Sea. Shell International Petroleum Mij, The Hague, pp. 41–60.

Goryainov, S.V., 1991. Orientation stability of the Yenisei ridge vergences. Soviet Geology and Geophysics, 32: 107–110.

Grew, E.S. and Manton, W.I., 1986. A new correlation of sapphirine granulites in the Indo-Antarctic metamorphic terrain: Late Proterozoic dates from the eastern Ghats province of India. Precambrian Research, 33: 123–137.

Hack, J.T., 1982. Physiographic divisions and differential uplift in the Piedmont and Blue Ridge. U.S. Geological Survey Professional Paper 1265, 49 pp.

Haller, J., 1971. Geology of the East Greenland Caledonides. Interscience Publishers, Switzerland, 413 pp.

Han, D.X. and Yang, Q. (Editors), 1980. Coal geology of China, vol. 2. Coal Industry Press, Beijing, 415 pp.

Hanna, W.F., Sweeney, R.E., Hildenbrand, T.G., Tanner, J.G., McConnell, R.K. and Godson, R.H., 1989. The gravity anomaly map of North America. In: A.W. Bally and A.R. Palmer (Editors), The geology of North America-an overview. Geology of North America series, vol. A. The Geological Society of America, Boulder, pp. 17–27.

Hatcher, Jr., R.D., Thomas, W.A., Geiser, P.A., Snoke, A.W., Mosher, S. and Wiltschko, D.V., 1989. Alleghanian orogen. In: R.D. Hatcher, Jr., T.A. Thomas, and G.W. Viele (Editors), The Appalachian-Ouachita orogen in the United States. The geology of North America series, vol. F-2. Geological Society of America, pp. 233–318.

Heafford, A.F., 1988. Carboniferous through Triassic stratigraphy of the Barents shelf. In: W.B. Harland and E.K. Dodswell (Editors), Geological evolution of the Barents Shelf region, Graham and Trotman, London, pp. 89–108.

Henderson, C.M., Bamber, E.W., Richards, B.C., Higgins, A.C. and McGugan, A., 1993. Permian. In: D.F. Scott and J.D. Aitken (Editors), Sedimentary cover of the craton in Canada. Geology of Canada series, vol. 5, Geological Survey of Canada, pp. 272–293.

Hoffman, P.E., 1989. Precambrian geology and tectonic history of North America. In: A.W. Bally and A.R. Palmer (Editors), The geology of North America-an overview. The geology of North America, vol. A. Geological Society of America, pp. 447–512.

Hofmann, J., 1987. Fault tectonics and magmatic ages in the Jetty Oasis area, Mac Robertson Land: a contribution to the Lambert rift development . In: M.R.A. Thomson, J.A. Crame, and J.W. Thomson (Editors), Geological evolution of Antarctica. Proceedings of the Fifth International Symposium on Antarctic Earth Sciences. Cambridge University Press, Cambridge, pp. 107–112.

Huang, K.N., 1988. Were the Emeishan basalts erupted as a consequence of back-arc spreading? Scientia Geologica Sinica, 3: 289–298.

Hulver, M.L., 1995, in progress. Post-orogenic evolution of the Appalachian Mountains and its foreland. Ph.D. dissertation, University of Chicago.

Hutchison, C.S., 1989. Geological evolution of South-east Asia. Clarendon Press, Oxford, 368 pp.

Il'yukhina, N.P. and Verbitskaya, N.G., 1977. Formation conditions and stratigraphy of carboniferous coal deposits of the Siberian platform. International Geology Review, 19: 429–440.

Isbell, J.L., Seegers, G.M. and Gelhar, G.A., this volume. Late Paleozoic glacial and post-glacial deposits, central Transantarctic Mountains, Antarctica.

Ishii, K.-I., Lui, X.-Y., Koichiro, I. and Huang, B.-H. (Editors), 1991. Pre-Jurassic geology of Inner Mongolia, China, a report of China-Japan Cooperative Research Group, 1987–1989. China-Japan Cooperative Research Group, Osaka, 402 pp.

Jensen, L.N., Riis, F. and Boyd, R. (Editors), 1992. Post-Cretaceous uplift and sedimentation along the western Fennoscandian Shield. Norsk Geologisk Tidsskrift, 72, pp. 223–338.

Jin, Y.-G., Glenister, B.F., Kotlyar, G.V. and Sheng, J.-Z., 1994. An operational scheme of Permian chronostratigraphy. In: Y.-G. Jin, J. Utting and B.R. Wardlaw (Editors), Palaeontology and stratigraphy. Palaeoworld, 4: 1–13.

Jubitz, K.-B., Znosko, J. and Franke, D. (Editors), 1985. Lithologic-palaeogeographical map. Saxonian. Southwest border of the East-European Platform. International Geological Correlation Program Project 86. 1: 1,500,000 map. Zentrales Geologisches Institut, Berlin.

Kale, V.S., 1991. Constraints on the evolution of the Purana basins of peninsular India. Journal of the Geological Society of India, 38: 231–252.

Kamenev, E.N., 1993. Structure and evolution of the Antarctic shield in Precambrian. In: R.H. Finlay, R. Unrug, M.R. Banks and J.J. Veevers (Editors), Gondwana eight: assembly, evolution, dispersal. A.A. Balkema, Rotterdam, pp. 141–151.

Kashik, D.S. (Editor), 1990. Permian key section of the Omolon massif. Leningrad, Nauka Publishers, 197 pp.

Khain, V.E., 1985. Old craton and Paleozoic fold belts, Geology of the USSR series, Part 1. Gebruder Borntraeger, Berlin, 272 pp.

Kluth, C.F., 1986. Plate tectonics of the ancestral Rocky Mountains. In: J.A. Peterson (Editor), Paleotectonics and sedimentation in the Rocky Mountain region, United States. American Association of Petroleum Geologists Memoir, 41: 353–369.

Kogbe, C.A., 1991. Stratigraphy and tectonic history of the Iullemmeden Basin in West Africa. South African Journal of Geology, 94: 19–32.

Kreuser, T., Wopfner, H., Kaaya, C.Z., Markwort, S., Semkiwa, P.M. and Aslanidis, P., 1990. Depositional evolution of Permo-Triassic Karoo basins in Tanzania with reference to their economic potential. In: C.A. Kogbe and J. Lang (Editors), Major African continental Phanerozoic complexes and dynamics of sedimentation. Journal of African Earth Sciences, 10: 151–167.

Kröner, A. and Williams, I.S., 1993. Age of metamorphism in the high-grade rocks of Sri Lanka. Journal of Geology, 101: 513–521.

Kutzbach, J.E. and Ziegler, A.M., 1993. Simulation of Late Permian climate and biomes with an atmosphere-ocean model: comparisons with observations. Philosophical Transactions of the Royal Society of London, Series B, 341: 327–340.

Langford, R.P., 1992. Permian coal and palaeogeography of Gondwana. Bureau of Mineral Resources (Australia) Record 1991/5, Palaeogeography 39, 136 pp.

Larsen, R.M., Brekke, H., Larsen, B.T. and Talleras, E. (Editors), 1992. Structural and tectonic modelling and its application to petroleum geology, Proceedings of Norwegian Petroleum Society Workshop, 18–20 October 1989, Stavanger, Norway. Norwegian Petroleum Society Special Publication 1, Elsevier, Amsterdam, 549 pp.

Lemosquet, Y. and Pareyn, C., 1983. Bechar basin. In: R.H. Wagner, C.F.W. Prins, and L.F. Granados (Editors), The Carboniferous of the World, Volume II: Australia, Indian subcontinent, South Africa, South America, and North Africa. IUGS Publication 20: 306–315.

Lerch, M.F., Xue, F., Kröner, A. Zhang, G.W. and Tod, W., 1995. A Middle Silurian-Early Devonian magmatic arc in the Qinling Mountains of central China. Journal of Geology: 103, 437–450.

Levell, B.K., Braakman, J.H. and Rutten, K.W., 1988. Oil-bearing sediments of Gondwana glaciation in Oman. American Association of Petroleum Geologists Bulletin, 72: 775–796.

Limarino, C.O. and Spalletti, L.A., 1986. Eolian Permian deposits in west and northwest Argentina. Sedimentary Geology, 49: 109–127.

Lindsay, J.F., this volume. Permian post-glacial environments of the Australian plate.

Liu, G.,1990. Permo-Carboniferous paleogeography and coal accumulation and their tectonic control in the North and South China continental plates. International Journal of Coal Geology, 16: 73–117.

López-Gamundí, O.R., this volume. Glacial-postglacial transition in the Late Paleozoic basins of southern South America.

López-Ramos, E., 1981. Geología de México, Tomo III. Secretaria de Educacion Publica, Mexico, 446 pp.

Lottes, A.L. and Rowley, D.B., 1990. Reconstruction of the Laurasian and Gondwanan segments of Permian Pangea. In: W.S. McKerrow and C.R. Scotese (Editors), Palaeozoic palaeogeography and biogeography. Geological Society Memoir 12, pp. 383–395.

Lottes, A.L. and Ziegler, A.M., 1994. World peat occurrence and the seasonality of climate and vegetation. Palaeogeography, Palaeoclimatology, Palaeoecology, 106: 23–37.

Lovera, O.M., Richter, F.M. and Harrison, T.M., 1989. The $^{40}Ar/^{39}Ar$ thermochronometry for slowly cooled samples having a distribution of diffusion domain sizes. Journal of Geophysical Research, 94: 17917–17935.

Lutz, T.M. and G. Omar, 1991, An inverse method of modeling thermal histories from apatite fission-track data. Earth and Planetary Science Letters, 104: 181–195.

Luo, Z., Jin, Y., Zhu, K. and Zhao, X., 1988. On Emei taphrogenesis of the upper Yangtze platform. Chengdu College Review of Geology, 34: 11–24 (in Chinese with English summary).

Machado, E.R., 1975. Gondwana coal in southern Brazil. In: K.S.W. Campbell (Editor), Gondwana geology. Australian National University Press, Canberra, pp. 271–293.

Manuilova, M.M., 1991. Precambrian geology and geochronology of the Baikalian mountain region, and the problem of the Baikalides. Soviet Geology and Geophysics, 32: 51–58.

Markwick, P.J. and Rowley, D.B., 1995, in press. The geological evidence for Triassic to Pleistocene glaciations: implications for eustacy. In: C. Pindell and C.L. Drake (Editors), SEPM Special Publication, 80 pp.

Martin, H., 1975. Structural and palaeogeographical evidence for an Upper Palaeozoic sea between southern Africa and South America. In: K.S.W. Campbell (Editor), Gondwana Geology. Australian National University Press, Canberra, pp. 37–51.

Martin, H., 1981. The late Palaeozoic Gondwana glaciation. Geologische Rundschau, 70: 480–496.

McKee, E.D., Oriel, S.S., et al., 1967. Paleotectonic maps of the Permian system. U.S. Geological Survey miscellaneous geologic investigations map I-450. 164 pp.

Mesner, J.C. and Wooldridge, L.C.P., 1964. Maranhýo Paleozoic basin and Cretaceous coastal basins, north Brazil. American Association of Petroleum Geologists Bulletin, 48: 1475–1512.

Metcalfe, I., 1993. Southeast Asian terranes: Gondwanaland origins and evolution. In: R.H. Finlay, R. Unrug, M.R. Banks and J.J. Veevers (Editors), Gondwana eight: assembly, evolution, dispersal. A.A. Balkema, Rotterdam, pp. 181–200.

Metcalfe, I., 1994. Gondwanaland origin, dispersion, and accretion of East and Southeast Asian continental terranes. Journal of African Earth Sciences, 7: 333–347.

Meyen, S.V., 1976. Carboniferous and Permian lepidophytes of Angaraland. Palaeontographica, Abteilung B, 157: 112–157.

Meyen, S.V., 1982. The Carboniferous and Permian floras of Angaraland (a synthesis). Biological Memoirs, 7: 1–110.

Micholet, J., Molinas, E. and Penet, B., 1970. Le systeme du Karroo au Gabon. In: S.H. Haughton (Editor), Second Gondwana symposium, South Africa, 1970, proceedings and papers. Council for Scientific and Industrial Research, Scientia, Pretoria, South Africa, pp. 371–377.

Miller, E.L., Miller, M.M., Stevens, C.H., Wright, J.E. and Madrid, R., 1992. Late Paleozoic paleogeographic and tectonic evolution of the western U.S. Cordillera. In: B.C. Burchfiel, P.W. Lipman and M.L. Zoback (Editors), The Cordilleran orogen: conterminus U.S. Geology of North America series, vol. G-3. Geological Society of America, pp. 57–106.

Mosmann, R., Falkenhein, F.U.H., Gonçalves, A. and Filho, F.N., 1984. Oil and gas potential of the Amazon Paleozoic basins. American Association of Petroleum Geologists Memoir, 40: 207–241.

Nagornyy, Y.N. and Nagornyy, V.N., 1976. Geological evolution of the Donets Basin. Geotectonics, 10: 45–52.

Narain, H. (Editor), 1975. Bouger gravity anomaly and regional geology map of India, NGRI/GPH-5, 1: 5,000,000 map. National Geophysical Research Institute, Hyderabad, India.

Neustroev, A.P., Parfenov, L. M. and Rodionov, V.P., 1993. Paleomagnetic data and nature of the Tas--Khayakhtakh terrane in the Verkhoyansk-Kolyma region. Russian Geology and Geophysics, 34: 19–30.

Nie, S., Rowley, D.B. and Ziegler, A.M., 1990. Constraints on the locations of Asian micro-continents in the Palaeo-Tethys during the Late Palaeozoic. In: W.S. McKerrow and C.R. Scotese (Editors), Palaeozoic palaeogeography and biogeography, Geological Society of London Memoir 12, pp. 397–409.

Nie, S.Y., Van der Voo, R., Li, M.-S. and Rowley, D.B., 1993. Paleomagnetism of Late Paleozoic rocks in the Tianshan, northwestern China. Tectonics, 12: 568–579.

Nott, J., 1995. The antiquity of landscape on the North Australian craton and the implications for theories of long-term landscape evolution. Journal of Geology, 103: 19–32.

Nyblade, A.A., Lei, Y., Shive, P.N. and Tesha, A., 1993. Paleomagnetism of Permian sedimentary rocks from Tanzania and the Permian paleogeography of Pangea. Earth and Planetary Science Letters, 118: 181–194.

O'Brien, P.E., 1986. Stratigraphy and sedimentology of Late Palaeozoic glaciomarine sediments beneath the Murray Basin, and their palaeogeographic and palaeoclimatic significance. BMR Journal of Australian Geology and Geophysics, 10: 53–63.

Oelofsen, B.W., 1987. The biostratigraphy and fossils of the Whitehill and Irati Shale Formations of the Karoo and Paraná Basins. In: G.D. McKenzie (Editor), Gondwana six: stratigraphy, sedimentology and paleontology. Geophysical Monograph, 41: 131–138.

Oesterlen, P.M. and Blenkinsop, T.G., 1994. Extension directions and strain near the failed triple point juncture of the Zambezi and Luangwa Rift zones, southern Africa. Journal of African Earth Sciences, 18: 175–180.

Orpen, J.L., Swain, C.J., Nugent, C. and Zhou, P.P., 1989. Wrench-fault and half-graben tectonics in the development of the Palaeozoic Zambezi Karoo Basins in Zimbabwe-the "Lower Zambezi" and "Mid-Zambezi" basins respectively-and regional implications. Journal of African Earth Sciences, 8: 215–229.

Palmer, A.R., 1983. The decade of North American geology time scale. Geology, 11: 503–504.

Paquette, J.-L., Nédélec, A., Moine, B. and Rakotondrazafy, M., 1994. U-Pb, single zircon Pb-evaporation, and Sm-Nd isotopic study of a granulite domain in SE Madagascar. Journal of Geology, 102: 523–538.

Parrish, J.T. and Peterson, F., 1988. Wind directions predicted from global circulation models and wind directions determined from eolian sandstones of the western United States-a comparison. Sedimentary Geology, 56: 261–282.

Petri, S. and Fulfaro, V.J., 1983. Geologia do Brasil. Biblioteca de Ciencias Naturais, 9, 631 pp.

Petri, S. and Mendes, J.C., 1983. Brazil. In: B. M. Moullade and A.E.M. Nairn (Editors), The Phanerozoic geology of the world, vol. II: The Mesozoic. Elsevier, Amsterdam, pp. 151–179.

Pitman, W.C., 1978. Relationship between eustacy and stratigraphic sequences of passive margins. Geological Society of America Bulletin, 89: 1389–1403.

Pitman, W.C. and Golovchenko, X., 1991. The effect of sea level changes on the morphology of mountain belts. Journal of Geophysical Research, 96: 6879–6891.

Poag, C.W. and Sevon, W.D., 1989. A record of Appalachian denudation in postdrift Mesozoic and Cenozoic sedimentary deposits of the U.S. middle Atlantic continental margin. In: T.W. Gardener and W.D. Sevon (Editors), Appalachian geomorphology. Elsevier, Amsterdam, pp. 119–158.

Powell, C.McA. and Veevers, J.J., 1994. Introduction. In: J.J. Veevers and C.McA. Powell (Editors), Permian-Triassic Pangean basins and foldbelts along the Panthalassan margin of Gondwanaland. Geological Society of America Memoir 184, pp. 1–4.

Radhakrishna, B.P., 1989. Suspect tectono-stratigraphic terrane elements in the Indian subcontinent. Journal of the Geological Society of India, 34: 1–24.

Rahaman, M.A., Van Breemen, O., Bowden, P. and Bennett, J.N., 1984. Age migrations of anorogenic ring complexes in northern Nigeria. Journal of Geology, 92: 173–184.

Rao, C.P., 1988. Paleoclimate of some Permo-Triassic carbonates of Malaysia. Sedimentary Geology, 60: 163–171.

Renne, P.R., and Basu, A.R., 1991. Rapid eruption of the Siberian Traps flood basalts at the Permo-Triassic boundary. Science, 253: 176–179.

Retallack, G.J., 1980. Late Carboniferous to Middle Triassic megafossil floras from the Sydney Basin. In: C. Herbert and R.J. Helby (Editors), A guide to the Sydney Basin. Geological Survey of New South Wales Bulletin, 26: 384–430.

Richter, F.M., Lovera, O.M., Harrison, T.M. and Copeland, P., 1991. Tibetan tectonics from $^{40}Ar/^{39}Ar$ analysis of a single feldspar sample. Earth and Planetary Science Letters, 105: 266–278.

Riis, F., and Fjeldskaar, W., 1992. On the magnitude of the Late Tertiary and Quaternary erosion and its significance for the uplift of Scandinavia and the Barents Sea. In: R.M. Larsen, H. Brekke, B.T. Larsen, and E. Talleras (Editors), Structural and tectonic modelling and its application to petroleum geology, Proceedings of Norwegian Petroleum Society Workshop, 18–20 October 1989, Stavanger, Norway. Norwegian Petroleum Society, Special Publication 1, Elsevier, Amsterdam, pp. 165–185.

Roberts, D., Sturt, B.A. and Furnes, H., 1985. Volcanite assemblages and environments in the Scandinavian Caledonides and the sequential development history of the mountain belt. In: D.G. Gee and B.A. Sturt (Editors), The Caledonide orogen-Scandinavia and related areas. John Wiley and Sons, Chichester, pp. 919–930.

Rowley, D.B., 1993. Regional tectonics of South China: is the Qinzhou trough succession a Taconic-style allochthon? In: Abstracts with Programs, Geological Society of America, 1993 Meeting, Boston, Massachusetts, October 25–28, 1993, vol. 25, pp. A-118.

Rowley, D.B. and Markwick, P.J., 1992. Haq et al. Eustatic sea level curve: implications for sequestered water volumes. The Journal of Geology, 100: 703–715.

Saleeby, J.B., 1992. Petrotectonic and paleogeographic settings of U.S. Cordilleran ophiolites . In: B.C. Burchfiel, P.W. Lipman and M.L. Zoback (Editors), The cordilleran orogen: conterminus U.S.. Geology of North America series, vol. G-3, Geological Society of America, pp. 653–682.

Scheibner, E., 1987. Paleozoic tectonic development of eastern Australia in relation to the Pacific region. In: J.W.H. Monger and J. Francheteau (Editors), Circum-Pacific orogenic belts and evolution of the Pacific Ocean Basin. Geodynamics series 18, American Geophysical Union, pp. 133–165.

Schoewe,W.H., 1951. Coal resources of the Permian system in Kansas. State Geological Survey of Kansas, Bulletin 90: 53–68.

Scotese, C.R., and Langford, R.P., 1995, in press. Pangea and the paleogeography of the Permian.

Sengör, A.M.C., Cin, A., Rowley, D.B., and Nie, S.Y., 1993. Space-time patterns of magmatism along the Tethysides: a preliminary study. Journal of Geology, 101: 51–84.

Sharief, F.A., 1982. Lithofacies distribution of the Permian-Triassic rocks in the Middle East. Journal of Petroleum Geology, 4: 299–310.

Sharma, M., Basu, A.R. and Nesterenko, G.V., 1992. Temporal Sr-, Nd-, and Pb-isotopic variations in the Siberian flood basalts: implications for the plume-source characteristics. Earth and Planetary Science Letters, 113: 365–381.

Shaw, R.D., Zeitler, P.K., McDougall, I. and Tingate, P.R., 1992. The Paleozoic history of an unusual intracratonic thrust belt in central Australia based on 40Ar-39Ar, K-Ar and fission track dating. Geological Society of London Journal, 149: 937–954.

Sheldon, R.P., 1964. Paleolatitudinal and paleogeographic distribution of phosphorite. U.S. Geological Survey Professional Paper, 501-C: 106–113.

Sheng, J.-Z. and Jin, Y.-G., 1994. Correlation of Permian deposits in China. In: Y.-G. Jin, J. Utting and B.R. Wardlaw (Editors), Palaeontology and stratigraphy. Palaeoworld, 4: 14–113.

Shiraishi, K., Ellis, D.J., Hiroi, Y., Fanning, C.M., Motoyoshi, Y. and Nakai, Y., 1994. Cambrian orogenic belt in East Antarctica and Sri Lanka: implications for Gandwana assembly. Journal of Geology, 102: 47–65.

Stagg, H.M.J., 1985. The structure and origin of Prydz Bay and MacRobertson Shelf, East Antarctica. Tectonophysics, 114: 315–340.

Stampfli, G., Marcoux, J. and Baud, A., 1991. Tethyan margins in space and time. In: J.E.T. Channell, E.L. Winterer and L.F. Jansa (Editors), Paleogeography and paleoceanography of Tethys. Palaeogeography, Palaeoclimatology, Palaeoecology, 87: 373–409.

Subrahmanyam, C., 1978. On the relation of gravity anomalies to geotectonics of the Precambrian terrains of the South Indian Shield. Geological Society of India Journal, 19: 251–263.

Sun, D.-L., 1993. On the Permian biogeographic boundary between Gondwana and Eurasia in Tibet, China as the eastern section of the Tethys. Palaeogeography, Palaeoclimatology, Palaeoecology, 100: 59–77.

Surlyk, F., 1990. Timing, style and sedimentary evolution of Late Palaeozoic-Mesozoic extensional basins of East Greenland. In: R.F.P. Hardman and J. Brooks (Editors), Tectonic events responsible for Britain's oil and gas reserves. Geological Society, Special Publication 55, pp. 107–125.

Szabo, F. and Kheradpir, A., 1978. Permian and Triassic stratigraphy, Zagros Basin, south-west Iran. Journal of Petroleum Geology, 1: 57–82.

Szatmari, P., Carvahlo, R.S. and Simoes, I.A., 1979. A comparison of evaporite facies in the Late Paleozoic Amazon and the Middle Cretaceous South Atlantic salt basins. Economic Geology, 74: 432–447.

Tankard, A.J., Jackson, M.P.A., Ericksson, K.A., Hobday, D.K., Hunter, D.R. and Minter, W.E.L., 1982. Crustal evolution of southern Africa. Springer-Verlag, New York, 523 pp.

Taylor, E.L., Taylor, T.N. and Cùneo, N.R., 1992. The present is not the key to the past: a polar forest from the Permian of Antarctica. Science, 257: 1675–1677.

Tiwari, R.S. and Tripathi, A., 1987. Palynological zones and their climatic inference in the coal-bearing Gondwana of peninsular India. Palaeobotanist, 36: 87–101.

Trompette, R., 1983. Les Dahomeyides au Benin, Togo, Ghana et au sud du Niger et de la Haute--Volta. In: J. Fabre (Editor), Afrique de l'ouest: introduction géologique et termes stratigraphiques. Pergamon Press, New York, pp. 43–45.

Ufimtsev, G.F., 1986. Neotectonics of the Baykal Rift and inferences as to the structure. International Geology Review, 28: 969–976.

Uflyand, A.K., Natapov, L.M., Lopatin,V.M. and Chernov, D.V., 1991. The tectonic nature of the

Taymyr Peninsula. Geotectonics, 25: 512–523.

Van der Voo, R., 1993. Paleomagnetism of the Atlantic, Tethys and Iapetus oceans. Cambridge University Press, Cambridge, 411 pp.

Veevers, J.J. (Editor), 1984. Phanerozoic earth history of Australia. Clarendon Press, Oxford, 418 pp.

Veevers, J.J., 1994. Case for the Gamburtsev Subglacial Mountains of East Antarctica originating by mid-Carboniferous shortening of an intracratonic basin. Geology, 22: 593–596.

Veevers, J.J., Powell, C.McA., Collinson, J.W. and López-Gamundí, O.R., 1994. Synthesis. In: J.J. Veevers and C.McA. Powell (Editors), Permian--Triassic Pangean basins and foldbelts along the Panthalassan margin of Gondwanaland. Geological Society of America Memoir, 184: 331–354.

Vellutini, P. and Vicat, J.P., 1983. Sur l'origine des formations conglomeratiques de base du geosynclinal ouest-congolien (Gabon, Congo, Zaire, Angola). Precambrian Research, 23: 87–101.

Venkatakrishnan, R. and Culver, S.J., 1988. Plate boundaries in west Africa and their implications for Pangean continental fit. Geology, 16: 322–325.

Vinogradov, A.P. (Editor), 1969. Atlas of the lithological-paleogeographical maps of the U.S.S.R., vol. 2: Devonian, Carboniferous, and Permian. Akademia Nauk SSSR, Moscow.

Visser, J.N.J., 1987. The palaeogeography of part of southwestern Gondwana during the Permo-Carboniferous glaciation. Palaeogeography, Palaeoclimatology, Palaeoecology, 61: 205–219.

Visser, J.N.J., 1990. The age of the late Palaeozoic glacigene deposits in southern Africa. South African Journal of Geology, 93: 366–375.

Visser, J.N.J., this volume. A review of the Permo-Carboniferous glaciation in Africa.

Walter, H., 1985. Vegetation of the Earth. Berlin, Springer-Verlag, 318 pp.

Wang, H.-Z. (Compiler), 1985. Atlas of the palaeogeography of China. Cartographic Publishing House, Beijing, 143 pp.

Wang, Z.Q., 1985. Palaeovegetation and plate tectonics: palaeophytogeography of North China during Permian and Triassic times. Palaeogeography, Palaeoclimatology, Palaeoecology, 49: 25–45.

Webb, J.A. and Fielding, C.R., 1993. Permo-Triassic sedimentation within the Lambert Graben, northern Prince Charles Mountains, East Antarctica. In: R.H. Finlay, R. Unrug, M.R. Banks and J.J. Veevers (Editors), Gondwana eight: assembly, evolution, dispersal. A.A. Balkema, Rotterdam, pp. 357–369.

Wellman, P., 1987. Eastern highlands of Australia; their uplift and erosion. BMR Journal of Australian Geology and Geophysics, 10: 277–286.

Wellman, P. and Tingey, R.J., 1982. A gravity survey of Enderby and Kemp Lands, Antarctica. In: C. Craddock (Editor), Antarctic geoscience. IUGS Series B, 4, pp. 937–940.

Wickham, S.M., Litvinovsky, B.A., Zanvilevich, A.N. and Bindeman, I.N., 1995. Geochemical evolution of Phanerozoic magmatism in Transbaikalia, East Asia: a key constraint on the origin of K-rich silicic magmas and the process of cratonization. Journal of Geophysical Research, 100, No. B8, pp. 15641–15654.

Wopfner, H., 1981. Development of Permian intracratonic basins in Australia. In: M.M. Cresswell and P. Vella (Editors),Gondwana five: selected papers and abstracts of papers presented at the Fifth International Gondwana Symposium. A. A. Balkema, Rotterdam, pp. 185–190.

Wopfner, H. and Cassyhap, S.M., this volume. Transition from subpolar to subtropical climates in the Permo-Carboniferous in East Africa and India.

Wopfner, H. and Kaaya, C.Z., 1991. Stratigraphy and morphotectonics of Karoo deposits of the northern Selous Basin, Tanzania. Geological Magazine, 128: 319–334.

Wopfner, H. and Kreuser, T., 1986. Evidence for Late Palaeozoic glaciation in southern Tanzania. Palaeogeography, Palaeoclimatology, Palaeoecology, 56: 259–275.

Yarmolyuk, V.V., Kovalenko, V.I., Odone, M., Del Moro, A. and Tsareva, G.M., 1990. Petrology and geochemistry of the basalt-trachyrhyolite-comendite association of northern Mongolia. International Geology Review: 1218–1234.

Yemane, K., 1993. Contribution of Late Permian paleogeography in maintaining a temperate climate in Gondwana. Nature, 361: 51–54.

Yemane, K. and Kelts, K., 1990. A short review of palaeoenvironments for Lower Beaufort (Upper Permian) Karoo sequences from southern to central Africa: a major Gondwana Lacustrine episode. Journal of African Earth Sciences, 10: 169–185.

Young, R. and McDougall, I., 1993. Long-term landscape evolution: Early Miocene and modern rivers in southern New South Wales, Australia. Journal of Geology, 101: 35–50.

Yuzvitskii, A.Z., Budnikov, V.I., Budnikov, I. V. and Androsov, B.N., 1991. Regularities in the distribution of Upper Paleozoic coals in Siberian (Angara) province. Soviet Geology and Geophysics, 32: 73–76.

Zalán, P.V., Wolf, S., Conceição, C.J.C., Astolfi, M.A.M., Vieira, I.S., Appi, V.T., Zanotto, O.A., and Marques, A., 1991. Tectonics and sedimentation in the Paraná Basin. In: H. Ulbrich and A.C. Rocha-

Campos (Editors), Gondwana seven proceedings: papers presented at the Seventh International Gondwana Symposium. Instituto de Geociências-USP, São Paulo, pp. 83–117.

Zeck, H.P., Andriessen, P.A.M., Hansen, K., Jensen, P.K. and Rasmussen, B.L., 1988. Paleozoic paleocover of the southern part of the Fennoscandian Shield-fission track constraints. Tectonophysics 149: 61–66.

Zheng, Q.Q., 1985. The basic features of Emei Mountain Basalt and the relation of its mineralization in Guizhou. Geology of Guizhou, 2: 1–10.

Ziegler, A.M., 1990. Phytogeographic patterns and continental configurations during the Permian Period. In: W.S. McKerrow and C.R. Scotese (Editors), Palaeozoic palaeogeography and biogeography, Geological Society of London Memoir 12: 363–379.

Ziegler, A.M., 1993. Models come in from the cold. Nature, 361: 16–17.

Ziegler, A.M., Barrett, S.F. and Kazmer, C.S., 1982. The geologic history of the upwelling zones off western South America. In: Geology of phosphorite deposits and problems of phosphorite genesis, Akademia Nauk Siberskoe Otdelenie, Novosibirsk, pp. 38–40.

Ziegler, A.M., Rees, P.McA., Rowley, D.B., Bekker, A., Li, Q., and Hulver, M.L., 1995, in press. Mesozoic assembly of Asia: constraints from fossil floras, tectonics and paleomagnetism.

Ziegler, A.M., Rowley, D.B., Lottes, A.L., Sahagian, D.L., Hulver, M.L. and Gierlowski, T.C., 1985. Paleogeographic interpretation: with an example from the Mid-Cretaceous. In: G.W. Wetherill, A.L. Albee and F.S. Stehli (Editors), Annual Review of Earth and Planetary Sciences, vol. 13. Annual Reviews, Inc., Palo Alto, pp. 385–425.

Ziegler, A.M., Scotese, C.R., McKerrow, W.S., Johnson, M.E. and Bambach, R.K., 1979. Paleozoic paleogeography. In: F.A. Donath, F.G. Stehli and G.W. Wetherill (Editors), Annual Review of Earth and Planetary Sciences, vol. 7. Annual Reviews, Inc., Palo Alto, pp. 473–502.

Ziegler, P.A., 1988. Evolution of the Arctic-North Atlantic and the western Tethys. American Association of Petroleum Geologists, Memoir 43, 198 pp.

Ziegler, P.A. (Editor), 1990. Geological atlas of western and central Europe. Shell International Petroleum Mij, The Hague, 239 pp.

Zonenshain, L.P., Kuzmin, M.I. and Kononov, M.V., 1985. Absolute reconstructions of the Paleozoic oceans. Earth and Planetary Science Letters, 74: 103–116.

Zonenshain, L.P., Kuzmin, M.I. and Natapov, L.M., 1990. Geology of the USSR: a plate tectonic synthesis. In: B.M. Page (Editor), Geodynamics series 21. American Geophysical Union, 242 pp.

Zonenshain, L.P. and Natapov, L.M., 1989. Tectonic history of the Arctic region from the Ordovician through the Cretaceous. In: Y. Herman (Editor), The Arctic Sea. Van Nostrand Reinhold, New York, pp. 829–862.

APPENDIX 8.1 *Permian Geographic Names (numbers following geographic names in the text correspond to numbered positions on figures 8.1 - 8.4 and this appendix).*

Continent	Number	Name	Latitude	Longitude	Country
N. America	1	McCloud Arc	45	-117	U.S.A. (west)
N. America	2	Antler Belt	41	-117	U.S.A. (west)
N. America	3	Golconda Belt	39	-117.5	U.S.A. (west)
N. America	4	Rocky Mts. (Ancestral)	37	-107	U.S.A. (west)
N. America	5	Delaware Basin	33	-102	U.S.A., (southwest)
N. America	6	Williston Basin	47	-102	U.S.A., Canada
N. America	7	Alleghanian Mts.	40	-79	U.S.A. (east)
N. America	8	Grenville Mts.	52	-70	Canada (eastern)
N. America	9	Sverdrup Basin	77	-90	Canada (northern)
Greenland	10	Eskimonia High	76	-20	Greenland
Europe	11	Barents Sea	30	75	Norway, Spitsbergen, etc.
Europe	12	Scandes Mts.	66	15	Norway, Sweden
Europe	13	Oslo Dome	60	10	Norway, Sweden
Europe	14	North Sea	55	5	Gt. Britain, Denmark, etc.
Europe	15	Hercynian Mts.	45	3	Spain, France, etc.
Europe	16	Chisos Arc	38.5	26	Greece
N. Asia	17	Caucasian Arc	43	43.5	Georgia
N. Asia	18	Kyzylkum Upland	42	65	Uzbekistan
N. Asia	19	Karpinsky Swell	47.5	40	Ukraina, Russia
N. Asia	20	Precaspian Depression	50	50	Russia, Kazakhstan
N. Asia	21	Russian Seaway	50	50	Russia
N. Asia	22	Ural Mts.	55	60	Russia
N. Asia	23	Byrranga Range	75	103	Russia
N. Asia	24	Zaisan Mts.	48	84	Kazakhstan
N. Asia	25	Yenesei Ridge	60	92	Russia (Siberia)
N. Asia	26	Tunguska Basin	65	95	Russia (Siberia)
N. Asia	27	Kuznetsk Basin	53	85	Russia (Siberia

Continent	Number	Name	Latitude	Longitude	Country
N. Asia	28	Patom Uplands	59	115	Russia (Siberia)
N. Asia	29	Transbaikalian Uplands	52	114	Russia (Siberia)
N. Asia	30	Vilyui Basin	65	130	Russia (Siberia)
N. Asia	31	Omolon Block	64	160	Russia (eastern)
N. Asia	32	Mongolian Arc	46	110	Mongolia
China	33	Yin Shan Arc	41	115	China (north)
China	34	Ordos Basin	38	110	China (north)
China	35	Pan Xi Mts.	36	104	China (west)
China	36	Tarim Block	40	85	China (west)
China	37	Omeishan Plateau	27	103	China (south)
China	38	Sichuan Basin	30	105	China (south)
China	39	Yunkai Arc	22	110	China (south)
S.E. Asia	40	Indochinese Arcs	14	108	Vietnam, Thailand, etc.
S.E. Asia	41	Sibumasu Arc	4	101	Malaya. Thailand, etc.
China	42	Qiangtang Block	34	90	China (Tibet)
China	43	Lhasa Block	29.5	91	China (Tibet)
Near East	44	Iranian Block	34	58	Iran
Near East	45	Pelagonian Block	40	22	Greece, Turkey
Near East	46	Adria-Turkish Block	37	30	Greece, Turkey, etc.
S. America	47	Merida Basin	10	-70	Venezuela, Colombia
S. America	48	Amazon Basin	-5	-65	Brazil
S. America	49	Andean Mts.	-32	-70.5	Chile, Argentina, etc.
S. America	50	Asunción Arch	-25	-57.5	Paraguay, Brazil
S. America	51	Parana Basin	-25	-50	Brazil, Paraguay, etc.
S. America	54	Diamantina Range	-13	-42	Brazil
S. America	52	Veadeiros Range	-14	-48	Brazil
S. America	53	Maranhao Basin	-5	-45	Brazil
Africa	55	Mauritanide Belt	-12.5	16	Mauritania

Continent	Number	Name	Latitude	Longitude	Country
Africa	56	Atlas Mts.	29	-9	Morocco
Africa	57	Ougarta Mts.	29	-2	Algeria
Africa	58	Djeffara Basin	34	10	Tunisia, Libya
Africa	59	Iforas Mts.	20	2	Mali, Algeria
Africa	60	Air Massif	19	9	Niger
Africa	61	Zinder Hot Spot	14	10	Niger
Africa	62	Atacora Mts.	9	1	Benin, Togo
Africa	63	Gabon Basin	-0.5	10	Gabon
Africa	64	Mayombe Range	-5	13	Gabon, Congo,etc.
Africa	65	Windhoek Highlands	-22.5	17	Namibia
Africa	66	Cargonian Highlands	-27	24	S.Africa, Namibia, etc.
Africa	67	Karoo Basin	-30	25	South Africa
Africa	68	Cape Mts.	-35	20	South Africa
Africa	69	Lomagundi Highlands	-18	30	Zimbabwe, Mozambique
Africa	70	Muchinga Scarp	-13	31	Zambia
Africa	71	Makutu Mts.	-10.5	33.5	Zambia
Africa	72	Ruhuhu Basin	-10.5	35	Tanzania
Africa	73	Mitumba Range	-2	28	Zaire
Africa	74	Tanzanian Uplands	-7	35	Tanzania, Uganda
Africa	75	Uluguru Uplands	-7.5	37.5	Tanzania
Arabia	76	Arabian Basin	25	50	Saudi Arabia, etc.
Arabia	77	Akhdar High	23	57	Oman
Arabia	78	Huqf Axis	20	57	Oman
India	79	Panjal High	34	75	India (north)
India	80	Mahanadi Basin	23	83	India
India	81	Eastern Ghats	18	83	India (southeast)
Antarctica	82	Lambert Rift Shoulders	-70	70	Antarctica
Antarctica	83	Wegener-Mawson Belt	25	-124	Antarctica

Continent	Number	Name	Latitude	Longitude	Country
Antarctica	84	Ross High	-78	165	Antarctica
Australia	85	Perth Basin	-29	115	Australia (southwest)
Australia	86	Hammersley Range	-22.5	118	Australia (northwest)
Australia	87	Canning Basin	-20	123	Australia (northwest)
Australia	88	Kimberley Plateau	-17	127	Australia (north)
Australia	89	MacDonnell Ranges	-23.5	132.5	Australia (central)
Australia	90	Musgrave Range	-26	130	Australia (central)
Australia	91	Gawlor Range	-33	137	Australia (south)
Australia	92	Flinders Range	-32	139	Australia (south)
Australia	93	Lachlan Highlands	-36	148	Australia (southeast)
Australia	94	Sydney Basin	-33	151	Australia (southeast)
Australia	95	New England Mts.	-27	151	Australia (southeast)
Australia	96	Bowen Basin	-24	149	Australia (northeast)

9. GLACIAL-POSTGLACIAL TRANSITION IN THE LATE PALEOZOIC BASINS OF SOUTHERN SOUTH AMERICA

Oscar R. López-Gamundí

The apparent polar wander path and the accompanying migration of ice centers are considered to be first-order controls on the late Paleozoic glaciation at a supercontinental scale (Fig. 9.1; Crowell, 1983; Caputo and Crowell, 1985). This glaciation lasted at least 90 Ma and affected all Gondwana subcontinents (Hambrey and Harland, 1981).

Evidence from the stratigraphic record and paleomagnetic data suggest that the pole moved across South America, southern Africa, Antarctica, and Australia during the late Paleozoic. Three glacial episodes within this paleoclimatic mega-event have been identified in western Gondwanaland (Figs. 9.2, 9.3): (1) the latest Devonian-earliest Carboniferous glacial sediments present in the Solimões and the Amazonas basins of Brazil and the Titicaca Lake region of western Bolivia, (2) the well-documented mid-Carboniferous glacial deposits along the basins of south

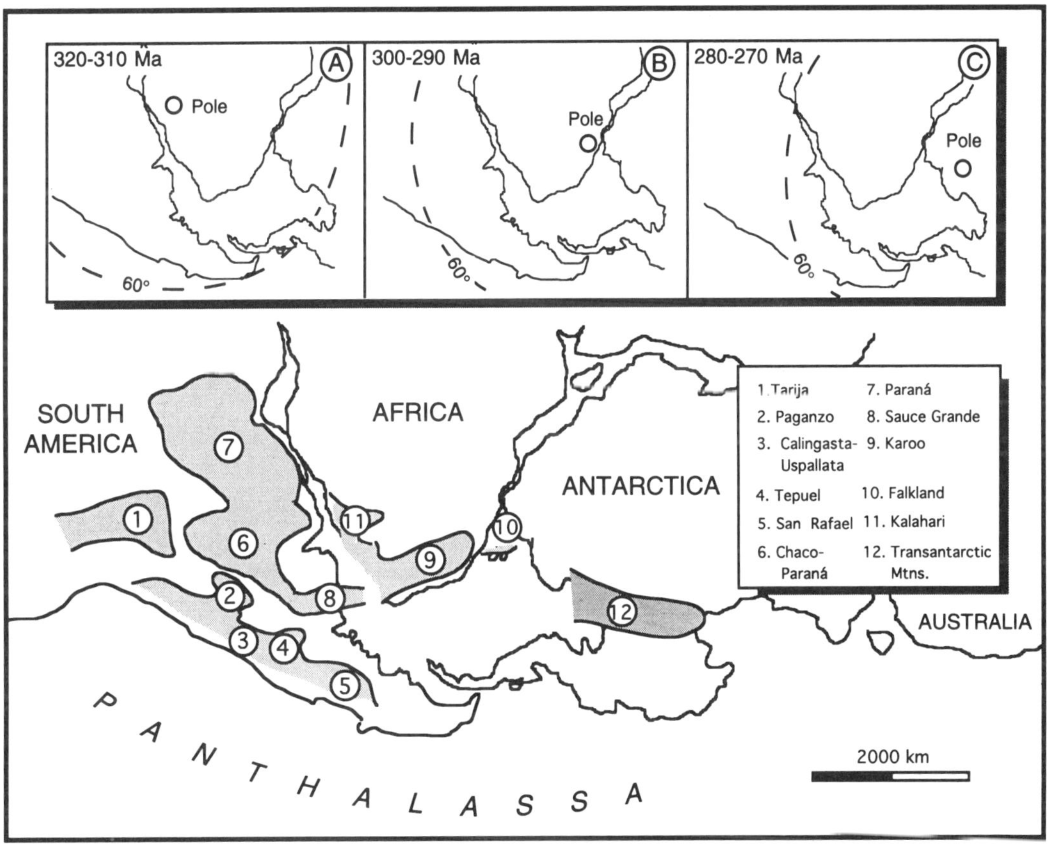

Fig. 9.1. Generalized late Paleozoic paleogeography of western Gondwanaland. Paleolatitudinal reconstructions and pole positions based on paleomagnetic data from (A) glacial deposits of the Tepuel basin, (B) glacial sediments of the Karoo basin, and (C) postglacial deposits of the Paganzo Basin (after Rapallini and Vilas, 1991; Rapallini , 1993).

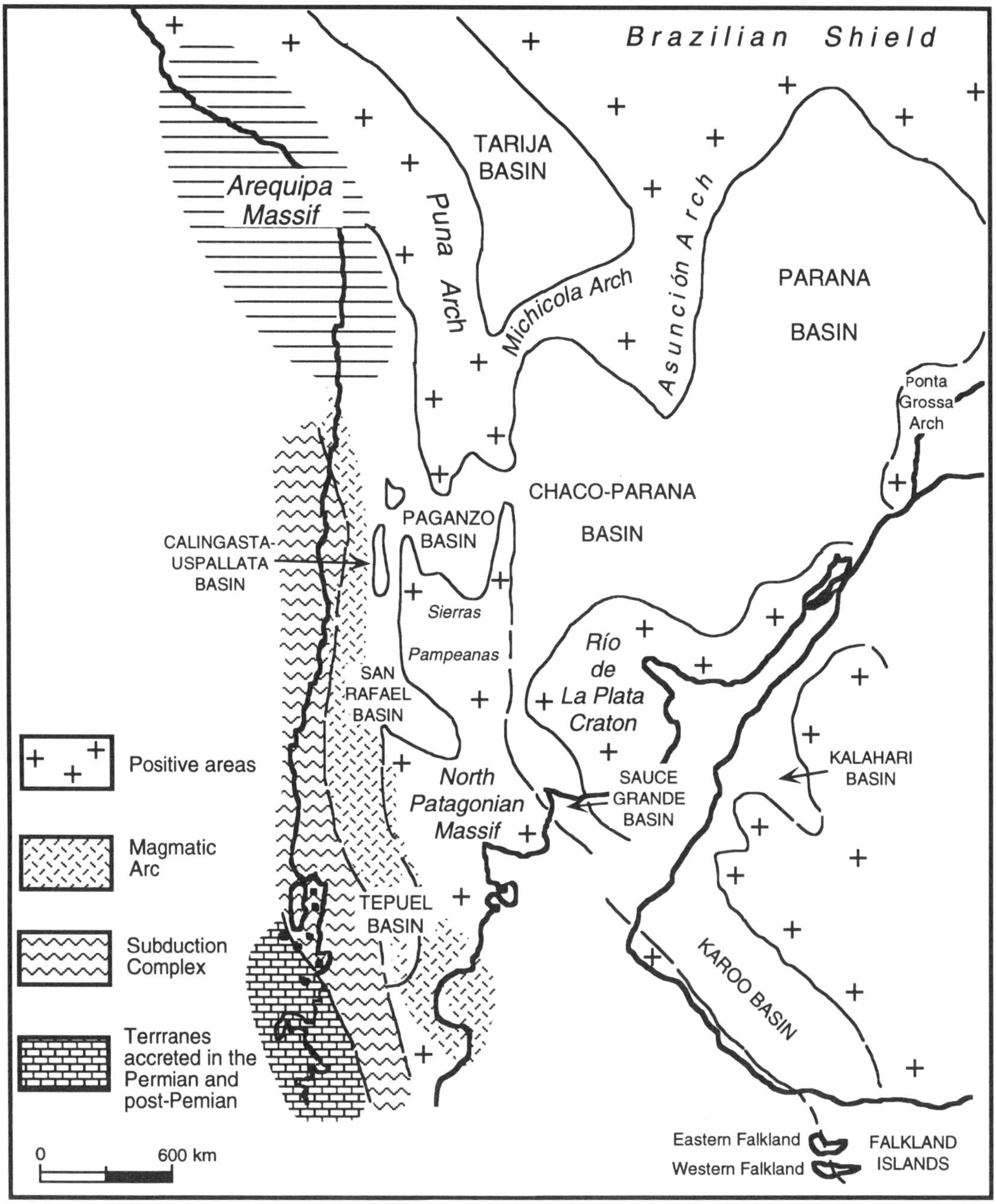

Fig. 9.2. Carboniferous-Early Permian paleogeography and tectonic setting of southern South America (After López-Gamundí and Rossello, 1993).

western South America and (3) the widespread Late Carboniferous (Stephanian)-Early Permian glacial sediments of eastern South America (the Paraná, Sauce Grande, and Falkland basins). The mid-Carboniferous glacial episode represents the onset of the Gondwana glaciation, which climaxed between the latest Carboniferous and the earliest Permian in eastern South America (Rocha-Campos, 1967; França and Potter, 1991) and South Africa (Visser, 1989) and ended in Australia by the latest Early Permian (Brakel and Totterdell, 1993).

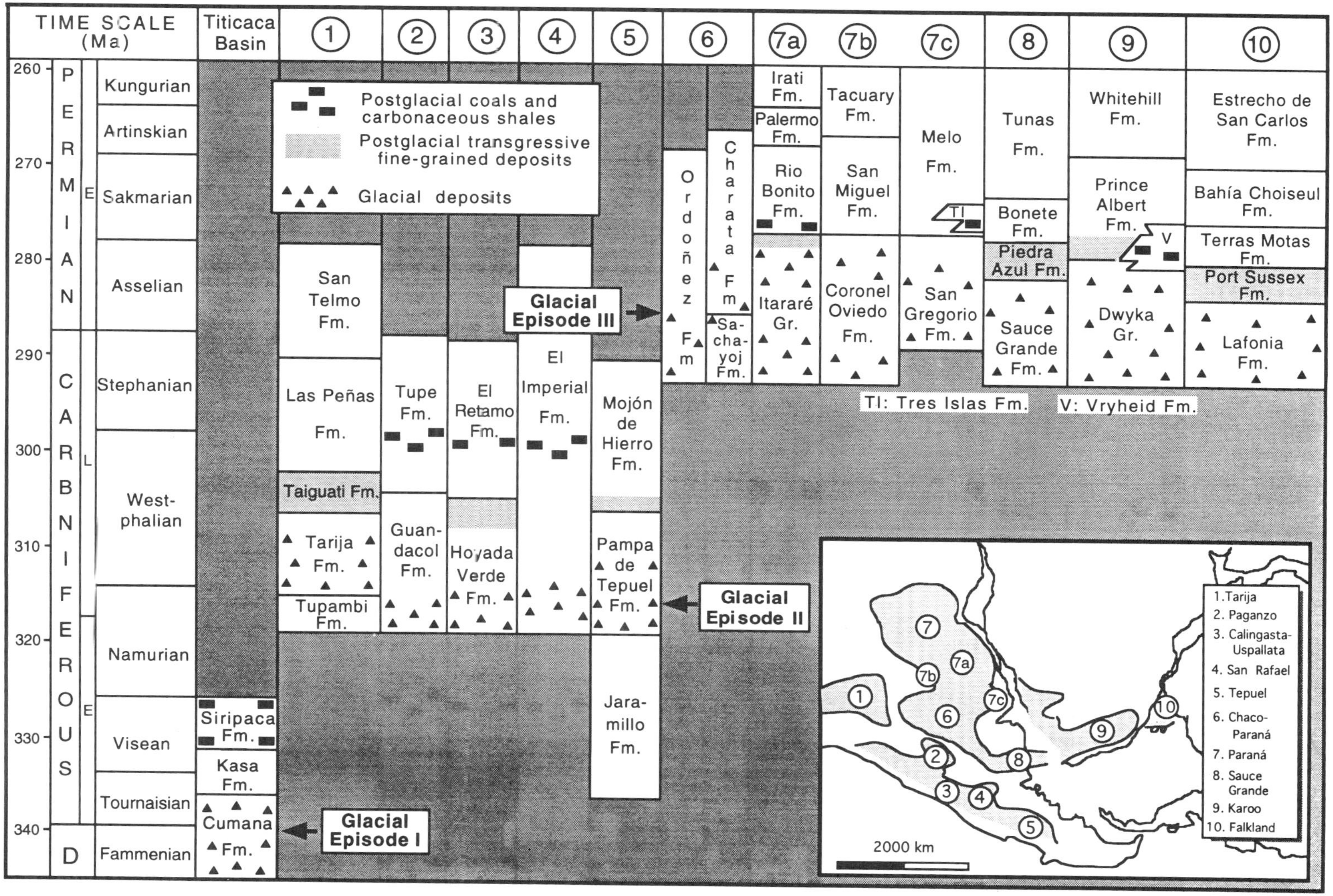

Fig. 9.3. Chronostratigraphic framework of the three glacial episodes across western Gondwanaland. Lithostratigraphic subdivisions of postglacial stratigraphy in the Titicaca, paleo-Pacific (1—5), Chaco-Paraná (6), and Karoo (9) basins have been simplified or omitted. Time scale from Palmer (1983).

In the last few years there has been substantial progress in characterizing paleogeographic and paleoclimatic aspects of the Gondwana glaciation. Chronocorrelation of glacial deposits has been improved as well, allowing a more accurate picture of the spatial and temporal evolution of this glaciation across Gondwanaland. This chapter provides an up-to-date review of the glacial-postglacial transition in the late Paleozoic basins of southern South America, presenting information not only from South American basins but also from South Africa. This permits the assessment of the influence of the Gondwana Ice Sheet, probably centered between South Africa and Antarctica during most of its existence, on the late Paleozoic basins of southern South America.

The Early Record of the Gondwana Glaciation: Early to Mid-Carbonifeous in the Andean Basins

Glacial Episode I (Late Devonian-Early Carboniferous)

The oldest record of the Gondwana glaciation in southern South America seems to be in the Titicaca Lake region of northwestern Bolivia (Fig. 9.3). Compressional tectonics seems to have triggered this first glacial episode. The latest Devonian (Late Famennian)-Early Carboniferous (Tournaisian-Viséan) Eo-Hercynian compression (Dalmayrac et el., 1980) partly coincided in time with this glacial episode in western Bolivia, where emersion was complete by the mid-Carboniferous (Namurian) (Sempere, 1989). This orogeny, locally known in Argentina as the Chañic phase, is represented by a strong angular unconformity in the "paleo-Pacific" or Andean basins of southern South America. This unconformity separates Late Devonian or older marine deposits from Early or, more commonly, mid-Carboniferous sediments. Despite the collision models that have been invoked to explain the Chañic orogeny (Ramos et al., 1986) and its influence on the onset of glaciation (Eyles, 1993), little evidence supports this hypothesis. The uplift of a large segment of the South American Gondwana margin could have been caused by orogenic movements related to the convergence of the Arequipa Massif and the South American continent by Late Devonian-Early Carboniferous times (Fig. 9.2; Dalmayrac et al., 1980; López-Gamundí and Rossello, 1993). The orogeny, accompanied by arc magmatism, resulted in deformation and uplift as shown by the scarcity of Early Carboniferous deposits.

Glaciomarine deposits have been recently described in the Late Devonian (Famennian)-earliest Carboniferous (Tournaisian) Cumaná Formation exposed in the Titicaca Lake region (Díaz Martínez et al., 1993a). The glacigenic section consists, from base to top, of laminated shales with dropstones, diamictites with striated clasts and deformed sandstone lenses, subordinate massive and graded conglomerates, and rippled and cross-bedded sandstones (Díaz Martínez et al., 1993a). The glaciomarine rocks of the Cumaná Formation are overlain by the predominantly deltaic deposits of the Kasa Formation. This unit consists of slope and delta front, often slumped, diamictites, shoreface to offshore laminated and bioturbated mudstones and sandstones with hummocky cross-stratification, and delta plain siltstones and channelized sandstones with unidirectional cross-bedding (Díaz Martínez et al., 1993b). The deposits have been interpreted as part of a fan delta complex fed by a proglacial outwash streams (Díaz Martínez et al., 1993b). The Kasa deposits are overlain transitionally by floodplain carbonaceous mudstones, coals, and thin sandstone beds of the Siripaca Formation. An Early Carboniferous (Viséan-Early Namurian) age is suggested for these coals (Rösler et al., 1989).

Although more detailed work is needed, this early glacial episode seems to be linked to similar glacially-influenced marine deposits of latest Devonian age described from the Amazonas and the Solimões basins of the Amazonan region of western Brazil (Rocha-Campos, 1981; Caputo, 1984) and included by Veevers and Powell (1987) in their glacial episode I. Sablock (1993) favors the presence of multiple but separate small, elevated glacial source areas, rather than sizable ice centers or ice sheets, for this glacial episode that followed the generally equable Middle to Late Devonian climatic conditions.

Glacial Episode II (Mid-Carboniferous)

A well-documented glacial episode waxed and waned along the southwestern margin of South America during the mid-Carboniferous (López-Gamundí, 1987; López-Gamundí and Espejo, 1993). The record of this glacial episode is present in the Tarija, Calingasta-Uspallata, Paganzo, San Rafael, and Tepuel basins (Figs. 9.1, 9.2, 9.3). Although all these basins formed along the active continental margin of southwestern Gondwanaland, they differ in geotectonic settings and subsidence regimes. The Tarija Basin has a thick late Paleozoic section and was developed as a foredeep between the Brazilian shield and the Arequipa Massif (Fig. 9.2). The Calingasta-Uspallata, Paganzo and San Rafael basins can be considered back-arc to foreland basins formed behind an active magmatic arc. Relatively thin (commonly less than 100m) glacigenic sequences characterize the late Paleozoic fill of these basins. In contrast, the Tepuel Basin hosts the thickest record of glacially-influenced marine deposits (about 2,000 m). High subsidence rates can be inferred for this basin, most probably developed in an intra- to fore-arc position (Fig. 9.2).

Paleogeographic reconstructions for the mid-Carboniferous indicate several possibly disconnected ice centers that supplied detritus into these basins. Paleo-ice flow directions suggest a general pattern from the continental interior toward the continental margin (Fig. 9.4A). Local changes to this pattern were provided by subsidiary, albeit independent, ice centers that dumped additional glacial detritus into the basins (Paganzo, Tarija, and Tepuel basins).

Tarija Basin

Glacial deposits are abundant in the Tarija Formation (Frakes and Crowell, 1969; Helwig, 1972). This unit, identified in surface and subsurface occurrences, ranges in thickness from 500 to 600 m and grades laterally into the sandstone-rich Chorro Formation in the northwestern sector of the basin in Bolivia.

The prevailing facies throughout the Tarija Formation is massive to poorly bedded diamictites. Subordinate facies are thin-bedded diamictites, laminated mudstones, intraformational conglomerates, and, increasingly abundant toward the top of the unit, channelized, trough cross-bedded, medium-sized sandstones (López-Gamundí, 1985). Dropstones in the basal shales of the formation (Itacuami Formation) suggest ice-rafting (Starck et al., 1993). The diamictites, with some scattered striated clasts (Frakes and Crowell, 1969; Helwig, 1972; Fernández Garrasino, 1981; López-Gamundí, 1985) contain several horizons of slumped sandstone beds.

Although the presence of a striated pavement carved into the Tupambi Formation sandstones indicates local grounded conditions (Starck et al., 1993), the bulk of the Tarija diamictites was deposited by gravity flows of glacially derived material triggered by steep depositional slopes and high rates of sediment supply. Such conditions existed especially along the eastern margin of the basin (Fig. 9.4; Schlatter and Nederloff, 1966).

Along the southern margin of the basin, the Tarija deposits are overlain by deltaic sand-rich deposits of the Las Peñas (=Escarpment) Formation. This contact has been interpreted as an erosional unconformity with evidence of subaerial exposure. Basinward, the Taiguati Formation mudstones, fine-grained sandstones, and subordinate diamictites onlap the upper part of the Tarija sediments. Fossils of the mid-Carboniferous (late Namurian-Early Westphalian) *Levipustula* zone are present in the Taiguati Formation (Rocha-Campos et al. 1977). Channels of the Las Peñas fluvio-deltaic complex downcut deeply the Tarija-Taiguati deposits forming extensive northward-draining paleovalleys identified in subsurface studies (Suárez-Soruco, 1986).

Calingasta-Uspallata and Paganzo Basins

A combination of high latitude and altitude enhanced the formation of ice centers along the margins of the Calingasta - Uspallata and the

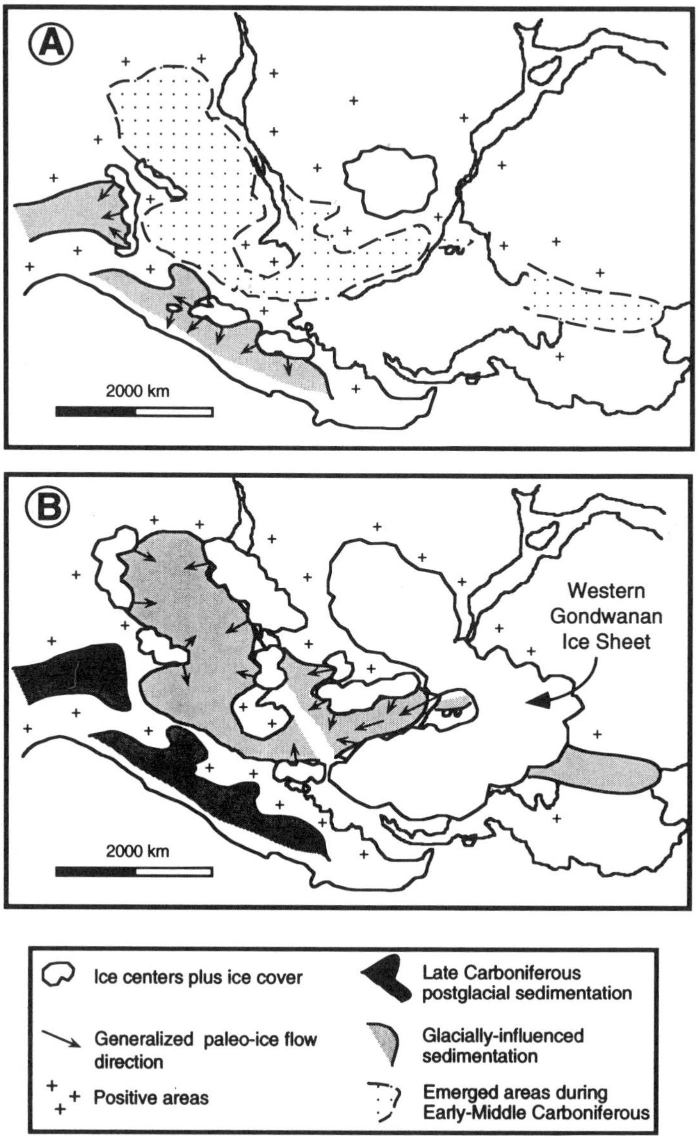

Fig. 9.4. A: Mid-Carboniferous (Namurian-Early Westphalian) paleogeography (glacial episode II). Glacially influenced sedimentation is confined to basins along the paleo-Pacific margin. Isolated ice spreading centers along the cratonic margins of these basins shed glacial detritus in mostly marine environments. Ice covers may have been present over elevated areas of central Africa. The rest of western Gondwanaland remained uplifted until sedimentation started by the end of the Carboniferous (Stephanian) (glacial episode III). Generalized paleo-ice flow directions from Frakes and Crowell (1969), Helwig (1972), López-Gamundí (1987). Western Antarctica (dashed line) not restored to its prebreakup position. B: Late Carboniferous (Stephanian)-earliest Permian paleogeography (glacial episode III). The growth and amalgamation of ice spreading centers in Africa and Antarctica led to the formation of the Western Gondwanan Ice Sheet. The Western Gondwana Ice Sheet and its satellite ice centers influenced sedimentation from the Karoo Basin of South Africa to the Chaco-Paraná and Sauce Grande basins of eastern Argentina. Glacially influenced sedimentation started in these basins when glacial conditions disappeared along the paleo-Pacific basins. Generalized paleo-ice flow directions from Frakes and Crowell (1969), França (1987) and Visser (1987). Western Antarctica (dashed line) not restored to its prebreakup position.

Paganzo basins of western Argentina. Mid-Carboniferous glacial deposits constitute the basal fill of both basins (Fig. 9.3). Detailed facies analyses suggest a complex glacial paleogeography strongly influenced by an irregular paleotopography. This paleorelief resulted from erosion subsequent to the uplift of the region during the latest Devonian-earliest Carboniferous and was later modified by glacial abrasion. Grounded ice conditions, evinced by a number of striated pavements carved in the basement (Martínez, 1993; Milana and Bercowski, 1993) and within the Carboniferous sequence (López-Gamundí and Rossello, 1995) and boulder pavements (González, 1981; López-Gamundí, 1983), suggest paleo-ice flow toward the northwest with some significant divergence from this general pattern indicative of local paleoslopes (Fig. 9.5).

The basal deposits of the Calingasta-Uspallata Basin range from proximal glaciomarine deposits to glacially influenced sediment gravity flows (Fig. 9.6). The principal facies are massive to poorly stratified sandy or muddy diamictites, well-stratified muddy diamictites, pebbly (dropstone) shales, and turbidites interpreted as the product of a complex interplay of glaciomarine sedimentation (sediment rain-out and ice-rafting) and gravity-induced reworking of glacial material (López-Gamundí, 1987). The areal extent of the outcrops along the eastern flank of the Calingasta-Uspallata Basin suggests an apron-like configuration along the eastern margin of the basin with glaciated areas located on the Proto-Precordillera high immediately to the east (Fig. 9.5). The predominant glacial association in the contiguous Paganzo Basin was deposited in a

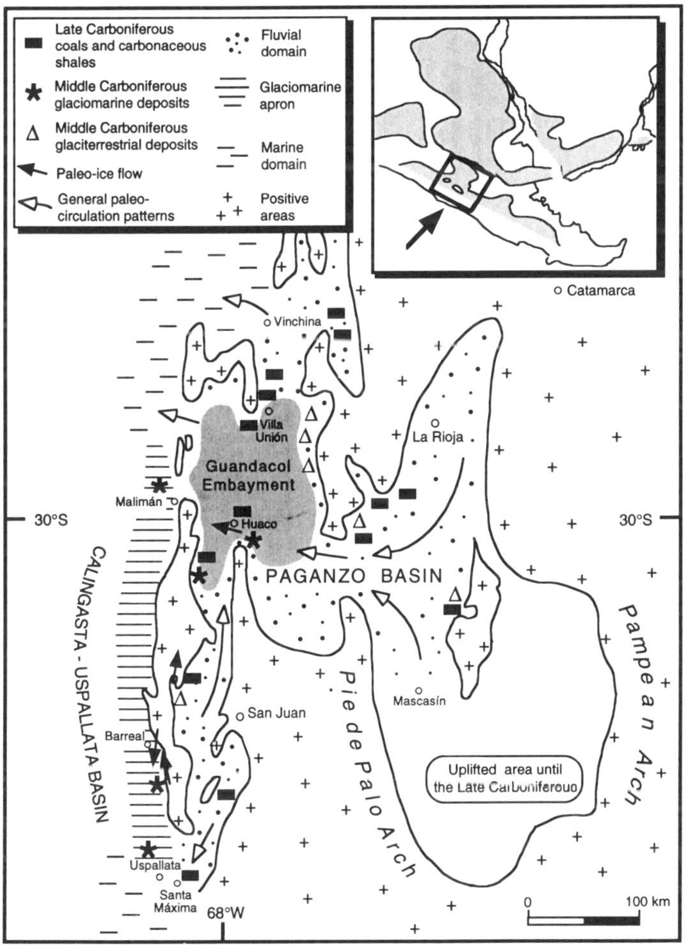

Fig. 9.5. Middle-Late Carboniferous paleogeography of Paganzo and Calingasta-Uspallata basins (western Argentina) (after López-Gamundí et al., 1989).

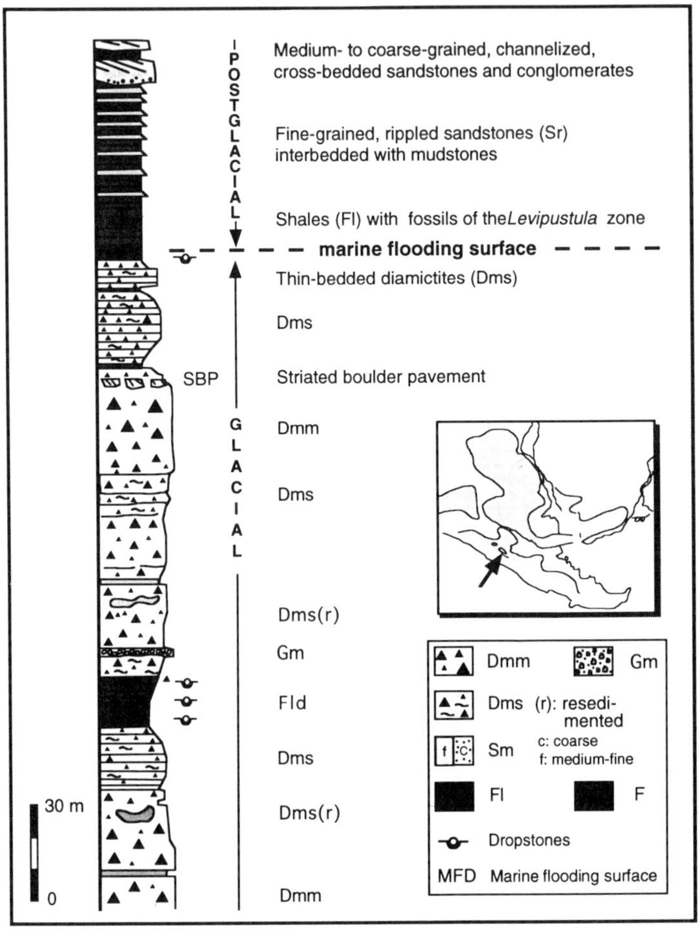

Fig. 9.6. Glacial Episode II. Glacial-postglacial transition in the Calingasta-Uspallata Basin near Barreal (see figure 9.5) (after López-Gamundí, 1983). Lithofacies code after Eyles et al. (1983).

brackish seaway (Guandacol embayment) open to the west in present coordinates (Fig. 9.4). It is characterized by massive and crudely bedded diamictites, thin-bedded diamictites, and shales with ice-rafted material (dropstone facies). The thin-bedded diamictites resulted from the combined action of subaqueous debris flows and ice-rafting. Intervening mudstones and calcareous shales contain a shelly fauna of brachiopods (Martínez, 1993) and marine microplankton (Ottone, 1991) of mid-Carboniferous age.

Mid-Carboniferous glaciterrestrial deposits have been identified in the north-central part of the Paganzo Basin (Limarino and Cesari, 1988; Limarino and Gutiérrez, 1990). Glaciolacustrine deposits have been identified only along a narrow strip of exposures adjacent to the northeast basin margin. These glaciolacustrine deposits contain laminated shales with dropstones, interbedded fine-grained turbidites, and diamictites. The massive diamictites and associated shales with dropstones are interpreted as deposited in a lacustrine environment from rain-out and settling of fines and ice-rafting respectively. These deposits are associated with clast-supported conglomerates and cross-bedded sandstones deposited in subaqueous fans (Limarino and Gutiérrez, 1990). Lacustrine dropstone shales seem to extend to the south in the western Paganzo Basin. In the southwest, several probably lacustrine horizons at or near the base of the late Paleozoic succession contain dropstones (Ortiz and Zambrano, 1981).

The glacigenic section in the Paganzo Basin is conformably overlain by a deltaic facies consisting of mouth bar sandstones, mudstones, carbonaceous shales and coals (Tupe Formation, Fig. 9.3). In the western Paganzo Basin coals are overlain by fluvio-deltaic facies of channelized cross-bedded sandstones, channel-margin rippled, fine-grained sandstones, and predominantly massive mudstones deposited in a floodplain paleoenvironment (Limarino and Cesari, 1988). The coals are thin (rarely thicker than 2 m), have low sulfur content (0.3–3% in weight) and range between sub-bituminous (brown coal) and bituminous (hard coal). Geochemical analyses indicate a predominantly woody-coaly kerogen type derived from terrestrial plants and a total organic content (TOC) up to about 54 percent (Villar and López-Gamundí, 1993). Both micro- and megafloristic assemblages of Late Carboniferous age (Late Westphalian-Stephanian) are associated with the coals. The megaflora, known as *NBG* (*Nothorhacopteris*, *Botrychiopsis* and *Ginkgophyllum*), is made up of lycophytes, sphenophytes, progymnosperms, pteridosperms, and Cordaitales.

San Rafael Basin

The glaciomarine facies of the San Rafael Basin are represented in the Lower Member of the El Imperial Formation (Espejo, 1990). Studies in progress indicate that the glacial section attains a maximum thickness of 60 m and consists of two principal facies: a diamictite facies and a pebbly shale (dropstone) facies. The diamictite facies consists of massive and stratified pebbly mudstones. The vertical transition from massive to stratified mudstones, the latter commonly with crude lamination and ice-rafted clasts, suggests a rain-out origin for these deposits. Intraclasts of diamictitic material are common in the stratified pebbly mudstones. Thin, isolated horizons of rippled sandstones indicate size segregation/sorting by sporadic traction bottom currents. The dropstone facies consists of laminated mudstones with abundant dropstones up to 30 cm in diameter. Thin-bedded fine-grained sandstones, often affected by soft-sediment deformation, are interbedded with the mudstones. The glacial association has a reduced thickness (20–30 m) toward the basin margin (Espejo, 1990). This condensed section consists of composite sandstone-conglomerate beds (probably outwash deltas) with intercalations of fine-grained sandstones with dropstones.

A mid-Carboniferous (Namurian) age has been assigned to the glacial section (Espejo, 1990) on the basis of the abundance of *Cristatiporites* sp. monosaccate pollen grains in the overlying shales, which in turn pass upward to fluvial facies with locally abundant coal seams of Late Carboniferous age (García, 1992). These coals have the same micro- and macrofloral assemblages, age, and stratigraphic position as those in the Paganzo Basin.

Tepuel Basin

The glacial deposits grouped in the Pampa de Tepuel Formation (Fig. 9.3) consist of diamictites (pebbly mudstones) and dropstone shales associated with laminated claystones and mudstones, fine-grained turbidites, medium- to coarse-grained sandstones, and clast-supported conglomerates. Although most diamictites have been interpreted as shallow-marine gravity flow deposits, the presence of striated clasts and dropstones in some shales suggest ice-rafting during sedimentation (Frakes and Crowell, 1969; López-Gamundí and Limarino, 1984). González Bonorino et al. (1988) have described 60 m-thick sequences of massive to crudely bedded diamictites in the Pampa de Tepuel Formation, interpreted as basal subglacial deposits followed by glaciomarine diamictites deposited in the inner shelf. Similar deposits of laterally extensive beds of massive diamictites up to 25 m thick have been alternatively interpreted as "line-source" debris flows, part of debris aprons developed in glaciated shelves (Eyles, 1990). The diamictite-bearing section is overlain by shales, mudstones, and fine-grained sandstones of the Mojón de Hierro Formation.

Glacial Episode III (Late Carboniferous-Early Permian)

Late Carboniferous-Early Permian glacial deposits are present in the Paraná, Chaco-Paraná, and Sauce Grande basins and in the Falkland (Malvinas) Islands (Figs. 9.1, 9.2). The glaciomarine deposits of the Sauce Grande Basin

(Sauce Grande Formation) of eastern Argentina and the Falkland Islands (Lafonia Formation) are overlain by Early Permian (Sakmarian) fine-grained deposits with fossils of the *Eurydesma* fauna and *Glossopteris* flora (Fig. 9.3) and correlatable in the Karoo and Kalahari basins of South Africa (Visser, 1989). This glacial-postglacial transition can be related to sudden sea-level rises associated with the decoupling of the Gondwana Ice Sheet from its substratum and its subsequent disintegration. The self-destructive collapse of this marine ice sheet started when high relative sea level peaked and ice retreat became irreversible (Visser, 1991a). The stratigraphic record of this climatic change is provided by the sharp contact between diamictites deposited by dense debris rain-out (topmost Dwyka Formation) and superjacent mudstones deposited by suspension settling and dilute turbidity flows (lowermost Prince Albert Formation; Visser, 1991a). This transgression indicates the transition from a marine ice sheet phase to a mountain ice cap phase characterized by the presence of tidewater glaciers at the heads of valleys along the basin margins, as documented for the northern margin of the Karoo Basin (Visser, 1991b).

Chaco-Paraná Basin

The glacial deposits of the Chaco-Paraná Basin are poorly known due to the lack of outcrops and the limited number of wells that have penetrated the glacigenic section. In contrast to the widespread marine influence evident during glacial sedimentation in the Paraná Basin, glaciterrestrial conditions seem to have been common in the Chaco-Paraná Basin during the Late Carboniferous and the earliest Permian. This basin was connected with the Paraná Basin to the east and the Sauce Grande Basin to the south (Fig. 9.2). Glaciterrestrial diamictites and lacustrine deposits, with subordinate marine intercalations, have been identified in several wells. Some associated facies found in the dominantly diamictitic section of a few wells indicate the presence of amalgamated sandstone beds interpreted as outwash-braidplain systems. Intercalated lacustrine mudstones contain land-derived woody material.

Recent biostratigraphic studies on pollen grains and spores indicate a Late Carboniferous-earliest Permian age for the diamictite-bearing Charata and Ordoñez formations (Russo and Archangelsky, 1987; Archangelsky and Vergel, 1991) and deposition under cold climatic conditions during the Late Carboniferous (Malizzia et al., 1993).

Paraná Basin

The Late Paleozoic glacigenic deposits of the Paraná Basin are the best known in South America. Outcrop belts along the basin margins and a dense exploratory well program have helped researchers define accurately the characteristics of the sedimentation in the Paraná Basin. Late Paleozoic glacially influenced sedimentation in this large intracratonic basin probably started by the Late Carboniferous (Westphalian). This glacial episode ended by the late Early Permian (Kungurian), according to palynological information (Daemon et al., 1991 in França, 1994), although data from invertebrates and megaflora indicate its demise by the Early Permian (Sakmarian) (Rocha-Campos and Rösler, 1978). Exposures of glacial sediments are common along the basin margins and have received extensive analysis (Rocha-Campos, 1967; Gravenor and Rocha-Campos, 1983). Only a brief summary of the salient features of the glacial and postglacial sedimentation in the central portion of the basin (southern Brazil) will be given in this chapter, Recent, subsurface information has provided a more comprehensive picture of the evolution of the glacial sedimentation in the basin (França, 1987; França and Potter, 1991; Eyles et al., 1993). The glacial-postglacial transition along the southern (Uruguay) and western (Paraguay) margins of the basin, only recently documented in the literature, is treated in some detail later in this chapter.

Central Paraná Basin. Glacial deposits are grouped under the Itararé Group (Fig. 9.3). These deposits reach 1,400 m in the deepest parts of the basin. On the basis of subsurface information, the Itararé Group has been subdivided into four formations: the Lagoa Azul, Campo Mourao, and Taciba formations in the central southern part and the Aquidauana Formation in the northwestern part of the basin (França and Potter, 1991). Each of the first three formations represents a deepening succession (Eyles et al., 1993) of basal turbidites overlain by debris flow and rain out diamictites.

The uppermost unit (Taciba Formation) has been subdivided into two members: the basal Rio Segredo Member sandstones and the Chapeu do Sol Member diamictites. The diamictites are the lateral equivalents of the Rio de Sul Member shales present in the central southern part of the basin (França, 1987). Grounded conditions in the form of striated pavements of bedrock (including roches moutonnée) and boulder pavements along the margins (Rocha-Campos et al., 1976; Santos, 1987) suggest that glaciers were probably restricted to the basin margins, although floating ice had extended further in the basin, as indicated by abundant dropstones in the fine-grained deposits (turbiditic, varve-like rhythmites and shales).

In the southern half of the basin, the glacigenic section of the Itararé Group is conformably overlain by deltaic sandstones of the Rio Bonito Formation. Thin (up to 50 cm) coal seams in the Itararé Group are considered as part of interglacial/interstadial sequences (Martini and Rocha-Campos, 1991). Thicker, early postglacial *Glossopteris*-bearing coals are found in the Rio Bonito deposits. They are derived from peatlands developed in paleovalleys or associated with deltaic and coastal plains (Martini and Rocha-Campos, 1991). The coals are sub-bituminous, ash-rich (40–45 %), and sulphur-poor and have a predominance of vitrinite compared to liptinite and inertinite (Mitzarek et al., 1993). Although type III kerogen is predominant, some analyses indicate a minor contribution from algae and bacteria (Henz et al., 1987).

Southern Margin of the Paraná Basin. The thick glacial section of the central part of the basin thins onto the Sur Riograndense Arch along the southern margin (Fig. 9.2). Known as the San Gregorio Formation in Uruguay (Fig. 9.3), it consists of a thin sequence of basal massive diamictites, slumped massive sandstones, rippled sandstones, shales with dropstones and scarce conglomerates (Ferrando and Andreis, 1990a). The presence of cephalopods, radiolaria and fish remains in the shales interbedded with the diamictites indicate a glaciomarine setting for this unit.

The basal diamictites show a variable thickness (10–40 m) due to the irregular bedrock topography. Striated pavements and roches moutonnées sculptured on crystalline basement indicate a general paleo-ice flow direction from east to west (Ferrando and Andreis, 1990a). The diamictites are massive, with poorly defined bed contacts. Associated massive sandstones show soft sediment deformation or ripples and lenticular beds. This basal suite grades upward to thin-bedded diamictites and shales, commonly with ice-rafted material. This association has been interpreted as lodgement and flow tills, predominating in the basin lows, which evolved upward to rain-out diamictites, gravity flow deposits, and settling from sediment plumes and ice rafting. The resedimented diamictites are crudely to well bedded and are commonly associated with channelized clast-supported conglomerates. Varve-like rhythmites overlie the diamictite-bearing section along the southernmost margin of the basin (Ferrando and Andreis, 1990a). They consist of shales, siltstones, and scarce fine-grained sandstones. The glacial section grades upward to the deltaic deposits of the Tres Islas Formation (Fig. 9.3), which in turn grade upward into the finer-grained, predominantly littoral (tidal) sediments of the Melo Formation. The deltaic Tres Islas deposits consist of coarse- to fine-grained sandstones with subordinate conglomerates and mudstones. Some of the shales are carbonaceous, although TOC values are generally below 5 percent (Andreis et al., 1991). Abundant trough cross-bedding in the coarse-grained sandstones indicates a paleocurrent flow near the west (Andreis et al., 1991). Sandstones tend to be common near the base of the Melo Formation and scarce to absent toward the top. This unit culminates with shales and heterolithic (sandstone-shale) facies. The fining upward trend has been interpreted as a response to a relative sea-level rise (Ferrando and Andreis, 1990b).

Western Margin of the Paraná Basin

Scattered outcrops on the eastern flank of the Asunción arch define the western margin of the Paraná Basin. Facies reconstructions and thickness variations indicate that this arch was active during glacial and postglacial sedimentation. The glacial deposits are known as the Coronel Oviedo Formation, equivalent to the Itararé Group in the Brazilian sector (Fig. 9.3).

The glacial deposits are overlain by the sandstone-dominated San Miguel Formation (Thornburg et al., 1993). The transition between the glacially influenced deposits and the postglacial San Miguel sediments is defined by the uppermost dropstone found in the muddy bioturbated upper section of the Coronel Oviedo, probably equivalent to fine-grained sediments of the Rio do Sul Member. Along the exposed margin of the basin, the San Miguel Formation consists of 40–100m-thick, progradational tide-dominated parasequences of offshore marine, distal to proximal marine-tidal bars, and estuarine distributary or tidal channel facies (Thornburg et al., 1993). Basinward, the lower part of the unit consists of bay mudstones and delta front and distributary channel sandstones (Thornburg et al., 1993). Coal-measures, common in the equivalent Rio Bonito Formation along the southeastern margin of the basin in Brazil, are absent in the San Miguel Formation. The San Miguel deposits pass upward transitionally to the mudstones of the Tacuary Formation, a basin margin equivalent of the Irati Formation (Fig. 9.3).

Sauce Grande Basin

Glacial deposits in the Sauce Grande Basin have been reported since the early years of this century (Keidel, 1916; Du Toit, 1927). Sedimentation started in the basin with the diamictite-bearing Sauce Grande Formation. This unit, ranging in thickness from 500–1,100 m, consists of massive to crudely stratified diamictites with soft-sediment deformation and subordinate sandstones and shales. The lower half to two thirds of the unit are dominated by massive diamictites with scarce sandstone lenses (Fig. 9.7). Striated and faceted clasts have been reported from the diamictites (Keidel, 1916; Harrington, 1947; Coates, 1969; Amos and López-Gamundí, 1981). Interbeds of rippled sandstones, conglomerates, and shales with dropstones become common toward the top of the unit (Andreis et al., 1987). The Sauce Grande deposits pass upward gradationally to shales and fine-grained sandstones of the Piedra Azul Formation, which in turn grade into bioturbated mudstones and rippled and cross-bedded sandstones with fossils of Early Permian *Eurydesma* fauna and *Glossopteris* flora. A similar succession has been encountered in an offshore well southeast of the Sierras Australes. There, the diamictite-bearing section is overlain by shales, siltstones, and claystones that correlate with the Piedra Azul and the Bonete formations. Spores found in the diamictite-bearing section of this well confirm a latest Carboniferous-earliest Permian age (Archangelsky et al., 1987).

The diamictites have been interpreted as glaciomarine deposits (mostly rain-out tills) subjected to downslope gravity remobilization (Coates, 1969; Andreis, 1984). The strong bimodal bipolar pattern of the paleocurrents in the Bonete Formation indicates a shallow marine environment, mostly below effective wave base with sporadic wave action (Andreis et al., 1979).

Falkland (Malvinas) Islands

Late Paleozoic rocks are exposed in both Falkland Islands. Detailed facies analyses of the glacial deposits (Lafonia Formation) have been carried out by Frakes and Crowell (1967), Scasso and Mendía (1985), and Bellosi and Jalfin (1984, 1987). Several lines of evidence confirm Adie's (1952) original proposal that the pre-breakup position of the Falkland Islands was immediately east of South Africa until at least the Early Jurassic (Fig. 9.2). Paleomagnetic (Mitchell et al., 1986) and radiometric (Rex and Tanner, 1982; Mussett and Taylor, 1994) data and paleogeographic inferences (Visser, 1987; Marshall, 1994) suggest that the late Paleozoic deposits of the Falkland Islands are intimately related to equivalent deposits in the Karoo Basin.

The basal Lafonian diamictites unconformably rest on Silurian or Devonian sedimentary rocks (Fig. 9.7). Most diamictites are massive or faintly stratified and are grouped in laterally continuous beds. Deformed sandstone bodies and intervals of shales with dropstones and diamictite clasts are present within the diamictites. Evidence of resedimentation (interbeds of conglomerate and cross-bedded or rippled sandstone) becomes especially common toward the top of the Formation (Bellosi and Jalfin, 1987).

The glacigenic deposits are overlain by marine shales with dropstones and well-bedded diamict

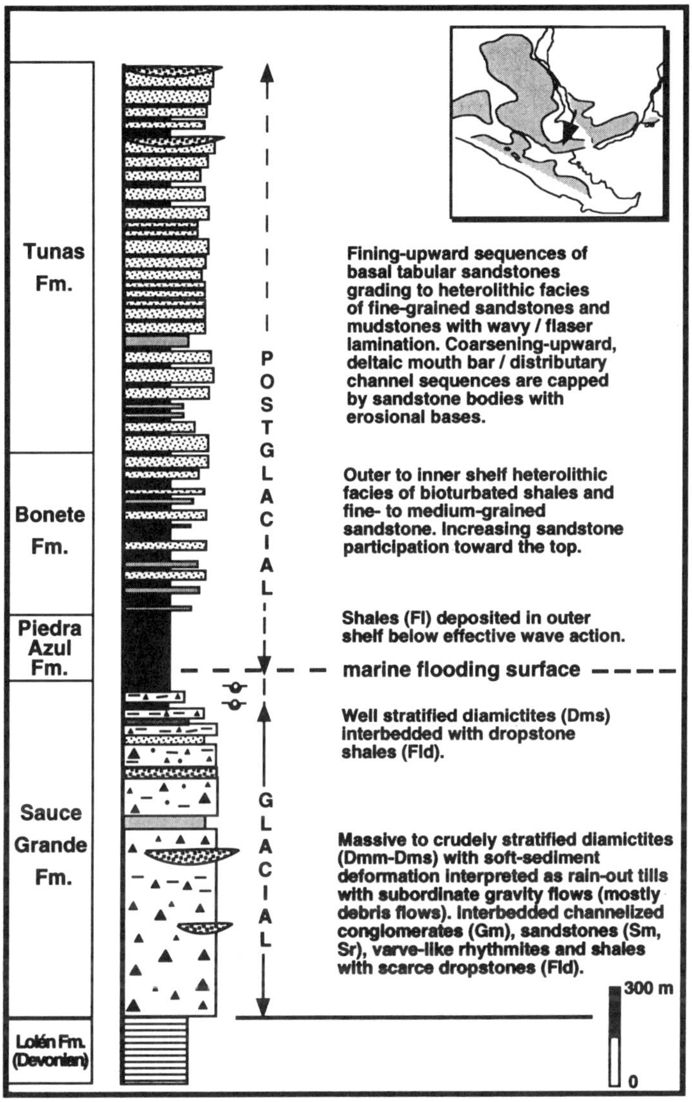

Fig. 9.7. Glacial Episode III. Glacial-postglacial transition in the Sauce Grande Basin in the Sierras Australes thrust-fold belt. Note overall coarsening upward trend of the late Paleozoic sequence from early postglacial transgressive deposits (Piedra Azul Formation) to deltaic sandstones (Tunas Formation).

ites (basal Black Rock Member of the Port Sussex Formation), which pass upward into marine shales, sandstones and thin chert levels of the uppermost Port Sussex Formation. This unit is covered by fluvial deposits of the Terras Motas Formation, which in turn are overlain by *Glossopteris*-bearing littoral deposits of the Bahía Choiseul Formation (Jalfin and Bellosi, 1983). The fine-grained deposits of the Port Sussex Formation have been interpreted as the result of a postglacial sea-level rise (Bellosi and Jalfin, 1984) that is related to the Sakmarian transgression (Dickins, 1984; López-Gamundí, 1989).

The Glacial-Postglacial Transition: Paleoclimatic and Sedimentologic Consequences

Sea-level fluctuations have been inferred in the major pre-Quaternary glaciations (Crowell, 1978; Bjørlykke, 1985; Proust et al., 1990). On the basis of the total area under glaciation and analogies with the Pleistocene glaciation, Crowley et al. (1991) estimated a sea-level lowering of around 100–120 m for the Gondwana glaciation. An equivalent sea-level rise can be consequently inferred during the final deglaciation phase.

Both glacioeustatic and glacioisostatic fluctuations are considered to be among those with the highest rates and, consequently, with high likelihood of being preserved in the stratigraphic record either as depositional transgressions (in the case of post- or interglacial eustatic sea-level changes coupled with moderate to high sediment supply) or as progradational wedges (in the case of glacioisostatic rebound and progradation of littoral-deltaic facies basinward). The facies architecture of several basin margins of southern South America during the glacial-postglacial transition indicates drastic changes that were at least partially caused by glacially induced changes in sea levels, probably coupled with other factors such as local tectonics.

Climatic changes during the glacial-postglacial transition were characterized by increasing humidity and temperature, resulting in cold to temperate humid conditions. These conditions led to the development of extensive peatlands and subsequent formation of coals in the Gondwana subcontinents. The relationship between these coals and the demise of the late Paleozoic glaciation is clearly expressed in the stratigraphic record by the stratigraphic proximity of glacial deposits and the postglacial coal seams, and by the diachroneity of the occurrence of the coals, which parallels the temporal trend of the glacial record across Gondwanaland (Fig. 9.3).

In basin margins under marine influence, the glacial deposits are commonly covered by open marine mudstones with poor to absent development of intervening shallow marine deposits. This sharp facies contrast indicates that the glacial-postglacial transition was characterized by sudden sea-level changes. These glacioeustatic sea-level rises had several paleoclimatic and sedimentologic consequences. They triggered transgression-related sediment starvation episodes that expanded the oxygen minimum zone, inducing a generalized euxinification and productivity concentration in some areas of the inundated shelves. They also induced a significant warming due to the reduced albedo of areas covered by water and brought humidity to coastal areas that contributed to the formation of peats landward. Thus, two glacial retreat facies associations ("valley glacier retreat" and "submarine glacial retreat") are defined on the basis of predominantly nonmarine and marine depositional conditions in different parts of the basin. Their sedimentologic characteristics and paleoclimatic implications are discussed in the following sections.

Valley Glacier Retreat Facies Association

This facies association reflects the retreat of the glaciation along basin margins under predominantly subaerial or shallow subaqueous conditions. Examples of this association are present in the Titicaca region of western Bolivia (glacial episode I) and in the Paganzo Basin of western Argentina (glacial episode II). Along the western margin of the Paganzo Basin, the late Paleozoic section exhibits evidence of grounded activity and the presence of basal massive muddy/sandy diamictites in discontinuous beds, probably lodgment tills. The glacial section is extremely thin (5–10 m), occasionally associated with lacustrine shales with dropstones, and invariably overlain by fluvial deposits with thin coal levels. This basal diamictitic facies shows extreme lateral thickness variability, which reflects an irregular bedrock topography. The overlying fluvial facies are predominantly coarse-grained and show a clear fining upward trend into the coal-bearing strata. Massive conglomerates, pebble to cobble conglomerates, and pebbly sandstones with crude trough cross-bedding and rippled coarse- to medium-grained sandstones have been grouped under a gravelly braided channel complex interpreted as deposited in a subaqueous proximal outwash fan (Martínez and Astini, 1992).

The development of these peat-forming conditions in western Argentina was coeval with the onset of glacially influenced sedimentation in southeastern South America (Chaco-Paraná, Paraná and Sauce Grande basins) (Fig. 9.3). The demise of peat-forming conditions in the Paganzo Basin may have resulted from the combined effect of low subsidence rates under a cool temperate climate coupled with a general trend toward seasonality and accompanying water fluctuations (López-Gamundí et al., 1993).

Facies associations related to valley glacier retreat facies association are also abundant in the South African sector of western Gondwanaland. Along the northeastern margin of the Karoo Basin, thin (seldom exceeding 5 m) glacial facies, interpreted as lodgment tills, are overlain by outwash plain pebbly sandstone/conglomerate facies which in turn pass upward to coal seams (Cairncross and Cadle, 1988). Emerging evidence that the postglacial marine shales of the Prince Albert Formation are at least partially coeval with these *Glossopteris* flora-bearing coals (Vryheid Formation), developed inland in the basin (Visser, 1992, pers. comm.), reinforces the possibility of a temporal and genetic linkage between the climatic amelioration during progressive recession of the marine ice sheet and the onset of peat-forming conditions. Similar facies arrangement has been identified north of the Karoo Basin in the mid-Zambesi basin (Oesterlen, 1990), where the glacial-postglacial transition is characterized from base to top by massive diamictites, stratified diamictites, massive sandstones, fine-grained clastics (siltstone, mudstones, rhythmites) and medium- to coarse-grained sandstones and *Glossopteris*-bearing carbonaceous mudstones. Local glacial conditions persisted during and after the marine ice sheet retreat in basin margin depressions incised in basement, such as those present along the northeastern margin of the Karroo Basin and some parts of the Kalahari Basin, fed with glacial material by valley glaciers (von Brunn, 1987; Visser, 1991b). In this case, fine-grained marine deposits assigned to the *Eurydesma* transgression (see next section) are mixed with, rather than immediately above, glacial deposits. The glacial debris and reworked glacial material with associated *Glossopteris*-bearing flora were released from retreating terrestrial ice, the latest phase of glacially influenced sedimentation in western Gondwanaland (mountain ice cap phase of Visser, 1991b).

An appealing analog for most Gondwana postglacial coals is the Holocene mires of northern Canada, characterized by significant amounts of rainfall in early postglacial times (Hazeldine, 1988). In recent years similar analogies have been proposed for specific cases of Gondwana postglacial coals (Martini and Glooschenko, 1985; Raja Rao and Mitra, 1985; Kreuser, 1987; Martini and Banks, 1989; Martini and Rocha-Campos, 1989).

Submarine Retreat (Transgressive) Facies Association

Marine fine clastics resting on glaciomarine deposits are present in several late Paleozoic basins of southern South America. Glacial period II ended late in the mid-Carboniferous (Westphalian) and was followed by a relative sea-level rise clearly identified in the Tarija (Taiguati Formation), Calingasta-Uspallata (upper section of the Hoyada Verde Formation and correlative units) and Tepuel (Mojón de Hierro Formation) basins, leading to an expansion of the marine domain. In these basins the glaciomarine sediments grade upward to postglacial open marine fine clastics with fossils of the *Levipustula* zone of mid-Carboniferous (Late Namurian-Early Westphalian) age. A younger glacioeustatic flooding event has been recognized for the glacial episode III in the Sauce Grande Basin where marine shales, mudstones and fine sandstones with fossils of the *Eurydesma* fauna and *Glossopteris* flora overlie glaciomarine deposits (Fig. 9.3). The presence of marine fine-grained deposits resting on the Lafonian diamictites and fossils of the *Glossopteris* flora in the Bahia Choiseul Formation suggests that the same flooding event was present in the Falkland Islands.

These transgressive fine clastics are considered to constitute the sedimentary record of a glacioeustatic sea-level rise that occurred in basin margins during the glacier retreat period subsequent to widespread glaciation. The examples studied fit the "submarine retreat" model proposed by Edwards (1986) when the rate of accommodation (Posamentier et al., 1988), in this case a function of glacioustasy and subsidence, exceeds the rate of deposition, resulting in a retrogradational facies pattern. Consequently, a drastic facies shift is apparent throughout the glacial-postglacial transition.

The postglacial facies association can be compared to a transgressive systems tract in sequence stratigraphic terms. These lithologies are characteristic of events of maximum flooding during transgressions, equivalent to the marine flooding surfaces (MFS) of sequence stratigraphy (Posamentier and Vail, 1988; Van Wagoner et al., 1990). Some of the cases treated here might represent the maximum landward extent of postglacial transgressions and be likened to maximum flooding surfaces. Examples of maximum flooding events within these postglacial transgressions have been tentatively identified in the Calingasta-Uspallata (Fig. 9.6), Sauce Grande (Fig. 9.7) and Falkland (Fig. 9.8) basins.

The postglacial facies association can be further subdivided into two facies: an early facies consisting of gravity flow deposits and fine-grained sediments with ice-rafted debris overlain by a late facies of ice-rafted debris-free claystones and mudstones deposited in open marine conditions mostly below effective wave action. The early facies (the early transgressive systems tract) is dominated by processes related to the withdrawal of the glacial source, such as gravity reworking of glacial material, ice-rafting, and background sedimentation dominated by settling of fines. Fine-grained deposits without evidence of wave or current reworking and thin beds of micritic limestones and chert, indicative of short periods of clastic starvation, characterize the deposition of the late facies (the late transgressive systems tract). Variations in this depositional motif were essentially regulated by the complex interplay of sediment supply and space of accommodation in these early postglacial shelves.

Conclusion

Although fragmentary, the present knowledge of the stratigraphic record of southern South America allows a definition of the temporal and spatial evolution of the Gondwana glaciation. Three glacial episodes are identified in western Gondwanaland. The oldest, glacial episode I (Famennian-Tournaisian), is present in the Titicaca region of western Bolivia and is related to age-equivalent glacial deposits described in the Solimões and the Amazonas basins. These glacial deposits probably resulted from a high elevation glaciation and not from ice sheets.

Glacial episode II (Namurian-Early Westphalian) affected basins along the paleo-Pacific margin of the Gondwana supercontinent. Glacial material was derived from isolated ice centers emplaced mostly along the cratonic margins of these basins. By the Late Carboniferous (Stephanian) glaciation ceased in the paleo-Pacific region and peat-forming conditions were favored by a rapid postglacial climatic amelioration as evidenced by the presence of *NBG* flora-bearing coal and carbonaceous shales in the Paganzo and the San Rafael basins.

The main phase of glacially influenced sedimentation in western Gondwanaland (glacial episode III) is documented in the Chaco-Paraná, Paraná, and Sauce Grande basins of southern South America and in the Karoo and Kalahari basins of southern Africa. This glacial episode started in the Late Carboniferous (Stephanian) and waned by the late Early Permian with the disintegration of the Gondwana Ice Sheet and its satellite ice centers. *Glossopteris*-bearing coals and carbonaceous shales are characteristic of early postglacial sedimentation in the basin margins (Paraná and Karoo basins). These organic-rich deposits are associated with fluvial sediments grouped under the valley glacier retreat facies association.

Basinward the early postglacial sedimentation was dominated by deposition of fines in open marine conditions created by the rapid glacio-eustatic sea-level rise. These fine-grained deposits, occasionally associated with chert horizons, are part of the submarine retreat facies association. These postglacial transgressions were characterized by two distinct marine faunas: the *Levipustula* fauna (glacial episode II) and the *Eurydesma* fauna (glacial episode III). After the collapse of the Gondwana marine ice sheet (glacial episode III), local glacial conditions persisted until the end of the early Permian (Artinskian) in depressions incised along some basin margins (the northern margin of Karoo Basin, some parts of the Kalahari Basin and probably the eastern margin of the Paraná Basin), where small ice caps provided glacial debris. In these cases, the fine-grained marine sediments with *Eurydesma* fauna lie between, rather than immediately above, glacially influenced deposits. In contrast, deposits of the *Eurydesma* transgression pass upward directly to open marine Prince Albert mudstones in the central part of the Karoo Basin (Visser, this volume).

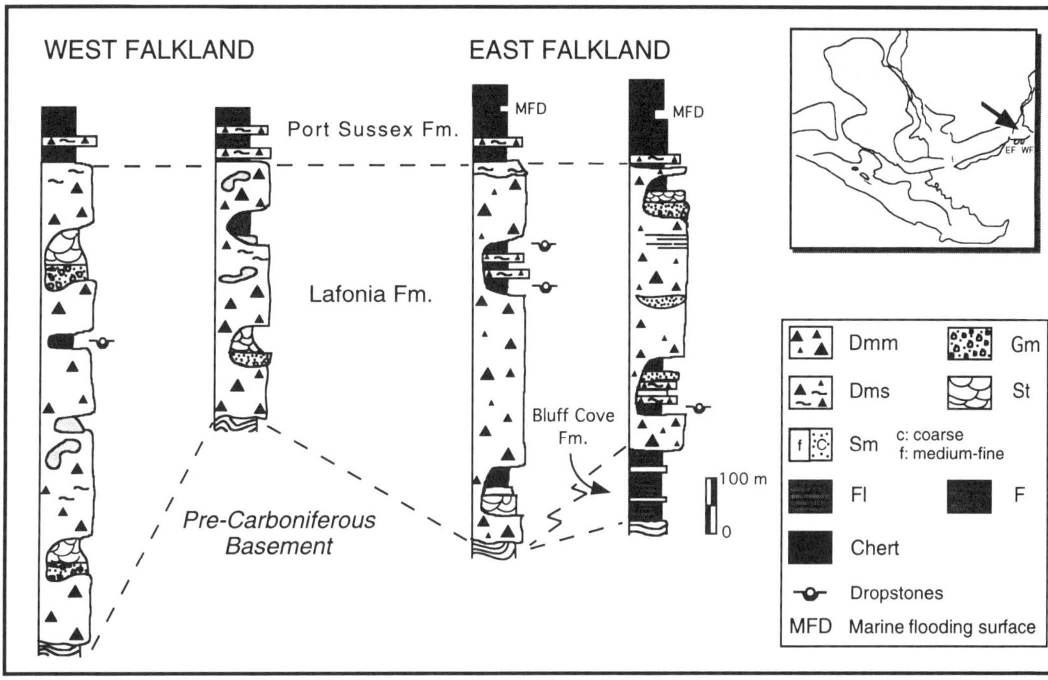

Fig. 9.8. Glacial Episode III. Glacial and early postglacial succession in Falkland Islands (after Frakes and Crowell, 1967 and Bellosi and Jalfin, 1987). Lithofacies code after Eyles et al. (1983). Note that subglacial facies such as esker-type deposits (St and Gm) are more common in the West Falkland outcrops, whereas more distal facies prevail in East Falkland. This facies trend is consistent with a prebreakup position of the Falkland microplate east of coast of South Africa, with East Falkland in a more basinal position (Gondwana reconstruction inset), translated eastward and rotated 180° from its present position.

References

Adie, R.J., 1952. The position of the Falkland Islands in a reconstruction of Gondwanaland. Geological Magazine, 89: 401–410.

Amos, A.J., and López-Gamundí, O.R., 1981. Late Paleozoic Sauce Grande Formation of eastern Argentina. In: M. Hambrey and W. Harland (Editors), Earth's pre-Pleistocene glacial record. Cambridge University Press, Cambridge, pp. 872–877.

Andreis, R.R., 1984. Análisis litofacial de la Formación Sauce Grande (Carbónico superior?), Sierras Australes, provincia de Buenos Aires. Annual Meeting Project 211-IGCP Late Paleozoic of South America, Abstracts, pp. 28–29.

Andreis, R.R. Lluch, L.L. and Iñíguez, A.M., 1979. Paleocorrientes y paleoambientes de la Formaciones Bonete y Tunas, Sierra Australes, provincia de Buenos Aires. 6th Congreso Geológico Argentino, Actas, 2, pp. 207–229.

Andreis, R.R., Amos, A.J., Archangelsky, S. and Conzález, C.R., 1987. Cuenca Sauce Grande. In: S. Archangelsky (Editor), El Sistema Carbonífero en la República Argentina. Academia Nacional de Ciencias de Córdoba, Córdoba, pp. 213–233.

Andreis, R.R., Ferrando, L.A. and Herbst, R., 1991. Terrenos carboníferos y pérmicos de la República Oriental del Uruguay. El Sistema Pérmico en la República Argentina y en la República Oriental del Uruguay, 12th International Congress on Stratigraphy and Geology of the Carboniferous and Permian and Academia Nacional de Ciencias de Córdoba (preprint), pp. 315–349.

Archangelsky, S., Azcuy, C.L., Conzález, C.R. and Sabattini, N., 1987. Edad de las biozonas. In: S. Archangelsky (Editor), El Sistema Carbonífero en la República Argentina. Academia Nacional de Ciencias de Córdoba, Córdoba, pp. 293–301.

Archangelsky, S. and Vergel, M., 1991. Cuenca Chacoparanaense: B. Paleontología, Bioestratigrafía y Paleoecología. In: El Sistema Pérmico en la República Argentina y en la República Oriental del Uruguay. 12th International Congress on Stratigraphy and Geology of the Carboniferous and

Permian and Academia Nacional de Ciencias de Córdoba (pre-print), pp. 40–44.

Bellosi, E. and Jalfin, G., 1984. Litoestratigrafía y evolución paleoambiental neopaleozoica de las Islas Malvinas. Argentina. IX Congreso Geológico Argentino, Actas, 5: 66–86.

Bellosi, E. and Jalfin, G., 1987. Area Islas Malvinas. In: S. Archangelsky (Editor), El Sistema Carbonífero en la República Argentina , Academia Nacional de Ciencias de Córdoba, Córdoba, pp. 211–223.

Bjørlykke, K., 1985. Glaciations, preservation of their sedimentary record and sea level changes. A discussion based on the Late Precambrian and Lower Paleozoic sequence in Norway. Palaeogeography, Palaeoclimatology, Palaeoecology, 51: 197–207.

Brakel, A.T. and Totterdell, J.M., 1993. The Sakmarian-Kungurian palaeogeography of Australia. In: R. H. Findlay, R. Unrug, M.R. Banks and J.J. Veevers (Editors), Gondwana Eight. A. A. Balkema, Rotterdam, pp. 385–396.

Cairncross, B. and Cadle, A.B., 1988. Depositional paleoenvironments of the coal-bearing Permian Vryheid Formation in the east Witbank Coalfield, South Africa. South African Journal of Geology, 91: 1–17.

Caputo, M.V., 1984. Glacição neodevoniana no continente Gondwana occidental. 38th Congresso Brasileiro de Geologia, vol. 2, pp. 725–739.

Caputo, M.V. and Crowell, J.C., 1985. Migration of glacial centers across Gondwana during Paleozoic Era. Geological Society of America Bulletin, 96: 1020–1036.

Coates, D.A., 1969. Stratigraphy and sedimentation of the Sauce Grande Formation, Sierra de la Ventana, Southern Buenos Aires Province, Argentina. In: A. J. Amos (Editor), Gondwana Stratigraphy, Unesco, Paris, pp. 799–816.

Crowell, J.C., 1978. Gondwanan glaciation, cyclothems, continental positioning and climate change. American Journal of Sciences, 278: 1345–1372.

Crowell, J.C., 1983. Ice ages recorded on Gondwanan continents. Geological Society of South Africa Transactions, 86: 237–262.

Crowley, T.J., Baum, S.K. and Hyde, W.T., 1991. Climate model comparison of Gondwanan and Laurentide glaciations. Journal of Geophysical Research, 96: 9217–9226.

Dalmayrac, B., Laubacher, G., Marocco, R., Martinez, R. and Romasi, P., 1980. La Chaine hercynienne d'Amerique du Sud, structure et evolution d'un orogene intracratonique. Geologische Rundschau, 69: 1–21.

Díaz Martínez, E., Isaacson, P.E., Lema, J.C., Christensen, K.L., Vavrdova, M. and Antelo, B., 1993a. Significance of a Late Devonian (Famennian) glacial marine diamictite, northern Altiplano, Bolivia. Compte Rendus XII Congrés International de la Stratigraphie et Géologie du Carbonifere et Permien, vol. I, pp. 293–304.

Díaz Martínez, E., Isaacson, P.E. and Sablock, P.E., 1993b. Late Paleozoic latitudinal shift of Gondwana: stratigraphic/sedimentologic and biogeographic evidence from Bolivia. Documents des Laboratoires de geologie de Lyon, vol. 125, pp. 119–138.

Dickins, J.M., 1984. Late Paleozoic glaciation. Bureau of Mineral Resources Journal, 9: 163–169.

Du Toit, A.L., 1927, A geological comparison of South America with South Africa. With a palaeontological contribution by F. R. Cowper Reed, Carnegie Institute Publication 381, pp. 1–157.

Edwards, M., 1986. Glacial Environments. In: H.G. Reading (Editor), Sedimentary environments and facies. Blackwell, London, pp. 445–469.

Espejo, I.S., 1990. Análisis estratigráfico, paleoambiental y de proveniencia de la Formación El Imperial en los alrededores de los ríos Diamante y Atuel (provincia de Mendoza). Ph.D. thesis, University of Buenos Aires, 315 p.

Eyles, C. Eyles, N. and Franca, A.B., 1993. Glaciation and tectonics in an active intracratonic basin; the late Paleozoic Itararé group. Sedimentology, 40: 1–25.

Eyles, N., 1990. Late Precambrian tillites of the Avalonian-Cadonian belt; marine debris flows in an active tectonic setting. Palaeogeography, Palaeoclimatology, Palaeoecology, 79: 73–98.

Eyles, N., 1993. Earth's glacial record and its tectonic setting. Earth Science Reviews, 35: 1–248.

Eyles, N., Eyles C.H. and Miall, A.D., 1983. Lithofacies types and vertical profile models; an alternative approach to the description and environmental interpretation of glacial diamict and diamictite sequences. Sedimentology, 30: 393–410.

Fernández Garrasino, C.A., 1981. Late Paleozoic Tarija Formation, southern Bolivia and northern Argentina. In: M. Hambrey and W. Harland (Editors), Pre-Pleistocene Glacial Record. Cambridge University Press, Cambridge, pp. 853–54.

Ferrando, L.A. and Andreis, R.R., 1990a. La glaciación gondwánica en Uruguay (Extremo Austral de la cuenca intracratónica de Paraná). Late Paleozoic of South America, Project 211 IUGS-UNESCO, Annual Meeting of the Working Group, Abstracts, pp. 72–79.

Ferrando, L.A. and Andreis, R.R., 1990b. Una nueva localidad de la Formación Melo (Depto. Rivera, Uruguay). Litofacies e interpretación paleoambiental preliminar. Late Paleozoic of South America, Project 211 IUGS-UNESCO, Annual Meeting of the Working Group, Abstracts, pp. 69–71.

Frakes, L.A. and Crowell, J.C., 1967. Facies and paleogeography of late Paleozoic Lafonian diamictite, Falkland Islands. Geological Society of America Bulletin, 78: 37–58.

Frakes, L.A. and Crowell, J.C., 1969. Late Paleozoic glaciation. Part I, South America. Geological Society of America Bulletin, 80: 1007–1042.

França, A.B., 1987. Stratigraphy, depositional environment and reservoir analysis of the Itararé Group (Permo-Carboniferous), Paraná Basin-Brazil. Ph. D. thesis, University of Cincinnati, Ohio, 188 pp.

França, A.B., 1994. Itararé Group: Gondwanan Carboniferous-Permian of the Paraná Basin, Brazil. In: M. Deynoux, J.M.G. Miller, N. Eyles, I.J. Fairchild and G.M. Young (Editors), The Earth's glacial record. Cambridge University Press, Cambridge, pp. 70–94.

França, A.B. and Potter, P.E., 1991. Stratigraphy and reservoir potential of glacial deposits of the Itararé Group (Carboniferous-Permian), Paraná basin, Brazil. American Association of Petroleum Geologists Bulletin, 75: 62–85.

García, G., 1992. Estudios paleoflorísticos y análisis bioestratigráfico de la Formación El Imperial y otras unidades equivalentes, Paleozoico superior de la cuenca San Rafael, provincia de Mendoza, Republica Argentina. Ph.D. thesis, University of Buenos Aires, Buenos Aires, 321 pp.

González, C.R., 1981, Pavimento glaciario en el Carbónico de la Precordillera. Revista Asociación Geológica Argentina, 36: 262–266.

González Bonorino, G., Rafine, G., Vega, V. and Guerin, D., 1988. Ambientes de plataforma nerítica dominada por tormentas en la seccion glacigénica del Grupo Tepuel, Chubut. Revista Asociación Geológica Argentina, 43: 239–252.

Gravenor, G.P. and Rocha-Campos, A.C., 1983. Patterns of late Paleozoic glacial sedimentation on the southeast side of the Paraná Basin, Brazil. Palaeogeography, Palaeoclimatology, Palaeoecology, 43: 1–39.

Hambrey, M.J. and Harland, W.B. (Editors), 1981. Earth's pre-Pleistocene glacial record. Cambridge University Press, Cambridge, 1004 pp.

Harrington, H.J., 1947. Explicación de las Hojas 33m y 34m, Sierras de Curamalal y de la Ventana, provincia de Buenos Aires. Boletín Dirección Nacional de Geología y Minería, 61, 43 pp.

Hazeldine, R.S., 1988. Coal reviewed: depositional controls, modern analogues and ancient climates. In: M.K. Whately and K.T. Pickering (Editors), Deltas: sites and traps for fossil fuels. Geological Society London, Special Publication No. 41, pp. 289–308.

Helwig, J., 1972. Stratigraphy, sedimentation, paleogeography and paleoclimates of Carboniferous ("Gondwana") and Permian of Bolivia. Bulletin of American Association of Petroleum Geologists, 56: 1008–1020.

Henz, G.I., Rodrigues, R. and Formoso, M.L.L., 1987. Organo-geoquímica dos carvões da jazida do Faxinal, Municipio Arroio dos Ratos, Rio Grande do Sul. Geochimica Brasiliensis 1: 161–175.

Jalfin, G. and Bellosi, E., 1983. Análisis estratigráfico de la Formacion Bahía Choiseul, Pérmico de la Isla Soledad, Islas Malvinas, Argentina. Revista Asociación Geológica Argentina, 38: 248–262.

Keidel, J., 1916, La geología de las sierras de la provincia de Buenos Aires y sus relaciones con las montañas de Sudáfrica y Los Andes. Anales del Ministerio de Agricultura, Sección Geologóa, Mineralogía y Minería, Buenos Aires, IX, vol. 3, 78 pp.

Kreuser, T., 1987. Late Paleozoic glacial sediments and transition to coal-bearing Lower Permian in Tanzania. Facies, 17: 149–158.

Limarino, C.O. and Cesari, S.N., 1988. Paleoclimatic significance of the lacustrine Carboniferous deposits in northwest Argentina. Palaeogeography, Palaeoclimatology, Palaeoecology, 65: 115–131.

Limarino, C.O. and Gutiérrez P., 1990. Diamictites in the Agua Colorada Formation. A new evidence of Carboniferous glaciation in northwest Argentina. Journal of South American Earth Sciences, 3: 9–20.

López-Gamundí, O.R., 1983, Modelo de sedimentación glacimarina para la Formación Hoyada Verde del Paleozoico superior de la provincia de San Juan. Revista Asociación Geológica Argentina, 38: 60–72.

López-Gamundí, O.R., 1985. Sedimentología de la Formación Tarija, Carbonífero de la Sierra de Aguaragüe, provincia de Salta. Revista Asociación Geológica Argentina, 41: 334–355.

López-Gamundí, O.R., 1987. Depositional models for the glacial-marine sequences of Andean late Paleozoic basins of Argentina. Sedimentary Geology, 52: 109–126.

López-Gamundí, O.R., 1989. Postglacial transgressions in late Paleozoic basins of western Argentina: a record of glacioeustatic sea level rise. Palaeogeography, Palaeoclimatology, Palaeoecology, 71: 257–270.

López-Gamundí, O.R., Alvarez L., Andreis, R.R., Bossi, G.E., Espejo, I.S., Fernandez Seveso, Kokogian, D., Legarreta, L., Limarino, C.O. and Sessarego, H., 1989. Cuencas intermontanas. In: G.A. Chebli and L.A. Spalletti (Editors), Cuencas Sedimentarias Argentinas. Serie Correlación Geológica No. 6, Universidad de Tucumán, pp. 123–168.

López-Gamundí, O.R. and Espejo, I.S., 1993. Correlation of a paleoclimatic mega-event: the Carboniferous glaciation in Argentina. Compte Rendus XII

Congrès International de la Stratigraphie et Géologie du Carbonifere et Permien, vol. I, pp. 313–324.

López-Gamundí, O.R. and Limarino, C.O., 1984. Facies de abanico submarino en el Grupo Tepuel (Paleozoico superior), provincia de Chubut. Revista Asociación Geológica Argentina, 39: 251–261.

López-Gamundí, O.R., Limarino, C.O. and Cesari, S.N., 1992. Late Paleozoic paleoclimatology of central west Argentina. Palaeogeography, Palaeoclimatology, Palaeoecology, 91: 305–329.

López-Gamundí, O. Limarino, C.O. and Cesari, S.N., 1993. Paleoclimatic significance and age constraints of the Carboniferous coals of Paganzo basin. In: R.H. Findlay, R. Unrug, M.R. Banks and J.J. Veevers (Editors), Gondwana Eight. A.A. Balkema, Rotterdam, pp. 291–298.

López-Gamundí, O.R. and Rossello, E.A., 1993. Devonian-Carboniferous unconformity in Argentina and its relation to the Eo-Hercynian orogeny in southern South America. Geologische Rundschau, 82: 136–147.

López-Gamundí, O.R. and Rossello, E.A., 1995. Pavimento glacial en la Formación Leoncito, provincia de San Juan. Revista de la Asociación Geológica Argentina, 50: 35-39.

Malizzia, D.C., Miller, E.E., Rosso, M. del R., Labayen, I. and Hernaez, S., 1993. Geochemical interpretatrion of the Carboniferous-Permian in the Alhuampa Sub-basin, North-east Argentina. Compte Rendus XII Congrès International de la Stratigraphie et Géologie du Carbonifere et Permien, vol. I, pp. 239–252.

Marshall, J.E.A., 1994. The Falkland Islands: A key element in Gondwana paleogeography. Tectonics, 13: 499–514.

Martínez, M., 1993. Hallazgo de fauna marina en la Formacion Guandacol (Carbonifero) en la localidad de Agua hedionda, San Juan, Precordillera Nororiental, Argentina. Compte Rendus XII Congrès International de la Stratigraphie et Géologie du Carbonifere et Permien, vol. 2, pp. 291–296.

Martínez, M. and Astini, R., 1992. Características del relleno sedimentario de una artesa glaciaria neopaleozoica en la Precordillera Central, Argentina. IV Reunión Argentina de Sedimentología, vol. III, pp. 161–167.

Martini, P. and Banks, M.R., 1989. Sedimentology of the cold-climate, coal-bearing, Lower Permian "Lower Freshwater Sequence" of Tasmania. Sedimentary Geology, 64: 25–41.

Martini, I.P. and Glooschenko, W., 1985. Cold climate peat formation in Canada, and its relevance to Lower Permian coal measures of Australia. Earth Science Reviews, 22: 1–107.

Martini, I.P. and Rocha-Campos, A.C., 1984. Interglacial and early post-glacial, lower Gondwana coal sequences in the Paraná basin, Brazil. In: H. Ulbrich and A.C. Rocha Campos (Editors), Gondwana Seven, Instituto de Geociencias, Universidade de São Paulo, pp. 317–336.

Milana, J.P. and Bercowski, F., 1993. Late Paleozoic glaciation in Paganzo Basin, western Argentina: sedimentological evidence. Compte Rendus XII Congrès International de la Stratigraphie et Géologie du Carbonifere et Permien, vol. I, pp. 325–335.

Mitchell, C., Taylor, G.K., Cox, K.G. and Shaw, J., 1986. Are the Falkland Islands a rotated microplate? Nature, 319: 131–134.

Mitzarek, G.L., Mendoça Filho, J.G. and Peralba, M. do C.R., 1993. Petrographical, Chemical Analyses of Coal Samples from Iruí Coalfield, Rio Grande do Sul, Brazil. Third Latin American Congress on Organic Geochemistry, Rio de Janeiro (Brazil), Extended Abstracts, pp. 104–106.

Musset, A.E. and Taylor, G. K., 1994. $^{40}Ar/^{39}Ar$ ages for dykes from the Falkland Islands with implications for the break-up of southern Gondwanaland. Journal of the Geological Society of London, 151: 79–81.

Oesterlen, P.M., 1990. A new occurrence of the Dwyka Formation in western Zimbabwe. South African Journal of Geology, 93: 382–388.

Ortiz, A. and Zambrano, J., 1981. La provincia geológica de Precordillera Oriental. 8th Congreso Geológico Argentino, Actas 3, pp. 59–74.

Ottone, E.G., 1991. Palinologie du Carbonifiere Superieure de la coupe de Mina Esperanza, bassin Paganzo, Argentine. Revue de Micropaleontologie, 34: 118–135.

Palmer, A.R., 1983. The decade of North American geology 1983 geologic time scale. Geology, 11: 503–504.

Posamentier, H.W., Jervey, M. and Vail, P.R., 1988. Eustatic controls on clastic deposition I- conceptual framework. In: C.K. Wilgus, B.S. Hastings, C.G. St.C. Kendall, H.W. Posamentier, C.A. Ross, and J.C. Van Wagoner (Editors.), Sea-level changes: an integrated approach. Society of Economic Paleontologists and Mineralogists, Special Publication 42, pp. 125–154.

Posamentier, H.W. and Vail, P.R., 1988. Eustatic controls on clastic deposition II–sequence and system tract models. In: C.K. Wilgus, B.S. Hastings, C.G. St.C. Kendall, H.W. Posamentier, C.A. Ross, and J.C. Van Wagoner (Editors.), Sea-level changes: an integrated approach, Society of Economic Paleontologists and Mineralogists, Special Publication 42, pp. 125–154.

Proust, J., Deynoux, M. and Guilocheau F., 1990. Effets conjugués de l'eustatisme et de l'isostasie sur les plates-formes stables en période glaciare. Exemple des dépôts glaciaires du Protérozoïque supérieur de l'Afrique de l'Ouest au Mali occidental. Bulletin Societé géologique de France, VI: 673–681.

Raja Rao, C.S. and Mitra, N.D.,1985. Late Paleozoic climate in the Indian segment of Gondwanaland. 9t[h] International Congress on Stratigraphy and Geology of the Carboniferous, vol. 5, pp. 415–422.

Ramos, V.A. Jordan, T., Allmendinger, R., Mpodozis, C., Kay, S.M., Cortés, J. and Palma, M., 1986, Paleozoic terranes of the central Argentine-Chilean Andes. Tectonics, 5: 855–880.

Rapallini, A.E., 1993. Paleomagnetismo del Grupo Tepuel: implicancias en la evolución tectónica de Patagonia y la paleogeografía del Gondwana en el Carbonífero. Reunión Grupo Argentino Proyecto Pangea, Comunicaciones, San Juan, pp. 48–51.

Rapallini, A. and Vilas, J.F., 1991. Paleomagnetismo. In: El Sistema Pérmico en la República Argentina y en la República Oriental del Uruguay, Academia Nacional de Ciencias, Córdoba, pp. 66–68.

Rex, D.C. and Tanner, P.W.G., 1982. Precambrian age for gneisses at Cape Meredith in the Falkland Islands. In: C. Craddock (Editor), Antarctic Geoscience Symposium on Antarctic Geology and Gophysics, University of Wisconsin Press, Madison, pp. 107–110.

Rocha-Campos, A.C., 1967, The Tubarão Group in the Brazilian portion of the Paraná basin. In: J.J. Bigarella, R. Becker and I. Pinto (Editors), Problems in Brazilian Gondwanan Geology. International Symposium on Gondwana Stratigraphy and Paleontology, Curitiba, pp. 27–102.

Rocha-Campos, A.C., 1981. Late Devonian Curuá Formation, Amazon Basin, Brazil. In: M.J. Hambrey and W.B. Harland (Editors), Earth's pre-Pleistocene glacial record. Cambridge University Press, Cambridge, pp. 888–891.

Rocha-Campos, A.C., de Carvalho, R. and Amos, A.J., 1977. A Carboniferous (Gondwana) fauna from Subandean Bolivia. Boletim Instituto de Geociencias, Universidade de São Paulo, 6: 181–191.

Rocha-Campos, A.C., Oliveira, M.E.C.B. de, Santos, P.R. dos and Saad, A.R., 1976. Boulder pavements and the sense of movement of late Paleozoic glaciers in central eastern São Paulo State, Paraná Basin, Brazil. Boletim Instituto de Geociências, Universidade de São Paulo, 7: 149–160.

Rocha-Campos, A.C. and Rösler, O., 1978. Late Paleozoic faunal and floral successions in the Paraná Basin, southeastern Brazil. Boletim Instituto de Geociências, Universidade de São Paulo, 9: 1–16.

Rösler, O., Iannuzzi, R. and Suárez, R., 1989. A flora Carbonífera da Formaíao Kasa em Belén, Penénsula de Copacabana, Altiplano boliviano, e a importância das formas trifoliadas. XI Congresso Brasileiro de Paleontologia, Resumo des Comunicaçoes, pp. 45–46.

Russo, A. and Archangelsky, S., 1987, Cuenca Chacoparanaense. In: S. Archangelsky (Editor), El Sistema Carbonífero en la República Argentina, Academia Nacional de Ciencias, Córdoba, pp. 139–184.

Sablock, P.E., 1993. A warm to cold paleoclimate change in Gondwana (Devonian through Carboniferous): a consequence of coupled continental shift and onlap. Compte Rendus XII Congrès International de la Stratigraphie et Géologie du Carbonifere et Permien, vol. I, pp. 355–367.

Santos, P.R. dos, 1987. Facies e evoluçao paleogeografica do Subgrupo Itararé/Grupo Aquidauana (Neopaleozoico) na Baçia do Paraná, Brasil. Ph.D. thesis, Instituto de Geociencias, Universidade de São Paulo, 128 pp.

Scasso, R.A. and Mendía, J.E., 1985. Rasgos estratigrãficos y paleoambientales del Paleozoico de las Islas Malvinas. Revista Asociación Geológica Argentina, 40: 26–50.

Schlatter, L.E. and Nederloff, M.H., 1966. Bosquejo de la geología y paleogeografía de Bolivia. Boletín Servicio Geológico de Bolivia, 8: 49.

Sempere, T., 1989. Paleozoic Evolution of Central Andes (10°–26° S). 28th International Congress, Washington D.C., Abstracts, 3, p. 73.

Starck, D., Gallardo, E. and Schulz, A., 1993. Neopaleozoic stratigraphy of the Sierras Subandinas Occidentales and Cordillera Oriental, Argentina. Compte Rendus XII Congrés International de la Stratigraphie et Géologie du Carbonifere et Permien, vol. 2, pp. 353–372.

Suárez-Soruco, M. 1986. The Carboniferous system of Bolivia: a brief summary. Late Paleozoic of South America, Project 211 IUGS-UNESCO, Annual Meeting of the Working Group, Abstracts, pp. 11–26.

Thornburg, J.B., Thomas, A.R. and Johansen, S.J., 1993. The San Miguel Formation: mid Permian, postglacial sedimentation in the western Paraná Basin, southeastern Paraguay. Symposium on Carboniferous to Triassic Pangea (Calgary), Canadian Society of Petroleum Geologists and Global Sedimentary Geology Program, Abstracts, p. 308.

Van Wagoner, J.C., Mitchum, R.M., Campion, K.M and Rhamanian, V.D., 1990. Siliciclastic sequence stratigraphy in well logs, cores, and outcrops. American Association of Petroleum Geologists, Methods in Exploration Series, 7, 55 pp.

Veevers, J.J. and Powell, C. McA., 1987. Late Paleozoic glacial episodes in Gondwanaland reflected in transgressive-regressive depositional sequences in Euramerica. Geological Society of America Bulletin, 98: 475–487.

Villar, H. and López-Gamundí O.R., 1993. Carbones y pelitas carbonosas del Carbonífero de la cuenca de Paganzo: contexto litofacial y potencial generador de hidrocarburos. XII Congreso Geológico Argentino and II Congreso de Exploración de Hidrocarburos, vol. 1, pp. 375–381.

Visser, J.N.J., 1987. The palaeogeography of part of southwestern Gondwana during the Permo-Carboniferous glaciation. Palaeogeography, Palaeoclimatology, Palaeoecology, 70: 377–391.

Visser, J.N.J., 1989. The Permo-Carboniferous Dwyka Formation of Southern Africa: deposition by a predominantly subpolar marine ice sheet. Palaeogeography, Palaeoclimatology, Palaeoecology, 70: 377–391.

Visser, J.N.J., 1991a. Self-destructive collapse of the Permo-Carboniferous marine ice sheet in the Karoo Basin: evidence from southern Karoo. South African Journal of Geology, 94: 255–262.

Visser, J.N.J., 1991b. The paleoclimatic setting of the late Paleozoic marine ice sheet in the Karoo Basin of southern Africa. In: J.B. Anderson and G.M. Ashley (Editors), Glacial marine sedimentation; Paleoclimatic significance, Geological Society of America, Boulder, Special Paper 261, pp. 181–189.

von Brunn, V., 1987. A facies analysis of Permo-Carboniferous glacigenic deposits along a paleoscarp in northern Natal, South Africa. In: G.D. McKenzie (Editor), Gondwana Six: Structure, Tectonics and Geophysics, Geophysical Monography 40, pp. 113–122.

10. A REVIEW OF THE PERMO-CARBONIFEROUS GLACIATION IN AFRICA

Johan N. J. Visser

The Permo-Carboniferous glaciation was a major event in the geological history of the Gondwana supercontinent, of which Africa, together with the Arabian Peninsula and Madagascar, formed about 35 percent of the surface area. Permo-Carboniferous glacigene deposits occur over southern and central Africa and extend north into the Arabian Peninsula, suggesting widespread cold conditions over the continent. The objective of this chapter is to give a brief review of the glacial deposits and to present a series of maps depicting the deglaciation on the continent.

The Dwyka Group in the Karoo and the Kalahari basins (Fig. 10.1) has been studied in detail and an age framework established (Visser, 1990a), but information on the glacial deposits in the rest of southern and in central Africa is scanty and in most cases the precise age of the late Paleozoic deposits is uncertain. Du Toit (1937) was perhaps the first to depict the distribution of the "Great Ice Cap" over Africa during the Late Carboniferous. He placed the northern margin of the ice just north of the equator in central Africa. Frakes and Crowell (1970) and Crowell and Frakes (1975) depicted ice lobes radiating from subcontinental ice centers to the fringes of the present African continent. They concluded that the whole of southern and central Africa was glaciated, with the peripheral regions characterized by subaqueous deposition and the center of the continent dominated by the reworking of scant glacial debris. In their reconstruction they showed several diachronous centers of glaciation that migrated over the continent as Gondwana drifted over the pole. They also concluded that mountain glaciers rather than a continental ice sheet occurred in central Africa.

Rust (1975) suggested an ice sheet centered over the continental highlands with ice spreading out into the surrounding areas. In the Karoo and the Kalahari basins the ice flowed into a marine environment, whereas in East Africa it terminated on land giving rise to fluvioglacial deposits. He concluded that the actual centers of glacial deposition in central Africa were most probably determined by preexisting drainage and the location of lowlands produced by ice scouring. According to Veevers and Powell (1987) the mid-Carboniferous to Early Permian glaciation was the most extensive, and ice centers located over the highlands of southern Africa, extended into central South America. Langford (1991) suggested the presence of a latest Carboniferous to earliest Permian ice sheet over southern and central Africa, Madagascar, and parts of South America and Antarctica. A smaller ice sheet was located over areas of high relief in the horn of Africa and the southern part of the Arabian Peninsula. According to Langford (1991) the Gondwana Ice Sheet collapsed during the Early Sakmarian transgression, and by the Late Sakmarian only small ice caps remained over elevated areas in south and central Africa and as in the horn of Africa. By the Artinskian all ice caps had melted except for a possible small one near the present coast on the border between Namibia and Angola.

Visser (1993a) depicted a continuous ice sheet over southern and central Africa during the Late Carboniferous. The major ice-spreading center was located over the uplands of south-central Africa, from which ice lobes extended into South America and Antarctica. During the Early Sakmarian transgression the Karoo and the Kalahari basins became largely ice-free; the ice sheet was melting away over central Africa, but an ice lobe was maintained in the Paraná Basin. During the Artinskian, small ice caps remained over the uplands of southern Africa. A Gondwana Ice Sheet complex over Antarctica and southern Africa with its paleo-eastern margin in central Africa was postulated by Eyles (1993). A smaller

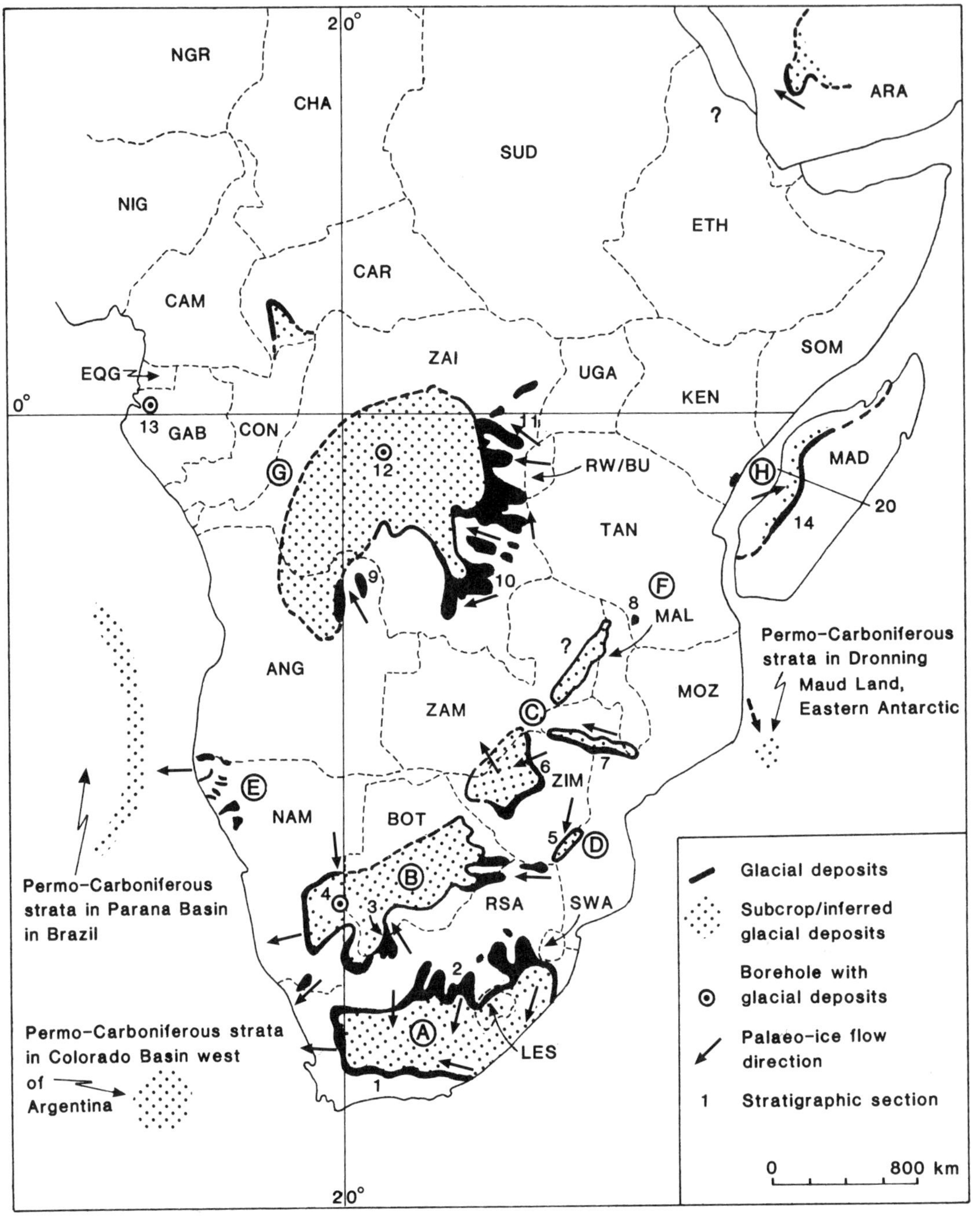

Fig. 10.1. Map showing the distribution of the Permo-Carboniferous glacigene deposits in Africa. Position of Arabia and Madagascar after De Wit et al. (1988). A = Karoo Basin; B = Kalahari Basin (including Warmbad Basin); C = Zambesi basins; D = Limpopo Basin; E = Huab Basin (including Kaokoveld valleys); F = Tanzanian basins; G = Congo Basin (including Gabon); H = Malagasy Basin (including coastal Kenya); ANG = Angola; ARA = Arabia; BOT = Botswana; CAM = Cameroon; CAR = Central African Republic; CHA = Chad; CON = Congo; EQG = Equatorial Guinea; ETH = Ethiopia; GAB = Gabon; KEN = Kenya; LES = Lesotho; MAD = Madagascar; MAL = Malawi; MOZ = Mozambique; NAM = Namibia; NGR = Niger; NIG = Nigeria; RSA = South Africa; RW/BU = Rwanda/Burundi; SOM = Somalia; SUD = Sudan; SWA = Swaziland; TAN = Tanzania; UGA = Uganda; ZAI= Zaire; ZAM = Zambia; ZIM = Zimbabwe

separate ice sheet occurred over India touching the horn of Africa.

Constraints on the Modeling of the Gondwana Ice Sheet over Africa

The differences in views on the reconstruction of the Gondwana Ice Sheet that I have outlined can be attributed to the scalar properties and dynamics of an ice sheet, a lack of age constraints on the glacigene deposits, factors controlling the preservation of glacial sediments, a lack of paleotopographic data, and a fragmented approach in the study of glacial sedimentation on Gondwana:

1. The maximum size at which an ice sheet can sustain itself over a supercontinent depends largely on climate and topography (Kerr, 1993), knowledge of the latter for Permo-Carboniferous Africa is limited. Frakes and Crowell (1970) and Eyles (1993) were in favor of different ice centers over Gondwana as the supercontinent drifted over the south pole. Dickins (1985) on the other hand found no evidence for the existence of a continental ice sheet over Gondwana. However, Langford (1991) comparing the surface area and Eyles (1993) the volume of the Gondwana Ice Sheet during the Late Carboniferous with that of the Pleistocene Ice Sheet in the northern hemisphere found no major differences in scale. Visser (1993a) calculated that the Gondwana Ice Sheet was about twice the size of the present Antarctic Ice Sheet during maximum glaciation. Thus, during maximum glaciation a continuous ice cover over a large part of Gondwana could have existed. An ice sheet is also a very dynamic system in which growth and decay are responses to climatic conditions over the supercontinent, land configuration, topography, and sea-level changes, so major in-phase as well as out-of-phase fluctuations in the shape and size of the ice sheet can be expected.

2. The lack of reliable age data for the glacigene deposits poses a major problem in ice sheet reconstruction. Glacial deposits commonly represent the deglaciation phase, whether final or interglacial. In the Karoo Basin, deposition by the marine ice sheet was triggered by changing environmental conditions during temporary (interglacial) as well as final collapses of the ice front (Visser, 1993b). Therefore, hardly any record exists for the onset of glaciation over Africa. It is postulated that glaciation could have started over elevated areas during the Namurian regression (Veevers and Powell, 1987; Visser, 1990a). In the western Argentinian basins, glacial deposition took place from the Namurian to the Early Westphalian and in the eastern basins from the Asselian to the Sakmarian (López-Gamundí, 1989; González, 1990), whereas in the Paraná Basin glacial deposition started during the Late Westphalian (Gravenor and Rocha-Campos, 1983; Dos Santos et al., 1988). Visser (1990a) suggested that glacial deposition in the Karoo Basin probably started during the Westphalian-Stephanian transition. There are also differences in opinion regarding the time of the collapse of the Gondwana Ice Sheet. Veevers and Powell (1987), Dickins (1985). and Langford (1991) assumed that the ice sheet completely collapsed during the Early Sakmarian transgression. In the Karoo, Kalahari, and Paraná basins and in eastern Australia evidence for Late Sakmarian and Artinskian glaciation is present (Visser, 1990a; Dos Santos et al., 1988; Banks, 1981). Therefore, the final collapse of the Gondwana Ice Sheet was a diachronous event controlled by local climate and topography. Time resolution within the glacial successions in Africa is also poorly controlled. However, up to three major depositional cycles can be recognized in the rock record, with the top of the middle one more or less at the Early Sakmarian (*Eurydesma*) sea-level highstand. Therefore, a "sequence stratigraphy" approach based on deglaciation sequences (DS) was used in correlating the glacial deposits in the various basins using the Dwyka Group in the Karoo Basin as a reference stratotype.

3. Factors controlling the preservation potential of glacial sediment are the base level of erosion, syndepositional subsidence and postglacial tectonics. Glacial material deposited on land has a very low preservation potential (less than 10%, according to Eyles, 1993) and will be reworked by postglacial processes unless subsidence within the region increases the preservation potential. On the other hand, glacial material deposited in a marine environment has a high preservation potential. However, isostatic rebound and later

tectonic uplift can result in the erosion of the glacial bedforms or deposits. Therefore reconstruction of the ice cover can be biased in favor of those regions where glacial deposits are preserved.

4. The fundamental effect of topography is to decouple the response of an ice sheet from the climate (Kerr, 1993). Although ice sheets are less constrained by bed topography, the latter controls the position of ice streams, whereas the margin topography controls the formation of glaciers and ice lobes. However, in regions where the ice sheet is thin, bed topography increasingly influences ice dynamics. Experience shows that glacial striae alone can be misleading for the regional interpretation of paleo-ice flow and should be evaluated together with the regional paleoslope (Visser, 1990b). The last phase of a glaciation is commonly characterised by ice caps over uplands and by valley glaciation with its characteristic lithofacies association (Visser, 1987a).

5. In modeling the glaciation of Africa, it is necessary to evaluate data from South America, Antarctica, and Madagascar, as these regions formed integral parts of Africa from the late Paleozoic to the Early Mesozoic. Therefore, tectonic or climatic events along the peripheral regions of southwestern Gondwana (Argentina, Brazil, the Weddell Sea region, and Dronning Maud Land) would have influenced glacial sedimentation in Africa.

Glacial Deposits

Descriptions of the glacial deposits in the Permo-Carboniferous African basins are based on fourteen stratigraphic sections (Fig. 10.2). However, to illustrate facies and age relationships additional sections from the Karoo Basin will be presented when describing the basin. Five lithofacies associations are recognized:

1. The *glaciotectonized bedrock-diamictite facies* is commonly thin (<10 m) and in composition matches that of the underlying bedrock. The clasts are commonly angular. The diamictite may show color banding due to the incorporation of bedrock material. These diamictites are interpreted as subglacial deposits preserved in bedrock lows.

2. The *massive homogeneous diamictite facies* consists of up to 250 m of diamictite without noticeable bedding. Small deformed as well as undeformed sandstone bodies may be present. Detailed studies of the diamictite show the presence of faint bedding planes on a meso-scale or planes defined by subtle changes in texture of the diamictite. The clasts show a poor fabric, with a high percentage of clasts having high ab-plane dips. These diamictites are interpreted as subaqueous deposits formed by either dense rain-out of glacial sediment, sediment gravity flow, or a combination of both.

3. The *stratified diamictite-rhythmite-shale facies* consists of alternating diamictite and mudrock forming sequences up to 300 m thick. The diamictites are bedded and may consist of stacked units. The rhythmite consists of alternating silty and clayey laminae. Lonestones are commonly present in the shale and the rhythmite. The facies was deposited subaqueously with the diamictites representing debris-flow deposits. The rhythmites formed either as distal silty turbidites interbedded with clay from suspension settling or as fall-out from under- or overflows (Visser, 1983a). Iceberg rafting of debris commonly occurred during deposition of the facies.

4. The *conglomerate-sandstone-diamictite facies* is thin (<60 m), and lithological units show rapid lateral as well as vertical facies changes. Diamictite commonly occurs at the base of the facies. The conglomerate fills erosional channels. The sandstone, which commonly contains pebbles or conglomerate lenses, varies from massive to cross-bedded. The facies is interpreted as an outwash deposit. The basal diamictite may have been deposited subglacially, but debris-flow diamictites interbedded with outwash material suggest subaerial outwash stream deposition or subaqueous debris-flow fan sedimentation where associated with subordinate mudrock.

5. The *conglomerate-sandstone-mudrock-coal* facies may vary from thin (<50 m) to several hundreds meters thick, building cyclic deposits. The internal arrangement of lithologies defines upward-fining as well as upward-coarsening sequences. The stones in the conglomerate may be faceted and striated. Cross-bedding is well developed in the sandstone. The upward-fining sequences represent fluvial deposits with coal

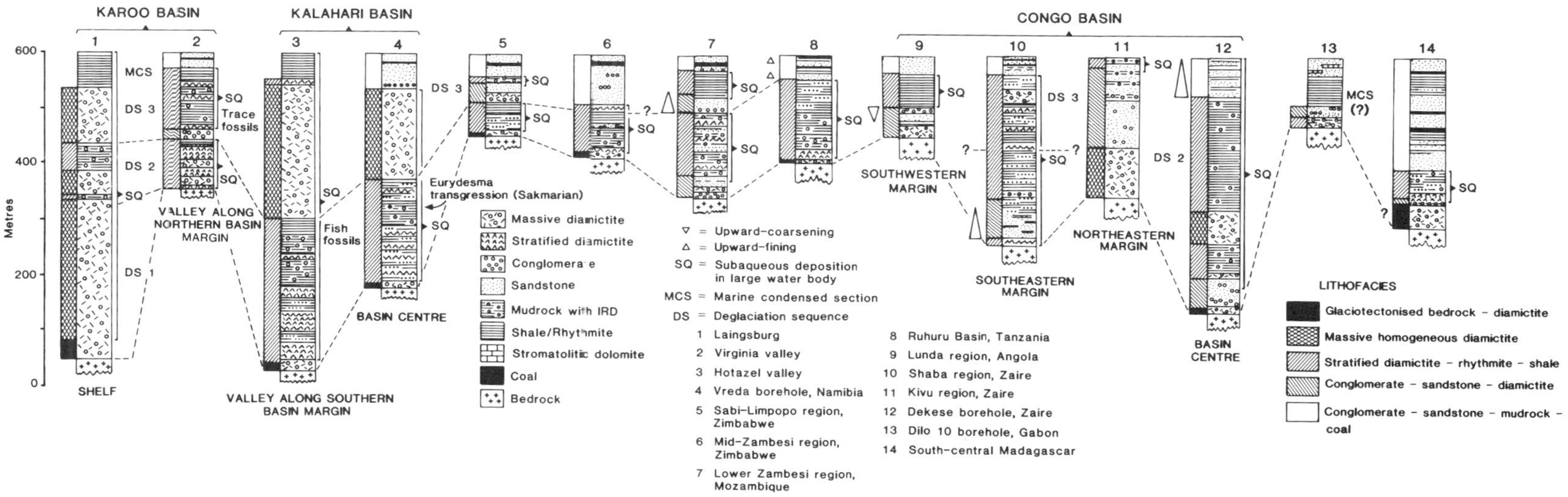

Fig. 10.2. Simplified stratigraphic sections of the most prominent Permo-Carboniferous basin fills in central and southern Africa. See Figure 10.1 for the location of the sections. Section 5 after Bond (1981a); 6 after Chappell and Humphreys (1970) and Bond (1981b); 7 after Rocha-Campos (1981a); 8 after Wopfner and Kreuser (1986); 9 after Rocha-Campos and Dos Santos (1981); 10, 11, and 12 after Cahen and Lepersonne (1981), 13 after Michelot et al. (1970), 14 after Rakotoarivelo (1970) and Frakes (1981).

formation during abandonment of the flood plain. The upward-coarsening sequences represent deltaic deposition with coal formation during abandonment of the delta plain.

Karoo Basin

The Karoo Basin lies entirely within South Africa and Lesotho (Figs. 10.1, 10.3). The basin evolved from a back-arc basin during the Late Carboniferous to a foreland one in the Late Permian (Visser, 1992). Outcrops of the glacigene Dwyka Group are present along the western, southern and eastern basin margins. Boreholes within the basin as well as along the northern margin supply additional information. The Dwyka Group along the southern basin margin is up to 800 m thick and consists of massive diamictite (± 80% of the rock volume) and subordinate mudrock (Fig. 10.2). Along the northern basin margin, a heterogeneous lithology consisting predominantly of mudrock and stratified diamictite is present within ancient glacial valleys up to 300 m deep. Visser (1986) therefore distinguished between a shelf facies assemblage in the south and an upland facies assemblage in the north.

North-south sections illustrate the influence of topography on the distribution of facies within the basin (Figs. 10.4, 10.5, 10.6). The uplands (Cargonian highlands) along the northern basin margin separated the Karoo and the Kalahari basins. Valleys were cut into the uplands from both the south and the north. Facies distribution shows that these uplands were a major source of glacial sediment, but large volumes of material were also transported into the basin from the east and the south.

The shelf deposits consist of the massive homogeneous diamictite and the stratified diamictite-rhythmite-shale facies which implies subaqueous deposition. However, ice was locally grounded as shown by the presence of the glaciotectonized bedrock-diamictite facies at the base of the glacial succession (Visser, 1989, 1990b). Most of the valley fills consist of the stratified diamictite-rhythmite-shale facies which also suggests subaqueous deposition in a typical fjord setting. In sections where a conglomerate-sandstone-diamictite facies is present (Koringhuis in Fig. 10.4, and section 2 in Fig. 10.2), those deposits represent subaqueous outwash facies. In the northeastern valleys the conglomerate-sandstone-mudrock-coal facies is present above the glacial deposits, suggesting postglacial deltaic infilling of the valleys. Deposition of the glacial

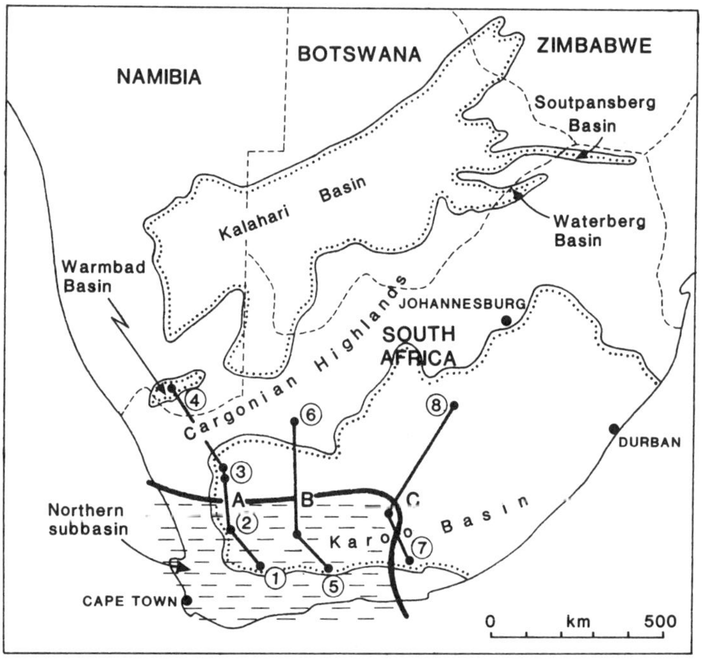

Fig. 10.3. Locality map showing the locations of the main basins containing glacial deposits in southern Africa. A, B, and C represent section lines. 1 – Laingsburg/Floriskraal area; 2 = Elandsvlei; 3 = Loeriesfontein; 4 = Aussenkjer; 5 = Klaarstroom; 6 = West of Prieska; 7 = Wolwefontein; 8 = Edenburg

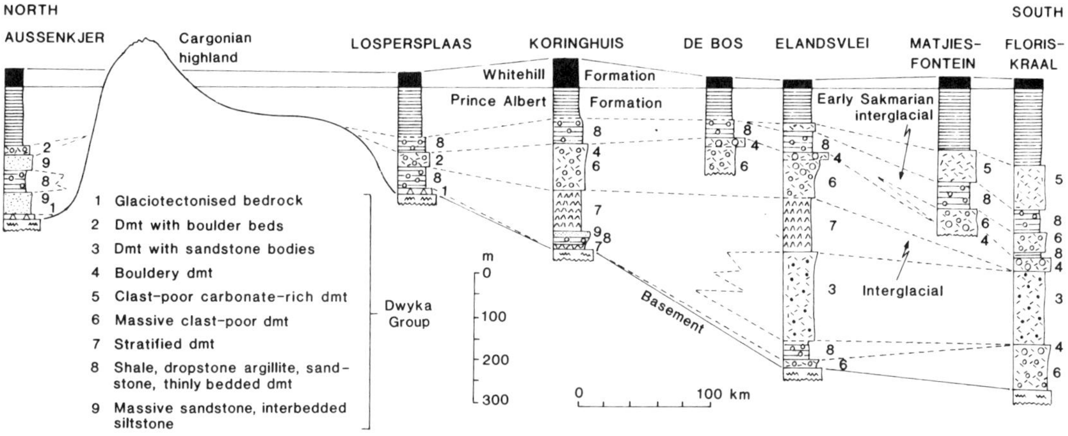

Fig. 10.4. North-south section along the western margin of the Karoo Basin. Section A in Figure 10.3.

deposits on the shelf was thus made by a marine ice sheet and those in the valleys by tidewater glaciers (Visser, 1989).

The shelf facies consists of three deglaciation sequences (Fig. 10.7). The lowest sequence (DS 1) was deposited only in the western part of the Karoo Basin (northern sub-basin in Fig. 10.3) and onlaps onto bedrock in the north and the east. The top of DS 1 consists of interglacial deposits that probably formed during the Early Stephanian. The top of DS 2 coincides with the Early Sakmarian sea-level highstand. Beds belonging to the upper part of DS 2 are present at the base of some valley fills. DS 3 is developed in the eastern and the southeastern parts of the basin and in the valleys (Figs. 10.4, 10.6). The Sakmarian-to-Artinskian age for the larger part of the valley fills is supported by palynology (Visser, 1990a).

It is speculated that ice flowed into the Karoo Basin after the Late Namurian (Visser, 1990a), but it was only since the sea-level highstand in the Westphalian (López-Gamundí, 1989; González, 1990) that deposition commenced in the Karoo Basin. Analysis of the three deglaciation

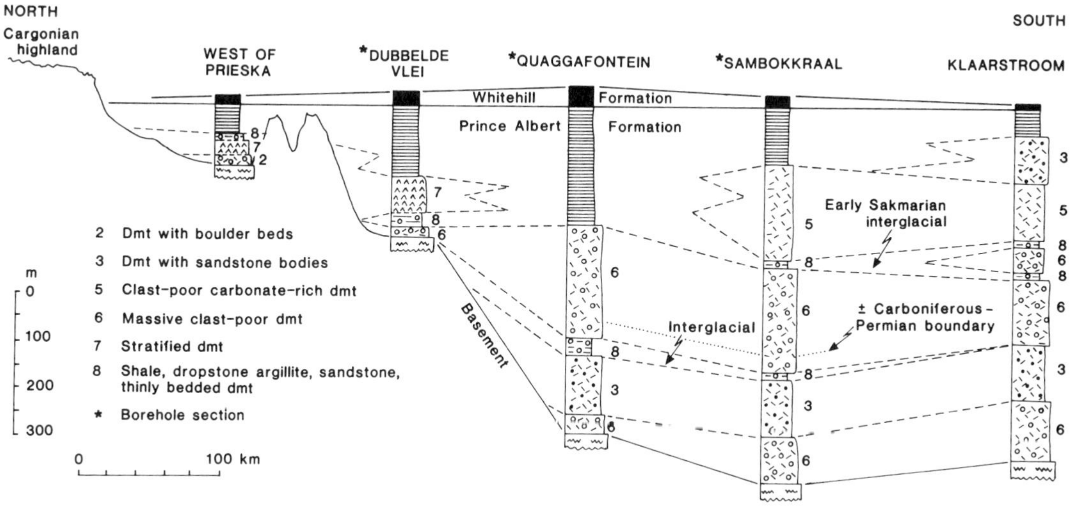

Fig. 10.5. North-south section along the western interior of the Karoo Basin. Section B in Figure 10.3. Carboniferous-Permian boundary after Anderson (1977).

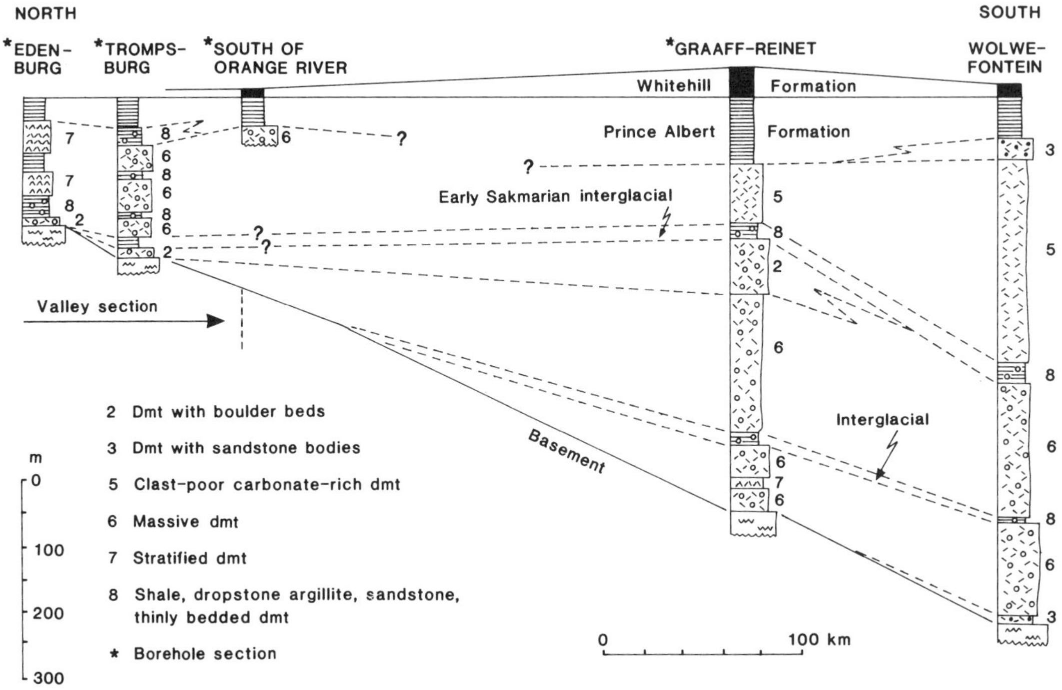

Fig. 10.6. North-south section along the center of the Karoo Basin. Section C in Figure 10.3.

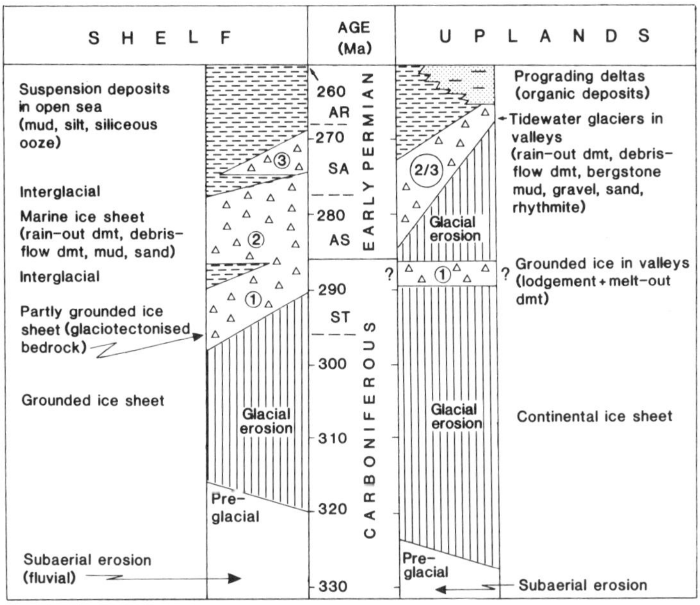

Fig. 10.7. The three deglaciation sequences (DS 1—3) in the Karoo Basin. ST = Stephanian; AS = Asselian; SA = Sakmarian; AR = Artinskian; dmt = diamictite. In only one valley along the northern margin of the Karoo Basin was possible evidence of DS 1 found at the base of the glacial succession. Time scale after Harland et al. (1982).

sequences suggets an evolutionary change in ice flow over southern Africa (Fig. 10.8). During the Late Carboniferous the uplands to the north of the Karoo Basin were a major ice-spreading center with ice streams flowing south into the Karoo Basin. The influence of this spreading center declined during the Sakmarian. A major ice-spreading center developed southeast of the Karoo Basin, supplying most of the sediments for DS 3. As the basin was open towards the west (Visser, 1987b), vast volumes of glacial sediment were transported beyond the western margin of the basin during the Late Carboniferous and deposited in the Colorado Basin (Fig. 10.1) where up to 1,000 m of massive diamictite is present (O. R. López-Gamundí, pers. commun., 1991).

Kalahari Basin and Outcrops in Northern Namibia

The western part of the Kalahari Basin lies within Namibia and the eastern part within Botswana (Fig. 10.1). On the southern border between Namibia and South Africa, the small Warmbad Basin (Karasburg Basin of Martin (1981b)) was originally part of the Kalahari Basin. Good outcrops are present only along the western basin margin. Elsewhere, postglacial mudrocks and younger deposits overlie the glacial succession so that borehole data are the main source of information. In the southeastern and the eastern sections of the basin, the Dwyka Group is thicker due to greater subsidence along fault-controlled troughs. The edges of the grabens near stable hinge zones were favorable sites for coal deposition (Langford, 1991).

In northern Namibia, glacial deposits are found in the Huab Basin, which formed the eastern margin of the large Paraná Basin in South America, in subsurface in Ovamboland, and in large glaciated valleys close to the border with Angola (Kaokoland).

Sedimentation in the Kalahari Basin as well as in northern Namibia was partly controlled by the paleotopography (Martin, 1981a; Visser, 1983b; Smith, 1984). Large valleys, some up to a 1,000 m deep, were cut along the southern, eastern, and northwestern margins of the Kalahari Basin (Fig. 10.1). Valleys in the east may have been up to 400 km long. The southern and the southeastern basin margins were defined by the Cargonian highlands, whereas the Windhoek highlands lay to the north of the basin. The basin was open to the west, partly via the Warmbad Basin (Visser, 1983b). The patchy distribution and the large variation in thickness of the basal diamictite unit also show the influence on sedimentation of bedrock topography. The basin floor apparently sloped towards the southwest. In northern Namibia preglacial relief was about 1,500 m (Martin, 1981a).

The internal arrangement of stratigraphic units shows great variation. However, a threefold subdivision can be recognized in the basin fill except along the northeastern margin where only two units are present (Fig. 10.2). A basal diamictite unit varying in thickness from 0 to 173 m, is present in bedrock lows (Smith, 1984). The basal unit is conformably overlain by an argillaceous unit consisting of shale and rhythmite with or without ice-rafted debris, thin diamictite beds and subordinate sandstones. To the east and in the Warmbad Basin, sandstones become more predominant at the base of this unit, whereas upvalley thin interbedded coals occur near the top of the mudrock sequence. The thickness of this unit is less influenced by bedrock topography and varies from about 10 to 230 m. Marine fossils (Eurydesma, Bryozoa, Asteroidean, Peruvispira, Orthoceras, radiolaria, foraminifera, goniatites, conularia, crinoid stems, and paleoniscoid fish) are present in the western part of the basin (Martin, 1981a).

The upper unit, which oversteps the lower units onto basement, consists of massive diamictite or pebbly mudstone with thin interbedded mudrocks attaining thicknesses of up to 250 m in the central and the southwestern parts of the basin (Visser, 1983b). To the east the unit becomes indistinguishable from the middle argillaceous unit which there forms the top of the glacial succession. Lateral facies changes to silty rhythmites with sparse clasts also occur in the west. Thin carbonate beds may be present within the pebbly mudstone. Several of the fossil species that occur in the middle argillaceous unit, as well as fossil wood, are present as transported fragments in the upper diamictite unit in the west (Martin, 1981a). Over the central and the western parts of the basin, the glacial beds are unconformably overlain by sandstone and pebbly sandstone.

Diamictite deposits in the Huab Basin have a maximum thickness of 7 m (Horsthemke et al., 1990). Frakes and Crowell (1970) described a 150m-thick sequence in a Kaokoveld valley consisting of alternating diamictite, sandstone, and conglomerate. Martin (1981a) found a boulder mudstone in a Kaokoveld valley that he correlated with the upper diamictite unit in the Kalahari Basin.

The basal diamictite unit consists, from the bottom up, of the glaciotectonized bedrock-diamictite, the massive homogeneous diamictite, and the shale-rhythmite-stratified diamictite facies. This implies grounded ice followed by resedimentation of glacial debris in bedrock lows that were probably flooded during onset of deglaciation. The middle argillaceous unit consists of the shale-rhythmite-stratified diamictite facies, suggesting subaqueous deposition during the Early Sakmarian sea-level highstand. Marine conditions prevailed in the west (Martin, 1981a; Visser, 1983b; Smith, 1984) and brackish conditions may have dominated in the east. The sandstones in the Aussenkjer section (Fig. 10.4) represent delta front deposits, whereas the conglomerate-sandstone-diamictite facies along the northeastern basin margin was deposited as glacial outwash. The upper unit consists of the massive homogeneous diamictite and the stratified diamictite-rhythmite-shale facies. This implies subaqueous rain-out of glacial sediment and resedimentation by sediment gravity flow. In the upper part of valleys and in the central and the western parts of the basin, a conglomerate-sandstone-mudrock-coal facies was deposited. These represent fluvial and coastal swamp deposits. The thin glacial deposits in the Huab Basin may have been deposited subglacially (Horsthemke et al., 1990). The conglomerate-sandstone-diamictite facies in a Kaokoveld valley represents a subaqueous outwash fan deposit which implies that some of the Kaokoveld valleys were typical fjords. Martin (1981a) has suggested that glacial deposition there occurred under marine conditions.

Two well-defined deglaciation sequences can be recognized in the central and the western parts of the basin (Fig. 10.2). The upper part of the lower sequence (middle argillaceous unit) contains Eurydesma fossils of Early Sakmarian age (Dickins, 1985). Thus during the Early Sakmarian transgression, muds were deposited in the basin, except along the eastern basin margin and in the

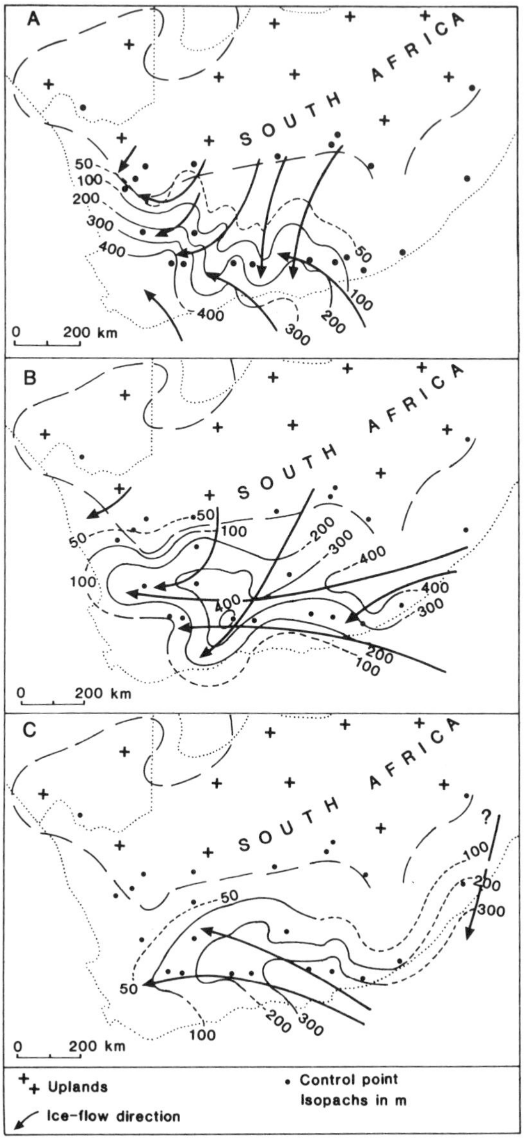

Fig. 10.8. Isopach maps of deglaciation sequences in the Dwyka Group. A = Ds 1 below the Early Stephanian interglacial; B = DS 2 below the Early Sakmarian interglacial; C = DS 3 postdating the Early Sakmarian interglacial. Palaeo-ice flow based on striated surfaces, clast composition, and facies changes.

Warmbad Basin, where glacial outwash gravels and sands built fan deltas. The upper deglaciation sequence, which is Sakmarian to Artinskian in age (Visser, 1990a), was deposited only in the western part of the basin. The boulder mudstone in northern Namibia can be correlated with a similar unit in the Paraná Basin that may be of Artinskian age (Dos Santos et al., 1988; Visser, 1990a). The lower deglaciation sequence can thus be correlated with DS 2 and the upper one with DS 3 in the Karoo Basin.

Ice flow patterns for DS 2 were toward the west and the southwest. Sediment was also transported into the basin along north-south striking valleys from the Cargonian and Windhoek highlands (Fig. 10.1). The Sakmarian-Artinskian DS 3 was deposited by ice flowing toward the basin from the Cargonian and the Windhoek highlands. In the Huab Basin and the Kaokoveld valleys ice flow was to the west, transporting sediment into the Paraná Basin of Brazil (Martin, 1961, 1981a; Horsthemke et al., 1990).

Zambesi-Limpopo Basins

Outcrops of glacial deposits occur in the mid-Zambesi and the Sabi-Limpopo regions of Zimbabwe and in the Zambesi-Limpopo region of Mozambique (Fig. 10.1). Frakes and Crowell (1970) believe that these are erosional remnants of a formerly much more extensive area of deposition because of the uneven floor and the absence of a coarse-grained marginal facies. According to Rust (1975), glacial sedimentation in the Zimbabwe terrain took place in separate downwarped basins (half-grabens), partly bounded by fault scarps. Rust (1975) shows a connection between the Kalahari Basin and the Zambesi basins during the Early Permian, but these latter basins were separated from the central African ones by the Nyasa rise. Outcrops within the basins show deformation and tilting, suggesting that postglacial faulting give rise to the preservation of the glacial beds. The paleotopography was uneven (Bond, 1970; Chappell and Humphreys, 1970) with paleoslopes towards the west and the south in the case of the Sabi-Limpopo region.

The glacial beds, which have a maximum thickness of about 160 m, consist of arenaceous to argillaceous diamictite, conglomerate, sandstone, and silty rhythmite with ice-rafted debris and mudstone. Diamictite or conglomerate is commonly present at the base of the succession and is overlain by mudrock and rhythmite with subordinate sandstone (Fig. 10.2). In the mid-Zambesi region, Bond (1970) described a facies change from east to west with an increase in mudrocks towards the basin center. Glacial deposits are absent in the west.

In the lower Zambesi region, a conglomerate-sandstone-diamictite facies formed by glacial outwash is present at the base of the succession. These deposits are overlain by the stratified diamictite-rhythmite-shale facies deposited subaqueously. Chappell and Humphreys (1970) and Bond (1981a) described a glacial pavement and injections of diamictite in the bedrock in the mid-Zambesi region that suggest that glaciers had been in contact with the bedrock prior to deposition of the overlying stratified diamictite-rhythmite-shale facies. Thus, west and south of the Lomagundi uplands, ice retreat was almost immediately followed by flooding and the deposition of debris-flow diamictites and turbidites and the suspension settling of mud in a possible freshwater lake (Bond, 1981b). In the lower Zambesi region, flooding was preceded by glacial outwash upon retreat of the ice. The conglomerate-sandstone-mudrock-coal facies occurs at the top of the succession, suggesting delta progradation into the lakes (Fig. 10.2). This delta progradation probably followed a renewed advance of the ice into the basin, evidence of which is less obvious in the lower Zambesi region than to the west and the south of the Lomagundi uplands.

Although Bond (1970, 1981a, 1981b) recognized up to three glacial events in the basins, only two deglaciation sequences are evident and can be tentatively correlated with the two sequences (DS 2, DS 3) in the Kalahari Basin (Fig. 10.2). DS 2 constitutes most of the succession. In the lower Zambesi region this correlation becomes uncertain as no evidence of glacial deposition is present in the upper third of the succession (Fig. 10.2). There are no proper age data for the glacial deposits. Pollen analyses of the rhythmites below the coals suggest an Early Permian age, with the result that Bond (1981a, 1981b) attributed a latest Carboniferous age to the underlying diamictites. If the correlation between the Kalahari and the Zambesi basins is correct, the top of DS 2 corresponds with the Early Sakmarian Eurydesma transgression (Fig. 10.2). This is in line with the suggestion by Bond (1970) that palynological evidence is consistent with a younger Early Permian age. The poorly developed DS 3 capped by the coal measures could be Sakmarian to Early Artinskian in age.

Ice flow in the mid-Zambesi region was to the west and the northwest, in the lower Zambesi region to the west-northwest and in the Sabi-Limpopo region to the south. The eastern margin of the mid-Zambesi basin consisted of the escarpment of the Lomagundi uplands with deep west-trending incised valleys (Bond, 1970). The present drainage has re-excavated these valleys floored with fluvio-glacial outwash. The lower Zambesi region, where fluvioglacial deposits also occur at the base of the succession, represents a major west-trending valley system also incised in the basement uplands of Mozambique. In the same way the glacial deposits of the Sabi-Limpopo region were preserved in a south-southwest-trending valley system incised into the Lomagundi-Mozambique uplands (Rocha-Campos, 1981a).

Although Rust (1975) inferred isostatic rebound for this part of Gondwana during deglaciation, a postglacial transgression is evident in the Karoo, Kalahari, and Zambesi basins, explaining the overlapping of postglacial mudrocks onto the bedrock. A rise in the inland lake levels as a result of the large meltwater input would explain the flooding of these basins and glacial valleys.

Tanzanian Basins

Paleozoic to Early Jurassic sedimentary rocks equivalent to the Karoo Supergroup of southern Africa cover a large area in eastern Tanzania. Rocks belonging to the basal unit are exposed only in the Ruhuhu Basin situated on the northeastern shore of Lake Nyasa (Wopfner and Kreuser, 1986). The preserved succession represents relics of the original basin fill that had been down-faulted into the basement. There may have been 300 to 450 m relief in the basin, with maximum sediment accumulation in depressions.

The glacial succession, which consists of a basal diamictite followed by bedded diamictite, rhythmite with ice-rafted debris, and finally black shale with carbonate concretions, has a thickness of about 100 m (Fig. 10.2). However, diamictite with a thickness exceeding 110 m was recorded in boreholes (Wopfner and Kreuser, 1986). In many instances the black shale unit overlaps the glacial beds and rests directly on crystalline basement. This succession is unconformably overlain by coal measures and by Late Permian strata with marine beds at the top. The black shale contains microflora indicating an Asselian age, whereas the coal measures are of Artinskian age (Wopfner and Kreuser, 1986). The glacial succession is thus tentatively correlated with DS 2 of the southern basins.

The glacial beds belong to the stratified diamictite-rhythmite-shale facies, suggesting subaqueous deposition. Wopfner and Kreuser (1986) interpreted the massive diamictite at the very base of the succession as a subglacial deposit, but the thick diamictites encountered in boreholes argue against such an interpretation. The diamictite formed by resedimentation of glacial debris in a freshwater lake fed by melting ice on the uplands. The coal measures belong to the sandstone-mudrock-coal facies forming upward-fining sequences. These suggest fluvial deposition.

No data on ice-flow directions are available for the Ruhuru Basin. North-northeast-trending uplands probably formed during the commencement of the East African rifting (Wopfner and Kreuser, 1986), thereby separating the Tanzanian basins from the large Congo Basin to the west. Wopfner and Kreuser (1986) are of the opinion that the rising rift shoulders could have formed ice-spreading centers during the glaciation so that one can infer ice flow into the basin from the west and possibly from the south. The strata within the Ruhuru Basin, including the Late Permian marine transgression, can be correlated with deposits in the Malagasy Basin to the northeast. This suggests a connection between the

Tanzanian and the Malagasy basins having a northeastern paleoslope. Regional ice flow out of the Ruhuru Basin was thus most likely to the northeast.

Congo Basin (including Gabon)

The Congo Basin underlies a large area in Zaire and the northeastern part of Angola and extends into the Central African Republic (Fig. 10.1). Outcrops near the coast in Gabon probably form part of the Congo Basin (Rust, 1975). Most of the basin is covered by younger rocks, outcrops are restricted to the eastern basin margin in Zaire (Kivu and Shaba regions) and the southwestern margin in northeast Angola. Outcrops of Karoo strata are present in Gabon, but the glacigene basal unit was intersected only in boreholes (Hambrey, 1981a). The same applies to the central part of the Congo Basin, where the thickest glacial succession was intersected in the Dekese borehole (Cahen and Lepersonne, 1981). The Congo Basin may represent an intracratonic sag basin (Langford, 1991).

The paleotopography is exposed along the eastern margin of the basin. Diamictite occurs in stepped U-shaped and in hanging paleovalleys exhibiting striated surfaces and roche moutonnées (Cahen and Lepersonne, 1981). The paleoslope was toward the west. The outcrops in northeastern Angola are structurally controlled, but deposition may have occurred in north-trending paleovalleys (Rocha-Campos and Dos Santos, 1981). The diamictite in western Gabon occurred in linear depressions with an east-west strike and may represent glacial valleys (Micholet et al., 1970).

The glacigene beds, which have a maximum thickness of about 450 m in the basin center, belong to the Lower Subgroup of the Lukuga Group in Zaire. Borehole cores in the basin center and outcrops in the northeast (sections 11, 12 in Fig. 10.2) show more or less the same stratigraphic succession, with diamictite and pebbly sandstone at the base and predominantly mudrock with ice-rafted debris at the top. The section in the southeast (section 10 in Fig. 10.2) consists primarily of mudrock and rhythmite with subordinate bedded diamictite. Mudrock and sandstone with interbedded coals follow directly on the glacials. The outcrops in northeastern Angola (Lutôe Series) show much more variation in lithology, but regionally also consist of basal diamictite overlain by pebbly sandstone and conglomerate (section 9 in Fig. 10.2). Mudrocks with no evidence of glaciation are common toward the top of the succession. The thickness of the glacial beds may be about 40 m, as the upper contact is difficult to establish. The glacial beds in Gabon are about 17 m thick and consist predominantly of stratified diamictite overlain by cross-bedded sandstone and conglomerate. Pyrite-bearing black shale with cherts and phosphate abruptly overlies the thin transition at the top of the glacials (Micholet et al., 1970).

The glacial succession (Lukuga Glacials) along the northeastern basin margin (Kivu region) has a thick basal diamictite overlain by a conglomerate-sandstone-diamictite facies. A conglomerate-sandstone-diamictite facies representing glacial outwash deposits is present at the base of the succession in the central part of the basin (section 12 in Fig. 10.2). Along the southeastern basin margin (Shaba region) two stratified diamictite-rhythmite-shale facies are separated by about 50 m of mudrock (section 10 in Fig. 10.2). In the Shaba region, the entire glacial succession, which is overlain by a fluvial sandstone-mudrock-coal facies, was thus deposited subaqueously. In the northeastern and the central parts of the basin the glacial outwash facies is overlain by a stratified diamictite-rhythmite-shale facies that was deposited subaqueously (sections 11, 12 in Fig. 10.2). The origin of the thick, apparently massive basal diamictite in section 11 from the Kivu region is uncertain. Studies from the Karoo Basin have shown that thicknesses of 100 m of diamictite cannot be deposited subglacially during a single deglaciation event, so the beds probably originated as stacked debris-flow diamictons that accumulated subaqueously (Visser, 1989).

The glacial sections in northeastern Angola consist of a stratified diamictite-rhythmite-shale facies overlain by an upward-coarsening conglomerate-sandstone-mudrock facies representing a regressive event (section 9 in Fig. 10.2). The basal diamictite contains sandstone and conglomerate bodies suggesting subaqueous deposition (Rocha-Campos and Dos Santos, 1981). In the Gabon section (13 in Fig. 10.2), a stratified diamictite-rhythmite-shale facies occurred at the base of the glacial succession and is overlain by a conglomerate and cross-bedded sandstone facies, suggesting infilling of the basin very much like that which occurred in the Angolan sections. Micholet et al. (1970) interpreted the overlying black shale as a lacustrine deposit, but the presence of pyrite, phosphorite and chert in the shale and the stromatolitic dolomites overlying the shale may indicate a marine incursion (Cook, 1984).

Along the eastern and the southern parts of the Congo Basin (Kivu and Shaba regions and northeastern Angola) and in the west, the ice front retreated in water, so glacial deposition occurred subaqueously. Along the paleo-northern basin margin (presently basin center) the ice retreated on land. The glacial outwash in this region was subsequently inundated due to rising lake levels. This implies that ice retreat started first in the north; when lake levels subsequently rose (perhaps as a result of the melting of the ice), it triggered the retreat of the ice along the eastern and southern basin margins. Infilling of the basin in the southeast resulted in the deposition of coal measures probably during the Late Sakmarian or the Early Artinskian.

Although there may be a hint of two deglaciation sequences in the southeast of the Congo Basin, all other sections represent only one glacial event. There are no proper age data available, and the glacials are described as Late Carboniferous to Early Permian in age (Hambrey, 1981; Rocha-Campos and Dos Santos, 1981). The upper part of the succession along the eastern basin margin may be Early Sakmarian in age (Cahen and Lepersonne, 1981). These data correspond with the age of DS 2 in southern Africa. Therefore, the age of the glacial deposits in the Congo Basin is taken as very Late Carboniferous to Early Sakmarian.

Reliable paleo-ice flow directions along the eastern basin margin suggest a paleoslope to the west (Fig. 10.1). A north-trending upland to the east of the basin must have formed a prominent ice-spreading center (Frakes and Crowell, 1970). Ice flow in northeastern Angola was to the north-northwest, probably along a major valley system (Fig. 10.1; Rocha-Campos and Dos Santos, 1981). The presence of glacial outwash along the northern basin margin suggests a possible south-dipping paleoslope at the site of the Dekese borehole. In Gabon no definite direction of ice flow can be established except for east-west linear depressions in the basement (Micholet et al., 1970). If the interpretation of a marine incursion is correct, then this event could have taken place only from the west. Ice could thus have flowed west out of the Congo Basin and deposited the glacial beds in the Sergipe-Algohas Basin of northeastern Brazil (Rocha-Campos, 1981b; Langford, 1991).

Malagasy Basin (Including Coastal Kenya)

The glacial beds occur at the base of the Sakoa Group, a thick succession of moderately deformed late Paleozoic to Mesozoic sedimentary rocks. The group forms an outcrop belt along the length of Madagascar and can be correlated with late Paleozoic strata in Kenya in East Africa. The glacial succession outcrops in a limited region in south-central Madagascar (Fig. 10.1; Frakes, 1981). In Kenya, where the preservation of the Karoo strata was fault-controlled, the glacial beds outcrop in the Kilifi area along the coast (Cannon et al., 1981).

Very little is known about the paleotopography of Madagascar. In coastal Kenya deposition occurred in a depression with a width of between 200 and 300 km and in a paleoslope to the north-northeast (Cannon et al., 1981).

The glacial beds, which have a thickness of up to 450 m (Rakotoarivelo, 1970), consist of a basal massive diamictite overlain by a conglomerate bed, a thin succession of stratified diamictite, and then pebbly mudstone with thin interbedded sandstones at the top (section 14 in Fig. 10.2). The glacial beds are overlain by a 1,200m succession of coal, sandstone, and shale. In coastal Kenya a basal diamictite is followed by feldspathic sandstones that Cannon et al. (1981) correlated with the coal measures in Madagascar. Frakes (1981), who ascribed a

Carboniferous to Permian age to the glacial deposits, suggested the presence of two glacial episodes in the Malagasy Basin, but sufficient evidence is lacking. The conglomerate bed probably represents reworked glacial sediment during a transgression. Only one deglaciation sequence that can be tentatively correlated with DS 2 in east and central Africa (Ruhuru Basin in Tanzania and Congo Basin in Zaire), is present. In the Malagasy Basin, limestones with a mid-Permian age overlie the fluvial beds. The age of the glacials can thus be constrained to very Late Carboniferous to Early Sakmarian, whereas the overlying coal measures may have a possible mid- to Late Sakmarian age.

The basal diamictite may have been deposited subglacially. The largest part of the succession consists of the stratified diamictite-rhythmite-shale facies, indicating subaqueous deposition in a possible freshwater lake (Frakes, 1981). The overlying sandstone-mudrock-coal facies represents infilling of the basin by deltaic and swamp deposits.

No direct evidence for the direction of paleo-ice flow is available. Frakes and Crowell (1970) measured paleocurrents directed to the northeast. The Malagasy Basin formed the distal part of the linear north-northeast-trending depression of coastal Kenya, so one can infer regional paleo-ice flow to the northeast (Fig. 10.1). The marine incursion into the basin was an arm of the Tethyan ocean (Kamen-Kaye, 1978; Cannon et al., 1981; Langford, 1991), supporting the inferred ice-flow pattern. Highlands that formed ice-spreading centers were thus located to the west and the southwest in Africa (Frakes and Crowell, 1970).

Other Related Areas

Reference to the deglaciation along the eastern margin of South America and its relation to events in Africa has already been made in discussions of the Karoo, Kalahari, and Congo basins. The Ogaden Basin in southeast Ethiopia contains a fairly complete succession of Karoo rocks, but glacial deposits are noticeably absent (Worku and Astin, 1992). The base of the Karoo succession there consists of fluvial fan deposits of Early to Late Permian age. Dow et al. (1971) described Paleozoic glacial rocks in northern Ethiopia, but Hambrey (1981b) considered them more likely to be of Late Ordovician age. Glacial deposits in southwestern Arabia, which, in a Gondwana reconstruction, formed part of Africa (Fig. 10.1), are of Stephanian to Sakmarian age (McClure and Young, 1981) and they can possibly be correlated with DS 1 and 2 in southern and central Africa. Descriptions of the strata, however, suggest the presence of only one deglaciation sequence, making a correlation with DS 2 more likely. The source of the glacial sediments was to the southeast (Fig. 10.1).

The late Paleozoic basin in which the lower part of the Beacon Supergroup in the Weddell Sea region of Antarctica (Ohio Range and Pensacola and Ellsworth mountains) was deposited continued in the Karoo Basin of South Africa (Elliot, 1975; Visser, 1989; Langford, 1991). Judging from deglaciation sequences in the Karoo Basin, most of the diamictites in those areas would be of Asselian to Artinskian age (DS 2, DS 3). To the north in Dronning Maud Land, thin diamictites are overlain by sandstone with plant remains of a possible Asselian to Sakmarian age (Larsson et al., 1990). The paleoslope was apparently towards the south (Fig. 10.1). The glacial deposits in the Weddell Sea region and Dronning Maud Land thus show a relationship similar to that between the Karoo Basin and central African basins in that DS 2 was the most widespread deglaciation sequence, and DS 3 was preserved only along the shelf regions.

Deglaciation History

Modeling the behavior of the Gondwana Ice Sheet during its existence over Africa of at least 50 Ma requires an integration of data pertaining to southwestern Gondwana, along with a fair amount of interpretation on the ages of glacial events, paleotopography, and postglacial tectonism. In some areas the deglaciation history can be modeled with greater confidence since there are more data, but in central Africa and in Madagascar extrapolation is required. Three time slices based on the deglaciation events in the Karoo Basin were selected, but maps depicting maximum and minimum ice cover are constructed for only the very Late Carboniferous to the Early Sakmarian and the Late Sakmarian to the Artinskian.

Late Westphalian-Stephanian

The oldest preserved diamictites of a possible Late Westphalian to Stephanian age are found at the base of the Dwyka Group in the northern sub-basin located in the western part of the Karoo Basin (unit 6 in Figs. 10.4, 10.5; Fig. 10.3). The extent of the predepositional ice cover over Africa is unknown because of the absence of glacial sediment. Glacial deposition in the Karoo Basin (base of DS 1) was apparently initiated by a relative sea-level rise during the long-lasting Late Westphalian interglacial period which is well recorded in the Argentinian basins (González, 1990; López-Gamundí et al., 1992). Glacial sedimentation in the western half of the Karoo Basin was terminated with the deposition of a thin mud unit (top of DS 1), suggesting open water during the Late Stephanian (Fig. 10.5).

Very Late Carboniferous-Early Sakmarian

Close to the end of the Carboniferous the Gondwana Ice Sheet expanded again to reach what was probably it greatest coverage of the supercontinent during the Asselian (Frakes and Crowell, 1970; Dickins, 1985). During the ensuing deglaciation sediments of DS 2 were widely deposited in the African basins as well as in the Paraná and Malagasy basins, Arabia, and Antarctica.

During maximum ice cover over Africa, the margin of the continental ice sheet was located over central Africa and extended into Arabia (Fig. 10.9). Langford (1991) suggested that Arabia was at such a low paleolatitude (± 40°) that it is doubtful whether an ice sheet could have been maintained. However, there were major uplands in the horn of Africa that could have produced favorable conditions for ice sheet formation with ice flowing northwest into Arabia (Fig. 10.9). Ice streams flowed into the Congo, Paraná, Kalahari and Karoo basins. Ice-spreading centers were located over uplands that more or less followed the structural trend in Africa. Marine sections of the ice sheet were present in the Paraná and Karoo basins. Part of the Paraná Basin and the area to the west of the Karoo Basin was inundated by an ice-free sea in which water depth and oceanic circulation controlled the position of the ice margin. In western Argentina a cold dry climate prevailed (López-Gamundí et al., 1992), whereas a marine embayment existed in the northern part of South America (Langford, 1991).

The Early Sakmarian transgression associated with an interglacial caused the collapse of part of the Gondwana Ice Sheet. Collapse was severe in the intermediate paleolatitudinal regions of Africa where the Congo and Malagasy basins became almost ice-free (Fig. 10.10). Meltwater lakes formed in the Zambesi, Malagasy, and Congo basins. Collapse of the marine sections of the ice sheet left large parts of the Paraná, Kalahari and Karoo basins also ice-free. The possible marine conditions in the Gabon region of the Congo Basin could be attributed to an eastward extension of the marine embayment in the north of South America. Deposition of DS 2 was terminated in the Karoo and the Kalahari basins by suspension settling of marine muds, in the Zambesi and the Tanzanian basins by lacustrine muds, and in the Congo and the Malagasy basins by lacustrine muds and fluvioglacial outwash. Coals were deposited in the Malagasy Basin (Fig. 10.10).

Late Sakmarian-Artinskian

During the Late Sakmarian the Gondwana Ice Sheet advanced from the paleo-south into the Karoo Basin, but over the rest of Africa the continental part of the ice sheet continued to retreat, with only ice caps eventually remaining over uplands (Fig. 10.11). Upon ice retreat, large volumes of meltwater were discharge, so in low-lying regions fluvial, lacustrine, and deltaic deposition prevailed. The cool-to-temperate climate enhanced the formation of peatlands in

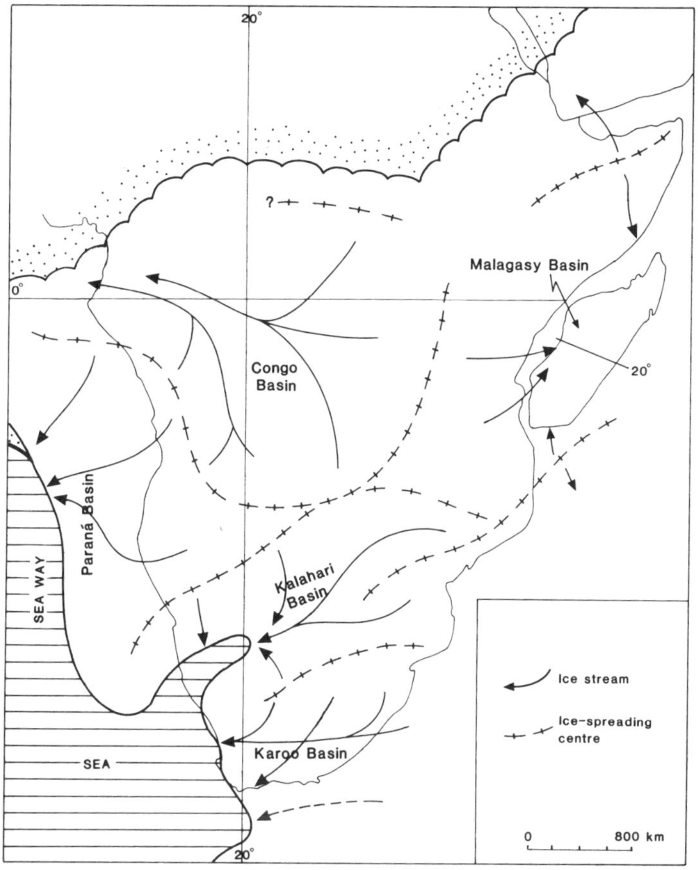

Fig. 10.9. A reconstruction of the Gondwana Ice Sheet over central and southern Africa during the latest Carboniferous to Asselian.

those regions. Glaciers flowed into the Paraná and the Kalahari basins, but polar conditions prevailed in the southern part of the Karoo Basin, which was covered by a marine ice sheet (Visser, 1991a). In southern Africa and eastern Brazil, the basal sediments of DS 3 were deposited under marine as well as freshwater conditions. The basins in northern South America were apparently cut off from the paleo-Pacific Ocean so evaporites formed in the western part of the Congo Basin during arid conditions (Langford, 1991).

All the ice had melted over central Africa by the Late Artinskian, and fluvial sedimentation had taken over the scene (Fig. 10.12). The Artinskian collapse of the marine ice sheet in the Karoo Basin was triggered by a relative sea-level rise due to basin subsidence associated with the transition from a back-arc to a foreland basin (Visser, 1992). This caused the abrupt transition from diamicton deposition to black marine mud (Visser, 1991b). Along the northern margin of the Karoo Basin tidewater glaciers fed by small ice caps over the uplands in Namibia-Angola, Zimbabwe and South Africa were maintained in some fjords. Infilling of the fjords ended with the formation of peatlands along the coastal regions and deltas (Fig. 10.12).

The glaciation model outlined for Africa is a key to our understanding of the paleo-northern and northeastern margins of the Gondwana Ice Sheet. However, the temporal and spatial behaviour of the ice sheet margin over the horn of Africa and Arabia (in the region of the Tethyan Sea) remains unresolved, as is clearly illustrated by recent reconstructions (Langford, 1991; Eyles, 1993). A study of the relationship between the Malagasy sea way and ice-sheet fluctuation over India and the East Antarctic shield may be of assistance in solving the problem.

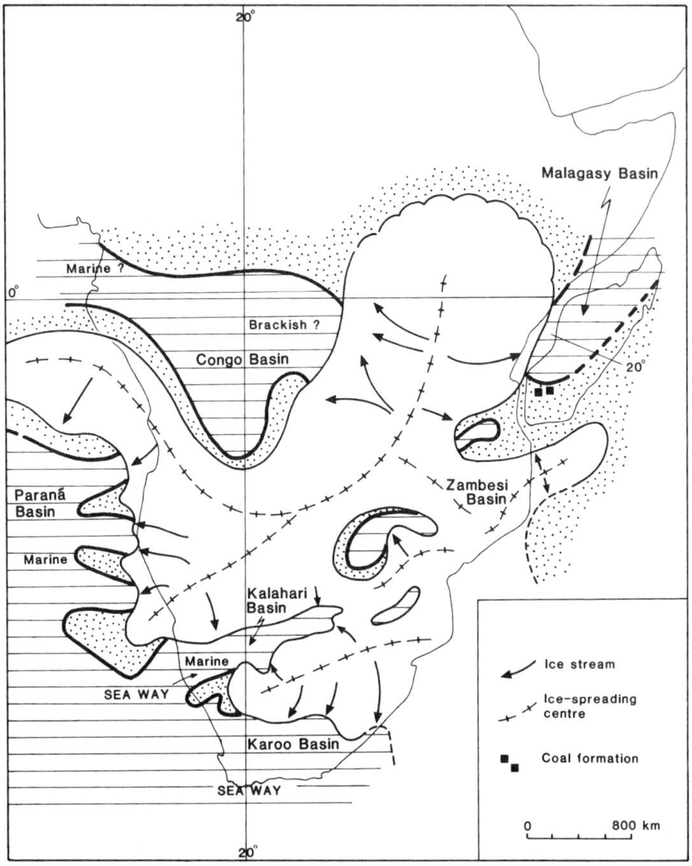

Fig. 10.10. A reconstruction of the Gondwana Ice Sheet over central and southern Africa during the Early Sakmarian (*Eurydesma*) transgression.

Conclusion

1. During the Permo-Carboniferous the interior of southern and central Africa consisted of a high-lying plateau region, except for the southern tip, where a marine shelf forming part of a large paleo-south-striking seaway in southwestern Gondwana was present. A marine embayment (sea way) extended from the Tethyan sea into the Malagasy Basin during the Early to mid-Permian, whereas part of the Congo Basin could have had a marine connection via northern South America in the Early Permian.

2. Rift basins in south-central and in east Africa were controlled by tectonic trends inherited from Proterozoic events in the African crust. Upstanding rift shoulders formed ice-spreading centers, whereas rift basin subsidence restricted ice cover after the Sakmarian.

3. Three deglaciation sequences (DS 1, 2, 3) were deposited in Africa. Sediments of the Stephanian DS 1 were preserved only in the shelf region; the Early Sakmarian DS 2 was the most widespread depositional event and glacigene deposits of the Artinskian DS 3 are confined to southern Africa. In central and in east Africa fluvial and lacustrine sediments accumulated simultaneously with the glacigene beds of DS 3.

4. The Gondwana Ice Sheet attained its maximum extent over Africa apparently from the Late Stephanian to the Asselian. Ice-spreading centers within the continental ice sheet were controlled by uplands. Ice flow moved largely via ice streams into low-lying regions, allowing modification (overdeepening) of the preglacial drainage patterns. The last phase consisted of a mountain glaciation, with valley glaciers and freshwater lakes in the interior basins. In the Karoo Basin a

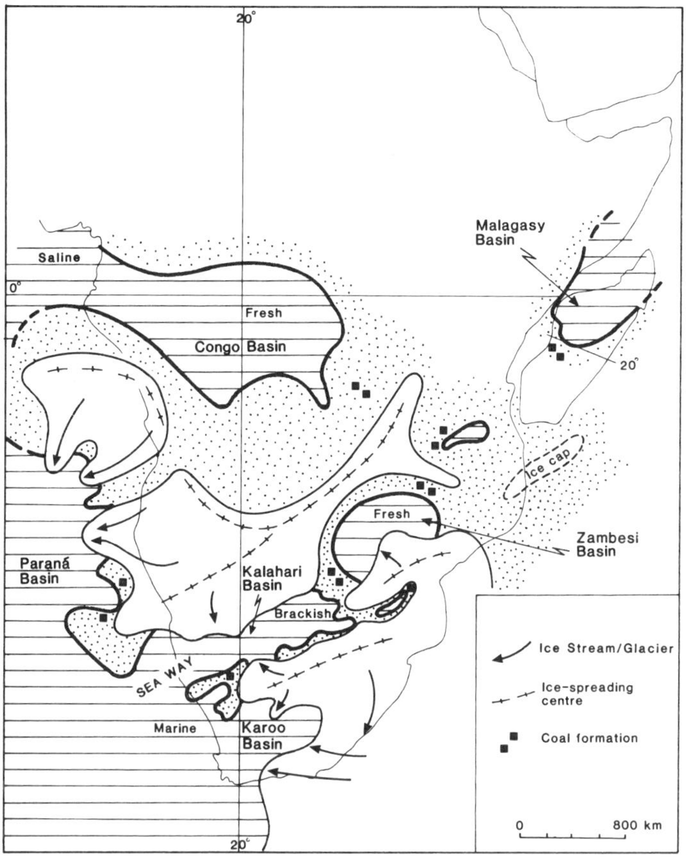

Fig. 10.11. A reconstruction of the Gondwana Ice Sheet over central and southern Africa during the Sakmarian and the Artinskian.

marine ice sheet was present but collapsed when tidewater glaciers were maintained for some time. Peatlands formed upon ice retreat during the Sakmarian and the Artinskian.

5. The marginal continental part of the Gondwana Ice Sheet over Africa collapsed initially as a result of climatic warming during the Early Sakmarian. The subsequent sea-level rise (*Eurydesma* transgression) also temporarily affected the marine ice sheet. The maritime sections of the ice sheet where moisture nursed the snow fields remained active into the Artinskian. Final collapse of the marine ice sheet took place when the rate of basin subsidence increased in the Artinskian, causing a relative sea-level rise.

6. The three major controls on the Permo-Carboniferous glaciation in Africa were geography, climate and relative sea-level. In modeling the late Paleozoic glaciation, Crowley and Baum (1992) emphasized the role of geography in initiating and maintaining the Gondwana Ice Sheet. Extensive areas of uplifted plateaux were the preferred sites for the accumulation of ice masses that evolved into a regional ice sheet through various feedback mechanisms. Large intracontinental sea ways funnelled moisture to interior ice-spreading centers. Globally the climate ameliorated from about the Sakmarian (Frakes, 1979), and regionally there was a climatic change as Africa moved into lower latitudes during the drift of Gondwana over the south pole (Caputo and Crowell, 1985). Relative sea-level change can be attributed to the collapse of the continental part of the Gondwana Ice Sheet or to basin subsidence, which enhanced rapid collapse of the ice sheet along seaways. A sea-level rise would have also influenced the relative elevation of the land mass and its configuration, thereby

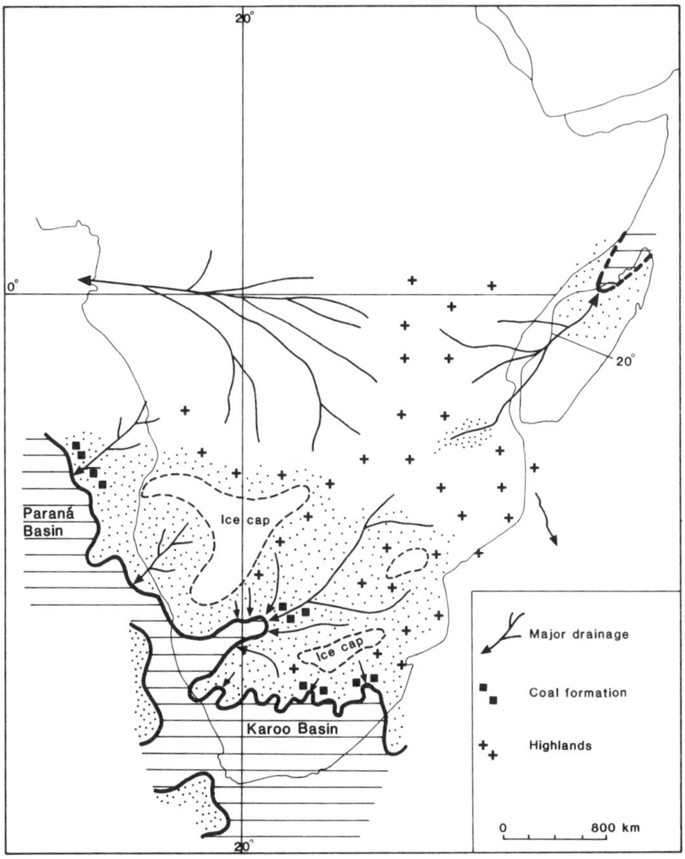

Fig. 10.12. A reconstruction of postglacial sedimentation over central and southern Africa during the Late Artinskian.

changing atmospheric and oceanic circulation, which were important feedback mechanism for regional climatic change.

7. The model for the Gondwana Ice Sheet over Africa holds some important clues for the understanding of cold climates on Earth. The products of ancient glaciations represent only short-lived deglaciation events in which sediment and water discharges were maximized, whereas the actual timing and duration of the glaciations are not constrained. Time scales for ancient glaciations are thus relative and may have consequences for our interpretation of the Pleistocene glaciation. The rock record is also biased in favor of subaqueous "glacial" facies (rain-out and debris-flow diamictites) and against ice-contact terrestrial sediments deposited directly from glaciers. This suggests a large-scale reworking and resedimentation of glacial debris, in sharp contrast to the Pleistocene sediment record which is biased in favor of terrestrial deposition. The discrepancy is attributed to the low preservation potential of terrestrial glacial deposits, and future reconstructions of ancient ice sheets must allow for continental sections for which no rock record may exist. Last, there is a relationship between plate tectonics and the late Paleozoic glaciation with regard to both cause and preservation of glacial products in sedimentary basins. In Africa the most complete record of the glaciation is preserved in the Karoo Basin the evolution of which was directly controlled by the active paleo-Pacific margin of Gondwana. This was also the period when the Gondwana margin experienced uplift, high plate-motion rates and a clockwise rotation all being forcing functions for climatic change (Veevers and Powell, 1987; Visser, 1992; Crowley and Baum, 1992). Past cold climates from the

rock record therefore cannot be interpreted as simple climatically driven cycles of ice sheet growth and decay.

References

Anderson, J.M., 1977. The biostratigraphy of the Permian and Triassic. Part 3, A review of Gondwana Permian palynology with particular reference to the northern Karoo Basin, South Africa. Mem. Bot. Surv. S. Afr., 41: 188 pp.

Banks, M.R., 1981. Late Palaeozoic tillites of Tasmania. In: M.J. Hambrey and W.B. Harland (Editors), Earth's pre-Pleistocene record. Cambridge University Press, Cambridge, pp. 495–501.

Bond, G., 1970. The Dwyka Series in Rhodesia. Proceedings of the Geological Associtaion, 81: 463–472.

Bond, G., 1981a. Late Palaeozoic (Dwyka) glaciation in the Middle Zambesi region. In: M.J. Hambrey and W.B. Harland (Editors), Earth's pre-Pleistocene record. Cambridge University Press, Cambridge, pp. 55–57.

Bond, G., 1981b. Late Palaeozoic (Dwyka) glaciation in the Sabi-Limpopo region, Zimbabwe. In: M.J. Hambrey and W.B. Harland (Editors), Earth's pre-Pleistocene record. Cambridge University Press, Cambridge, pp. 58–60.

Cahen, L. and Lepersonne, J., 1981. Late Palaeozoic tillites of the Congo Basin in Zaire. In: M.J. Hambrey and W.B. Harland (Editors), Earth's pre-Pleistocene record. Cambridge University Press, Cambridge, pp. 43–47.

Cannon, R.T., Simiyu Siambi, W.M.N. and Karanja, F.M., 1981. The proto-Indian Ocean and a probable Paleozoic/Mesozoic triradial rift system in East Africa. Earth Planetary Science Letters, 52: 419–426.

Caputo, M.V. and Crowell, J.C., 1985. Migration of glacial centers across Gondwana during Paleozoic Era. Geological Society of America Bulletin, 96: 1020–1036.

Chappell, J. and Humphreys, M., 1970. Glacial sedimentation in the Lower Karroo, mid-Zambesi Valley, Rhodesia. Proceedings of the Second Gondwana Symposium, South Africa, pp. 501–509.

Cook, P.J., 1984. Spatial and temporal controls on the formation of phosphate deposits - a review. In: J. Nriagu and P.B. Moore (Editors), Phosphate minerals. Springer-Verlag, New York, pp. 242–274.

Crowell, J.C. and Frakes, L.A., 1975. The Late Palaeozoic glaciation. In: K.S.W. Campbell (Editor), Gondwana geology. Australian National University Press, Canberra, pp. 313–346.

Crowley, T.J. and Baum, S.K., 1992. Modeling late Paleozoic glaciation. Geology, 20: 507–510.

De Wit, M.J., Jeffery, M., Bergh, H. and Nicolaysen, L., 1988. Geological map of sectors of Gondwana reconstructed to their disposition-150 Ma. American Assocociation of Petroleum Geologists and University of Witwatersrand.

Dickins, J.M., 1985. Late Palaeozoic glaciation. BMR J. Austr. Geol. Geophys., 9: 163–169.

Dos Santos, P.R., Rocha-Campos, A. and Canuto, J.R., 1988. Paleogeographic evolution of the late Paleozoic glaciation in the Paraná Basin, Brazil. Abstr. 7th Gondwana Symp. São Paulo, Brazil, pp. 131.

Dow, D.B., Beyth, M. and Hailu, T., 1971. Palaeozoic glacial rocks recently discovered in northern Ethiopia. Geol. Mag., 108: 53–60.

Du Toit, A.L., 1937. Our wandering continents: an hypothesis of continental drifting. Oliver and Boyd, Edinburgh, 366 pp.

Elliot, D.H., 1975. Gondwana basins of Antarctica. In: K.S.W. Campbell (Editor), Gondwana geology. Australian National University Press, Canberra, pp. 493–536.

Eyles, N., 1993. Earth's glacial record and its tectonic setting. Earth-Science Review, 35: 1–248.

Frakes, L., 1979. Climates throughout geological time. Elsevier, Amsterdam, 310 pp.

Frakes, L.A., 1981. Late Palaeozoic diamictites of Madagascar. In: M.J. Hambrey and W.B. Harland (Editors), Earth's pre-Pleistocene record. Cambridge University Press, Cambridge, pp. 80–81.

Frakes, L.A. and Crowell, J.C., 1970. Late Paleozoic glaciation, vol. II. Africa exclusive of the Karroo Basin. Geological Society of America Bulletin, 81: 2261–2286.

González, C.R., 1990. Development of the Late Paleozoic glaciations of the South American Gondwana in western Argentina. Palaeogeography, Palaeoclimatology, Palaeoecology, 79: 275–287.

Gravenor, C.P. and Rocha-Campos, A.C., 1983. Patterns of late Paleozoic glacial sedimentation on the southeast side of the Paraná Basin, Brazil. Palaeogeography, Palaeoclimatology, Palaeoecology, 43: 1–39.

Hambrey, M.J., 1981a. Late Palaeozoic Karoo tillites in Gabon. In: M.J. Hambrey and W.B. Harland (Editors), Earth's pre-Pleistocene record. Cambridge University Press, Cambridge, pp. 41–42.

Hambrey, M.J., 1981b. Palaeozoic tillites in northern Ethiopia. In: M.J. Hambrey and W.B. Harland (Editors), Earth's pre-Pleistocene record. Cambridge University Press, Cambridge, pp. 38–40.

Harland, W.B., Cox, A.V., Llewellyn, P.G., Pickton,

C.A.G., Smith, A.G. and Walters, R., 1982. A geological time scale. Cambridge University Press, Cambridge, 131 pp.

Horsthemke, E., Ledendecker, S. and Porada, H., 1990. Depositional environments and stratigraphic correlation of the Karoo sequence in northwestern Damaraland. Communications of the Geological Survey of Namibia, 6: 63–73.

Kamen-Kaye, M., 1978. Permian to Triassic faunas and paleogeography: Somalia, Kenya, Mozambique, Madagascar, South Africa. Journal of Petroleum Geology, 1: 79–101.

Kerr, A., 1993. Topography, climate and ice masses: a review. Terra Nova, 5: 332–342.

Langford, R.P., 1991. Permian coal and palaeogeography of Gondwana. BMR Record 1991/95, 136 pp.

Larsson, K., Lindström, S. and Guy-Ohlson, D., 1990. An Early Permian palynoflora from Milorgfjella, Dronning Maud Land, Antarctica. Antarctica Sci., 2: 331–344.

López-Gamundí, O.R., 1989. Postglacial transgressions in Late Paleozoic basins of western Argentina: a record of glacioeustatic sea level rise. Palaeogeography, Palaeoclimatology, Palaeoecology, 71: 257–270.

López-Gamundí, O.R., Limarino, C.O. and Cesari, S.N., 1992. Late Paleozoic paleoclimatology of central west Argentina. Palaeogeography, Palaeoclimatology, Palaeoecology, 91: 305–329.

Martin, H., 1961. The hypothesis of continental drift in the light of recent advances of geological knowledge in Brazil and in South West Africa. Alex Du Toit Memorial Lecture 7, Annex. Transactions of the Geological Society of South Africa, 64: 47 pp.

Martin, H., 1981a. The Late Palaeozoic Dwyka Group of the South Kalahari Basin in Namibia and Botswana and the subglacial valleys of the Kaokoveld in Namibia. In: M.J. Hambrey and W.B. Harland (Editors), Earth's pre-Pleistocene record. Cambridge University Press, Cambridge, pp. 61–66.

Martin, H., 1981b. The Late Palaeozoic Dwyka Group of the Karasburg Basin, Namibia. In: M.J. Hambrey and W.B. Harland (Editors), Earth's pre-Pleistocene record. Cambridge University Press, Cambridge, pp. 67–70.

McClure, H. and Young, G.M., 1981. Late Palaeozoic glaciation in the Arabian Peninsula. In: M.J. Hambrey and W.B. Harland (Editors), Earth's pre-Pleistocene record. Cambridge University Press, Cambridge, pp. 275–277.

Micholet, J., Molinas, E. and Penet, B., 1970. Le Système du Karroo an Gabon. Proceeding of the Second Gondwana Symposium, South Africa, pp. 371–380.

Rakotoarivelo, H.J., 1970. Palynostratigraphic review of the coal basin of the Sakoa-Sakamena (Madagascar). Proceeding Papers Second Gondwana Symp., South Africa, pp. 537–540.

Rocha-Campos, A.C., 1981a. Late Palaeozoic "Série Tilitica" of Mozambique. In: M.J. Hambrey and W.B. Harland (Editors), Earth's pre-Pleistocene record. Cambridge University Press, Cambridge, pp. 52–54.

Rocha-Campos, A.C., 1981b. Late Palaeozoic tillites of the Sergipe-Algoas basin, Rondônia and Matogrosso, Brazil. In: M.J. Hambrey and W.B. Harland (Editors), Earth's pre-Pleistocene record. Cambridge University Press, Cambridge, pp. 838–841.

Rocha-Campos, A.C. and Dos Santos, P.R., 1981. Late Palaeozoic tillites of the Lutôe Series, Angola. In: M.J. Hambrey and W.B. Harland (Editors), Earth's pre-Pleistocene record. Cambridge University Press, Cambridge, pp. 48–51.

Rust, I.C., 1975. Tectonic and sedimentary framework of Gondwana Basins in southern Africa. In: K.S.W. Campbell (Editor), Gondwana geology. Australian National University Press, Canberra, pp. 537–564.

Smith, R.A., 1984. The lithostratigraphy of the Karoo Supergroup in Botswana. Bulletin of the Geological Survey Department Botswana, 26: 239 pp.

Veevers, J.J. and Powell, C.McA., 1987. Late Paleozoic episodes in Gondwanaland reflected in transgressive-regressive depositional sequences in Euramerica. Geolological Society America Bulletin, 98: 475–487.

Visser, J.N.J., 1983a. The problems of recognizing ancient subaqueous debris flow deposits in glacial sequences. Transactions of the Geological Society of South Africa, 86: 127–135.

Visser, J.N.J., 1983b. An analysis of the Permo-Carboniferous glaciation in the marine Kalahari Basin, southern Africa. Palaeogeography, Palaeoclimatology, Palaeoecology, 44: 295–315.

Visser, J.N.J., 1986. Lateral lithofacies relationships in the glacigene Dwyka Formation in the western and central parts of the Karoo Basin. Transactions of the Geological Society of South Africa, 89: 373–383.

Visser, J.N.J., 1987a. The influence of topography on the Permo-Carboniferous glaciation in the Karoo Basin and adjoining areas. In: G.D. McKenzie (Editor), Gondwana Six: stratigraphy, sedimentology, and paleontology. Geophysical Monograph, American Geophysical Union, 41: 123–129.

Visser, J.N.J., 1987b. The palaeogeography of part of southwestern Gondwana during the Permo-Carboniferous glaciation. Palaeogeography, Palaeoclimatology, Palaeoecology, 61: 205–219.

Visser, J.N.J., 1989. The Permo-Carboniferous Dwyka

Formation of southern Africa: deposition by a predominantly subpolar marine ice sheet. Palaeogeography, Palaeoclimatology, Palaeoecology, 70: 377–391.

Visser, J.N.J., 1990a. The age of the late Palaeozoic glacigene deposits in southern Africa. South African Journal of Geology, 93: 366–375.

Visser, J.N.J., 1990b. Glacial bedforms at the base of the Permo-Carboniferous Dwyka Formation along the western margin of the Karoo Basin, South Africa. Sedimentology, 37: 231–245.

Visser, J.N.J., 1991a. The paleoclimatic setting of the late Paleozoic marine ice sheet in the Karoo Basin of southern Africa. In: J.B. Anderson and G.M. Ashley (Editors), Glacial marine sedimentation; paleoclimatic significance. Geological Society of America, Special Paper 261, pp. 181–189.

Visser, J.N.J., 1991b. Self-destructive collapse of the Permo-Carboniferous marine ice sheet in the Karoo Basin: evidence from the southern Karoo. South Africa Journal of Geology, 94: 255–262.

Visser, J.N.J., 1992. Basin tectonics in southwestern Gondwana during the Carboniferous and Permian. In: M.J. De Wit and I.G.D. Ransome (Editors), Inversion tectonics of the Cape Fold Belt, Karoo and Cretaceous basins of southern Africa. A.A. Balkema, Rotterdam, pp. 109–115.

Visser, J.N.J., 1993a. A reconstruction of the Late Palaeozoic ice sheet on southwestern Gondwana. In: R.H. Findlay, R. Unrug, M.R. Banks and J.J. Veevers (Editors), Gondwana 8: assembly, evolution and dispersal. A.A. Balkema, Rotterdam, pp. 449–458.

Visser, J.N.J., 1993b. Sea-level changes in a back-arc-foreland transition: the late Carboniferous-Permian Karoo Basin of South Africa. Sedimentary Geology, 83: 115–131.

Wopfner, H. and Kreuser, T., 1986. Evidence for Late Palaeozoic glaciation in southern Tanzania. Palaeogeography, Palaeoclimatology, Palaeoecology, 56: 259–275.

Worku, T. and Astin, T.R., 1992. The Karoo sediments (Late Palaeozoic to Early Jurassic) of the Ogaden Basin, Ethiopia. Sedimentary Geology, 76: 7–21.

11. TRANSITION FROM FREEZING TO SUBTROPICAL CLIMATES IN THE PERMO-CARBONIFEROUS OF AFRO-ARABIA AND INDIA

Helmut Wopfner and Satyendra M. Casshyap

In discussions on global warming and climatic catastrophes sight is frequently lost of the fact that the climate of Earth was never stable. Even within the record of our modern calendar, climatic changes have influenced human societies and contributed to the advance and the decline of civilizations. In the course of the geologic history of our planet, periods of warm climate, when even polar regions were free from ice, alternated with cold regimes, during which large regions were covered by glaciers, thick ice sheets, and permafrost. Such "ice house" periods (Veevers, 1990) existed in the Late Precambrian (around 600 Ma) and in the late Paleozoic (around 280 Ma). Short-lived manifestations of cold climate, when temperatures were apparently low enough for surface water to freeze and for local ice sheets to develop, have been recorded from the Late Ordovician of Africa (South Africa and Algiers/Morocco), from the Late Devonian of Brazil, and form the Early Carboniferous of North Africa (Caputo and Crowell, 1985; Veevers and Powell, 1987). Whereas these episodes were of only local significance at best, the late Paleozoic "ice house" period left its traces not only on all of Gondwana (Crowell and Frakes, 1975; Casshyap, 1979; Hambrey and Harland, 1981; Veevers and Powell, 1987; Visser, 1989, 1990; Wopfner, 1989; 1991) but also on the northernmost parts of Laurasia (Chumakov, 1985). The cold climatic regime of this period is recorded by deposits laid down by glaciers such as tillites, containing striated and faceted clasts, commonly resting on grooved and polished glaciated pavements. Other sediments attesting to the proximity of glaciers are fine-grained laminated siliciclastic deposits with dropstones, ice-rafted clasts up to boulder size that were released when the ice melted.

In Afro-Arabia the northern limit of glaciation is well known (Wopfner, 1991; 1993). Glaciation also affected the whole of India (Casshyap and Qidwai, 1974; Casshyap, 1979; Casshyap and Tewari, 1987), and glaciers shed their debris into the adjoining sea. The effects of glacial transport are still evident on exotic terranes in Tibet (Liang et al., 1983) and in western Yunnan (Wang, 1983; Wopfner and Jin, 1993). As climate ameliorated and glaciers retreated, different kinds of sediments, such as coals and redbeds, record the climatic amelioration during the terminal period of the Paleozoic and the beginning of the Mesozoic. The transition from the ice-bound Gondwana of the Late Carboniferous to the warm and even hot subtropical climate of the Early Triassic is well documented by lithological and biological climatic indicators, observed in well-dated sections from a number of locations widely distributed in both the Afro-Arabian and the Indian parts of Gondwana. They permit a well-founded reconstruction of the multifaceted scenario of climatic transition and its global repercussions in late Paleozoic times. As the interpretation of lithologic indicators may not always be unequivocal, we present a concise, although generalized, description of the actual succession of the sedimentary rocks. This, we hope, will demonstrate the basis for our interpretation.

Geotectonics

In any discussion of paleoclimatic changes, the relative positions of Gondwana fragments have to be considered. We generally accept the classical reconstructions of Smith and Hallam (1970) and De Wit et al. (1988) with some modifications, as shown in Figure 11.1. Based on structural, lithofacies and biological evidence, Malagasy (Madagascar) is placed adjacent to East Africa to

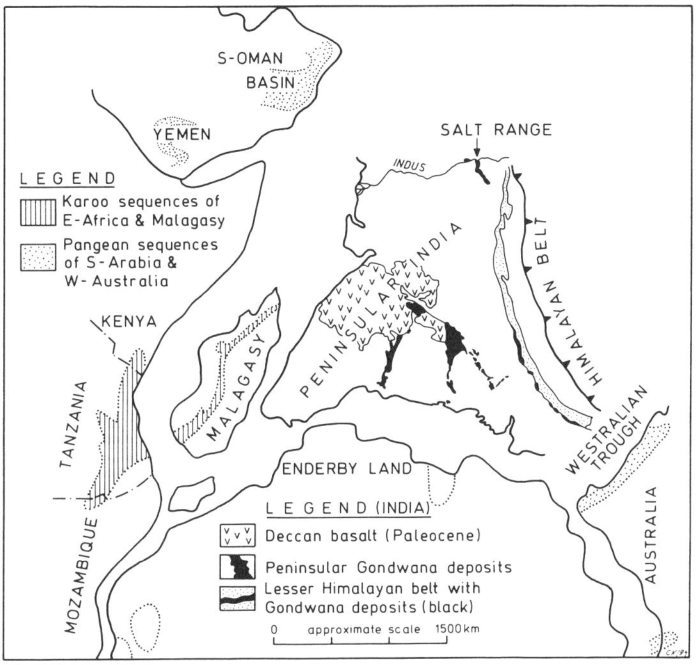

Fig. 11.1. Assembly of NE-Gondwana in late Early Permian and areas of exposure of Pangean depositional sequences.

form the eastern margin of the Karoo rift system and the southern continuation of the Malagasy Trough, as suggested by Wopfner (1991; 1994a). India is placed further northeast to accommodate observed changes in lithofacies and to provide a marine pathway from the Tethys to Madagascar. In order to depict the factual situation on the northern margin of the Indian subcontinent, we have excluded from Figure 11.1 Gondwana fragments north and east of the Himalayas, such as Lhasa Block, Baoshan and Tengchong Terranes (Wopfner and Jin, 1993), and Sibumasu Block (Metcalfe, 1988). The reconstructed position of the Salt Range as shown in Figure 11.1 may be debatable, as the underlying Cambrian halite beds may demand a closer association with the Cambrian salt basin of Oman in Early Permian times.

The Karoo basins of eastern Africa were essentially intracratonic rifts (Wopfner, 1990). The deposits were laid down in generally north-northeast-trending syngenetic graben systems, extending from Lake Nyasa to southeastern Kenya. To the south the graben structures continue up to the Zambesi River (such as the Metangula and Lulangu basins; Utting, 1978), where the structural pattern changes to the foreland basin configuration of the Karoo Basin in South Africa. Structural and stratigraphic relationships between them and time-equivalent tectonostratigraphic units of Madagascar have been well established (Wopfner, 1994a) but their northward continuation and their connection with Permian basins in southern Arabia is less clear. Buried graben structures in Somalia may have been part of that link (Bosellini, 1989).

Gondwana sediments in India are developed in two distinct geographic and tectonostratigraphic domains, namely the Peninsular part to the south of the Ganges/Brahmaputra line and the Extra-Peninsular part to the north. In Peninsular India the Gondwana sediments occur in linear, frequently fault-bounded intracratonic basins. The Extra-Peninsular region includes the Lesser Himalayan belt and the Tethyan or Greater Himalayan belt (Fig. 11.1). In the Extra-Peninsular Himalayan region, Gondwana sediments extend from Arunachal Pradesh, via Bhutan, Sikkim, Darjeeling, and Nepal, to Kashmir. In its tectonostratigraphic sense, Extra-Peninsular India forms the transition from the continental domain of the Gondwana craton to the progressively marine environment of the Tethys. A rift environment accompanied by basic volcanism controlled deposition in the Kashmir region (Gaetani and Garzanti, 1991).

The Pangea Mega-Succession

The sedimentary succession which was initiated by the Late Paleozoic glacial event is here referred to as the Pangea Mega-Succession. In Africa it is termed the Karoo Supergroup; in India it is known as Gondwana Supergroup. Its deposition was preceded by a major hiatus, so its sediments generally rest unconformably on older rocks.

The time of the onset of glaciation is still very much under debate. Tillites and other glacial deposits are usually unfossiliferous, and the few fossils that have been found generally indicate a Late Carboniferous (Stephanian) age (Kemp, 1975; Truswell, 1980; Besems and Schuurman, 1987). Veevers and Powell (1987) and Visser (1990) have argued that central Gondwana was covered by glaciers much earlier than the peripheral parts. They suggested that the hiatus preceding the Pangea succession represents the time during which Gondwana was scoured by continental glaciers that left no trace of their activities. From this hypothesis they concluded that glaciation commenced in Namurian times. Such negative evidence is difficult to accept, especially as the presumed period of glaciation is not matched by appropriate sedimentary records. In South Africa the youngest rocks disconformably underlying the Dwyka Group are those of the Lake Mentz Subgroup. As this unit is of Namurian age (Crowell, 1983) the onset of glaciation must postdate Namurian times. Similarly in northwest-Australia (Bonaparte Gulf Basin), Stephanian fluviatile sandstones of the basal Kulshill Group rest unconformably on a sandstone-carbonate succession of Namurian age (Foster, 1986). As glacigene deposits do not occur until higher in the Kulshill Group, commencement of glaciation as early as Namurian is very unlikely.

For East Africa we suggest a Stephanian age for the main period of glaciation, preceded perhaps by a period of decreasing temperatures during the Westphalian (Wopfner, 1989; 1991; Diekmann, 1993). This is supported by corrected fission track ages of 300 Ma at Lake Rukwa, indicating the time of thermal uplift which created the highlands where glaciations started from (Mbede, 1993).

The depositional regime of the Pangea Mega-Succession was terminated by the transition from the intracratonic rift to the pericratonic passive margin environment in middle to late Triassic times, when India and Madagascar separated from Africa (Bosellini, 1989; Wopfner, 1994a). In southern Africa, deposition of the Pangea Mega-Succession (Karoo Supergroup) was terminated by the extrusion of basaltic lavas of the Jurassic, the Drakensberg volcanics.

In India the Pangea Mega-Cycle corresponds to the Gondwana Supergroup. It commenced with deposition of the Talchir Tillite and associated glacigene deposits. The Talchir glaciation is generally considered to be of earliest Permian age. However, an earlier beginning of the actual glaciation, perhaps in Stephanian times, is indicated by the Permo-Carboniferous ages of some postglacial marine faunas (P. Singh, 1973; T. Singh, 1983; 1987; Joshi et al., 1990.).

A Middle Triassic hiatus is also recognizable in most basins of Peninsular India, followed by the Late Jurassic-Early Cretaceous Bagra/Jabalpur succession (where it terminates the Pangea Mega-Succession). However, in Godavari Valley sedimentation continued apparently without interruption into the Cretaceous (Chikiala Formation) and was terminated by the extrusion of the Deccan Lavas (dated at 65 Ma ± 5). This essentially Jurassic-Cretaceous succession, which should heneceforth be designated as Upper Gondwana (or Younger Gondwana), is not considered in this chapter.

The stratigraphic record

East Africa

One of the best and most complete of the Pangea sections in East Africa is that of the Ruhuhu Basin in southwestern Tanzania, where the succession has been exposed along the rift shoulder of the modern Nyasa Rift (Fig. 11.2). This classical succession, originally described by

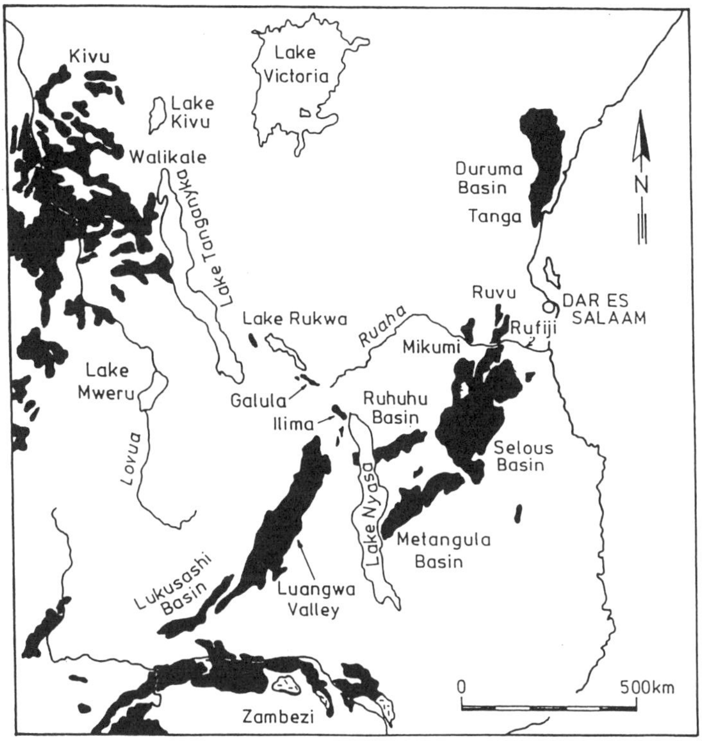

Fig. 11.2. Generalized distribution of Karoo basins (in black) in eastern and part of central Africa. Sediment fill of these basins corresponds to the Pangea Mega-Sequence.

Stockley (1932) as the Songea Series, is now referred to as the Songea Group. It comprises about 3000 m of Late Carboniferous to Early Triassic sedimentary rocks, which are entirely of terrestrial origin. Its formal lithostratigraphic subdivision is shown in Figure 11.3. Deposition was controlled by climatic and tectonic events and resulted in the development of five distinctive sub-successions (Wopfner, 1989; 1994a; Kreuser et al., 1990). As basin geometry was controlled by an expanding rift system, sedimentation was continuous within the central grabens, but depositional overlap and unconformities developed across collapsed rift margins and on

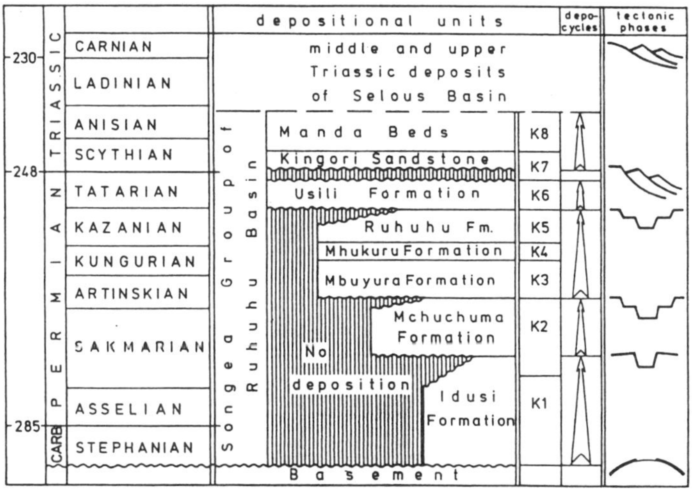

Fig. 11.3. Simplified stratigraphy and depositional cycles of the Pangea Mega-Sequence in southern Tanzania. The symbols K1 to K8 are the old stratigraphic units used by Stockley (1932) (after Wopfner, 1994a).

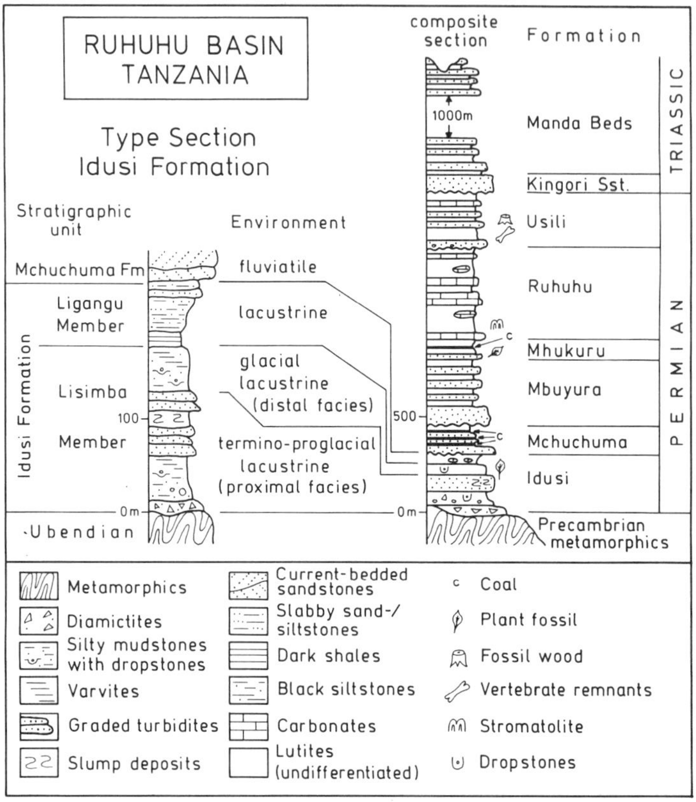

Fig. 11.4. Generalized lithology of the Pangea Mega-Sequence in southwest Tanzania and detailed lithofacies and environment of the glacigene Idusi Formation (after Diekmann, 1993).

tilted intrabasin horsts (Wopfner 1990; 1994a). The generalized lithologic succession of the Songea Group is shown in figure 11.4.

The sedimentary succession of the Songea Group commenced with the deposition of the glacigene Idusi Formation. This unit comprises basal tillites with striated and faceted clasts of up to boulder size (Wopfner and Kreuser, 1986), succeeded by various kinds of reworked and slurried glacial debris, slumped diamictites, and coarse sandstones, deposited in water bodies at the front of glaciers.

The overlying rhythmites with dropstones and turbidites were deposited in a lake environment adjacent to glaciers (Wopfner, 1989; 1993). The uppermost unit of the Idusi Formation, now termed the Lilangu Member (Diekmann, 1993), consists of black kaolinitic, more or less bituminous, pyritic and plant-bearing lutites with typical carbonate concretions. In the deeper parts of the basin the lutites are interspersed by sandstone-turbidites.

The succession of the Idusi Formation shows an overall threefold subdivision that can be attributed to climatic control, reflecting the change from glacial to periglacial and finally postglacial conditions (Wopfner, 1989; 1993; 1994b; Wopfner and Diekmann, 1992; Diekmann, 1993). This is illustrated by the detailed profile in Figure 11.4.

The second sub-succession is formed by the coal-bearing Mchuchuma Formation. The basal part of this formation is formed by the Mpera Sandstone Member (Semkiwa, 1992), a coarse, arkosic, kaolinitic, and pyritic sandstone. The coal-bearing part of the Mchuchuma Formation contains up to seven coal seams, which appear at the top of point-bar successions or lacustrine swamps (Casshyap et al., 1987; Semkiwa, 1992).

The third depo-cycle comprises, in ascending order, the Mbuyura, Mhukuru, and Ruhuhu formations (Fig. 11.3), reaching a cumulative thickness of almost 1,000 m (Fig. 11.4). The Mbuyura Formation is essentially a red-bed succession with current-bedded, coarse-grained, and conglomeratic sandstones, consisting of up to 70 percent fresh feldspars. It is interbedded with red lutites, containing nodular calcretes (Wopfner, 1989; 1993; Kaaya, 1992). It is followed by the coal-bearing Mhukuru Formation, a lacustrine succession of black, *Glossopteris*-bearing lutites and thin coal seams. The Ruhuhu Formation consists of pale green, calcareous siltstones, marls, and thin, stromatolitic marly limestones. The presence of the pelecypod *Palaeomutela* attests to a freshwater environment. Early diagenetic barite concretions suggest at least a temporary increase of water alkalinity. Desiccation cracks near the top indicate a shallowing of the lake and beginning aridity.

The fourth depositional cycle, represented by the Usili Formation, was introduced by a marked erosional event, witnessed by thick basal conglomerates. The remainder of the Usili Formation comprises greenish grey and red mudstones with interspersed fine-grained sandstones and nodular limestones of a semiarid environment (Wopfner, 1989). In places there is an abundance of silicified wood as well as remnants of *Pareiasauria, Gorgonopsia* and other reptiles. A newly discovered skull of *Dicynodon lacerticeps* attests to the Late Permian age of this unit (Kaaya, 1992).

The deposits of the fifth mega-cycle, which are essentially of Early Triassic age, are separated from the preceding rock unit by a regional unconformity (Fig. 11.3). Sedimentation commenced with the Kingori Sandstone, a coarse-grained, current-bedded, supermature quartz sandstone with bedload-shear and dewatering deformations. Syngenetic homoaxial quartz growth indicates low pH values for interstitial fluids. The basal 2 m of the sandstone contains large rip-up clasts of the underlying Usili Formation. These features indicate a warm pluvial climate.

The overlying Manda Beds, with a thickness of up to 1,300 m, consist of a repetitive succession of fining-upward cyclothems. Each unit begins with a thick bed of feldspathic bioturbated sandstone, commonly cemented by laumontite (Wopfner et al., 1991), fining upward to be concluded by red micaceous siltstones, indicative of a hot and semihumid climate (Wopfner, 1989; 1993). Vertebrate fossils (anomodont and diapside reptiles), including a remnant of *Parotosuchus*, indicate a late Skythian to early Anisian age (Markwort, 1991).

The Pangean succession of Tanzania is concluded by fluviatile sandstones of Early to Middle Triassic age of the Selous Basin (Hankel, 1987), situated northeast of the Ruhuhu Basin (Fig. 11.2).

Madagascar

Depositional successions equivalent to the Karoo Supergroup of continental Africa form an almost continuous belt along the entire west and northwest coasts of the island. The occurrences are assigned to the Morondava Basin along the west coast and to the Diego (Mahajanga) Basin on the northwest coast. During Permo-Carboniferous and Early Triassic times, these basins were opposite to Tanzania (Fig. 11.1) and formed the eastern part of the Malagasy Trough (Wopfner, 1994a).

The succession in Madagascar is subdivided into three major lithostratigraphic units-in ascending order, the Sakoa Group, the Sakamena Group, and the Isalo Group. Their cumulative thickness in the southern Morondava Basin may exceed 6,000 m (Bésairie, 1972). The Sakoa Group and the Sakamena Group together are time equivalents of the East African Pangean Mega-Succession. They were deposited in individual graben structures, draped over by sediments of the Isalo Group.

The Sakoa Group commences with glacigene sediments with a threefold subdivision identical to that of the Idusi Formation in Tanzania, including the black, organic rich lutite beds at the top. The glacigene deposits are overlain by a coal-measure succession, which contains the same microflora as the Mchuchuma Formation in southern Tanzania. The strata above the coal-bearing succession commence with a basal conglomerate that is probably a time equivalent of the Mbuyura Formation. The succeeding units are marine and contain brachiopod and nautiloid faunas comparable to those identified in the Salt Range in Pakistan (Teichert, 1970).

Arabian Peninsula

Typical tillites resting on glaciated pavements and other glacigene rocks have been described from Yemen by Kruck and Thiele (1983). In southern Oman a Late Carboniferous to Early

Permian glacigene succession showing a threefold subdivision similar to that of the Idusi Formation has been reported from exposures and from many petroleum exploration wells. The Oman succession, termed the Haushi Group, consists of basal diamictites succeeded by periglacial sandstones, turbidites and diamictites and finally by the black lutite facies of the Rahab Shale. The latter forms the cap rock for important Omani oil deposits contained within the underlying periglacial and glacial sediments (Lewell et al., 1988).

In central Saudi Arabia, the fluviodeltaic deposits of the Permo-Carboniferous Unayzah Formation (Al-Laboun, 1987) are exact time equivalents of the Haushi Group further south. They are an important reference point establishing the northern limit of the late Paleozoic glaciation on the Afro-Arabian fragment of Gondwana (Wopfner, 1991).

Marine carbonates above the Rahab Shale contain a prolific brachiopod fauna (Hudson and Sudbury, 1959) of Artinskian age. It has been compared with brachiopod faunas in northern Madagascar and in the Salt Range (Kummel and Teichert, 1966; 1970; Teichert, 1970). This glacigene succession is then overlain by carbonates and evaporites of the Khuff Group (Khalifa, 1993) which extends north onto the Arabian platform and is succeeded by the Early Triassic Sudair Formation.

In northern Oman the Pangean succession commences with the glaciolacustrine Al Khlata Formation, which, at least in parts, appears to be an equivalent of the Rahab Shale of southern Oman. It is overlain by the marginal marine deposits of the Early Permian Gharif Formation which in turn is blanketed by the shelf carbonates of the Late Permian Khuff Formation (Lee, in Robertson et al., 1990; Robertson and Searle, in Robertson et al., 1990).

Salt Range, Pakistan

The succession exposed in the Salt Range and in the Trans-Indus Ranges documents yet another transition from the typical Gondwana facies to a Tethyan environment (Kummel and Teichert, 1966, 1970). It may be assumed that prior to the opening of the Malagasy Trough, the Cambrian salt basin underlying the Pangean deposits of the Salt Range and the Cambrian salt basin of southern Oman belonged to the same geological structure. Similarly, as in South Oman, there are no coal deposits present in the Salt Range, except for a few lenticles of coal within the basal part of the postglacial Warchha Formation.

The basal, glacigene Tobra Formation which reaches a maximum thickness of up to 120 m, comprises tillites and freshwater lutites (with *Glossopteris, Gangamopteris*, and *Samaropsis*) in the eastern and the central Salt Range regions and a mixed diamictite-boulder bed facies in the western Salt Range and in the Trans-Indus Ranges. Perfectly soled, faceted, and striated clasts, illustrated by Schindewolf (1967), from the basal tillite clearly indicate intraglacial transport. The glaciation is equated with the Talchir glaciation of Peninsular India (Schindewolf, 1967). The overlying Dandot Formation is a periglacial marine facies containing *Eurydesma* and a typical bryozoa/mollusc fauna. It resulted from the eustatic rise of sea level caused by melting of the glaciers. Gaetani and Garzanti (1991) indicate the existence of a hiatus separating the Asselian Dandot Formation from the overlying reddish to purplish fluviatile deposits of the Warchha and the Sarthai formations.

A marine transgression in Artinskian time introduced the succession of the "*Productus* Limestone" which is now subdivided in ascending order into the Amb, Wargal, and Chhidru formations. Marine conditions, some quite marginal, continued uninterruptedly across the Permian-Triassic boundary (Kummel and Teichert, 1970). Shale interbeds of the Amb Formation also contain *Glossopteris* and *Gangamopteris*. Sedimentation ceased in the Middle Triassic, and a major hiatus separates the Middle Triassic Kingriali Dolomite from the Early Jurassic Datta Formation (Gaetani and Garzanti, 1991).

The marine fauna contained within the "*Productus* Beds" resembles the marine faunas of Madagascar but also marine faunas of the Tethyan margin of Australia (Teichert, 1970; Archbold, 1983).

Peninsular India

In Peninsular India the Gondwana sediments occur in linear intracratonic basins as shown in Figure 11.5. The basins have a long history of

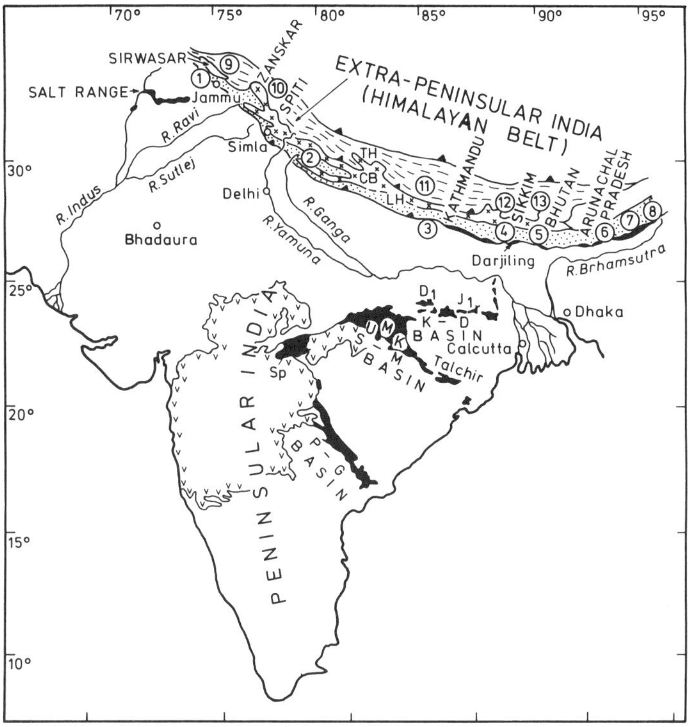

Fig. 11.5. Map of Indian subcontinent showing occurrences of Gondwana strata in the Peninsular and Extra-Peninsular (Himalayan) regions. Numerals in Extra-Peninsular region (Lesser and Tethyan Himalayan belts) refer to localities of the generalized lithologic columns, (see Figures 11.7 and 11.8). Important Peninsular Gondwana basins are identified by letters: Sp = Satpura Basin; K-D = Koel-Damodar Basin; S-M = Son-Mahanadi Basin; P-G = Godavari Basin. For symbols see Figure 11.1.

continental sedimentation. The deposits contain a typical Gondwana flora but interbeds with a marine fauna are developed locally near the base of the succession (Casshyap, 1979). The Gondwana Supergroup has been investigated and reviewed by Ghosh and Mitra (1972), Casshyap (1973; 1979), Dickins and Sha (1979), and Singh (1987) among others. Figure 11.6 illustrates a generalized Permian Gondwana succession, representing a lithological record corresponding to glacial and glaciofluvial deposits (Talchir Formation) and to postglacial events (Karharbari and Barakar formations, Barren Measures, Raniganj Formation). The succession reflects the climatic change from cold to temperate to warm, humid conditions as sedimentation progressed from Early Permian to the Triassic period. Although not necessarily time equivalent, depositional successions similar to those observed in the Pangea Mega-Succession of Africa and Madagascar can be identified.

Gondwana sedimentation was initiated by the deposition of glacial and glacigene sediments on an uneven Precambrian basement, forming the lower Talchir Formation. The formation crops out as thin, discontinuous patches on the floor and the margin of exhumed valleys and the thickness rapidly increases toward the low-lying interior parts of the basin (Banerjee, 1966; Casshyap and Qidwai, 1974; Frakes et al., 1975; Datta et al., 1979; Dickins and Sha, 1979; Casshyap and Srivastava, 1987). Figure 11.6A summarizes the vertical distribution of facies generally occurring in the Talchir Formation. In sections where Talchir rocks rest unconformably on Archaean basement, the succession invariably starts with a basal diamictite that in turn passes upward into either mudstone facies containing dropstones or an interbedded shale and siltstone-sandstone facies that also contains dropstones. Both these facies pass upward into a massive, faintly bedded sandstone.

Massive diamictite which commonly occurs in the basal parts, abounds in poorly sorted clasts, some showing pentagonal shape and striations. They are lodgement tills deposited on basement rocks by grounded ice.

Stratified diamictite occurs above the massive diamictite with a gradational contact. It represents subglacial melt water till, deposited terrestrially or in shallow water below grounded or floating ice (Boulton and Eyles, 1979; Dreimanis, 1979).

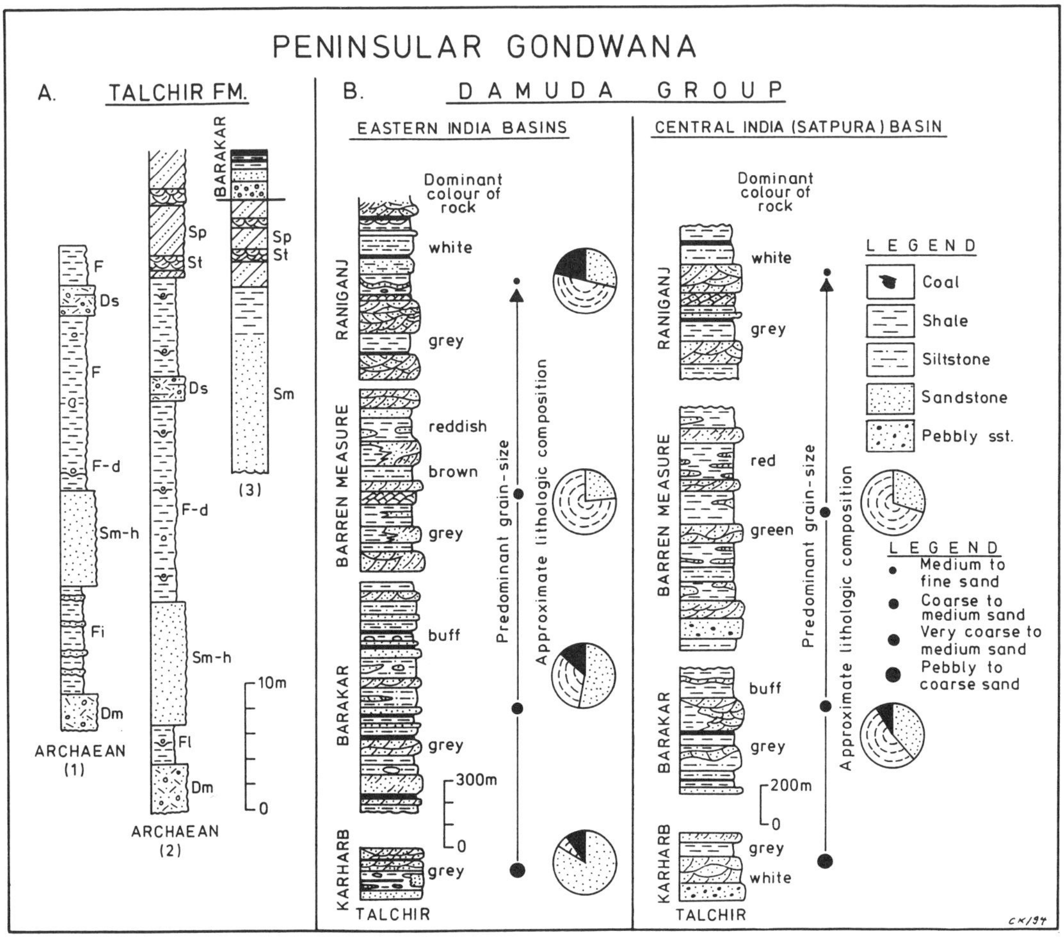

Fig. 11.6. Generalized stratigraphic sections of Permian Gondwana of Peninsular India. A. Glacial and glacigene sequences of Talchir Formation: measured stratigraphic sections at different localities in the Son-Mahanadi Basin: 1. Bansjhol section, east of Korba; 2. Ambikapur Road-Hado River section, northeast of Korba; 3. Kharpari Creek section, near Korbi, north of Korba. Facies codes are: Dm-Ds: Diamictite massive/stratified; Sm-h: Sandstone massive and faintly bedded; Sp-St: Planar and trough-type cross-bedded sandstone; S-pb: Pebbly sandstone; Fl: Laminated lutites with or without dropstones; Fi: Interbedded shale-sandstone facies. B. Generalized stratigraphic sections of postglacial Permian fluviatile deposits of the Damuda Group of Eastern India (Koel-Damodar and Son-Mahanadi Basins) and central India (Satpura Basin) showing dominant lithology, colour, grain size and sedimentary structures.

Reworked till also occurs as thin sheets of diamictite alternating with thin sandstone, related to seasonal mass flow of till and sand due to high discharge of meltwater streams in proglacial outwash plains (Boulton and Eyles, 1979). Isolated diamictite bodies represent melt-out tills (Casshyap and Srivastava, 1987) or frozen glacial debris slumped from a shelf into deeper parts of the basin.

Massive conglomerate bodies may be related to catastrophic discharges of natural meltwater dams (Collinson and Thompson, 1982) or to longitudinal proglacial outwash plains (Boothroyd and Ashley, 1975). The fossiliferous basal conglomerate facies in the Hasdo section near Manendragarh represents a facies of submarine outwash material of delta fans (Casshyap and Srivastava, 1987). Cross-bedded pebbly sandstones originated in braided channels of proglacial outwash plains. Englacial esker deposits are also present.

The middle and the upper parts of the Talchir Formation are dominated by fine-grained lithofacies (Fig. 11.6). They consist of alternating shale and fine-grained sandstones, interpreted as levee deposits of sinuous periglacial stream channels, mudstone-siltstone facies with laminations and cross-laminations, deposited by suspension in standing water bodies which developed distally to outwash complexes of temperate valley glaciers (Shaw, 1975). Laminites, commonly with dropstones, reflect seasonal variations of meltwater discharge.

The marine intercalation in the Hasdo River section is characterized by an abundance of *Eurydesma* along with *Deltopecten*. In Umaria the bivalves are rare, but the fauna abounds in brachiopods, including *Linoproductus* and *Spiriferidae*. Similarly, the Daltonganj fauna in Bihar in the east and the Bhaduara fauna of Rajasthan in the west also suggests cool environment.

The second depositional cycle is introduced by the postglacial, fluvial, coarse sandstones and conglomerates with interbeds of shale and coal of the Karharbari Formation. Slopes with increased gradients produced by periodic uplift became the site for braided streams which transported abundant bed load. The occurrence of a distinct conglomeratic horizon of about 40 m thickness in the upper Karharbari coalfields of the Mahanadi Basin suggests yet another episode of tectonic uplift at the end of the Karharbari sedimentation.

Gradients decreased thereafter, and on this subdued landscape the coal-bearing Barakar Formation developed. Consequently the streams underwent a progressive change from braided to moderately sinuous to meandering channels, with recurring fining upward cycles. At this stage the expanded basins aerially became unified into longitudinal alluvial plains, apparently larger in size than their present limits. The sinuous to meandering Barakar rivers, loaded with abundant mixed and suspended load, flowed on a gentle paleoslope from southeast to northwest. The resultant sediments are characterized by a progressive increase in fine clastics through time from lower to upper Barakar and along the length of the basin in the downcurrent direction.

The bulk of the overlying Barren Measures and Raniganj sediments were deposited largely by meandering streams, flowing on a distinctly gentle paleoslope directed northwest and west. The overall dominance of red beds with intercalated green layers indicates a hot, semiarid to arid climate. Owing to the near-flat topography and the proximity of the shoreline in the north, occasional marine ingressions of the low-lying terrain during Barren Measures and Raniganj sedimentation occurred.

Evidence from the lithofacies of Triassic Gondwana rocks suggests a new episode of tectonic uplift and a readjustment of the paleoslope. The Triassic sediments record a sudden increase in coarse clastics, including pebble-beds, and less fine grained clastics than are found in the underlying succession. An increase in annual precipitation may also be indicated.

Extra-Peninsular India

The Pangean successions of Extra-Peninsular India are broadly subdivided into those of the Lesser Himalaya and those of the Tethyan Himalaya. The former have many features in common with the cratonic facies, including a coal-bearing succession, whereas the latter show a stronger marine influence and, in places, are associated with basic volcanism (Acharyya et al., 1979; Tripathi and Singh, 1987; Gaetani and Garzanti, 1991).

Lesser Himalaya. Occurrences of the Late Paleozoic succession form an almost continuous belt along the eastern Lesser Himalayas, extending from Arunachal Pradesh to central Nepal via Bhutan and Darjeeling. Further important sections are exposed in the Rangit tectonic window in Sikkim. Generalized stratigraphic columns of the Lesser Himalaya are shown in Figure 11.7.

The Gondwana beds of the Lesser Himalayas have been investigated by Acharyya (1971), Acharyya et al. (1979), Tripathi and Chowdhury (1983), Singh (1983), and Joshi et al. (1990) among others. The succession may be correlated confidently with the Gondwana Supergroup of Peninsular India; the Rangit Pebble Slate is equivalent to the Talchir Formation, and the overlying succession is coeval to the formations of the Damuda Group (Barakar, Barren Measures, Raniganj).

The Rangit Pebble Slate consists of a basal diamictite with subrounded to angular clasts of up to boulder size, supported in a slatey matrix. The fact that some of the blade-shaped clasts of

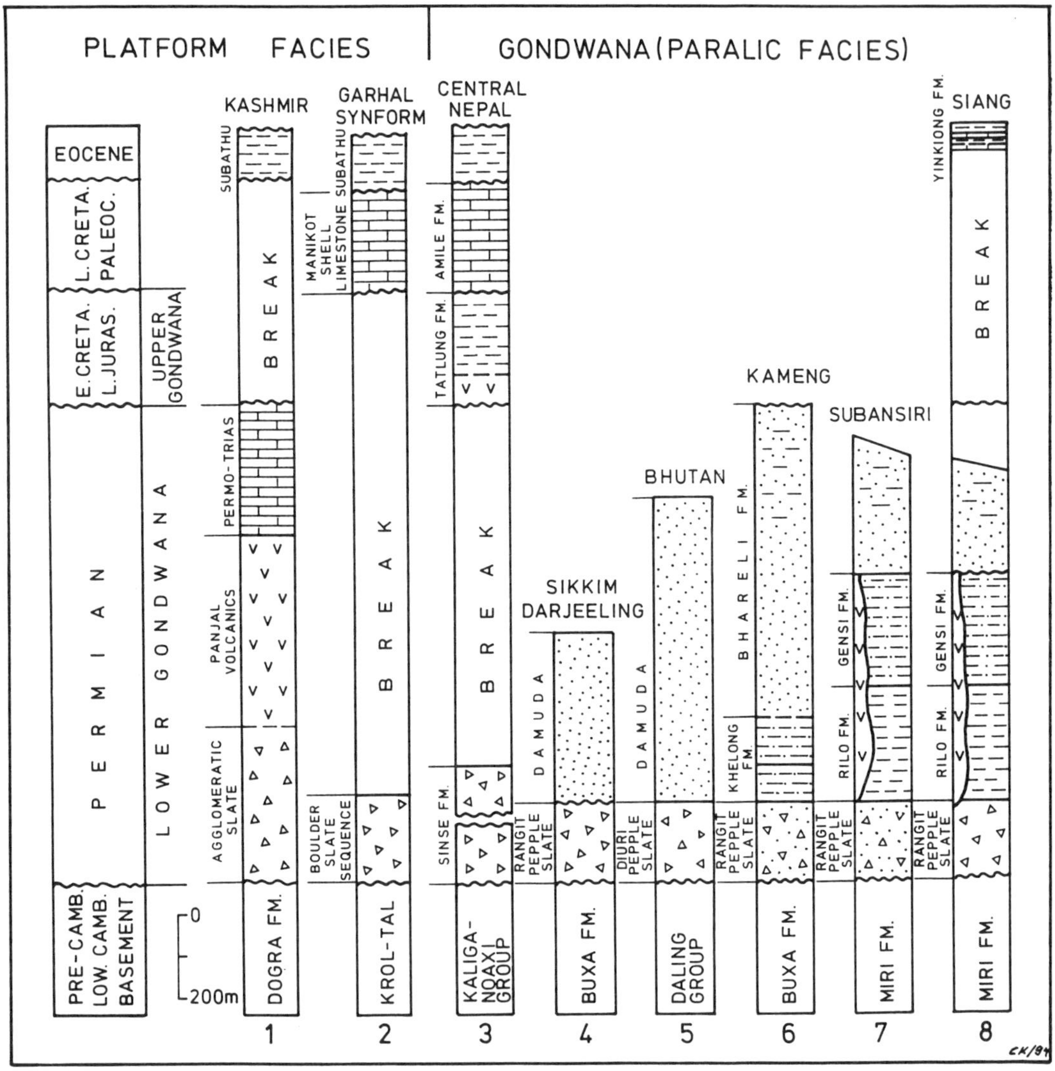

Fig. 11.7. Generalized stratigraphic columns of Gondwana sequences of the Lesser Himalaya (after Tripathi and Singh (1987). For identification, see encircled numbers in Figure 11.5.

nearly 1 m length are placed in an almost vertical position indicates a glacial origin. The overlying sandstones, siltstones, and greywackes represent proglacial outwash and glaciomarine deposits. The thick succession of black slates, sometimes laminated in the basal parts and containing fossiliferous calcareous concretions, like the Garu Formation (Singh, 1983; 1987), originated in a distal periglacial to postglacial environment.

The parallel-bedded, thick sandstone succession above the black slates represents a regressive, marginal marine facies, whereas the succeeding coal-bearing unit derived from a fluviodeltaic environment.

Tethyan Himalaya. The bulk of the sediments of the Tethyan Himalayan belt are marine (Fig. 11.8). Reliable correlations between the marine part of Tethys and the Gondwana of Peninsular India and Arunachal have been established (Nakazawa and Kapoor, 1977). This zone extends from Kashmir to Spiti and finally into Tibet and southwest China. For the purpose of this paper only the first locality is discussed.

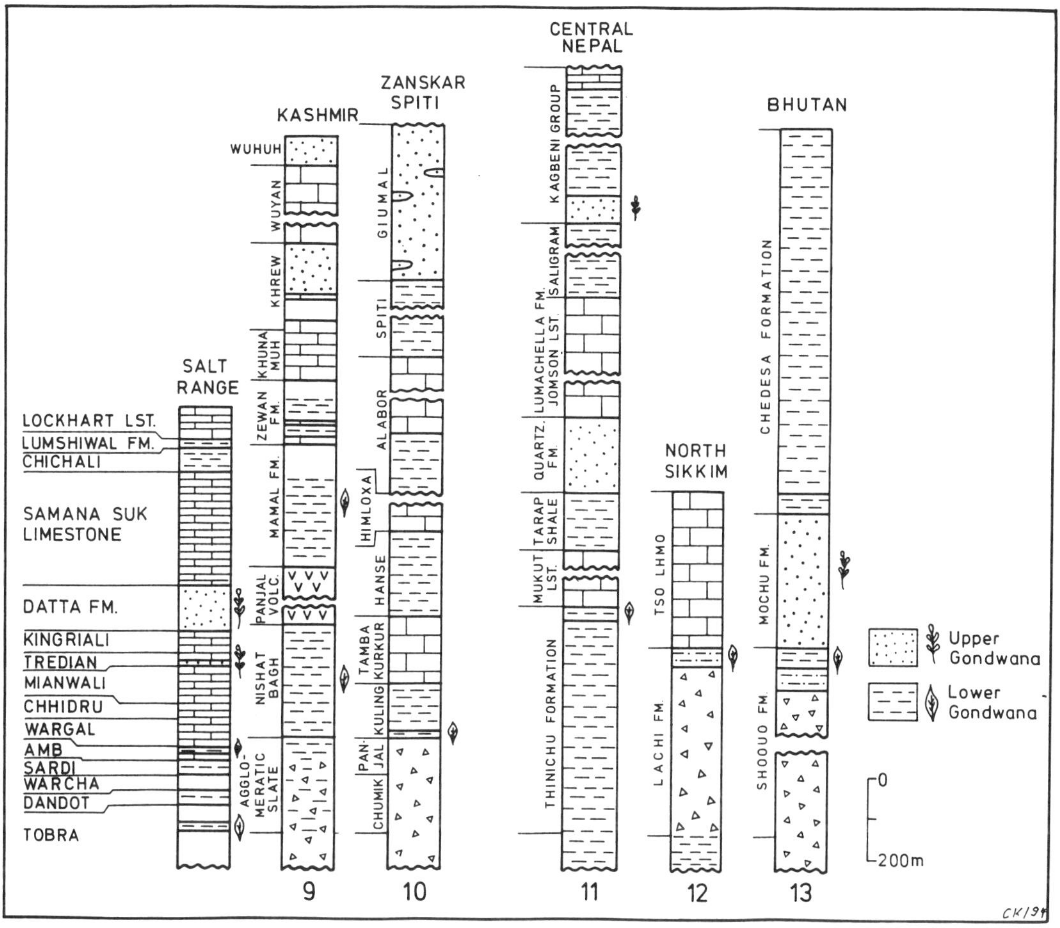

Fig. 11.8. Generalized lithologic columns of Gondwana sequences of the Tethys Himalaya, including that of the Salt Range in Pakistan (after Tripathi and Singh, 1987). For location, see encircled numbers in Tethyan Himalayan Belt in Figure 11.5.

Kashmir features a Gondwana succession dominated by thick continental basalts succeeding the glacigene succession. It is quite different from all other sections described so far and more like the Pangean sections of Tibet or the Baoshan terrane of west Yunnan. Gaetani and Garzanti (1991) interpreted the succession in Kashmir as products of an Early Permian rift.

The succession commences with Asselian diamictites of glacial origin, which rest unconformably on the Carboniferous *Fenestella* Beds. The diamictites are overlain by greywacke and black siltstones, which in turn are succeeded by the Agglomerate Slate. The latter contains an Artinskian faunal assemblage. The Nishatbag Formation, located conformably above the Agglomerate Slate and below the Panjal Volcanics (Panjal Traps), has yielded *Gangamopteris kashmirensis* and *Glossopteris nishatbaghensis*. According to Stampfli et al. (1991), the thick basalt accumulations of the Panjal Traps are of Late Kazanian to Tatarian age. The Mamal Formation (Kumar et al., 1980) overlies the Panjal Volcanics and in turn is overlain conformably by the Djulfian Zewan Formation (Gaetani and Garzanti, 1991). It has yielded a rich floral assemblage, including *Gangamopteris kashmirensis, Glossopteris cf. communis*, and *G. cf. feistmanteli* and is characterized by the presence of many typical Cathaysian forms, such as *Lobatannularia esfolia*, and *Rajahia mamalensis*.

Paleoclimate

Afro-Arabia

Glaciation and deglaciation. The onset of the Pangean depositional succession in the Late Carboniferous is synonymous with the late Paleozoic glaciation. In the Afro-Arabian region the glaciation did not extend beyond the geographical limit of the supercontinent, as it did in most other Gondwana fragments. In several cases, transitions from glaciated to nonglaciated terrains may be recognized. The best example exists in Arabia, where the glacigene deposits of the Haushi Group in Oman find their exact time equivalent in the fluviodeltaic succession of the Unayzah Group in central Saudi Arabia (Al Laboun, 1987; Besems and Schuurman, 1987). Other transition zones between glacial and periglacial regions have been identified by Wopfner (1991) between Gabon and Niger and East Africa and Libya. Thus, the northern limit of the late Paleozoic glaciers may be reconstructed with a fair degree of confidence (Wopfner, 1991). Such a reconstruction is attempted in Figure 11.9. It demonstrates that a climatic gradient must have existed between polar regions and lower latitudes.

Visser (1989) demonstrated that in South Africa glacial deposits (Dwyka Group) resulted from a subpolar ice sheet that had developed in proximity to the late Paleozoic South Pole. Tanzania and its neighbouring regions in Zambia and Madagascar were at not much lower latitudes, but ice transport directions (Diekmann, 1993) indicate that the glacial deposits of the Idusi Formation resulted from plateau-type glaciers. Lodgement tills were less prominent and the dominant facies were reworked glacial debris and periglacial deposits (such as slurried tills, slump masses, outwash gravels, laminites) (Wopfner and Kreuser, 1986; Wopfner, 1989; 1993; Wopfner and Diekmann, 1992; Diekmann 1993). Similar conditions prevailed in neighboring Zambia (Utting, 1978; Huebner, 1965) where in some areas, ancient glaciated valleys with west-flowing glaciers have been identified (Butakoff, 1948). Highlands that fed these glaciers were formed by a northeast-trending salient emanating from South Africa (Rust, 1975) and highlands to the east of Lake Tanganyika (Butakoff, 1948; Wopfner, 1991; 1993).

Madagascar was situated at about the same latitude as the Ruhuhu Basin (Fig. 11.1) and experienced comparable climatic and depositional conditions, except that the glaciers apparently entered the Morondava Basin from the south or southeast (Rust 1975).

Southern Arabia was situated at a lower latitude. Consequently, the glacial deposits were apparently controlled by marked fluctuation of ice movements, resulting in repeated intertonguing of flow tills, turbidites, and slurried

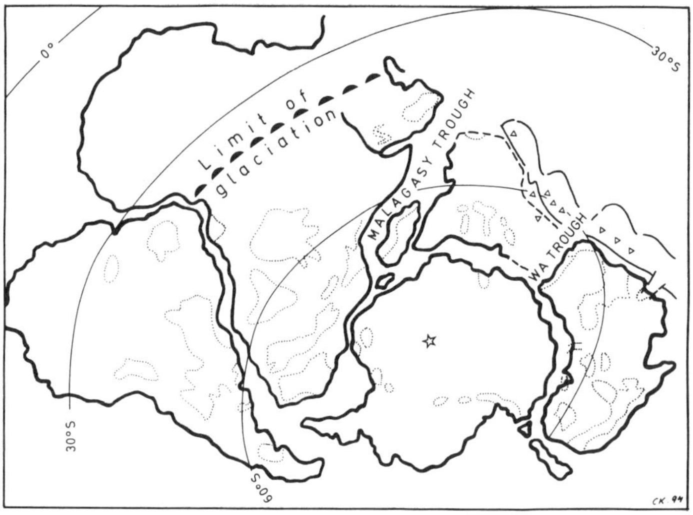

Fig. 11.9. Late Carboniferous/Early Permian reconstruction of Gondwana (after Smith and Hallam, 1970), showing Pangean basins with glacial deposits (Truswell, 1980) and the northern limit of glaciation in Afro-Arabia (Wopfner, 1991, 1993). A South Pole position (star) similar to that chosen by Levell et al. (1988) but adjusted for the glacial demarcation line, places Indian and East African glacial deposits at about 55°S. Tibet, Baoshan, and Sibumasu are purely schematic.

material (Levell et al., 1988). The same authors suggest a southerly source for the glaciers, whereas Kruck and Thiele (1983) proposed a movement of ice from the northwest (Sudan). Saudi Arabia and Niger were occupied by large delta and braid plains akin to urstromtaeler, fed by meltwaters of the glaciers. They were situated within the boreal climatic belt, which, at the beginning of the Permian, was situated below 40°S Gondwana latitude (Fig. 11.9).

The retreat of the glaciers in Afro-Arabia is marked by the deposition of lacustrine laminites. The regular repetitious nature of the laminites suggests a seasonal control of deposition, but only few of them have been identified as true varves (Diekmann, 1993). The common occurrence of dropstones demonstrates the proximity of glaciers flowing into proglacial lakes (Fig. 11.4). Interspersed with the laminite facies may be bedded siltstones, sandstone wedges, slump masses, and turbidites. Locally developed rain-out tills may indicate a temporary advance of the ice or increased calving activity. All these deposits are characterized by their "arkosic" composition, typical for deposits derived from glacial abrasion.

The beginning of the final deglaciation is marked by black, usually pyritic, kaolinitic lutites with an appreciable kerogene content, as described by Wopfner (1994b) and represented by the Lilangu Member of the Idusi Formation of East Africa (Fig. 11.4). Turbidites, which may be irregularly interspersed with the black lutites, indicate high sediment intake and instability of nearshore conditions. Plant fossils near the top of the Lilangu Member in the Ruhuhu Basin are mainly *Glossopteris* and *Gangamopteris* (Diekmann, 1993). The Lilangu Member corresponds to the "schistes noirs" of the Morondava Basin of western Madagascar (Bésarie, 1972) and the black bituminous Rahab Shale of southern Oman.

The kerogenous lutite facies clearly demonstrates the termination of the late Paleozoic glaciation. There is no unequivocal evidence for any glacial action succeeding the black lutite facies in Northeast Africa, Arabia, or Madagascar. The very content of organic matter, fossil plants and a sudden diversification of microflora suggest a much increased bioproduction associated with climatic amelioration. As in modern fjords (see Wopfner, 1994b for literature), the mass of organic debris caused anoxic conditions within the sediment column, thus producing the typical lithofacies association. Although demonstrable not in East Africa but in many other Gondwana basins (such as Madagascar, Oman, North India, West and South Australia), this event was accompanied by a marine transgression caused by the eustatic rise in sea level and the appearance of *Eurydesma*.

Dating the deglaciation succession thus provides a reasonable time for the beginning of the climatic amelioration. Microfloral assemblages obtained from the upper Lilangu Member put it within the *Granulatisporites confluence* Oppel-Zone (Foster and Waterhouse, 1988) placing the deglaciation succession within the Asselian to the Early Sakmarian. On the basis of palynological data, Besems and Schuurman (1987) suggested a middle-late Stephanian to Sakmarian age for the whole glacigene succession of South Oman, placing the Rahab Shale in the Late Asselian/Early Sakmarian. This is in keeping with the Sakmarian to Artinskian age of the overlying marine limestone (Hudson and Sudbury, 1959). Of additional interest is the appearance of Euramerican taxa, suggesting that the boundary between the Gondwana and the Euramerican floral province coincides roughly with the northern limit of glaciation, as shown in Figure 11.9.

The situation in the Salt Range and in Madagascar is similar to that described for Afro-Arabia.

Transition to the warmer climates. In eastern Africa and Madagascar the glacigene succession is succeeded by coal-bearing deposits. The basal part of the Mchuchuma Formation is formed by the Mpera Sandstone Member (Semkiwa, 1992), a fluviatile sandstone with early diagenetic pyrite and kaolin, indicating acid and reducing pore water conditions of the underflow. This, together with the overlying coal seams, suggests high precipitation rates and prolific plant growth, providing the biomass for the coal accumulation and sufficient organic acids to produce low-pH pore waters. Temperatures must have been high enough to produce the abundant vegetation. Although climate may still have been seasonal, a climate at least temperate to warm is suggested. Assuming a paleo-latitude of about 50°S and using the paleo-temperature curves of Frakes (1979), an average annual temperature of 10°C is indicated for the Sakmarian, rising to 15°C at the end of the Permian (Fig. 11.10).

There are no coal deposits in Arabia. The glacigene Haushi Group is succeeded by the marginal marine dolomites and evaporites of the Khuff Group, indicating a warm, seasonally hot, and mildly evaporitic climate.

The warm climatic phase. This phase, which was characterized by elevated temperatures and, at its beginning, by high CO_2 concentrations (Fig. 11.10), is exemplified by the lithofacies development of the Upper Songea Group of southwest Tanzania. Practically identical climatic histories are indicated by the time-equivalent successions in Zambia (Utting, 1978) and Madagascar (Bésarie, 1972).

The red beds of the Mbuyura Formation are interpreted as the product of a hot and arid to semiarid climate. The abundance of fresh feldspars in the sandstones and conglomerates indicates the absence of any chemical weathering which can be expected only if the crystalline basement disintegrated to grus in an arid environment. The current bedded sandstones were laid down by episodic flash floods in braided wadi systems or by sheet flooding. The intervening red micaceous siltstones with calcretes were deposited in overbank claypans or inland floodouts.

At the end of deposition of the Mbuyura Formation precipitation increased leading to the development of a large swamp basin that received the wadi run-off from the elevated regions in the west and the northwest (Kaaya, 1992), thus creating the environment for the coal

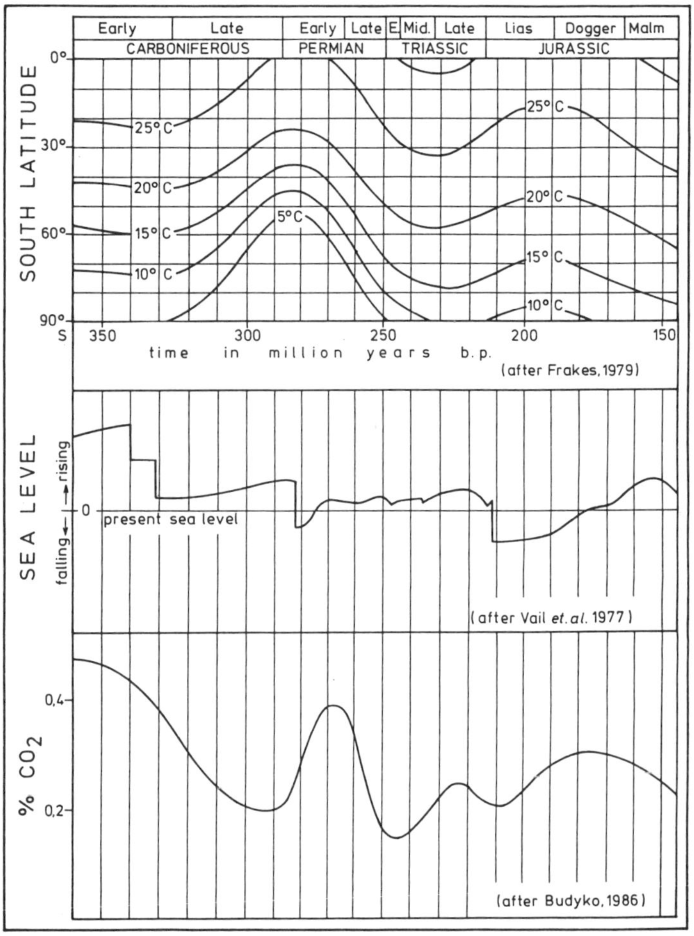

Fig. 11.10. Compilation of mean annual temperatures for the Southern Hemisphere, relative changes of sea level, and variations of CO^2 concentrations in the atmosphere for the time interval Carboniferous to Jurassic (after Wopfner, 1993).

accumulation of the Mhukuru Formation around the Kungurian/Kazanian boundary.

Elevated precipitation rates persisted throughout the Kazanian and, combined with renewed subsidence, led to the extensive inland lake in which the Ruhuhu Formation, with its stromatolites and oolitic limestone lenses, was laid down. Temperatures were warm to hot, leading to occasional mildly alkaline conditions in the lake waters (Kaaya, 1992). Toward the end of the Permian, temperatures remained high, but precipitation may have decreased slightly, producing the shallow lake and the mudflat landscape of the Usili Formation. The abundance of silicified tree trunks suggests local pH-gradients within the ground water and sufficient plant growth to sustain a varied tetrapod population.

A dramatic change in depositional environment occurred at the beginning of the Triassic. In part this change was due to renewed rift activity (Wopfner, 1994a) but it was also affected by marked climatic modifications. Sedimentary structures of the Kingori Sandstone indicate high-volume transport of abundant bed load material requiring humid to hyperhumid conditions (Wopfner, 1989; 1993). The almost pure quartz composition of the Kingori Sandstone is explained by its derivation from a mantle of deeply weathered basement created by tropical climate (Markwort, 1991). Deposition of the succeeding Manda Beds was controlled by high temperatures and reduced precipitation rates (Wopfner et al., 1991).

In southern Arabia the climate was hot and semiarid, as indicated by the dolomite and gypsum interbeds of the Khuff Group (Khalifa, 1993). A warm climate is also confirmed by the parautochthonous Upper Permian limestones of northern Oman (Blendinger et al., in Robertson et al., 1990; Lee, in Robertson et al., 1990). The red and green shales of the Lower Triassic Sudair Formation were still the product of a semiarid climate, but thereafter aridity increased to form the late Early Triassic evaporites (Schlumberger, 1976).

India

The Gondwana Basins of Peninsular India were at about the same latitude as those of East Africa and Madagascar (Fig. 11.9). Consequently they experienced a similar climatic history.

The glacial and transitional phases. Climatic development during the depositional event of the Talchir Formation is exemplified by the succession in the central parts of the Son-Mahanadi Gondwana Basin. The massive basal diamictite shows the characteristic textural, structural and compositional features of a basal till. In addition, the presence of ice-contact features such as crescent marks and subparallel striations on the bedrock such as those seen in the Tan River section and roches moutonnées in neighboring areas (Datta et al.,1979) confirm the concept that the basal diamictite was deposited directly as a lodgment till without any subsequent reworking. It falls under the category of true till (tillites) considered to have formed during glacial retreat (Boulton and Eyles, 1979).

The retreat of the Talchir glaciers was followed immediately by the development of variably sized meltwater lakes. They occupied depressions of the unevenly shaped basin floor that resulted from the scouring action of the glaciers. In these impounded water bodies, the mud released from glacial debris was deposited as finely laminated mudstone devoid of all current structures. Clasts that now occur as dropstones embedded in the laminites were released from floating ice. The occurrence of cross-laminated lenses of siltstones in the upper part of the mudstone facies is indicative of weak currents and the consequent reworking of fine clastics. More effective reworking and shallowing of the basin caused the mudstone to pass upward gradationally into a massive or faintly bedded fine to very fine sandstone.

The Extra-Peninsular glacial and proglacial successions were deposited at lower latitudes and close to the Tethyan margin of Gondwana (Fig. 11.9). There are some lodgement tills, but wash-out and rain-out tills appear to dominate. The thick succession of black lutites with marine-fossiliferous carbonate concretions (Acharyya et al., 1979; Singh, 1983; 1987) reflects the deglaciation process in a distal, marginal marine position created by the eustatic rise in sea level (Fig. 11.10).

The presence of *Eurydesma* and a bryozoa/mollusc fauna in several parts of Peninsular and Extra-Peninsular India (Arunachal Pradesh, Sikkim, Bhutan) and in the Salt Range in Pakistan, together with the occurrence of a *Gangamopteris* flora and monosaccate pollen, are all suggestive of a cool climate throughout the period of Talchir sedimentation up to the top layers, which date to the early Sakmarian.

The postglacial phase. This phase was marked by a cool, temperate climate during deposition of alluvial fan and braided river facies of the Karharbari and Barakar Formations, which abound with fresh to slightly weathered feldspar. The mega- and micro-floras are similar to those of the Talchir Formation, although recurrence of coal seams may indicate at least seasonal warming of the climate. The red colored Barren Measures provide the first signs of a warm, humid climate during the Middle Permian throughout Peninsular India, followed by the coal-bearing Raniganj Coal Measures with a flora dominated by *Glossopteris*.

The coastline facies of Lesser Himalaya and the shelf facies of Tethyan Himalayas, which succeed the basal tillites and other glacigene deposits, were formed in lower latitudes and abound in carbonate and fine clastic facies. They are indicative of a warm temperate climate.

The Triassic period represented by the Panchet/Pachmarhi Formations, comprising pebbly sandstone and reddish and yellow-colored shale, suggests a seasonally arid, warm climate.

Conclusion

The concept that Paleozoic glaciation centers coincided with pole positions as Gondwana drifted across the geographical South Pole (Crowell and Frakes, 1975; Caputo and Crowell, 1985) at first glance appears to be a convenient hypothesis. It seems to explain the formation of apparent centres of glaciation at different parts of Gondwana at different times, such as the Ordovician glaciation in Africa or the Devonian in South America. However, as Chumakov (1985) has pointed out, manifestations of warm climates along the same apparent polar wander path are just as common as those for phases of glaciation.

In the case of the late Paleozoic glaciation, the areal distribution of which exceeded that of any of the preceding ones by an order of magnitude, an almost synchronous initiation of the glaciation in the Stephanian in central and eastern Africa, southern Arabia, and India is indicated. We accept, however, the possibility of a slightly earlier beginning in localities nearer to the pole. The northern limit of the glacial front is well constrained in Afro-Arabia, as shown in figure 11.9. In India it certainly extended beyond the Himalayas and is present on Gondwana terranes in Tibet and Yunnan. Sedimentological features of glacial and proglacial deposits demonstrate that most of them were laid down during the final retreat of wet-based glaciers. In East Africa, Madagascar and southern Arabia they were deposited under terrestrial conditions, generally followed by fluctuating lacustrine environments. Well-exposed glaciated pavements in Oman and Yemen (Kruck and Thiele, 1983) and glacial valleys in eastern Africa (Butakoff, 1948; Diekmann, 1993) attest to such terrestrial environments. Marine incursions occurred only in Arabia and Madagascar and only after deglaciation caused a substantial eustatic sea-level rise.

Paleobiological and lithofacies data indicate that deglaciation was an almost synchronous event in central and eastern Africa, Madagascar, Arabia, and Peninsular and Extra-Peninsular India, as well as on the terranes of Tibet and West Yunnan. This was a comparatively rapid change, most strikingly demonstrated by the marked transition from the deglacation facies (Rahab Shale) to the warm-water carbonates and evaporites in Oman and in the Salt Range. A similar situation also exists in Gabon and southwest Yunnan. Synchronous deglaciation is further corroborated by the massive rise in eustatic sea level in the Sakmarian, accompanied by a substantial increase in CO_2 concentration, as demonstrated in Figure 11.10.

This rapid and synchronous deglaciation event is not compatible with the drift of Gondwana across the South Pole unless one accepts unreasonably high drift rates for the whole of the supercontinent. The evidence presented in our study indicates a rapid modification of climatic zones in the Early Permian, the warm, subtropical belt expanding at the expense of the boreal/temperate zone.

The time required for deglaciation is difficult to estimate. Palynological data from East Africa and Oman suggest that the deglaciation process took up most of the Asselian and part of the Sakmarian, amounting to a time span of about 5 Ma. Accepting the beginning of the late Paleozoic glaciation at around 300 Ma, as suggested earlier, the total duration of the glacial period up to the eustatic sea-level rise would have been about 15 to 20 Ma, depending which absolute time scale is used. Using these values as a general time frame, two thirds of the total time span would have been taken up by glacial and interglacial conditions and about one third by the final elimination of the ice cover.

The dramatic increase of the CO_2-concentration, the peak of which was reached at about the Artinskian (Fig. 11.10), was associated with the rapid warming of the climate. Although some of that increase was no doubt due to increased volcanic activity in the Early Permian (Fischer, 1982), a climatic component must also be considered. The establishment of the postglacial vegetation, especially the vast coal swamps on Gondwana and the development of extensive carbonate platforms led to the consumption of the excess CO_2. Despite the decline in CO_2 in the atmosphere, average temperatures apparently increased up to a maximum around the Middle Triassic. Thus, it appears that climatic development was controlled not so much by CO_2 concentrations as by other global changes.

Cyclical merging of land masses, their subsequent disintegration, and changing oceanic currents (Crowell, 1978 and elaborated on by Veevers, 1990) may produce important back-coupling effects. Veevers (1990) suggested a 400Ma cycle in which the earth alternates between a stage of unified crust (Pangea) and a stage of dispersed continental fragments. Periods of a single continental mass experienced an ice house climate, exemplified by long-lasting glaciations, such as those of the latest Proterozoic and the late Paleozoic. In contrast, periods of continental dispersion and slow sea-floor spreading experienced high expulsion rates of CO_2 resulting from abundant magma emplacements and leading to greenhouse conditions. The short-lived Ordovician and Quaternary glaciations would have occurred during such greenhouse stages.

Diverging views were expressed by Fischer (1982), who proposed an ice house stage for both the late Paleozoic and the Quaternary glaciation periods.

Purely "geocentric" views may not be able to solve such discrepancies, and we propose that much closer attention be given to astronomical events, as suggested by Berger and Tricot (1986) and others. Steiner (1967) and Steiner and Grillmair (1973), for instance, suggested a recurrence of geological events at about 300 Ma intervals, governed by changes of mass distribution within the Milky Way System in the course of a galactic year. In the scheme of these authors, both the late Paleozoic and the Quaternary glaciations would have occurred during the same "season" of the galactic year. They also point out that both glaciations were preceded by major orogenic pulses, an observation that also holds true for the Late Ordovician glaciation.

Assuming that the Quaternary glaciation is part of an icehouse phase, it may be permissible to transpose the course of the Paleozoic glaciation to the Pleistocene/Holocene climatic history. Accordingly, the earth would still be within the waxing and waning stage of glaciation and the next glacial advance would be predetermined. An increase of the green house effect by increased atmospheric CO_2 may therefore be beneficial in softening the impact of the next glacial advance and climatic deterioration.

References

Acharyya, S.K., 1971. Rangit Pebble Slate - a new formation from Darjeeling Hills. Indian Minerals. Geological Survey of India, 25: 61–64.

Acharyya, S.K., Sha, S.C., Ghosh, S.C. and Ghosh, R.N., 1979. Gondwana of Himalaya and its Stratigraphy. IV International Gondwana Symposium Calcutta, India, vol. 2, pp. 420–430.

Al-Laboun, A.A., 1987. Unayzah Formation: a new Permian-Carboniferous unit in Saudi Arabia. American Association of Petroleum Geologists Bulletin, 71: 29–38.

Archbold, N.W., 1983. Permian marine invertebrate provinces of the Gondwana realm. Alcheringa, 7: 59–73.

Banerjee, Indranil, 1966. Turbidites in a glacial sequence: a study from the Talchir Formation, Raniganj coalfield, India. Journal of Geology, 74: 593–606.

Berger, A. and Tricot, C., 1986. Global climatic changes and astronomical theory of paleoclimates. In: A. Cazenave (Editor), Earth rotation: solved and unsolved problems. Reidel, Dordrecht, pp. 111–129.

Besems, R.E. and Schuurman, W.M.L., 1987. Palynostratigraphy of Late Paleozoic glacial deposits of the Arabian Peninsula with special reference to Oman. Palynology, 11: 37–53.

Bésairie, H., 1972. Geologie de Madagascar. Imprimerie Nationale, Tananarive, 463 pp.

Boothroyd, J.C. and G.M. Ashley, 1975. Process, bar morphology and sedimentary structures on braided outwash fans, northeastern Gulf of Alaska. In:

A.V.Jopling and B.C. McDonald (Editors), Glaciofluvial and glaciolacustrine sedimentation. Society of Economic Paleontologists and Mineralogists, Special Publication 23, pp. 193–222.

Bosellini, A., 1989. The continental margins of Somalia: their structural evolution and sequence stratigraphy. Mem. Sci. Geol. Istituto Geol. e Mineral, Università di Padova, XVI: 373–458.

Boulton, G.S. and Eyles, N., 1979. Sedimentation by valley glaciers: A model and genetic classification. In: C. Schuchter (Editor), Moraines and varves. A.A. Balkema, Rotterdam, pp. 11–23.

Budyko, M.J., 1986. The evolution of the biosphere. Reidel, Dordrecht, 528 pp.

Butakoff, N., 1948. Les formationes glaciaires et post glaciaires fossilifères d'age permo-carbonifère (Karroo inferieur) de la Région Walikale (Kivu, Congo belge). Mem. Inst. Geol. Univ. Louvain, 9: 1–124.

Caputo, M.V. and Crowell, J.C., 1985. Migration of glacial centres across Gondwana during Paleozoic Era. Geological Society of America Bulletin, 96: 1020–1036.

Casshyap, S.M., 1973. Paleocurrents and paleogeographic re-construction in the Barakar (Lower Gondwana) sandstones of peninsular India. Sedimentary Geology, 9: 283–303.

Casshyap, S.M., 1979. Patterns and sedimentation in Gondwana basins. IV. International Gondwana Symposium, Calcutta, India, vol. 2, pp. 525–551.

Casshyap S.M. and Qidway. H.A., 1974. Glacial sedimentation of Late Paleozoic Talchir diamictite, Pench Valley coalfield, central India. Geological Society of America Bulletin, 85: 749-760.

Casshyap S.M. and Srivastava, V.K., 1987. Glacial and proglacial Talchir sedimentation in Son-Mahanadi Gondwana basin: paleogeographic reconstruction. In G.D. McKenzie (Editor), Gondwana Six, American Geophysical Union, Washington, DC, Monograph 41, pp. 167–182.

Casshyap, S.M. and Tewari, R.S., 1987. Depositional models and tectonic evolution of Gondwana basins. Paleobotanist, 36: 59–66.

Casshyap, S.M., Kreuser, T. and Wopfner, H., 1987. Analysis of cyclic sedimentation in the Lower Permian Mchuchuma coalfield (Southwest Tanzania). Geologische Rundschau, 76: 869-883.

Chumakov, N.M., 1985. Glacial events of the past and their geological significance. Palaeogeography, Palaeoclimatology, Palaeoecology, 51: 319–346.

Collinson, J.D. and Thompson, D.B., 1982. Sedimentary structures. George Allen and Unwin, London, 194 pp.

Crowell, J.C., 1978. Gondwanan glaciation, cyclothems, continental positioning and climatic change. American Journal of Science, 278: 1245–1312.

Crowell, J.C., 1983. Ice ages recorded on Gondwanan continents. Transcriptions of the Geological Society of South Africa, 86: 237–262.

Crowell, J.C. and Frakes, L.A., 1975. The Late Palaeozoic glaciation. In: K.S.W. Campbell (Editor) Gondwana Geology, III. International Gondwana Symposium, Canberra, pp. 313–331.

Datta, N.R., Dutta, A. and Majumdar, A.K., 1979. Talchir glaciation in the Son-Mahanadi and Koel valleys and adjoining areas. IV International Gondwana Symposium, Calcutta, India, vol. 1, pp. 360–367.

De Wit, M., Jeffery, M., Bergh, H. and Nicolaysen, L., 1988. Geological map of sectors of Gondwana. American Association of Petroleum Geologists, Tulsa, OK.

Dickins, J.M. and Sha, S.C., 1979. Correlation of the main sequences of India and western Australia. IV. International Gondwana Symposium, Calcutta, India, vol. 2, pp. 387–407.

Diekmann, B., 1993. Paläoklima und glazigene Karoo-Sedimente des späten Paläozoikums in SW-Tansania. Geologisches Institut der Universität zu Köln, Sonderveröffentlichung 90, pp. 183.

Dreimanis, A., 1979. The problems of waterlain tills. In: C. Schuchter (Editor), Moraines and Varves. A.A. Balkema, Rotterdam, pp. 167–177.

Fischer, A.G., 1982. Long-term climatic oscillations recorded in stratigraphy. In: W.H. Berger and J.C. Crowell (Editors), Climate in earth history. National Academic Press, Washington, DC, pp. 97–104.

Foster, C.B., 1986. A review of Early Carboniferous-Early Permian selected samples from Kulshill Nos 1 & 2, Leseur No. 1, Lacrosse No. 1, Pelican Island No.1 and Berkley No. 1 of the Bonaparte Basin. Australian Geological Survey Organisation, Open File Report, 56 pp.

Foster, C.B. and Waterhouse J.B., 1988. The Granulatisporites confluence Oppel-zone and early Permian marine faunas from the Grant Formation on the Barbwire Terrace, Canning Basin, Western Australia. Australian Journal of Earth Sciences, 35: 135–157.

Frakes, L.A., 1979. Climates throughout geologic time. Elsevier, Amsterdam, 310 pp.

Frakes, L.A., Kemp, E.M. and Crowell, J.C., 1975. Late Paleozoic glaciation (V) Asia. Geological Society of America Bulletin, 86: 454–464.

Gaetani, M. and Garzanti, E., 1991. Multicyclic history of the northern India continental margin (Northwestern Himalaya). American Association of Petroleum Geologists Bulletin 75: 1427-1446.

Ghosh, P.K. and Mitra, N.D., 1972. A review of recent progress in the studies of the Gondwanas of India. II. International Gondwana Symposium, Pretoria, South Africa, pp. 23–47.

Hambrey, M.J. and Harland, W.B., 1981. Earth's pre-Pleistocene glacial record. Cambridge University Press, Cambridge, 1003 pp.

Hankel, O., 1987. Lithostratigraphic subdivision of the Karoo rocks of the Luwegu Basin (Tanzania) and their biostratigraphic classification based on microfloras, macrofloras, fossil woods and vertebrates. Geologische Rundschau, 76: 539-565.

Huebner, H., 1965. Permokarbonische glazigene und periglaziale Ablagerungen aus dem zentralen Teil des Kongobeckens. Stockholm Contributions Geology, 13: 41–61.

Hudson, R.G.S. and Sudbury, M., 1959. Permian brachiopoda from south-east Arabia. Museum National D'Historie Naturelle, VII, Paris, pp. 19–55.

Joshi, A., Wangda, D. and Mamgain, V.D., 1990. Marine Permian sequence in foothills of Bhutan, Lesser Himalaya. Indian Minerals, 44: 151–162.

Kaaya, C.Z., 1992. Depositional environment of Late Permian Karoo Beds in the Ruhuhu Basin and Mikumi area of Tanzania. Geologisches Institut der Universität zu Köln, Sonderveröffentlichung 83, 126 pp.

Kemp, E.M., 1975. The palynology of Late Palaeozoic glacial deposits of Gondwanaland. In: K.S.W. Campbell (Editor), Gondwana Geology. Australian National University Press, Canberra, pp. 397–413.

Khalifa, M.A., 1993. Lithostratigraphic classification and depositional history of the Permian rocks in Al Qasim Province, Saudi Arabia. Journal of African Earth Sciences, 16: 329–340.

Kreuser, T., Wopfner, H., Kaaya, C.Z., Markwort, S., Semkiwa, P. and Aslanidis, P., 1990. Depositional evolution of Permo-Triassic Karoo basins in Tanzania with reference to their economic potential. Journal of African Earth Sciences, 10: 151–167.

Kruck, W. and Thiele, J., 1983. Late Palaeozoic glacial deposits in the Yemen Republic. Geologisches Jahrbuch, 46: 3–29.

Kumar, G., Singh, G. and Srivastava, G.S., 1980. Paleozoic stratigraphy of Kashmir Basin with special reference to Lidar Valley, Kashmir. Symposium on Three Decades of Development in Paleontology, Hyderabad.

Kummel, B. and Teichert, C. 1966. Relations between Permian and Triassic formations in the Salt Range and Trans-Indus Ranges, West Pakistan. Neues Jahrbuch Geol. Paläont. Abhandlungen 125: 297–333, (Festband Schindewolf).

Kummel, B. and Teichert, C., 1970. Stratigraphy and Paleontology of the Permo-Triassic Boundary Beds, Salt Range and Trans-Indus Ranges, West Pakistan. Department of Geology Special Publication 4, University Press of Kansas, 110 pp.

Levell, B.K., Braakman, J.H. and Rutten, K.W., 1988. Oil-bearing sediments of Gondwana glaciation in Oman. American Association of Petroleum Geologists Bulletin, 72: 775–796.

Liang Dingyi, Nie Zetong, Guo Tieyin, Xu Baowen, Zhang Yizhi and Wang Weipin, 1983. Permo-Carboniferous Gondwana-Tethys facies in southern Karakoran Ali, Xizang (Tibet). Journal of the Wuhan College of Geology, 1: 9–27 (in Chinese with English abstract).

Markwort, S., 1991. Sedimentation und Diagenese der triassischen Karoo-Schichten des Ruhuhu-Beckens, Tansania. Geologisches Institut der Universität zu Köln, Sonderveröffentlichung 80, 118 pp.

Mbede, E.I., 1993. Tectonic development of the Rukwa rift basin in SW Tanzania. Berliner Geowissenschaftliche Abhandlungen, Reihe A, 152: 92 pp.

Metcalfe, I., 1988. Origin and assembly of southeast Asian continental terranes. In: M.G. Audley Charles and A. Hallam (Editors), Gondwana and Tethys. Geological Society of London, Special Publication, 37: 101–118.

Nakazawa, K. and Kapoor, H.M., 1977. Correlation of the marine Permian in the Tethys and Gondwana. IV. International Gondwana Symposium, Calcutta, India, vol. 2, pp. 409–416.

Robertson, A.H.F., Searle, M.P. and Ries, A.C. (Editors), 1990. The geology and tectonics of the Oman region. Geological Society London, Special Publication 49, pp. 845.

Rust, I.C., 1975. Tectonic and sedimentary framework of Gondwana Basins in southern Africa. Gondwana Geology. Australian National University Press, Canberra, pp. 537–564.

Schindewolf, O.H., 1967. Zur permischen Vereisung der Salt Range (W. Pakistan). Geologische Rundschau, 56: 914–918.

Schlumberger, 1976. Well Evaluation Conference, Iran, Services Techniques. Schlumberger, Paris, 180 pp.

Semkiwa, P.Z., 1992. Depositional environment and coal petrography of Permian coal deposits in Karoo Basins of SW-Tanzania. Geologisches Institut der Universität zu Köln, Sonderveröffentlichung 84, 184 pp.

Shaw, J., 1975. Sedimentary successions in Pleistocene ice marginal lakes. In: A.V. Jopling and B.C. McDonald (Editors), Glaciofluvial and glaciolacustrine sedimentation. Society of Economic Paleontologists and Mineralogists, Special Publication 23: 281–303.

Singh, P., 1973. A note on the fossiliferous formation in Lesser Himalaya of Nepal and Bhutan. Himalayan Geology, 3: 373–380.

Singh, T., 1983. Age and faunal affinity of the Garu Formation, Arunachal Pradesh. Himalayan Geology, 11: 271–286.

Singh, T., 1987. Permian biogeography of Indian Subcontinent with special reference to marine fauna. In: G.D. McKenzie (Editor), Gondwana Six. American Geophysical Union, Monograph, 41, pp. 239–249.

Smith, A.G. and Hallam, A., 1970. The fit of the southern continents. Nature, 225: 139–144.

Stampfli, G., Marcoux, J. and Baud, A., 1991. Tethyan margins in space and time. Palaeogeography, Palaeoclimatology, Palaeoecology, 87:373–409.

Steiner, J., 1967. The sequence of geological events and the dynamics of the Milky Way Galaxy. The present cosmic year - a preliminary study. Journal of the Geological Society of Australia, 14: 99–131.

Steiner, J. and Grillmair, E., 1973. Possible galactic causes for periodic and episodic glaciation. Geological Society of America Bulletin, 84: 1003–1018.

Stockley, G.M., 1932. The geology of the Ruhuhu Coalfields, Tanganyika Territory. Geological Society London, Quarterly Journal, 88: 610–622.

Teichert C., 1970. Marine fossil invertebrate faunas of the Gondwana region. II. International Gondwana Symposium, Pretoria, South Africa, pp. 125–138.

Tripathi, C. and Chowdhury, J.R., 1983. Gondwana of Arunachal Himalaya. Himalayan Geology, 11: 73–90.

Tripathi, C. and Singh, G., 1987. Gondwana and associated rocks of the Himalaya and their significance. In: G.D.McKenzie (Editor), Gondwana Six, American Geophysical Union, Monograph, 41: 195–205.

Truswell, E.M., 1980. Permo-Carboniferous palynology of Gondwanaland: progress and problems in the decade to 1980. BMR Journal of Australian Geology and Geophysics, 5: 95–111.

Utting, J., 1978. The Karroo stratigraphy of the northern part of the Luangwa Valley. Geological Survey of Zambia, Memoir 4, 64 pp.

Vail, P.R., Mitchum, R.M. and Thompson, S., 1977. Seismic stratigraphy and global changes of sea level, Part 4. Global cycles of relative changes of sea level. American Association of Petroleum Geologists, Memoir 26, pp. 83–97.

Veevers, J.J., 1990. Tectonic-climatic supercycle in the billion-year plate-tectonic era: Permian Pangean icehouse alternates with Cretaceous dispersed-continents greenhouse. Sedimentary Geology, 68: 1–16.

Veevers, J.J. and Powell, C.McA., 1987. Late Paleozoic glacial episodes in Gondwanaland reflected in transgressive-regressive depositional sequences in Euramerica. Geological Society of America Bulletin, 98: 475–487.

Visser, J.N.J., 1989. The Permo-Carboniferous Dwyka Formation of southern Africa: Deposition by a predominantly subpolar marine ice sheet. Palaeogeography, Palaeoclimatology, Palaeoecology, 70: 377–391.

Visser, J.N.J., 1990. The age of the late Palaeozoic glacigene deposits in southern Africa. African Journal of Geology, 93: 366–375.

Wang Yizhao, 1983. The characteristics and significance of Carboniferous gravel beds in Tengchong and Baoshan area, Western Yunnan. Contributions to the geology of the Qinghai-Xizang (Tibet) Plateau. Geological Publishing House Beijing, pp. 71–76 (in Chinese with English abstract).

Wopfner, H., 1989. Palaeoclimate and depositional control of the Karoo sequence in southern Tanzania. XIe Congres International de Stratigraphie et de Géologie Carbonifère Beijing 1987, Compte Rendu, 4: 89–95.

Wopfner, H., 1990. Rifting in Tanzanian Karoo basins and its economic implications. In: G. Rocci and M. Deschamps (Editors), Etudes récentes sur la Géologie l'Afrique. 15ᵉ Colloque de Géologie Africaine, Nancy 1990. CIFEG, Publication occasionelle 1990/22, Orleans, pp. 217–220.

Wopfner, H., 1991. Extent and timing of the Late Palaeozoic glaciation in Africa. Geologisches Institut der Universität zu Köln, Sonderveröffentlichung, 82: 447–453.

Wopfner, H., 1993. Paläoklimatische Kontrolle der Karoo-Ablagerungen im Südwesten Tanzanias. Geologisches Institut der Universität zu Köln, Sonderveröffentlichung 70: 57–76.

Wopfner., H., 1994a. The Malagasy Rift, a chasm in the Tethyan margin of Gondwana. Journal of Southeast Asian Earth Sciences, 9: 451–461.

Wopfner, H., 1994b. Early Permian deglaciation sequences reflect end of "ice house" conditions in northern Gondwana. Permophiles, 24: 22–23.

Wopfner, H. and Diekmann, B., 1992. Neue Ergebnisse aus der spätpaläozoischen Abfolge in der Karoo Tansanias. Zentralblatt für Geologie und Paläontologie, Teil I, 1991, pp. 2689–2701.

Wopfner, H. and Jin X., 1993. Baoshan and Tengchong Blocks of western Yunnan (China) in the Late Paleozoic mosaic of the eastern Tethys. Third International Symposium IGCP 321–Gondwana Dispersion and Asian Accretion, Kuala Lumpur, pp. 47–48.

Wopfner, H. and Kreuser, T., 1986. Evidence for Late Palaeozoic glaciation in southern Tanzania. Palaeogeography, Palaeoclimatology, Palaeoecology, 56: 259–275.

Wopfner, H,., Markwort, S. and Semkiwa, P., 1991. Eary diagenetic laumontite in the Lower Triassic Manda Beds of the Ruhuhu Basin, Southern Tanzania. Journal of Sedimentary Petrology, 61: 65–72.

12. PERMIAN POSTGLACIAL ENVIRONMENTS OF THE AUSTRALIAN PLATE

John F. Lindsay

Australia was a part of eastern Gondwana during the Permian (Fig. 12.1). The Lord Howe Rise and New Zealand lay further to the east and southeast respectively, while Antarctica adjoined the continent's southern margin (Lawver and Scotese, 1987). India lay to the west, extending north along the western Australian coast as far as North West Cape (Klootwijk et al., 1985; Powell et al., 1988).

During the Permian Australia was structurally complex and provided a variety of depositional settings in which both glacial and postglacial sediments could accumulate (Fig. 12.2; Eyles, 1993). In simplest terms, the Permian basins can be grouped into three broad categories: the northern and northwestern extensional settings, the intracratonic settings, and the eastern Australian active paleo-Pacific margin settings.

The Northern and Northwestern Extensional Settings

Along the northern and the northwestern margins of the Australian plate, basins formed by extensional processes in the Late Carboniferous-Early Permian and contain thick Permian successions. Perhaps, in part at least, some of these basins are related to the rifting of Sibumasu from the edge of Gondwana in the late Early Permian (Metcalfe, 1988). The northern and northwestern areas include some of the continent's major Permian depocenters especially the Bonaparte, Browse, Canning and Carnarvon basins (Fig. 12.2). The Bonaparte in particular contains a thick Permian succession that reaches 6,000 m at least, and the Browse and the Carnarvon basins each contain in excess of 3,500 m of Permian. Little is known of the offshore extent of the Permian succession in any of these basins. North of these basins, the Permian is present in Timor, Irian Jaya, and the Arafura Basin. Coal measures are present in Irian Jaya in particular (Pigram and Panggabean, 1983; Archbold et al., 1982), however, the nature of these basins is not well understood.

The Intracratonic Settings

Intracratonic basins containing Permian succes sions are spread across the width of the Australian continent. Most are relatively shallow features and appear to be related to the reactivation of preexisting structures (Fig. 12.2). The basins fall into two broad types, small localized features with relatively thick successions such as the Collie, Wilga, and Boyup basins (Playford et al., 1976; Backhouse, 1990) and broad, shallow depressions, such as the Cooper and the Galilee basins. Coal is present in postglacial settings in a number of these basins but the Cooper Basin is by far the most important, not only it is widespread, but it contains up to 1,000 m of Permian rocks including coal and, most significant, associated hydrocarbon production (Smyth and Cameron, 1982).

The Eastern Australian Active Margin Settings

During the Late Carboniferous and the earliest Permian, an active margin with a volcanic belt paralleled the east coast or the paleo-Pacific margin of Australia from the Sydney Basin to Townsville (Fig. 12.2). Late Paleozoic glaciation appears earlier on this part of the Australian continent than elsewhere, perhaps in response to uplift (Eyles, 1993). Subduction ceased in the Early Permian, when tectonic movement changed from essentially convergent to dominantly strike-slip (Cawood, 1982) and deep-water sediments collected in the region. This extensive basin

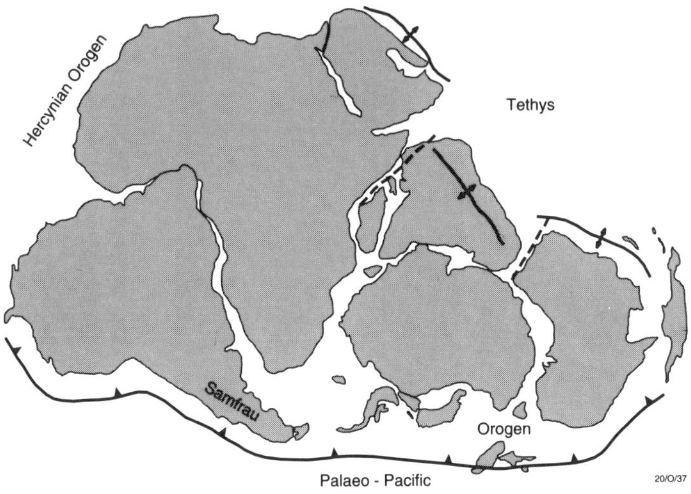

Fig. 12.1. Gondwana and its major tectonic elements in the Permian (after Langford, 1991).

system bordering the orogenic belt in eastern Australia was probably formed by back-arc extension in the Early Permian and became a foreland basin during mid-Late Permian compression (Schiebner 1973; Korsch, 1982; Murray, 1985; Mallet et al., 1988). The Sydney-Gunnedah-Bowen Basin system is by far the most important part of the basin complex and forms an extensive trough, more than 1,500 km in length, in front of the New England Orogen. The basins formed along this active margin contain perhaps the most extensive record of the Australian Permian and are certainly the most important in an economic sense, as they contain the continent's major exploitable coal resources.

The New England Orogen includes large areas of Permian rocks but is an extremely complex and still somewhat enigmatic region. During the Devonian-Carboniferous, it was the site of an east-facing, Andean-type convergent margin, with ignimbritic volcanism in the arc and fore-arc and accretionary wedge to the east. The New England Orogen appears to consist of a number of allochthonous terranes, and its evolution during the Permian is not well understood (Korsch et al., 1991).

New Zealand, located on the eastern margin of Gondwana, has a complex structural history. It consists of an older part (Western Province), where the basement is Ordovician or older, and a younger part (Eastern Province), where the basement is no older than the Carboniferous. The cratonic Western Province consists of neritic to marginal marine Permian quartzose-clastics rocks that are similar to those found in Tasmania. The Western Province appears to be closely related to Australia, Antarctica, and adjoining platforms, such as the Lord Howe Rise, and was probably contiguous with the Tasman Geosyncline (Crook and Feary, 1982).

Paleogeography

The global paleoclimate of the Permian was dominated by monsoonal atmospheric circulation (Parrish and Barron, 1986; Parrish et al., 1986). The size and the large latitudinal extent of Pangea caused the surface air pressure to deviate strongly from the normal zonal pattern. Since land areas respond more strongly to seasonal changes in solar radiation, the large Pangean megacontinent became warm in summer, establishing a low pressure cell, and cool in winter, developing a high-pressure cell, resulting in increased seasonality. This effect was greatest during the Late Permian-Triassic, when global plate movements caused the land area to be almost symmetrically disposed around the equator. Since large areas of Gondwana were located

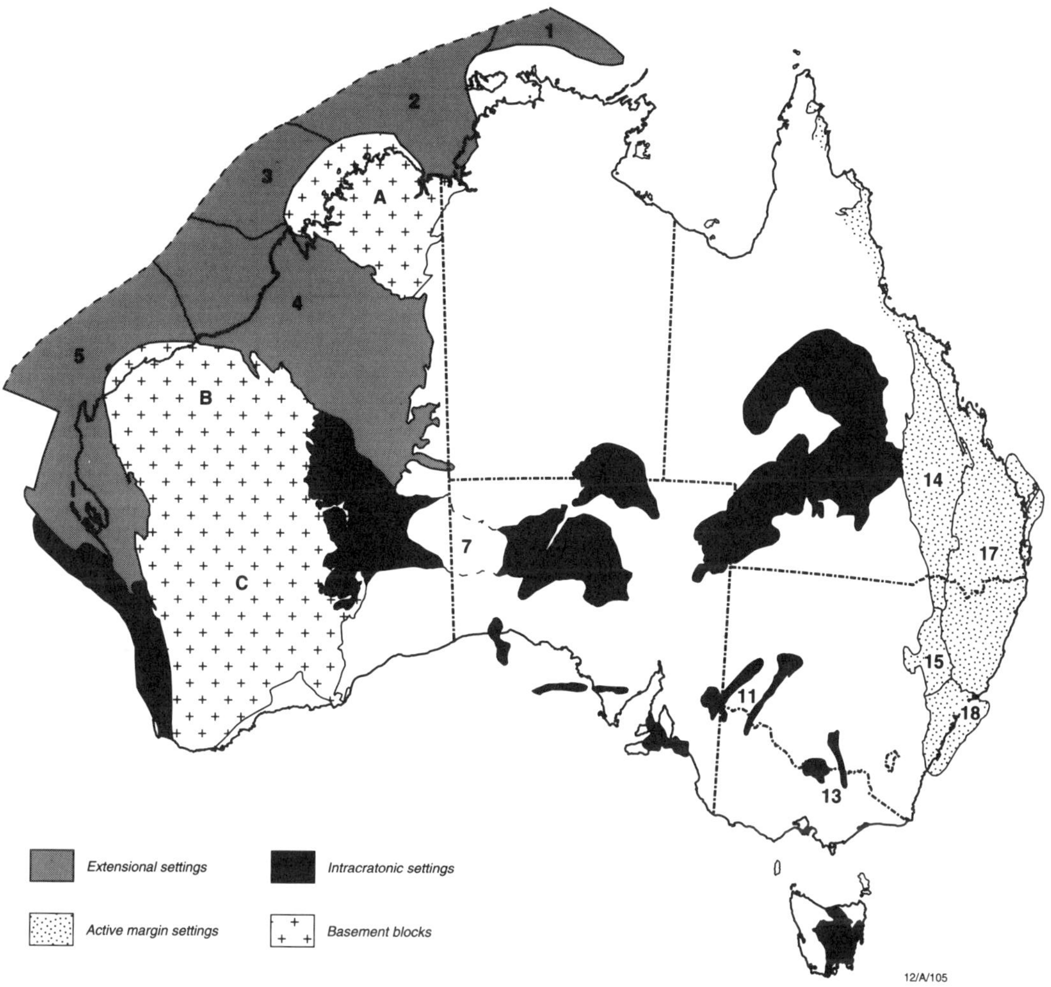

Fig. 12.2. Australian sedimentary basins containing Permian successions include: 1. Arafura, 2. Bonaparte, 3. Browse, 4. Canning, 5. Carnarvon, 6. Perth, 7. Officer, 8. Pedirka, 9. Arckaringa, 10. Cooper, 11. Murray Infrabasin, 12. Galilee, 13. Oaklands, 14. Bowen, 15. Gunnedah, 16. Tasmania, 17. New England Orogen, 18. Sydney. The following basement blocks were important centers for ice dispersal during the Permian: A. Kimberley Block, B. Pilbara Block, C. Yilgarn Block.

in polar regions during the Permian, the effect of the monsoonal atmospheric climate was probably considerably less than that of mid- to low latitudes.

The Gondwana supercontinent was thus influenced by a highly differentiated climate that included glacial, cold temperate, cool temperate, warm temperate, and arid to subtropical zones. Following the extensive latest Carboniferous-earliest Permian glaciation, the climate in the Early Permian was humid and temperate, while relative sea level was high. The western regions of Gondwana lay in warm temperate, semiarid, and arid to subtropical climatic zones throughout the Permian. In contrast, the central and the eastern regions of Gondwana lay within the warm, temperate-cool, temperate-cold, temperate, and glacial zones throughout the Early Permian and rotated into warm temperate to cool temperate zones during the Late Permian. Continentality increased considerably from the Early to Late Permian, in part due to the relative lowering of sea level in many regions of Gondwana (Langford, 1991).

The Ice Sheets

Late Paleozoic glacial successions are widespread in Australia and extend over a considerable period of time. Their distribution and timing are discussed by Crowell and Frakes (1971a, b) and Dickins (1985a, b, 1993) and summarized by Eyles (1993). The dramatic cooling that ultimately led to the Permian glaciations began in the latest Carboniferous in Australia in association with uplift and the local formation of mountain glaciers along the active paleo-Pacific margin of the continent. At this time there was only a limited glaciation in western Australia with glaciomarine sediments accumulating in the Carnarvon Basin and coarse outwash material accumulating in the Canning Basin to the north. Ice also advanced into southeastern Australia from ice centers in adjacent Antarctica.

Because of climatic cooling, by the beginning of the Permian Australia was covered by extensive ice sheets (Fig. 12.3), especially the southern and the western parts of the continent. The largest of these sheets occupied much of western Australia and extended into central Australia. The Yilgarn Block (Fig. 12.2) was a major center for ice dispersal, resulting in deposition of lodgement tills and other glacigenic sediments in the Collie and the Perth basins, and directing streams of ice east on the margins of the Officer Basin to the north and the northwest. The ice cap probably extended onto Antarctica to the south and on to India to the west (Struckmeyer and Totterdell,1992).

The Pilbara Block was also the focus for another major ice center, with ice flowing to the west-southwest, the south, and locally the southeast into the Carnarvon Basin. A floating ice sheet extended offshore. Along the southern margin of the Canning Basin, ice flowed to the north and the northwest, a pattern that, when taken in conjunction with northerly directions along the eastern Pilbara, suggests that stronger ice flow from the Yilgarn center was bypassing the less active Pilbara sheet. Grounded ice was also present on the Broome Arch and in the Fitzroy Trough in the Canning Basin. The ice thus came across the basin from the Precambrian hinterland to the south. The glacigenic deposits along the southern Canning Basin margin are ice-retreat successions, postdating the maximum advance. Icebergs or seasonal ice was also present in the Bonaparte Basin and may have reached the Kimberley and the Stuart blocks (Grey et al., 1977; Redfern, 1991; O'Brien and Christie-Blick, 1992).

The central Australian ice center radiated southwest into the Officer Basin, north onto the craton, and southeast into the Pedirka Basin. Ice from the Yilgarn and the central Australian centers converged in the Officer Basin and was deflected to the north and the south. An erosional high along the eastern side of the Pedirka Basin shed detritus westward and may have supported a transient ice cover during colder intervals that connected the great western and southern ice sheets.

A major ice sheet radiated from Victoria Land in Antarctica and encroached on southeast Australia. The south Australian portion of the ice sheet formed an irregular lobe with apparently no radial flow directions. To the west, ice flowed directly from the mountains into the sea, where its sediment load was deposited as turbidites in a fault-bounded trough (Townsend, 1976). An ice shelf also fed from the Gawler Craton to the south into local troughs. The glaciofluvial sediments of the southern Cooper Basin (Williams and Wild, 1984) must have been supplied by glaciers on the basin edge. Southwest of Adelaide the ice flowed north and northwest.

The ice lobe over Victoria was an extension of the Antarctic Ice Sheet. Floating ice extended only a short distance into the Murray infrabasin area. During the coldest periods, a floating ice sheet was probably connected to the glacial lobes of southeastern and southcentral Australia, but in warmer times an open seaway may have been present, except for a cover of seasonal winter ice.

Western Tasmania experienced an ice advance

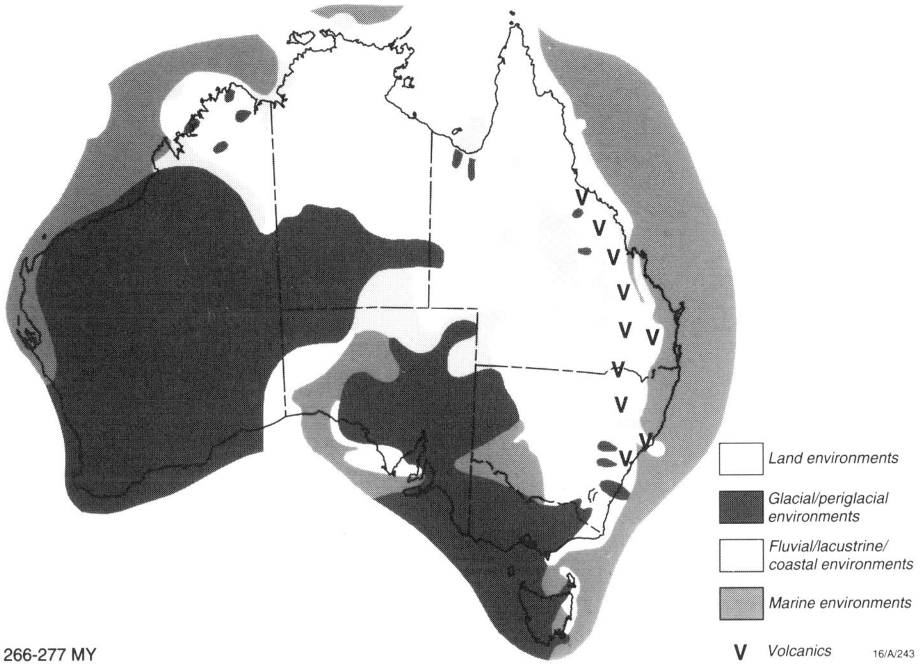

Fig. 12.3. The southern and southwestern parts of Australia were covered by vast areas of ice flowing from three independent ice centres by the beginning of the Permian. The east was dominated by a major volcanic belt that fed large fluvial systems to the west and relatively deep ocean depocenters to the east (after Struckmeyer and Totterdell, 1992).

from the west at the same time but the flow was deflected to the northeast and the southeast by local topography. The ice sheet extended to the ocean in central Tasmania developing another ice shelf. Local ice caps developed on islands to the east.

Eastern Australia developed only minor valley glaciers (Dickins and Sutherland, 1979). In the Galilee Basin alpine glaciers may have developed in highlands to the east and the north (Crowell and Frakes, 1971a; Hawkins, 1978) and local ice may have existed as far north as the Gulf of Carpentaria (Burgess, 1984).

Unglaciated Regions

Sedimentation was widespread along the active eastern margin of Australia during the glacial period. In northeastern Australia a shallow continental shelf extended north from the Yarrol Basin (Fig. 12.2). Further south, in the Sydney Basin, deposition began in a cold-water, shallow marine setting, accompanied by basaltic volcanism. Eastern Tasmania also experienced cold shallow-marine conditions.

The Murray and the Arckaringa basins of the southern interior were also marine but contain very impoverished faunas (Townsend and Ludbrook, 1975; O'Brien, 1986) suggesting that most marine organisms were under stress because of large influxes of meltwater into restricted embayments from the adjacent ice caps.

In the Bonaparte Basin, to the northwest, sea level fell, so that nearshore and high-energy barrier and beach facies grade up to fluvial and lacustrine deposits. There is also evidence of ice-rafting, by seasonal or glacial ice. Locally some deltas developed, and fluvial fans formed due to uplift (Mory and Beere, 1988). Sea level also fell in the Browse Basin at this time (Crostella, 1976).

Onshore fluvial deposition probably took place over much of the continent, especially as outwash aprons along the margins of ice sheets, but most of these deposits have not survived later erosion. However, in the Officer, southern Canning, Pedirka, and Cooper basins, glaciofluvial units have been preserved. The eastern Galilee Basin received fluvial and fluviolacustrine sediments from surrounding erosional highlands (Hawkins, 1978). The Denison Trough of the Bowen Basin contains a great thickness of terrestrial sedimentary rocks. Fluvial facies are also preserved in the smaller basins in eastern New South Wales.

Postglacial era

Early-Middle Sakmarian (277–271 Ma). Most of the ice had withdrawn from Australia by the end of the Asselian (277 Ma) (Fig. 12.4), changing the environment of the continent dramatically. Vast volumes of sediment were released over much of the southern half of the continent, and sea level began to rise because of the melting of the ice caps, especially in parts of Victoria and Tasmania, where the massive ice load had depressed the crust. As a consequence, marine transgressions occurred in southern Victoria, Tasmania, the eastern Murray infrabasinal area, and the Bonaparte, Carnarvon, northern Perth, and Temi basins, and to a lesser extent in the Sydney Basin. But in the Canning and the Arckaringa basins, the influx of sediment on their northeastern sides kept pace with the rising sea level and the shorelines prograded. In the Pedirka Basin, initially low-energy, fine-grained fluvial deposits gave way to higher-energy, sand-dominated fluvial aggradation by the end of the time internal. In the southern Perth Basin, a graben between the Yilgarn Block and Greater India, peatlands evolved in a fluvial setting. A sedimentary basin occupied a substantial portion of the

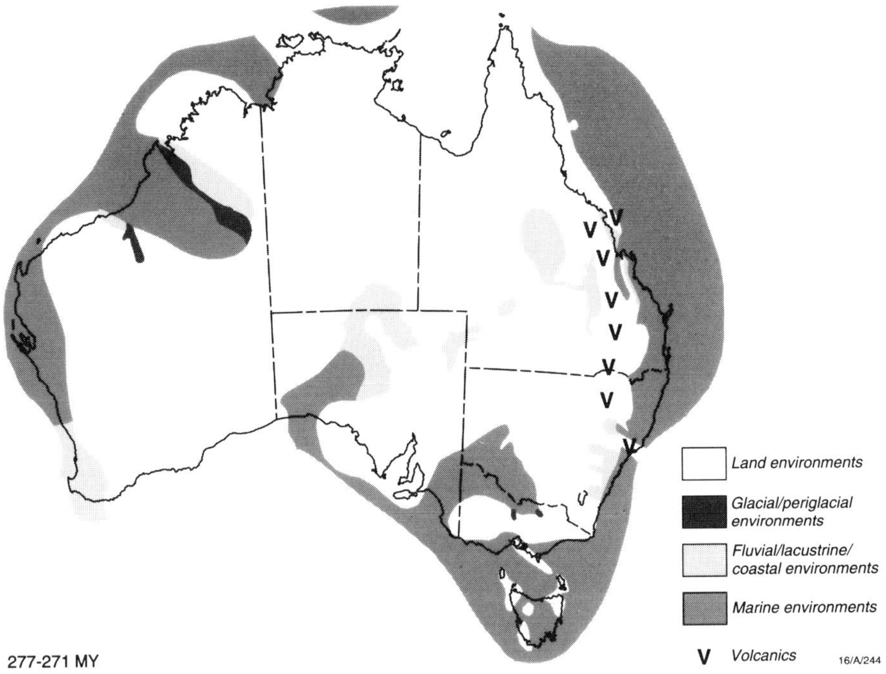

Fig. 12.4. The ice sheets had retreated completely from the Australian continent by the beginning of the Sakmarian (277 Ma) leaving behind large areas of unconsolidated sediment that were to dominate sedimentary processes for much of the remainder of the Permian. Due to the melting ice caps, sea level rose and shallow marine conditions expanded onto the continental margins (after Struckmeyer and Totterdell, 1992).

southwestern Yilgarn Block, but only remnants such as the Collie Basin are preserved (Wilson, 1989).

Further east, on the continent, carbonates were deposited within the generally clastic successions of the shallow marine Yarrol Basin. In the Denison Trough, an alluvial, coal-forming setting, including alluvial fan and lacustrine facies, developed (Draper and Beeston, 1985). The eastern Galilee Basin was partly occupied by rivers and some lakes but lacked peat development (Hawkins, 1978). In the Cooper Basin, a fluvial succession containing some coal developed.

Along the east coast, the deep marine area rapidly filled to become shallow marine, while further north, deep water turbidite sedimentation continued. Meanwhile, the volcanic belt in eastern Australia remained active.

The major erosional areas were broadly similar to those of the unglaciated regions of the earliest Permian, while the deglaciated central Australian highlands, after rising by isostatic rebound, were subjected to stream erosion, and remained so for the rest of the Permian and beyond.

Late Sakmarian-Early Artinskian (271–266 Ma). By the Late Sakmarian (271 Ma), the continent was no longer affected by glaciation or associated isostatic effects (Fig. 12.5). Deposition ended in the Murray Basin and Victoria province at this time, while in Tasmania fluvial sediments, including coal measures, extended across the state toward the southeast. The Arckaringa Basin had been totally filled by clastics derived from the east and from the large central Australian erosional area via the Pedirka Basin. Both basins began to develop coal measures by the Early Artinskian (Struckmeyer and Totterdell, 1992).

Coal measures, mostly fluvial, are prominent at this time, as they were also deposited in the Denison Trough (Bowen Basin) and in the Galilee, southern Cooper, Collie, and southern Perth basins. The Denison Trough and the Cooper basins also contained lakes. Coal was generally not forming at this time in the Sydney Basin.

In the shallow marine Yarrol Basin, limestone now dominated over clastic deposition, and a wide range of marine fauna flourished. Carbonates also began to dominate along the eastern plate margin. The marine region of eastern New England shallowed to become a carbonate-clastic province with vast areas of crinoidal limestone accumulating along the north coast of New South Wales (Lindsay, 1969, 1991). To the north, however, deeper water conditions still prevailed.

In the west, the northern Perth Basin accumulated a shallow marine carbonate-lutite succession, followed by a shallow-marine sandstone which may have been partly littoral in origin during the Late Sakmarian (Playford et al., 1976). Coal-measure sedimentation extended across the region by the Early Artinskian. During the same time period the eastern Carnarvon Basin experienced a mixture of marine and paralic conditions. Initially a low clastic influx on a wide shelf gave rise to the highly fossiliferous carbonate-lutite associations (Lavering, 1985). However, as sea level fell, the southeastern basin was exposed to a period of karst-forming subaerial weathering, followed by a renewed sea-level rise and transgression that set the stage for the outbuilding of a large delta complex.

Meanwhile, the Canning Basin began the interval with an erosional hiatus and then a major transgression with widespread nearshore to very shallow marine, dominantly sandy units derived from the reworking of vast volumes of glacial debris. In the northwest and the south, calcareous and coquinitic units contain a warm to temperate open ocean faunal assemblage (Forman and Wales, 1981; Towner et al., 1983). The northern margin differed again; a migrating barrier bar and lagoonal zone were constructed between the sea and an apron of coalescing fan deltas at the distal end of braided fluvial systems draining the Kimberley highlands to the north. A prograding shoreline of sandy deltas also developed along the eastern Bonaparte Basin (Dickins et al., 1972). Most of the basin continued to be shallow marine, receiving fine, carbonaceous, prodelta clastics and some limestone (Laws and Brown, 1976).

Late Artinskian-Kungurian (266–258 Ma). This was a time of rising sea level, and terrestrial sedimentation was replaced by marine settings. In the eastern part of the continent cooling climate locally brought the return of glaciation to western Victoria and to the uplands along the western

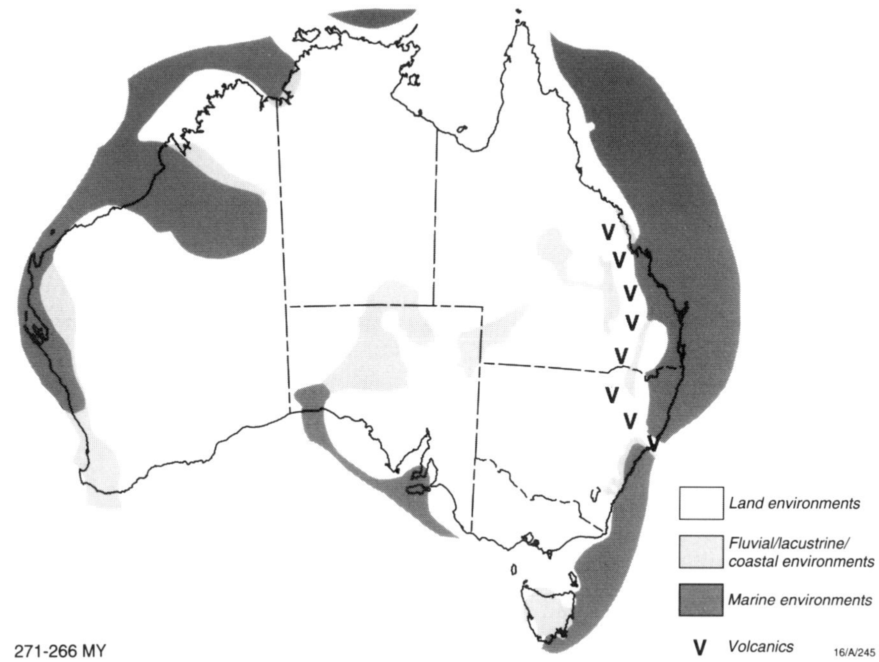

Fig. 12.5. As the effects of the glaciation became less important sea level rose and the subglacial region rebounded, creating many shallow basins, dominated by swamps and wetlands, that were ideal for the accumulation of peat. Thus, by the end of the Sakmarian, from around 271 Ma, the first of Australia's major coal-forming episodes began in the postglacial setting (after Struckmeyer and Totterdell, 1992).

margin of the Sydney Basin. Alpine glaciers developed amid 6,000m topography along the western Sydney Basin (Dulhunty, 1964). In Tasmania, the marine conditions initially advanced toward the north and the west (Banks and Clarke, 1987; Clarke, 1989); later marine conditions extended across the eastern half of the state. At the same time an apron of coalescing fan deltas formed in the northern Sydney Basin, while a similar line of alluvial fans and distal alluvial plains extended northwest, into the Gunnedah Basin. Shallow marine conditions continued in the Yarrol Basin, and marine conditions spread across the Bowen Basin. On the western flank of the basin a large delta developed (Brakel, 1989a, 1989b). The southern Cooper Basin twice saw the development of large lakes, about 250 by 150 km in size, in both cases extending diachronously westward. The first lake was infilled by deltas and the reclaimed territory converted to peatlands, which were in turn inundated when the second lake formed and encroached westwards (Thornton, 1979).

Marine advances also are evident on the western side of the continent, in the Bonaparte Basin and along the northern side of the Canning Basin. Most of the Canning Basin remained an extensive shallow sea, with clastic and some carbonate deposition in water temperatures of 24°C–26°C (Forman and Wales, 1981). Shallow marine conditions in the Carnarvon Basin varied from those of quiet lower offshore to wave-dominated upper shoreface zones, with storm activity prominent at various times (Moore et al., 1980a; Hocking et al., 1983; Moore and Hocking, 1983). Evidence of floating ice in the area is consistent with the reappearance of glaciers in southeast Australia, in marked contrast to the warm conditions prevailing in the nearby Canning Basin. Shallow marine conditions persisted in the Perth Basin. Fluvial coal measures continued to collect in the Collie and the southern Perth basins.

A major episode of uplift and folding affected a large portion of eastern Australia, generating hiatuses from the Sydney Basin to the Bowen

Basin. Highlands still existed in central Australia and in the region between the Sydney-Bowen trough and the eastern sea. Highlands also persisted from west of the Bowen Basin to north Queensland, as remnant of earlier igneous and tectonic activity.

Early Kungurian (Ufimian) (c.256–258 Ma). The Late Permian of Australia is distinguished by an important marine transgression in many basins. Most notable is the establishment of a seaway along the full length of the Sydney Bowen trough, isolating the New England-Yarrol highlands from the mainland. The northwestern margin area of the Bowen Basin was claimed by a Permian sea for the first time, but to the south, a fluvial facies prevailed. In Tasmania the transgression reached its maximum extent, with only the northeast and northwest corners of the state and some islands in the east remaining above sea level.

In western Australia transgression also continued in the Bonaparte and the Carnarvon basins. Most of the Carnarvon Basin was characterized by a broad sandy shelf with a very low paleoslope (Moore et al., 1980b; Kopsen and McGann, 1985; Bentley, 1988). The Canning Basin by contrast was dominated by tectonic emergence of the southern and the central portions, which were to last as dry land until the Late Jurassic. However, the Fitzroy Tough subsided to become the locus of a barrier lagoonal complex, mostly with fringing deltas in the southeast and very shallow marine water to the northwest. Paralic environments, including coal swamps (Playford et al., 1976) dominate the north Perth Basin. Coal-prone fluvial aggradation remained the feature of the Collie Basin and the southern Perth Basin.

Middle Kazanian - Middle Tatarian (250-256 Ma). Although many marine areas continued as such, a general progradation characterized the rest of the Permian, and nonmarine and paralic depositional sites increased at the expense of marine settings due to a eustatic drop in sea level (Fig. 12.6). The Middle Kazanian marks the

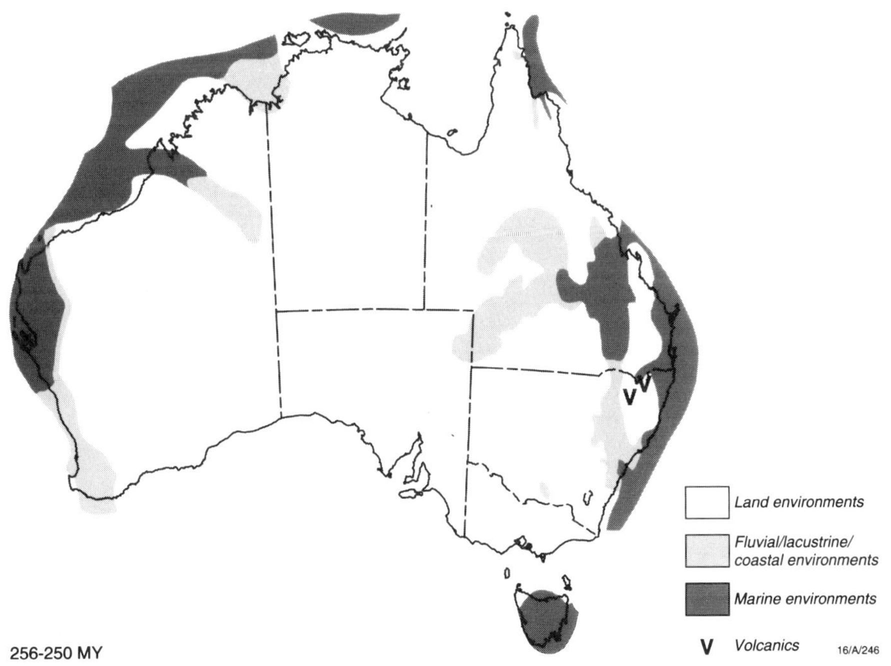

Fig. 12.6. The close of the Permian, from approximately 250 Ma on, saw the development of Australia's most important coal measures. The continent's interior basins gradually shallowed, and sea level fell, resulting in the deposition of coal in almost every nonmarine basin. At the close of the Permian at approximately 240 Ma, most of the basins had filled completely (after Struckmeyer and Totterdell, 1992).

beginning of the regression in the Sydney and the Gunnedah basins and the deposition of the upper coal measures succession in eastern Australia. The Sydney and the Gunnedah basins filled with deltaic and fluvial coal measures as clastics poured in from adjacent uplifted areas in New England and, to a lesser extent, from the western Lachlan highlands. The main progradation direction was southerly along the basin axes, but most detritus was introduced into the depositional system from the sides via alluvial fans (Struckmeyer and Totterdell, 1992).

On the eastern side of the continent, the Bowen Basin remained marine at first, with continuing transgression in the west and onto the Springsure Shelf leading into the southern Galilee Basin. Late coal-prone deltaic facies prograded from the north, to be interrupted by the same regression that was occurring in the Sydney Basin. Extensive fluvial coal measures deposition resumed in the Galilee Basin and later in the Cooper Basin, and the basins reached their maximum areal extent.

On the western side of the continent, in the Bonaparte Basin, an early marine shelf setting was succeeded by the construction of a very large delta and lagoon complex (Bhatia et al., 1984; Mory, 1988). Similarly, the eastern Fitzroy Trough received large volumes of deltaic and fluvial fill, while the western portion remained marine. The rest of the continent remained unchanged.

Late Tatarian (250–248 Ma). As the Permian came to an end, general progradation left very few marine domains within the present-day coastlines. The Late Permian was characterized by the most widespread coal deposition the continent has ever known, for every nonmarine basin from Cape York to Tasmania and across to Perth was accumulating coal (Brakel and Totterdell, 1988; Harrington and Brakel, 1989). Initially, peatlands evolved in a deltaic setting in the Sydney Basin, but with continuing southward progradation, alluvial fans advanced into the depositional site from the north and the northeast, burying the peat (Brakel, 1989a, 1989b). In the northern proximal areas, raised peat bogs formed on the fan surfaces during quiescent periods, and the coal seams are interbedded with thick conglomerates. The coal measures were succeeded by braided fluvial units. The Gunnedah Basin has a similar depositional history (Hamilton and Beckett, 1984). The progradational, coal-forming event of the main eastern Australian basins also involved Tasmania.

The Bowen Basin during this time was infilled with marine and deltaic facies, succeeded by fluvial systems including widespread peatlands, which evolved in a cool temperate climate (Rigby, 1972). The Galilee and the Cooper basins continued as large expanses of coal-prone fluvial deposition, draining east into the Bowen Basin. Differential subsidence in the Cooper Basin troughs produced meandering streams and lacustrine settings in the lower areas, some bordered by erosional highs (Thornton, 1979; Wells and O'Brien, 1989).

Coal measures also accumulated in the Perth Basin. However, most of the Carnarvon Basin, except for its northern end, remained erosional. A marine incursion in the southwestern Canning Basin led to the formation of coastal marshes at the head of the embayment (Forman and Wales, 1981). In the Fitzroy Trough, the earlier lagoonal setting was replaced by a large fluvial lobe sourced from the Kimberley uplands. The large delta in the Bonaparte Basin was flooded by a sea-level rise, after which a smaller delta was reestablished in a lagoonal tract behind a prograding barrier bar. This portion of the continent had a warmer climate than any other region at any time during the Permian, becoming warm temperate to tropical (Dickins, 1978).

Permian energy resources

During the early stages of the late Paleozoic glaciation of Gondwana, the *Botrychiopsis* floral assemblage was present in ice-free areas. This flora, a tundra-like association, was followed by a gymnosperm flora represented by monosaccate microflora similar to modern taiga or conifer-dominated woodlands. Reeds grew in open areas or along the rivers in a climate that was too harsh to support herbaceous plants (Retallack, 1980; Falcon, 1986).

As Gondwana began to warm following the Asselian glacial maxima, the *Botrychiopsis* and gymnosperm-type floras were replaced by new floral assemblages. The first arborescent floras to appear were the stunted *Gangamopteris* flora and early conifers that first appeared in the milder coastal climates and only later penetrated inland to develop forests. The *Glossopteris* flora evolved during the Sakmarian in water-logged swamps as the climate warmed and permafrost conditions ceased (Gould, 1975; Retallack, 1980).

In the mid-Early Permian, the taiga-like woodlands began to be replaced by extremely diversified and mixed *Glossopteris-Gangamopteris*-dominated swamp forests growing on swampy levee banks and in abandoned channels. They were large deciduous trees with an understorey of pteridophytes, ferns, club-mosses and lycopods, horsetails and early gymnosperms (Retallack, 1980; Falcon, 1986), while articulates dominated the open swamps.

Glossopteris dominated the coal swamps during the Late Permian, but other trees, including conifers, were present, growing in a strongly seasonal climate. Conifers probably grew on the upland and nondepositional areas beyond the basins (Retallack, 1980; Ziegler, 1990). In the Late Permian, dense, high-diversity *Glossopteris* forests dominated Australian coal swamps, but *Glossopteris* disappeared from floral assemblages at the end of the Permian.

In central Gondwana the latest Permian climax vegetation was dominated by high-order plants (gymnosperms, or pteridosperms), with smaller proportions of lower-order spore-producing plants (for example, ferns and horsetails) that required water for reproduction. The vegetation was possibly similar to that of a savanna-type system resulting from warmer conditions (Falcon, 1986).

Coal measures

Peat forms where there is sufficient organic growth and swampy conditions. Coal type is determined principally by the kind of flora that formed the precursor peat but also by the extent of alteration during deposition and early diagenesis. The kind of flora and the depositional setting were determined by complex interactions of geography and climate. Inland sedimentary basins have the greatest diversity of successions and are particularly susceptible to periodic flooding.

As elsewhere, the Permian floral succession in Australia was climatically controlled (Retallack, 1980), and some species migrated polewards as the Permian ice cap receded. Following the waning of glaciation early in the Permian, a generally cool-temperate climate prevailed for the remainder of the Permian period. The Permian flora consisted of a low diversity pteridosperm-dominated assemblage (Gould and Shibaoka, 1980). A few varieties of the lycopods, equisitaleans, and ferns were present (Gould, 1975). An important implication is that the peat-forming components of the flora may be regarded as a more or less homogeneous input to the Permian coal swamps.

Permian coal occurs widely across the continent (Fig. 12.7; Harrington, 1989). The major economic coal basins of eastern Australia, the Sydney, Gunnedah, and Bowen basins, were developed on the eastern margin of Australia during the Permian (Fig. 12.8), although the nature of the basins is still problematic (Harrington, 1982). The more westerly Permian basins represent large, sometimes interconnected basins, formed on broad and in places fault-bounded downwarps in the crust. The greatest production by far was in the Sydney and Bowen basins.

Coal formed in the Sydney Basin as the ice sheets retreated and the *Gangamopteris*-dominated flora spread (Retallack, 1980). Glacial influence was lowest in the Bowen Basin, which was 12° farther north in paleolatitude than the Sydney and the Gunnedah basins. The extent of marine influence on coals was least in the intracratonic basins, but reached a maximum in the Sydney, Gunnedah, Bowen, and Arckaringa basins. Marine conditions were established throughout most of the marginal basins between late in the Early Permian and early in the Late Permian. Following a regression in the Late Permian and further climatic amelioration, coal measures composed of a *Glossopteris*-dominated flora formed throughout the Sydney, Gunnedah, Bowen, Galilee and Cooper basins.

Although coal accumulated throughout the Permian postglacial period, most of it was formed in two time intervals, during the Early Permian, from about 271–266 Ma, and during the Late Permian, from 256–248 Ma (Fig. 12.8; Brakel and Totterdell, 1990a, b). The second time interval was by far the most important both in terms of the total accumulated volume of organic

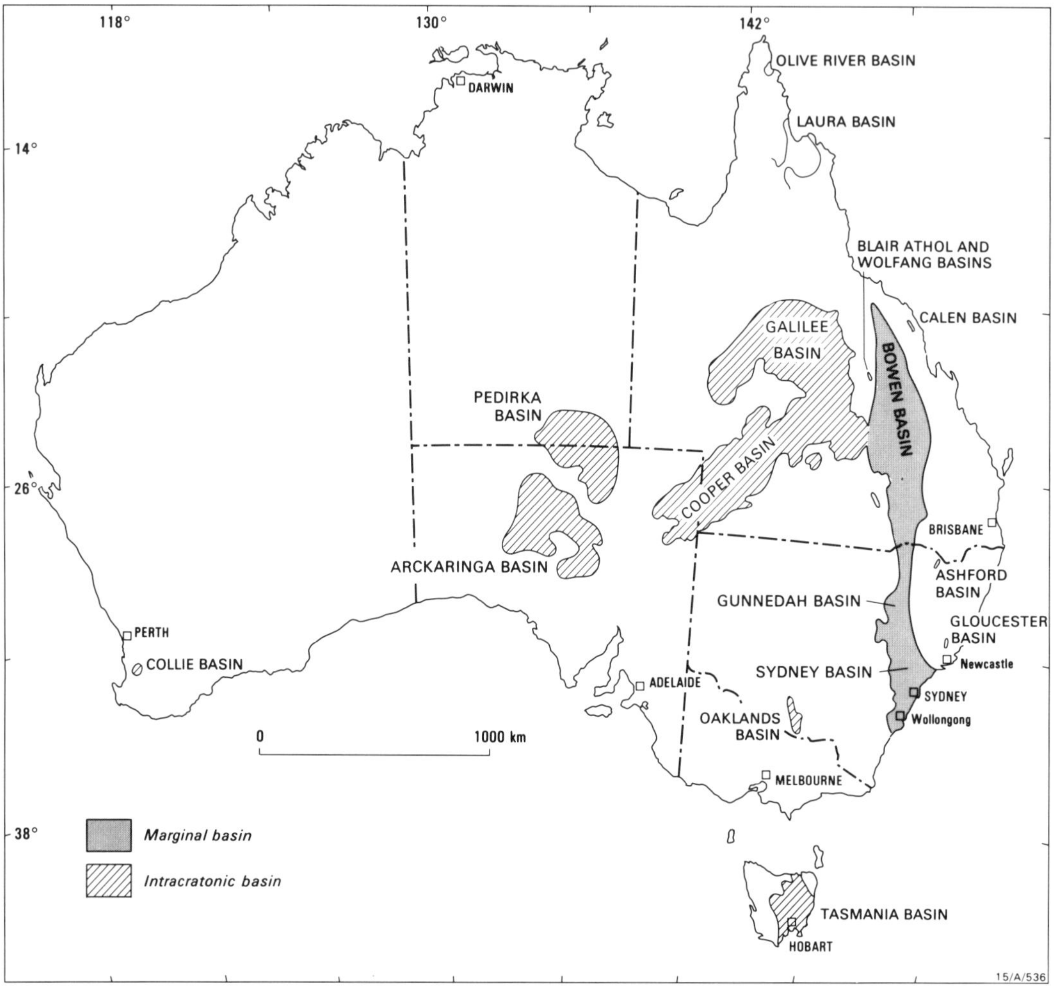

Fig. 12.7. Location of Permian coal basins in Australia (after Harrington, 1989).

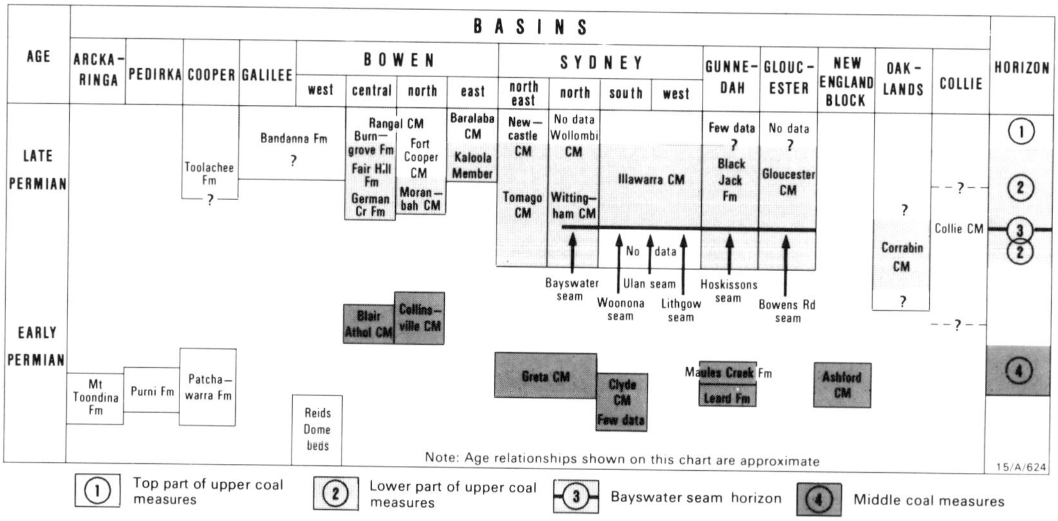

Fig. 12.8. Simplified correlation chart for Australian Permian coal basins (after Hunt, 1989).

material and in an economic sense. All humid depositional environments on land were potentially capable of preserving coal, given appropriate tectonic conditions. Deltaic and meandering fluvial systems were the most productive, but braided fluvial, fluviolacustrine, and back-barrier settings were also significant. Climate played a key role in peat preservation. Low paleo-temperatures greatly retarded organic decomposition, allowing vegetable matter to accumulate in situations and to thicknesses otherwise not possible. However, the large late Permian delta complexes in the Bonaparte Basin and the Fitzroy Trough developed in the warmest and driest climate of the Australian Permian (Dickins, 1978) and as a consequence preserved no economic coal.

Those coal basins with marine access in the Early Permian all ultimately regressed to coal measure deposition. Basins dominated by rift volcanism have a thick volcanic pile substituting for most of the basal marine succession, while basins beyond the reach of the sea have fluvial successions. The widespread accumulation of coal in earlier Permian basins can be attributed in part to a eustatic sea-level lowstand. However, other factors also played a role as the coal measures were not all synchronous and coal did not form in some basins. The coals of the Collie and the southern Perth basins were, by contrast, derived from peats that formed directly over a thin basal glacigenic layer deposited by the retreating Yilgarn Ice Cap. The Denison Trough succession began directly with coal measures, at least on the trough margins, although the base of the unit in the deep trough may overlie a fluvial or volcanic substrate.

In the Bowen Basin two regressive sea-level cycles are apparent. The first was fluviodeltaic, yielding very little coal in the west, while in the north a coal-bearing alluvial fan prism was built on basal Permian volcanics and older rocks. In the Cooper Basin, the large lakes were replaced by coal-forming peatlands as they were infilled; a hiatus occurred before coal deposition resumed toward the end of the Permian. Mid-Permian hiatuses occurred in the Galilee, Gloucester, and northern Perth basins and on the Roma Shelf before the onset of paralic and/or nonmarine deposition.

Hydrocarbon systems

Basins that are similar in age and have similar source-rock types and structural histories are linked into broad petroleum systems (Magoon and Dow, 1991). The Australian Permian postglacial period is contained largely within the Gondwana System (Bradshaw, 1993), which is critical with regard to the distribution of the continent's petroleum resources (Fig. 12.9). The system, which is the most diverse and widespread of the Australian petroleum systems, is mostly derived from Permian postglacial coal measure settings. Economic and subeconomic oil and gas accumulations occur in the Bonaparte, Cooper, and Bowen basins, but, even more important, many of the larger accumulations of the Mesozoic successions of the Eromanga and Surat basins were sourced from the underlying postglacial Permian coal measures.

In eastern Australia, the Gondwana system source rocks are predominantly nonmarine and are associated with the coal measure successions (Gilby and Mortimore, 1989; Jenkins, 1989; Heath, 1989). In the west hydrocarbon activity has mostly focused on the Canning and the Perth basins, although subeconomic fields have been discovered in the Bonaparte Basin. The Perth Basin has produced gas from Permian reservoir units in three fields. The gas was generated from

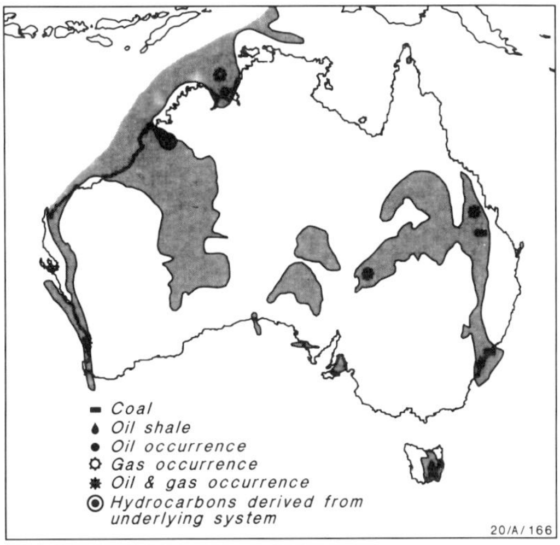

Fig. 12.9. The Gondwana petroleum system (after Bradshaw, 1993).

the associated coals and carbonaceous shales, all, ultimately, the product of the Permian postglacial environment.

Reservoir facies include Permian glacial to fluvial sandstones in the Cooper and other eastern Australian basins and Late Permian deltaic to shallow marine sands in the Petrel Sub-basin and the Perth Basin. Permian limestone is the gas reservoir in the Perth Basin (Lane and Watson, 1985).

Conclusion

The ice sheets that formed on the Australian continental block during the latest Carboniferous and the earliest Permian had a major effect on the broad-scale sedimentology of the continent. As the ice retreated, it left behind a newly sculpted landscape and vast volumes of unconsolidated sediment, an ideal setting for the development of vast peatlands. These changes, along with sea-level fluctuations resulting from the melting of the ice sheets, were to have a dramatic effect on the nature of the sediment fill in most onshore basins for the rest of the Permian and in some cases well beyond. Perhaps the most important outcome, at least in an economic sense, was the development of the continent's most important coal measures, many of which later generated hydrocarbons. The Permian sedimentary basins of Australia and much of Gondwana thus acted as enormous sinks for atmospheric carbon. Today, as a major source of energy, these same basins are a significant source of carbon dioxide and contribute to the planet's environmental problems.

References

Archbold, N.W., Pigram, C.J., Ratman, N. and Hakim, S., 1982. Indonesian Permian brachiopod fauna and Gondwana-South-East Asia relationships. Nature, 296: 556–558.

Backhouse, J., 1990. Permian palynostratigraphic correlations in southwestern Australia and their geological implications. In: E.M. Truswell and J.A. Owen (Editors), Proceedings of the 7th International Palynological Congress. Review of Palaeobotany and Palynology, 65: 229–237

Banks, M.R. and Clarke, M.J., 1987. Changes in the geography of the Tasmania Basin in the late Paleozoic. In: G.D. McKenzie (Editor), Gondwana six: Stratigraphy, sedimentology, and paleontology. American Geophysical Union, Geophysical Monograph, 41: 1–14.

Bentley, J., 1988. The Candace Terrace. a geological perspective. In: P.G., Purcell and R.R., Purcell (Editors), The North West Shelf, Australia. Proceedings of the Petroleum Exploration Society of Australia Symposium, Perth, pp. 157–171.

Bhatia, M.R., Thomas, M. and Boirie, J.M., 1984. Depositional framework and diagenesis of the Late Permian gas reservoirs of the Bonaparte Basin. Australian Petroleum Exploration Society Journal, 24: 299–313.

Bradshaw, M., 1993, Australian petroleum systems. PESA Journal, 21: 43–53.

Brakel, A.T., 1989a. Permian coal measures in the exposed part of the Bowen Basin. In: H.J. Harrington, A.T. Brakel, J.W. Hunt, A.T. Wells, M.F. Middleton, P.E. O'Brien, D.S. Hamilton, J. Beckett, C.R. Weber, S. Radke, J.M. Totterdell, D.J. Swaine and P.W. Schmidt (Editors), Permian coals of eastern Australia. Bureau of Mineral Resources, Australia, Bulletin 231: 69–91.

Brakel, A.T., 1989b. Sydney Basin: Permian coal measures stratigraphy and sedimentation. In: H.J. Harrington, A.T. Brakel, J.W. Hunt, A.T. Wells, M. F. Middleton, P.E. O'Brien, D.S. Hamilton, J. Beckett, C.R. Weber, S. Radke, J.M. Totterdell, D.J. Swaine and P. W. Schmidt (Editors), Permian coals of eastern Australia. Bureau of Mineral Resources, Australia, Bulletin 231: 9–41.

Brakel, A.T. and Totterdell, J.M., 1988. The Late Permian palaeogeography of the Australian continent. Proceedings of the 22nd Newcastle Symposium on Advances in the study of the Sydney Basin, Department of Geology, University of Newcastle, pp. 59–67.

Brakel, A.T. and Totterdell, J.M., 1990a. The Permian Palaeogeography of Australia. Bureau of Mineral Resources Geology and Geophysics, Record 1990/60, 126 pp.

Brakel, A.T. and Totterdell, J.M., 1990b. Glaciation and palaeogeography of Australia at the start of the Permian. In: Gondwana: terranes and resources. 10th Australian Geological Convention, Hobart. Geological Society of Australia, Abstracts, 25: 63–64.

Burgess, I.R., 1984. Carpentaria Basin: a regional analysis with reference to hydrocarbon potential. Australian Petroleum Exploration Society Journal, 24: 7–18.

Cawood, P.A., 1982. Structural relations in the subduction complex of the Paleozoic New England Fold Belt, eastern Australia. Journal of Geology, 90: 381–392.

Clarke, M.J., 1989. Lower Parrneer Supergroup. In: C.F. Burrett and E.L. Martin (Editors), Geology and mineral resources of Tasmania. Geological Society of Australia, Special Publication 15: 295–309.

Crook, K.A.W. and Feary, D. A., 1982. Development of New Zealand according to the fore-arc model of crustal evolution. Tectonophysics, 87: 65–107.

Crostella, A., 1976. Browse Basin. In: R.B. Leslie, H.J. Evans and C.L. Knight (Editors), Economic geology of Australia and Papua New Guinea. 3. Petroleum. Australasian Institute of Mining and Metallurgy, Monograph 7: 194–199.

Crowell, J.C. and Frakes, L.A., 1971a. Late Palaeozoic glaciation of Australia. Journal of the Geological Society of Australia, 17: 115–155.

Crowell, J.C. and Frakes, L.A., 1971b. Late Paleozoic glaciation: Part IV, Australia. Geological Society of America Bulletin, 82: 2515–2540.

Dickins, J.M., 1978. Climate of the Permian in Australia: the invertebrate faunas. Palaeogeography, Palaeoclimatology, Palaeoecology, 23: 33–46.

Dickins, J.M., 1985a. Late Palaeozoic glaciation. BMR Journal of Australian Geology and Geophysics, 9: 163-169.

Dickins, J.M., 1985b. Late Paleozoic climate with special reference to invertebrate fauna. Neuvieme Congres International de Stratigraphie e de Geologie du Carbonifere, Compte Rendu, 5: 394-402.

Dickins, J.M., 1993. Climate of Late Devonian to Triassic. Palaeogeography, Palaeoclimatology, Palaeoecology, 100: 89-94.

Dickins, J.M. and Sutherland, W.A., 1979. A glacial moraine in the Permian Lochinvar Formation of the Sydney Basin, New South Wales. Search, 10: 190–191.

Dickins, J.M., Roberts, J. and Veevers, J.J., 1972. Permian and Mesozoic geology of the northeastern part of the Bonaparte Gulf Basin. In: Geological Papers. Bureau of Mineral Resources, Australia, Bulletin 125: 75–102.

Draper, J.J. and Beeston, J.W., 1985. Depositional aspects of the Reids Dome beds, Denison Trough. Queensland Government Mining Journal, 86: 200–210.

Dulhunty, J.A., 1964. Our Permian heritage in central eastern New South Wales. Royal Society of New South Wales, Journal and Proceedings, 97: 161–166.

Eyles, N., 1993. Earth's glacial record and its tectonic setting. Earth-Science Reviews, 35: 1–248.

Falcon, R.M.S., 1986. A brief review of the origin, formation, and distribution of coal in southern Africa. In: C.R. Anhaeusser and S. Maske (Editors), Mineral deposits of Southern Africa, Volume IL. Geological Society of South Africa, Johannesburg, 1879–1898.

Forman, D.J. and Wales, D.W. (Compilers), 1981. Geological evolution of the Canning Basin, western Australia. Bureau of Mineral Resources, Australia, Bulletin, 210, pp. 91.

Gilby, A.R. and Mortimore, I.R., 1989. The prospects for Eromanga oil accumulations in the northern Cooper Basin region. In: B.J. O'Neil (Editor), The Cooper and Eromanga basins, Australia. Proceedings of the Petroleum Exploration Society of Australia, Society of Petroleum Engineers, and Australian Society of Exploration Geophysicists Conference, Adelaide, pp. 391–403.

Gould, R. E., 1975. The succession of Australian pre-Tertiary megafossil floras. Botanical Reviews, 41: 453–483.

Gould, R. E. and Shibaoka, M., 1980. Some aspects of the formation and petrographic features of coal members in Australia, with special reference to the Tasman orogenic zone. Australian Coal Geology, 2: 1–29.

Grey, K., Van De Graaff, W. J. E. and Hocking, R.M., 1977. Precambrian stromatolites as provenance indicators in the Permian Lyons Formation, Carnarvon Basin. Geological Survey of Western Australia, Annual Report for 1976, pp. 70–72.

Hamilton, D.S. and Beckett, I. J., 1984. Permian depositional systems in the Gunnedah region. Geological Survey of New South Wales, Quarterly Notes, 55: 1–19.

Harrington, H.J., 1982. Tectonics and the Sydney Basin. 16th Newcastle Symposium on Advances in the study of the Sydney Basin. Department of Geology, University of Newcastle, Abstracts, 15–19.

Harrington, H.J., 1989. Tectonic history. In: H.J. Harrington, A.T. Brakel, J.W. Hunt, A.T. Wells, M.F. Middleton, P.E. O'Brien, D.S. Hamilton, J. Beckett, C.R. Weber, S. Radke, J.M. Totterdell. D.J. Swaine and P.W. Schmidt (Editors), Permian coals of eastern Australia. Bureau of Mineral Resources, Australia, Bulletin 231: 377–394.

Harrington, H.J. and Brakel, A.T., 1989. Synopsis of geological evolution of the Permian coal basins. In: H.J. Harrington, A.T. Brakel, J.W. Hunt, A.T. Wells, M.F. Middleton, P.E. O'Brien, D.S. Hamilton, J. Beckett, C.R. Weber, S. Radke, J.M. Totterdell. D.J. Swainepp and P.W. Schmidt (Editors), Permian coals of eastern Australia. Bureau of Mineral Resources, Australia, Bulletin 231: 205–210.

Hawkins, P.J., 1978. Galilee Basin: a review of petroleum prospects. Queensland Government Mining Journal, 79: 96–112.

Heath, R., 1989. Exploration in the Cooper Basin. Australian Petroleum Exploration Society Journal, 29: 366–378.

Hocking, R.M., Williams, S.J., Moore, P.S., Denman, P.D., Lavering, I.H. and van de Graaff, W.J.E.,

1983. Explanatory notes on the Kennedy Range 1:250 000 geological sheet, Western Australia. Geological Survey of Western Australia, Record 1983/005, pp. 55.

Hunt, J.W., 1989, Coal-type trends in the Permian basins. In: H.J. Harrington, A.T. Brakel, J.W. Hunt, A.T. Wells, M.F. Middleton, P.E. O'Brien, D.S. Hamilton, J. Beckett, C.R. Weber, S. Radke, J.M. Totterdell. D.J. Swaine and P.W. Schmidt (Editors), Permian coals of eastern Australia. Bureau of Mineral Resources, Australia Bulletin 231: 212–234.

Jenkins, C.C., 1989. Geochemical correlation of source rocks and crude oils from the Cooper and Eromanga basins. In: B.J. O'Neil (Editor), The Cooper and Eromanga basins, Australia. Proceedings of the Petroleum Exploration Society of Australia, Society of Petroleum Engineers, and Australian Society of Exploration Geophysicists Conference, Adelaide, pp. 525–540.

Klootwijk, C.T., Conaghan, P.J. and Powell, C. McA., 1985. The Himalayan Arc: a large-scale continental subduction, oroclinal bending and back-arc spreading. Earth and Planetary Science Letters, 75: 167–183.

Kopsen, E. and McGann, G., 1985. A review of the hydrocarbon habitat of eastern and central Barrow-Dampier sub-basin, Western Australia Australian Petroleum Exploration Society Journal, 25: 154–176.

Korsch, R.J., 1982. Early Permian events in the New England Orogen. In: P.G. Flood and B.N. Runnegar (Editors), New England Geology. Department of Geology and Geophysics, University of New England, pp. 35–42.

Korsch, R.J., Harrington, H.J., Murray, C.G., Fergusson, C.L. and Flood, P.G., 1991. Tectonics of the New England Orogen. In: D.M. Finlayson (Editor), The Eromanga-Brisbane geoscience transect: a guide to basin development across Phanerozoic Australia in southern Queensland. Bureau of Mineral Resources, Australia, Bulletin 232: 35–52.

Lane, C.M. and Watson, R.J., 1985. Development of the Woodada gas field. Australian Petroleum Exploration Society Journal, 25: 316–328.

Langford, R.P., 1991. Permian coal and palaeogeography of Gondwana. Bureau of Mineral Resources Geology and Geophysics, Record 1991/95, 136 pp.

Lavering, I.H., 1985. Permian environments and palaeogeography of the onshore Carnarvon Basin and implications for petroleum prospectivity. PESA Journal, 7: 40–46.

Laws, R.A. and Brown, R.S., 1976. Bonaparte Gulf Basin. Southeastern part. In: R.B. Leslie, H.J. Evans and C.L. Knight (Editors), Economic geology of Australia and Papua New Guinea. 3. Petroleum. Australasian Institute of Mining and Metallurgy, Monograph 7: 200–208.

Lawver, L.A. and Scotese, C.R., 1987. A revised reconstruction of Gondwanaland. In: G.D. McKenzie (Editor), Gondwana six: Structure, tectonics and geophysics. American Geophysical Union, Geophysical Monograph, 40: 17–24.

Lindsay, J.F., 1969. Stratigraphy and structure of the Palaeozoic sediments of the lower Macleay region, north-eastern New South Wales. Royal Society of New South Wales, Journal and Proceedings, 102: 41–55.

Lindsay, J.F., 1991. A sequence analysis model for crinoidal limestones, Permian Yessabah Limestone, Manning-Macleay Basin, eastern Australia. Bureau of Mineral Resources Journal of Australian Geology and Geophysics, 12: 145–157.

Magoon, L.B. and Dow, W.G., 1991. The petroleum system-from source to trap. American Association of Petroleum Geologists Bulletin, 75: 627.

Mallet, C.W., Hammond, R.L. and Sullivan, T.D., 1988. Implications for the Sydney Basin of upper crustal extension in the Bowen Basin. Proceedings of the 22nd Newcastle symposium on advances in the study of the Sydney Basin, Department of Geology, University of Newcastle, pp. 1–8.

Metcalfe, I., 1988. Origin and assembly of south-east Asian continental terranes. In: M.G. Audley-Charles and A. Hallam (Editors), Gondwana and Tethys. Geological Society, Special Publication 37: 101–118.

Moore, P.S. and Hocking, R.M., 1983. Significance of hummocky cross-stratification in the Permian of the Carnarvon Basin, western Australia. Journal of the Geological Society of Australia, 30: 323–331.

Moore, P.S., Denman, P.D. and Hocking, R.M., 1980a. Sedimentology of the Byro Group (Lower Permian), Carnarvon Basin, western Australia. Geological Survey of Western Australia, Annual Report for 1979, pp. 55–64.

Moore, P.S., Hocking, R.M. and Denman, P.D., 1980b. Sedimentology of the Kennedy Group (Permian), Carnarvon Basin, western Australia. Geological Survey of Western Australia, Annual Report for 1979, pp. 65–72.

Mory, A.J., 1988. Regional geology of the offshore Bonaparte Basin. In: P.G. Purcell and R.R. Purcell (Editors), The North West Shelf, Australia. Proceedings of the Petroleum Exploration Society of lia Symposium, Perth, pp. 287–301.

Mory, A.J. and Beere, G.M., 1988. Geology of the

onshore Bonaparte and Ord basins in Western Australia. Geological Survey of Western Australia, Bulletin 134, pp. 184.

Murray, C.G., 1985. Tectonic setting of the Bowen Basin. In: Bowen Basin Coal Symposium. Geological Society of Australia, Abstracts, 17: 5–16.

O'Brien, P. E. and Christie-Blick, N., 1992. Glacially grooved surfaces in the Grant Group, Grant Range, Canning Basin and the extent of Late Palaeozoic Pilbara ice sheets. Bureau of Mineral Resources Journal of Australian Geology and Geophysics, 13: 87–92.

O'Brien, P.E., 1986. Stratigraphy and sedimentology of Late Palaeozoic glaciomarine sediments beneath the Murray Basin, and their palaeogeographic and palaeoclimatic significance. Bureau of Mineral Resources Journal of Australian Geology and Geophysics, 10: 53–63.

Parrish, J.M. and Barron, E.J., 1986. Paleoclimates and Economic Geology. Society of Economic Paleontologists and Mineralogists, Short Course 18, pp. 162.

Parrish, J.M., Parrish, J.T. and Ziegler, A.M., 1986. Permian-Triassic paleogeography and paleoclimatology and implications for Therapsid distribution. In: M.N. Hotten, P.D. MacLean, J.J. Roth, and E.C. Roth (Editors), The ecology and biology of mammal-like reptiles. Smithsonian Institution Press, Washington, DC, and London, pp. 109–131.

Pigram, C.J. and Panggabean, H., 1983. Geological data record Waghete (Yapekopra) 1:250,000 sheet area, Irian Jaya Geological Research and Development Centre, Indonesia.

Playford, P.E., Cockbain, A.E. and Low, G.H., 1976. Geology of the Perth Basin Western Australia. Geological Survey of Western Australia, Bulletin 124, pp. 310.

Powell, C.McA., Roots, S.R, and Veevers, J.J., 1988. Pre-breakup continental extension in East Gondwanaland and the early opening of the eastern Indian Ocean. Tectonophysics, 155: 261–283.

Redfern, J., 1991. Subsurface facies analysis of Permo-Carboniferous glaciogenic sediments: Canning Basin, Western Australia. Proceedings of 7th Gondwana Symposium, São Paulo, 1988, pp. 71.

Retallack, G., 1980. Late Carboniferous to Middle Triassic megafossil floras from the Sydney Basin. In: C. Herbert and R. Helby (Editors), A guide to the Sydney Basin. Geological Survey of New South Wales, Bulletin 26: 384–431.

Rigby, J.F., 1972. The flora of the Kaloola Member of the Baralaba Coal Measures, central Queensland. Geological Survey of Queensland, Publication 352, Palaeontological Papers 26: 1–16.

Schiebner, E., 1973. A plate tectonic model of the Palaeozoic tectonic history of New South Wales. Journal of the Geological Society of Australia, 20: 405–426.

Smyth, M. and Cameron, M., 1982. Organic petrology and source potential of sediments in the Eromanga Basin, south Australia. International Journal of Coal Geology, 1: 263–281.

Struckmeyer, H.I.M. and Totterdell, J.M., 1992, Australia, evolution of a continent. Australian Bureau of Mineral Resources, Canberra, 97 pp.

Thornton, R.C.N., 1979. Regional stratigraphic analysis of the Gidgealpa Group, southern Cooper Basin, Australia. Geological Survey of South Australia, Bulletin 49, pp. 140.

Towner, R.R., Crowe, R.W.A. and Yeates, A.N., 1983. Notes on the geology of the southern part of the Canning Basin. Bureau of Mineral Resources, Australia, Record 1976/95, pp. 116.

Townsend, I.J., 1976. Stratigraphic drilling in the Arckaringa Basin, 1969–1971. Geological Survey of South Australia, Report of Investigations 45, pp. 30.

Townsend, I.J. and Ludbrook, N.H., 1975. Revision of Permian and Devonian nomenclature of four formations in and below the Arckaringa Basin. Geological Survey of South Australia, Quarterly Geological Notes 54: 2–12.

Wells, A.T. and O'Brien, P.E., 1989. Coal measures of Early and Late Permian Gidgealpa Group in the Cooper Basin, south Australia and Queensland. In: H.J. Harrington, A.T. Brakel, J.W. Hunt, A.T. Wells, M.F. Middleton, P.E. O'Brien, D.S. Hamilton, J. Beckett, C.R. Weber, S. Radke, J.M. Totterdell. D.J. Swaine, and P.W. Schmidt (Editors), Permian coals of eastern Australia. Bureau of Mineral Resources, Australia, Bulletin 231: 149–170.

Williams, B.P.J. and Wild, E.K., 1984. Tirrawarra Sandstone and Merrimelia Formation of the southern Cooper Basin, south Australia, the sedimentation and evolution of a glaciofluvial system. Australian Petroleum Exploration Society Journal, 24: 377–392.

Wilson, A.C., 1989. Palaeocurrent patterns in the Collie Coal Measures. the implications for sedimentation and basin models. Professional Papers, Geological Survey of Western Australia Report 25: 1–85.

Ziegler, A.M., 1990. Phytogeographic patterns and continental configurations during the Permian period. In: W.S. McKerrow, and C.R. Scotese (Editors), Palaeozoic palaeogeography and biogeography. Geological Society of London, Memoir 12: 363–379.

13. UPPER PALEOZOIC GLACIAL AND POSTGLACIAL DEPOSITS, CENTRAL TRANSANTARCTIC MOUNTAINS, ANTARCTICA

John L. Isbell, Gina M. Seegers and Greg A. Gelhar

Upper Paleozoic rocks are exposed throughout the Transantarctic Mountains of Antarctica from northern Victoria Land to the Ohio Range, in the Ellsworth, Pensacola, Theron, and Prince Charles Mountains and in Dronning Maud Land. The most complete exposures of these rocks occur in the central Transantarctic Mountains between the Byrd Glacier and the Ohio Range (Fig. 13.1). Within this outcrop belt, upper Paleozoic strata attain a thickness of 1.5 km. These strata are relatively undeformed; they rest disconformably on Precambrian and Paleozoic basement and sedimentary rocks, and are disconformably overlain by nonmarine Triassic rocks. The upper Paleozoic rocks are laterally continuous and can be traced throughout the central Transantarctic

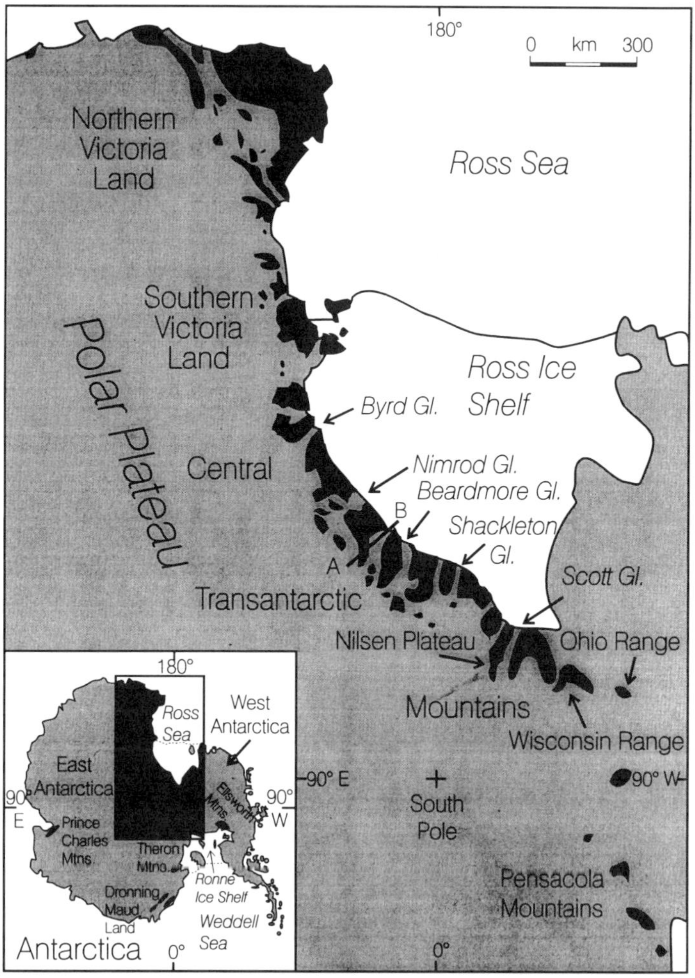

Fig. 13.1. Location map of the central Transantarctic Mountains. Line A-B near the Beardmore Glacier is the line of cross-section for Figure 13.6.

Mountains. These rocks provide an excellent spacial and temporal view of late Paleozoic depositional conditions in Antarctica.

Within the central Transantarctic Mountains, both glaciomarine and terrestrial glacial deposits occur. Basinal shales overlay these units, which in turn are overlain by deltaic and fluvial sandstone. Thick coal-bearing fluvial deposits occur at the top of the upper Paleozoic section. These rocks provide insight into deglaciation in marine and terrestrial settings and record a change from glacial to postglacial to peat-forming conditions as Antarctica drifted over the south pole in the Late Carboniferous and the Permian. This review outlines the depositional framework during glacial and postglacial times and speculates on the controls that influenced deposition of these rocks. Other important reviews include Barrett (1991) and Collinson et al. (1994).

Depositional Framework

Upper Paleozoic strata in the central Transantarctic Mountains range in age from Late Carboniferous to Late Permian, a conclusion based on palynomorph correlation with Australian palynologic zones (Kyle and Schopf, 1982), and on the presence of *Glossopteris* leaves. Additional palynomorph biostratigraphy is needed on these rocks, as time control within the upper Paleozoic rocks is presently provided by lithostratigraphy. Strata consist of, in ascending order: (1) interstratified glacial diamictite, sandstone, and shale, (2) basinal shale and deltaic sandstone, (3) thick fluvial sandstone, and (4) interstratified fluvial sandstone, shale, and coal (Fig. 13.2). These strata have been extensively studied in the area between the Byrd and the Beardmore glaciers (Byrd-Beardmore area). In this area, these

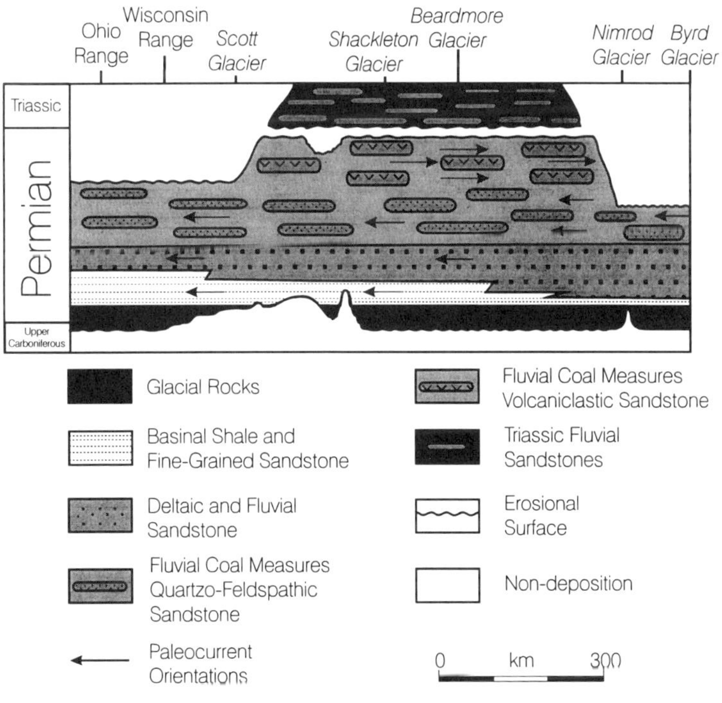

Fig. 13.2. Time-space diagram for the upper Paleozoic strata in the central Transantarctic Mountains (data from Long, 1964, 1965; Barrett, 1965; Minshew, 1967; Bradshaw et al., 1984; Coates, 1985; Barrett et al., 1986; Aitchison et al., 1988; Isbell, 1990; Isbell et al., in press).

rocks correspond to the Pagoda, Mackellar, Fairchild, and Buckley formations respectively (Barrett et al., 1986; Isbell et al., in press). Within these rocks, and laterally equivalent units, paleocurrent orientations indicate a paleoslope that dipped away from the present position of the Byrd Glacier toward the Ohio Range. These directions suggest an elongate basin oriented oblique to the present trend of the central Transantarctic Mountains. This depositional basin formed along the margin of the East Antarctic Craton, on the eroded core of the Ross Orogenic Belt (Collinson et al., 1994). Rocks of the East Antarctic Craton are located under the present Polar Plateau and extend to the present coastline along the western side of the Ross Sea. A late Paleozoic-early Mesozoic orogenic belt along the paleo-Pacific margin was located in present West Antarctica. Lindsay (1970) suggested that initial subsidence within the depositional basin began as glacial loading. Dalziel and Elliot (1982) proposed a tectonic origin for the basin, while Collinson et al. (1994) summarized the evidence for development of the area as a foreland basin. Foreland basins form adjacent to orogenic belts as a result of crustal flexure beneath a tectonic load. Such loading produces a moat-shaped basin in front of the orogenic belt. Storey (1992) listed some of the problems with this hypothesis.

Within the depositional basin, nonmarine conditions occurred in the area between the Byrd and the Beardmore glaciers, while marine to marginal-marine conditions may have occurred in the lower portion of the Ohio Range section (Bradshaw et al., 1984; Barrett et al., 1986; Aitchison et al., 1988; Isbell 1990). Younger Permian rocks in the Ohio Range were deposited under fluvial conditions (Bradshaw et al., 1984), suggesting progradation of the strandline within the basin. Rocks exposed between the Shackleton Glacier and the Ohio Range have not been studied as extensively or as recently as rocks in other parts of the central Transantarctic Mountains.

The Antarctic sector of Gondwana drifted across the south pole during the Late Carboniferous and the Permian. On the basis of data summarized by Powell and Li (1994), we believe that the central Transantarctic Mountains were located within several degrees of the late Paleozoic pole from 320 (Late Carboniferous) to 260 Ma (Late Permian; Fig. 13.3). Changing paleogeographic, climatic, and tectonic conditions influenced late Paleozoic deposition in the central Transantarctic Mountains.

Glacial Deposits

Upper Paleozoic glacial rocks in the central Transantarctic Mountains were deposited along the margins of a continental ice sheet. Lithofacies analysis indicates that both terrestrial and glaciomarine conditions occurred along the margins. Deposits of terrestrial-based temperate ice occur in the area between the present Byrd and Beardmore glaciers (Miller, 1989; Isbell et al., in press), while Aitchison et al. (1988) identified glaciomarine deposits in the Ohio Range. Paleocurrent orientations suggest that ice flowed away from the Byrd area toward the Ohio Range (Fig. 13.4). Deposits of these two areas are contrasted later in this chapter.

In the area between the Byrd and the Beardmore glaciers, glacial rocks of the Pagoda Formation rest unconformably on Precambrian and

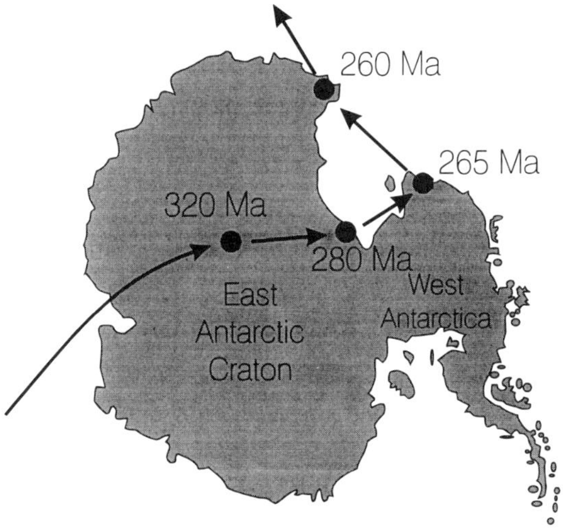

Fig. 13.3. Polar Wandering path for the Late Paleozoic (data from Powell and Li, 1994).

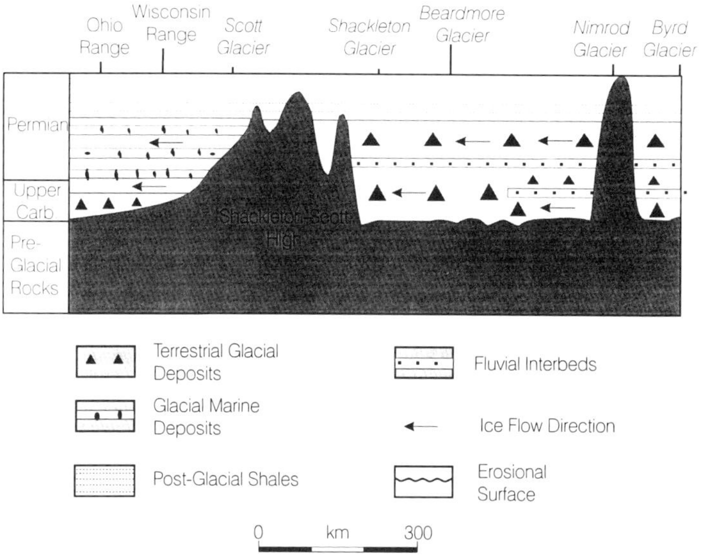

Fig. 13.4. Cross-section of glacial strata in the central Transantarctic Mountains, showing areas of glaciomarine and terrestrial glacial deposits. Note that the Shackleton-Scott High separates areas of glaciomarine from terrestrial glacial deposits (data from Long, 1964, 1965; Barrett, 1965; Minshew, 1967; Bradshaw et al., 1984; Coates, 1985; Barrett et al., 1986; Aitchison et al., 1988; Isbell, 1990; Isbell et al., in press).

lower Paleozoic rocks. Pagoda strata are laterally continuous across the area and attain a maximum thickness of 390 m just south of the Nimrod Glacier (Barrett et al., 1986). Diamictites are 5 to 60 m thick, characteristically rest on erosional and striated surfaces, and consist of unstratified diamictite and crudely stratified diamictite. These rocks are interpreted as lodgement and melt-out tills respectively (Miller, 1989; Isbell et al., in press). Sandstone bodies surrounded by diamictite suggest deposition in subglacial and englacial meltwater channels, while coarse-grained cross-bedded sandstone and laminated shale were deposited in outwash streams and in lacustrine settings. These lithofacies and the presence of numerous striated surfaces suggest deposition from wet-based terrestrial glaciers (Barrett et al., 1986; Miller, 1989; Isbell et al., in press).

At the opposite end of the central Transantarctic Mountains, in the Ohio Range, Aitchison et al. (1988) and Bradshaw et al. (1984) identified an up-section change from glacial to glaciomarine deposits in the Buckeye Formation. Evidence that suggests direct deposition from grounded ice includes massive diamictites resting on eroded and grooved surfaces and nonoriented boulder horizons (Bradshaw et al., 1984). Higher in the formation, stratified diamictites occur. Diamictites with gradational bases and pebbles with long axes parallel to bedding characterize these beds. Such features suggest deposition of material from floating glacial ice (Aitchison et al., 1988). Throughout the formation, diamictites are separated by thin interbeds of laminated siltstone and fine-grained, graded and rippled sandstone. Such rocks suggest deposition by underflow currents. Isolated pebbles, which pierce underlying laminae, occur within the fine-grained beds. Bradshaw et al. (1984) and Aitchison et al. (1988) interpreted these pebbles as ice rafted dropstones. The presence of hummocky cross-stratification (Aitchison et al., 1988), fossil marine acritarchs (Kemp, 1975), and low iron content (Frakes and Crowell, 1975) provide further evidence for marine conditions. Low iron content reflects a reducing environment, which Frakes and Crowell (1975) believed characterized glaciomarine settings. The presence of diamictites resting on striated/grooved surfaces and of stratified diamictites resting on gradational bases in the Ohio Range suggests deposition from an ice sheet grounded below sea level.

The interstratification of diamictite, conglomerate, sandstone, and shale in the central Transantarctic Mountains records numerous advances and retreats of the late Paleozoic ice front (Lindsay, 1970; Frakes et al., 1971; Miller, 1989). It is uncertain whether interstratification of these rocks represents minor fluctuations in the position of the ice front, or complete withdrawal of the ice from this area. In the Ohio Range, massive diamictites of the Buckeye Formation resting on striated surfaces indicate initial grounded glacial advance of a marine ice sheet. Because marine ice sheets are grounded below sea level, they are inherently unstable and are susceptible to rapid collapse. A relative rise in sea level will result in detachment of the ice from the sea floor, formation of an ice shelf, and retreat of the ice upflow to a point where the grounding line is held in place by basement highs. Upsection in the Buckeye Formation, stratified diamictites overlying gradational contacts suggest establishment of glaciomarine conditions. These conditions probably represent decoupling of the ice from the underlying bed and development of an ice shelf (Aitchison et al., 1988; Anderson and Thomas, 1991). Rising sea level from melting and retreat of the ice may have controlled the position of the grounding line on the paleoslope. In the Byrd-Beardmore area, rocks interbedded with the diamictites are entirely nonmarine. These deposits are located upflow from Ohio Range rocks and indicate substantial retreat of the ice front; however, there is no evidence that glaciomarine conditions ever extended into this area. The marine-nonmarine transition zone must have been located in the area between the Ohio Range and the Beardmore Glacier, as suggested by the change from marine/glaciomarine to terrestrial deposits up the direction of ice flow.

An area of thin diamictite occurs between the Shackleton and the Scott glaciers (Coates, 1985). This area separates terrestrial glacial deposits of the Byrd-Beardmore area from glaciomarine strata in the Ohio Range. In the Shackleton-Scott area, glacial rocks are locally absent and younger rocks rest directly on the basal unconformity (Fig. 13.4; Barrett, 1965, Minshew, 1967; La Prade, 1972). These features suggest that this area may have been a subglacial high (Shackleton-Scott High). If present, such a high may have prevented flooding in the Byrd-Beardmore area during minor retreats of the ice front. Frakes et al. (1971) identified two late Paleozoic glacial basins in the area of the present central Transantarctic Mountains, that correspond to the areas located on opposite sides of the Shackleton-Scott high.

Glacial to postglacial transition

The demise of Gondwana glaciation in Antarctica can be inferred by identifying large-scale trends within the deposits and by examining the contact between the glacial and the postglacial strata. In the Ohio Range, clast content and clast size within diamictites decrease upsection as does the abundance of dropstones within interbedded lithologies. These features led Bradshaw et al. (1984) and Aitchison et al. (1988) to suggest a gradual demise of the ice sheet in this area. With the migration of the grounding line toward the Byrd-Beardmore area, glaciomarine conditions were replaced by marine deposition. Glaciomarine conditions probably never reached the Byrd-Beardmore area, however, since no glaciomarine deposits are known from this area. The proposed high located between the Shackleton and the Scott glaciers could have prevented flooding into the Byrd-Beardmore area during glacial times.

Throughout the central Transantarctic Mountains, a sharp contact separates glacial deposits below from postglacial deposits above (Bradshaw et al., 1984; Barrett et al., 1986; Isbell et al., in press). In the Ohio Range, silty shale of the Discovery Ridge Formation rest on a sharp to disconformable contact with the underlying glaciomarine deposits of the Buckeye Formation (Long, 1964; Bradshaw et al., 1984). In the Byrd to Beardmore area, black shale of the Mackellar Formation is in sharp contact with diamictite of the underlying Pagoda Formation (Fig. 13.5). Dropstones are extremely rare in the shales, and where present, occur only in the lower 1.5 m.

The sharp contact between diamictite of the Pagoda Formation and black shale of the Mackellar Formation is a flooding surface that appears to extend throughout the 1,300 km length of the central Transantarctic Mountains. This surface records the final disappearance of glacial ice from the area and represents at least a 1,300 km shift in the position of the ice front up the paleoslope. Collapse of the ice sheet and the resulting rise in sea level could have submerged the Shackleton-Scott High. Flooding across the high would then have established basinal conditions throughout the area of the present central Transantarctic Mountains.

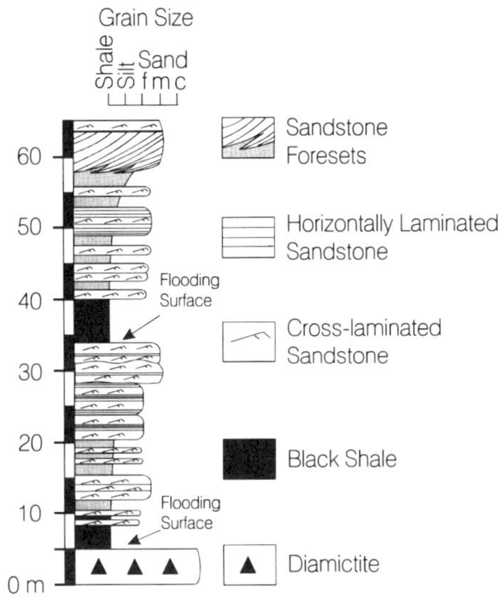

Fig. 13.5. Stratigraphic section through the Mackellar Formation in the area between the Byrd and the Nimrod glaciers. Note the two flooding surfaces overlain by coarsening upward sequences.

Comparison with the Cenozoic Record of Antarctica

The Cenozoic history of glaciation in Antarctica is similar to the glacial history recorded by upper Paleozoic strata in the central Transantarctic Mountains. Both are represented by terrestrial ice sheet, marine ice sheet, and ice shelf environments; large scale advances and retreats of the ice front; and marine flooding of deglaciated areas.

Antarctica currently contains two ice sheets: the East and the West Antarctic ice sheets. The East Antarctic Ice Sheet is the larger of the two and is a dry-based terrestrial glacier. This ice sheet is predominantly grounded above sea level; therefore, small-scale changes in sea level do not affect the position of its margins (Clapperton and Sugden, 1990). Advance and retreat of the ice probably results from long-term climatic changes. Although much of the ice is grounded above sea level, several depressions that extend below sea level occur in the sub-ice topography (Webb, 1990). Cenozoic marine microfossils found within glacial deposits high in the Transantarctic Mountains suggest that these depressions were inundated by marine waters during major ice retreat (Webb and Harwood, 1991). Advancing ice then eroded the marine sediment and deposited it in the Transantarctic Mountains. Fossil wood and leaves within these deposits also support retreat of the ice during the Cenozoic (Webb and Harwood, 1991).

The West Antarctica Ice Sheet is smaller and is grounded primarily below sea level. This ice sheet is probably wet based, and is ablating faster than the East Antarctic Ice Sheet (Anderson and Molnia, 1989). Much of the West Antarctic Ice Sheet drains into the Ross and the Weddell seas where it is presently detached from its underlying bed and exists as the Ross and the Ronne ice shelves respectively (Anderson and Molnia, 1989). During the late Wisconsinan, grounded ice extended to the edge of the continental shelves in the Ross and the Weddell seas. During the post-Wisconsinan rise in sea level, ice in the Ross and the Weddell embayments detached from their underlying beds, formed ice shelves, and retreated to their present positions, where highs (pinning points) have prevented further retreat of the grounding line (Anderson et al., 1991). Studies of the West Antarctic Ice Sheet suggest that it is unstable and susceptible to further collapse with a rise in sea level (Anderson and Thomas, 1991).

Late Paleozoic glaciation in the Ohio Range may have been analogous to the present West Antarctic Ice Sheet, although on a smaller scale. Both represent marine ice sheet conditions with grounding of ice below sea level. A slight rise in sea level probably facilitated formation of ice shelves and rapid retreat of ice fronts up the paleoslope to a point where the glaciers were stabilized by basement highs. In such a scenario, numerous advances and retreats of an ice front are possible. Late Paleozoic terrestrial glaciation in the area between the present Byrd and Beardmore glaciers corresponds to the present East Antarctic Ice Sheet. Collapse of these two ice sheets may have allowed for inundation of marine waters into interior basins.

Postglacial Shales

Black shales overlie glacial strata throughout the central Transantarctic Mountains and represent sediment-starved conditions following postglacial flooding of the basin. Dropstones are absent from these rocks, except for rare occurrences near the glacial/black shale contact. Upward, these shales contain thin beds of graded and cross-laminated sandstone (0.01–0.1 m thick). These beds coarsen upward into thick sandstone beds (1 to 10 m thick). Such sands were introduced into the basin by underflow currents in front of prograding deltaic complexes (Miller and Collinson, 1994; Isbell et al., in press).

In the area between the Byrd and the Nimrod glaciers, two coarsening-upward successions occur (Fig. 13.5). A sharp contact separates sandstone at the top of one succession from shale in the overlying succession. Isbell et al. (in press) interpreted this contact as a regional flooding surface and suggested that rapid fluctuations in water levels within the basin may indicate further stages in the disintegration of the remaining late Paleozoic ice sheet elsewhere in Antarctica and in Gondwanaland.

The salinity of water in the basin and the basin's relationship to the open ocean are unknown. A low-diversity nonmarine ichnofossil fauna and high carbon/sulfur ratios suggest deposition in fresh to brackish waters in the Beardmore Glacier area (Mackellar Formation; Miller and Collinson, 1994). Farther down the paleoslope, in the Ohio Range, Bradshaw et al. (1984) interpreted equivalent strata (Discovery Ridge Formation) as marine due to their location between glaciomarine strata below and possible marine strata above (lower Mount Glossopteris Formation). Salinity within the basin may have been affected by the shape of the basin, a high influx of fresh water, and/or circulation patterns within the basin. The proposed Shackleton-Scott High may have persisted into postglacial time. Such a high could have restricted circulation and inhibited the mixing of fresh water in the Byrd-Beardmore area with normal seawater in the Ohio Range. An opening between the Thurston Island and the Marie Byrd Land crustal blocks (Grunow et al., 1991) may have connected late Paleozoic waters in the paleo-Pacific Ocean with basinal waters in the area of the present Transantarctic Mountains (Miller and Collinson, 1994).

Deltaic and Fluvial Sandstones

Thick sandstone deposits (250 m thick; Fairchild Formation) conformably overlie the postglacial shales in the area between the Byrd and the Beardmore glaciers. In the area between the Byrd and the Nimrod glaciers, basal beds consist of large (5–10 m thick), low-angle dipping medium-grained sandstone beds that interfinger with postglacial shales. These foreset beds represent "Gilbert deltas" that prograded into the postglacial basin. Such deltas are produced by flow of river waters into unstratified basins, usually freshwater lakes where shallow depths inhibit thermal stratification. "Gilbert deltas" have not been reported farther down the paleoslope (Miller and Collinson, 1994).

Upward, deltaic beds are replaced by large sandstone channel-form bodies, 2 to 20 m thick and 100 to 1000 m wide. These sandstones probably represent braided streams (Barrett et al., 1986; Isbell, 1990, 1991). Paleocurrent orientations define a paleoslope dipping uniformly toward the Ohio Range (Barrett et al., 1986; Isbell 1990, 1991). There, Bradshaw et al. (1984) interpreted equivalent rocks (the lower portion of the Mount Glossopteris Formation) as marine to marginal marine on the basis of the presence of the trace fossil *Monocraterion (?)*. The distribution of this trace fossil may have been ubiquitous, as Zawiskie et al. (1983) described *Monocraterion* within fluvial deposits in northern Victoria Land. The change from "Gilbert deltas" in the Byrd-Nimrod area to deltaic deposits lacking well-developed foreset beds in strata exposed from the Beardmore to the Ohio Range supports a lateral change in salinity down the paleoslope. Nonetheless, lateral and vertical facies changes within the basin suggest progradation of a deltaic platform upon which fluvial sedimentation developed.

In the Beardmore area, facies patterns and paleocurrent orientations display only slight variations across the outcrop belt (Fig. 13.6; Isbell, 1990). This indicates the existence of a broad cratonic basin with a uniformly dipping paleoslope. Stacked channel sandstone bodies

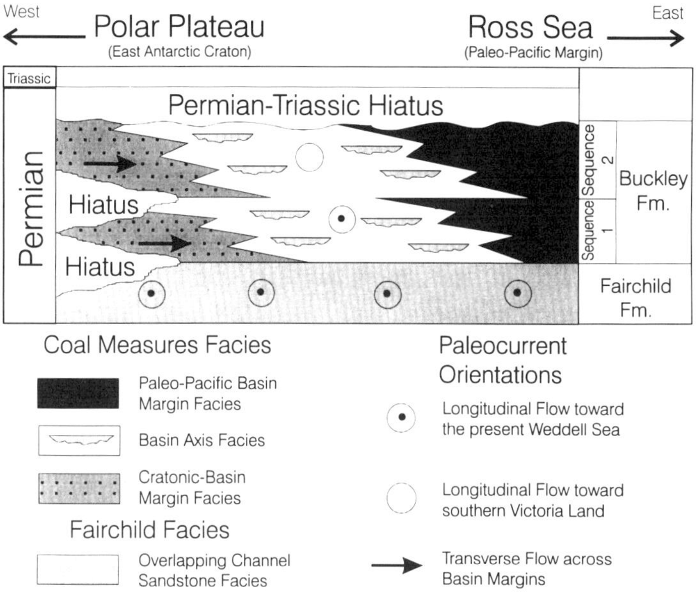

Fig. 13.6. Time-space diagram for fluvial rocks in the Beardmore Glacier area. Note the change in facies patterns and paleocurrent orientations between the Fairchild and the Buckley formations.

such as those described earlier typically occur in areas of slow subsidence. Under such conditions, laterally shifting channels erode their flood plains, leaving behind only channel deposits (Isbell, 1990).

Coal Measures

Interstratified fluvial sandstone, shale, and coal deposits occur at the top of the upper Paleozoic section in the central Transantarctic Mountains. These rocks record changing depositional, tectonic, and climatic conditions near the end of the Paleozoic in Antarctica. These strata are unconformably overlain by Lower Triassic fluvial sandstone and noncarbonaceous shale (Barrett et al., 1986; Collinson et al., 1994).

Paleocurrent orientations within coal measure strata in the Beardmore Glacier area define the existence of a large trough-shaped basin oriented oblique to the present trend of the central Transantarctic Mountains. Basin margins were located on the East Antarctic craton along what is now the Polar Plateau, and in the direction of the paleo-Pacific margin of Antarctica along what is now the eastern side of the Ross Sea (Fig. 13.1). The change from a uniformly dipping paleoslope to a trough-shaped basin suggests the formation of a foreland basin at the beginning of coal measure deposition (Isbell, 1990).

Coal-bearing strata of the Buckley Formation, exposed in the area between the present Byrd and Beardmore Glaciers, are entirely nonmarine (Barrett, et al., 1986, Isbell et al., in press). Paleocurrent orientations from rocks in the lower portion of the coal measures reveal that regional drainage was away from this area and toward the location of the present Weddell Sea. Farther down the paleoslope, in the Ohio Range, Bradshaw et al. (1984) reported laterally equivalent rocks of the middle and upper portions of the Mount Glossopteris Formation as marine/marginal marine deposits that grade up section into nonmarine deposits. Their marine interpretation is based on the presence of *Monocraterion,* a questionable marine trace fossil.

A regional paleoslope reversal occurs near the middle of the coal measure strata in the Beardmore Glacier area (Fig. 13.6). Streams flowing regionally toward Victoria Land deposited the rocks above this reversal (Isbell 1990, 1991). This reversal coincides with a change from quartzo-feldspathic sandstone to volcaniclastic sandstone. In the Ohio Range, Long (1964, 1965) reported paleocurrents oriented toward the present Weddell Sea. This is in contrast to directions taken from rocks in the Beardmore area. The nature of the strata in the area between these two sites is largely unknown.

Rocks of the more than 750m thick Buckley Formation in the Beardmore Glacier region are the most extensively studied coal measure rocks in the central Transantarctic Mountains. Two major depositional successions occur within these rocks (Fig. 13.6). Each succession is bounded below by disconformities along the western (Polar Plateau) basin margin and by conformities near the center of the depositional basin. These disconformities are recognized within measured sections by (1) an abrupt change from quartzo-feldspathic sandstone to volcaniclastic sandstone, (2) an abrupt change from dominantly fine-grained flood-plain lithofacies to dominantly coarse-grained channel-sandstone lithofacies, (3) an abrupt change in paleocurrent orientations, and by (4) the presence of an intraformational conglomerate just above the contact (Fig. 13.6).

Within each depositional succession, three geographically distributed lithofacies assemblages occur: (1) a thick succession of overlapping sandstone bodies along the Ross Sea coastline (paleo-Pacific basin margin lithofacies assemblage), (2) thick interstratified mudstone and sandstone along the center of the outcrop belt (basin axis lithofacies assemblage), and (3) a thin succession of overlapping sandstone bodies along the polar plateau side of the outcrop belt (cratonic basin margin lithofacies assemblage). Upward within each depositional succession, the overlapping sandstone lithofacies assemblage along the Ross Sea expands basinward displacing the other two lithofacies assemblages toward the present Polar Plateau. These shifts result in two large-scale coarsening-upward sequences on the Ross Sea side of the outcrop belt. Two large-scale fining-upward sequences occur on the Polar Plateau side of the area (Fig. 13.6). These patterns suggest the progradation of large clastic wedges into the basin from the paleo-Pacific margin (Isbell, 1991). The lower depositional succession is bounded above by a disconformity along the western basin margin and by a correlative conformity within the depositional basin. An abrupt shift in the position of facies toward the Ross Sea marks the base of the overlying succession. The upper succession is bounded above by the Permian-Triassic regional unconformity. Sandstone composition in the lower depositional succession is quartzo-feldspathic, while upper-succession sandstones are dominated by volcaniclastic framework grains.

Lithofacies analysis of the two depositional successions reveals the occurrence of braided-stream deposits along the margins of this trough-shaped basin. Away from the margins and toward the axis of the basin, the ratio of flood-plain shale to channel sandstone increases. Along the axis of the basin, braided-stream lithofacies are isolated within thick flood-plain deposits (Isbell, 1990). Paleocurrent analysis suggests that the rivers flowed transversely across the basin margins toward the basin axis. There, these streams joined a major trunk stream that flowed longitudinally down the axis of the basin. Upward within each depositional succession, the position of this longitudinal drainage axis shifted cratonward concurrent with displacement of the geographically distributed lithofacies assemblages described earlier. An abrupt shift in the position of this drainage axis toward the Ross Sea coincides with succession boundaries.

Modeling by Flemings and Jordan (1989, 1990) have produced foreland-basin depositional packages similar to those in the Beardmore Glacier area. Flemings and Jordan (1990) suggested that abrupt shifts in depositional patterns toward the orogenic belt occur during the onset of thrust loading on elastic lithosphere. Rapid subsidence during thrusting traps coarse-grained sediment next to the orogenic belt (Blair and Bilodeau, 1988). Later, as subsidence decreases, progradation of a clastic wedge with cratonward displacement of depositional patterns occurs. Combined paleocurrent, lithofacies, and petrologic analyses suggest that tectonic loading and the progradation of clastic wedges into the basin controlled deposition of strata in the Buckley

Formation. The presence of two depositional successions in the Buckley Formation suggests that two loading events occurred during the Late Permian.

Coal seams and fossil plants occur throughout the coal measure strata, but are abundant only in the basin axis lithofacies assemblage. The abundance of coal beds along the basin axis suggests that clastic influx and subsidence rates controlled the formation and preservation of peat. High topographic relief and high rates of clastic influx along the basin margins were probably unfavorable for peat accumulation, whereas high subsidence rates coupled with lower rates of clastic influx along the basin axis may have allowed for both the accumulation and the preservation of peat/coal deposits.

Fossil plants within coal measures strata in the central Transantarctic Mountains include *in situ* tree stumps, silicified logs, *Glossopteris* and *Gangamopteris* leaves, and *Vertebraria* roots. Fossil leaf mats and tree rings within fossil logs suggests seasonal growth of these plants. Taylor et al. (1992) reported an absence of frost rings within Upper Permian fossil stumps, which strongly suggest warm summer temperatures (well above 0°C), despite the areas near polar position (Fig. 13.3). Winter temperatures are less constrained by tree ring data, however, as ring growth ceased at the end of the growing season when the trees became dormant (Taylor, personal communication).

Permian rocks and fossils in the central Transantarctic Mountains suggest warm (above 0°C summer temperatures), humid conditions during deposition of the coal measures. Such conditions are difficult to explain, as many climate modelers have argued for cold arid climates within the polar interior of Gondwanaland (Fawcett et al., 1994). The postglacial seaway in the central Transantarctic Mountains, with its possible connections to the paleo-Pacific Ocean, may have played a major role in moderating climatic conditions. Although the seaway was reduced in size by delta progradation, it probably existed at least until the middle of coal measures deposition. At that time, a drainage reversal in the Byrd-Beardmore area suggests termination of the basin's connection with the paleo-Pacific Ocean; however, warm humid conditions appear to have continued until the end of the Permian. Noncarbonaceous shale in the overlying rocks suggests a change to an arid climate by the Early Triassic.

Coal measures strata and plant fossils in the central Transantarctic Mountains record changing depositional, climatic, and tectonic conditions near the end of the Paleozoic in Antarctica. Extensive study of coal-measure rocks in the Byrd-Beardmore area suggests that tectonic activity controlled the distribution of lithofacies, including coal, in the depositional basin. Tectonic loading and formation of a foreland basin was perhaps the single most important factor that influenced Late Permian sedimentation in this area.

Conclusion

Changing paleogeographic, climatic and tectonic conditions controlled the deposition of upper Paleozoic rocks in the central Transantarctic Mountains. These rocks record the presence of subglacial highs, fluctuations in the position of ice margins and ice centers as Antarctica drifted over the South Pole, plate convergence along the paleo-Pacific continental margin, and the formation of a foreland basin.

Paleocurrent data suggest that during much of the late Paleozoic, strata in the Ohio Range were deposited downstream from strata in the Byrd-Beardmore area. Differences in depositional environments characterized the two areas. Terrestrial glacial and nonmarine deposition prevailed in the Byrd-Beardmore area, while glaciomarine, marine, marginal marine, and deltaic conditions may have occurred toward the Ohio Range. By the Late Permian, the shoreline appears to have prograded beyond the Ohio Range and fluvial conditions prevailed throughout the area occupied by the present central Transantarctic Mountains.

Late Carboniferous to Early Permian glaciation occurred during Antarctica's drift over the South

Pole. Glacial rocks in the central Transantarctic Mountains record numerous advances and retreats of the ice front. In the Ohio Range, ice advance occurred in a glaciomarine setting. During retreat of the ice front, decoupling of ice from the sea bed resulted in migration of the grounding line upflow and development of an ice shelf. The area between the Shackleton and the Scott glaciers could have been a subglacial high. This high, if it existed, could have served as a pinning point that prevented further migration of the grounding line. The high also may have served as a physical barrier to marine incursions into the Byrd-Beardmore area during larger-scale fluctuations in the position of the ice front. Rocks in the area between the present Byrd and Beardmore glaciers were deposited entirely by terrestrial glaciers and in lacustrine and fluvial environments. A sharp contact between diamictites and black shales marks the transition from glacial to postglacial conditions. This contact is a flooding surface and represents rapid disintegration of glacial ice, followed by a major rise in basinal water levels. Destruction of glacial ice in the area of the present central Transantarctic Mountains may have been facilitated by migration of the pole toward northern Victoria Land/Australia.

Flooding across the hypothetical Shackleton-Scott High resulted in sediment-starved basinal conditions throughout the central Transantarctic Mountains. This structure may have continued to influence sedimentation by restricting the mixing of waters on opposite sides of the high. Initial sediment-starved conditions were replaced by sediment gravity flows, prograding deltas, and braided river systems. Deposition at that time occurred in a broad, slowly subsiding cratonic basin.

Major changes in facies patterns and paleocurrent orientations at the base of the coal measure rocks in the Beardmore Glacier area signal the change from a cratonic to a foreland basin. These rocks were deposited on flood plains and by braided streams flowing transversely across basin margins and longitudinally down the basin axis. High subsidence rates and low rates of clastic influx characterized the longitudinal drainage axis. Coal was preferentially deposited and preserved along the basin axis.

References

Aitchison, J.C, Bradshaw, M.A. and Newman, L., 1988. Lithofacies and origin of the Buckeye Formation: Late Paleozoic glacial and glacio-marine sediments, Ohio Range, Transantarctic Mountains, Antarctica. Palaeogeography, Palaeoclimatology, Palaeoecology, 64: 93–104.

Anderson, J.B., Kennedy, D.S., Smith, M.J. and Domack, E.W., 1991. Sedimentary facies associated with Antarctica's floating ice masses. In: J.B. Anderson and G.M. Ashley (Editors), Glacial marine sedimentation; paleoclimatic significance. Geological Society of America, Special Paper 261: pp. 1–26.

Anderson, J.B. and Molnia, B.F., 1989. Glacial-marine sedimentation. American Geophysical Union, Short Course in Geology 9, 127 pp.

Anderson, J.B. and Thomas, M.A., 1991. Marine ice-sheet decoupling as a mechanism for rapid, episodic sea-level change: the record of such events and their influence on sedimentation. Sedimentary Geology, 70: 87–104.

Barrett, P.J., 1965. Geology of the area between the Axel Heiberg and Shackleton Glaciers, Queen Maud Range, Antarctica. Part 2. Beacon Group. New Zealand Journal of Geology and Geophysics, 8: 344–363.

Barrett, P.J., 1991. The Devonian to Triassic Beacon Supergroup of the Transantarctic Mountains and correlatives in other parts of Antarctica. In: R.J. Tingey (Editor), The Geology of Antarctica. Clarendon Press, Oxford, pp. 120–152.

Barrett, P.J., Elliot, D.H. and Lindsay, J.F., 1986. The Beacon Supergroup (Devonian-Triassic) and Ferrar Group (Jurassic) in the Beardmore Glacier area, Antarctica. In: M.D. Turner and J.F. Splettstoesser (Editors), Geology of the Central Transantarctic Mountains. American Geophysical Union, Antarctic Research Series, 36: 339–428.

Blair, T.C. and Bilodeau, W.L., 1988. Development of tectonic cyclothems in rift, pull-apart, and foreland basins: sedimentary response to episodic tectonism. Geology, 16: 517–520.

Bradshaw, M.A., Newman, J. and Aitchison, J.C., 1984. Preliminary geological results of the 1983-84 Ohio Range Expedition. New Zealand Antarctic Record, 5: 1–17.

Clapperton, C.M. and Sugden, D.E., 1990. Late Cenozoic glacial history of the Ross Embayment, Antarctica. Quaternary Science Reviews, 9: 253–272.

Coates, D.A., 1985. Late Paleozoic glacial patterns in the central Transantarctic Mountains, Antarctica. In: M.D. Turner and J.F. Splettstoesser (Editors), Geology of the Central Transantarctic Mountains. Antarctic Research Series, American Geophysical Union, 36: 325–338.

Collinson, J.W., Isbell, J.L., Elliot, D.H., Miller, M.F. and Miller, J.M.G., 1994. Permian-Triassic Transantarctic Basin. In: J.J. Veevers and C. McA. Powell (Editor), Permian-Triassic Pangean Basins and foldbelts along the Panthalassan Margin of Gondwanaland. Geological Society of America, Memoir 184: pp. 173–222.

Dalziel, I.W.D. and Elliot, D.H., 1982. West Antarctica: problem child of Gondwanaland. Tectonics, 1: 3–19.

Fawcett, P.J., Barron, E.J., Robison, V.D. and Katz, B.J., 1994. The climatic evolution of India and Australia from the Late Permian to mid-Jurassic: a comparison of climate model results with the geologic record. In: G.D. Klein (Editor), Pangea: paleoclimate, tectonics, and sedimentation during accretion, zenith, and breakup of a supercontinent. Geological Society of America, Special Paper 288: pp. 139–157.

Flemings, P.B. and Jordan, T.E., 1989. A synthetic stratigraphic model of foreland basin development. Journal of Geophysical Research, 94: 3851–3866.

Flemings, P.B. and Jordan, T.E., 1990. Stratigraphic modeling of foreland basins: interpreting thrust deformation and lithosphere rheology. Geology, 18: 430–434.

Frakes, L.A. and Crowell, J.C., 1975. Characteristics of modern glacial marine sediments: application to Gondwana glacials. In: K.S.W. Campbell (Editor), Gondwana geology. Australian National University Press, Canberra, pp. 373–380.

Frakes, L.A., Matthews, J.L. and Crowell, J.C., 1971. Late Paleozoic glaciation: Part III, Antarctica. Bulletin of the Geological Society of America, 82: 1581–1604.

Grunow, A.M., Kent, D.V. and Dalziel, I.W.D., 1991. New paleomagnetic data from Thurston Island: implications for the tectonics of West Antarctica and Weddell Sea Opening. Journal of Geophysical Research, 96: 17,935–17,954.

Isbell, J.L., 1990. Fluvial sedimentology and basin analyses of the Permian Fairchild and Buckley Formations, Beardmore Glacier region, and the Weller Coal Measures, Southern Victoria Land, Antarctica. Ph.D. thesis, Ohio State University, Columbus, Ohio, 347 pp.

Isbell, J. L., 1991. Evidence for a low-gradient alluvial fan from the palaeo-Pacific margin in the Upper Permian Buckley Formation, Beardmore Glacier region, Antarctica. In: M.R.A. Thomson, J.A. Crame and J.W. Thomson (Editors), Geological evolution of Antarctica. Cambridge University Press, Cambridge, pp. 215–217.

Isbell, J. L., Seegers, G.M. and Gelhar, G., in press. Stratigraphy of Upper Carboniferous and Permian rocks exposed between the Byrd and Nimrod Glaciers. Antarctic Journal of the U.S., 29.

Kemp, E.M., 1975. The palynology of Late Paleozoic glacial deposits of Gondwanaland. In: K.S.W. Campbell (Editor), Gondwana Geology. Australian National University Press, Canberra, pp. 397–413.

Kyle, R.A. and Schopf, J.M., 1982. Permian and Triassic palynostratigraphy of the Victoria Group, Transantarctic Mountains. In: C. Craddock (Editor), Antarctic Geoscience. University of Wisconsin Press, Madison, WI, International Union of Geological Sciences, Series B-4, pp. 649–659.

La Prade, K.E., 1972. Permian-Triassic Beacon Supergroup of the Shackleton Glacier Area, Queen Maud Range, Transantarctic Mountains. In: R.J. Aide (Editor), Antarctic geology and geophysics. Universitetsforlaget, Oslo, pp. 373–378.

Lindsay, J.F., 1970. Depositional environments of Paleozoic glacial rocks in the central Transantarctic Mountains. Geological Society of America Bulletin, 81: 1149–1172.

Long, W.E., 1964. The stratigraphy of the Horlick Mountains. In: R.J. Adie (Editor), Antarctic geology. North-Holland Publishing Company, Amsterdam, pp. 352–363.

Long, W.E., 1965. Stratigraphy of the Ohio Range, Antarctica. In: J.B. Hadley (Editor), Geology and Paleontology of the Antarctic. American Geophysical Union, Antarctic Research Series, 6: 71–116.

Miller, J.M.G., 1989. Glacial advance and retreat snequences in a Permo-Carboniferous section, central Transantarctic Mountains. Sedimentology, 36: 419–430.

Miller, M.F. and Collinson, J.W., 1994. Late Paleozoic post-glacial inland sea filled by fine-grained turbidites: Mackellar Formation, central Transantarctic Mountains. In: M. Deynoux, J.M.G. Miller, E.W. Domack, N. Eyles, I.J. Fairchild and G.M. Young (Editors), Earth's glacial record. Cambridge University Press, Cambridge, pp. 215–233.

Minshew, V.H., 1967. Geology of the Scott Glacier and Wisconsin Range areas, Central Transantarctic Mountains, Antarctica. Ph.D. thesis, Ohio State University, Columbus, Ohio, 268 pp.

Powell, C. McA. and Li, Z. X., 1994. Reconstruction of the Panthalassan Margin of Gondwanaland. In: J.J. Veevers and C. McA. Powell (Editor), Permian-Triassic Pangean Basins and foldbelts along the Panthalassan Margin of Gondwanaland. Geological Society of America, Memoir 184: pp. 5–10.

Storey, B.C., 1992. The palaeo-Pacific margin of Antarctica: an active boundary from Carboniferous to Jurassic times?. Geological Society of America, Abstracts with Programs, 24: p. 192.

Taylor, E.L., Taylor, T.N. and Cúneo, N.R., 1992. The present is not a key to the past: a polar forest from the Permian of Antarctica. Science, 257: 1675–1677.

Webb, P.N., 1990. The Cenozoic history of Antarctica and its global impact. Antarctic Science, 2: 3–21.

Webb, P.N. and Harwood, D.M., 1991. Late Cenozoic glacial history of the Ross Embayment, Antarctica. Quaternary Science Review, 10: 215–223.

Zawiskie, J.M., Collinson, J.W. and Hammer, W.R., 1983. Trace fossils of the Permian-Triassic Takrouna Formation, northern Victoria Land, Antarctica. In: R.L. Oliver, P.R. James and J.B. Jago (Editors), Antarctic Earth Science. Australian Academy of Science and Cambridge University Press, Cambridge, pp. 215–220.

14. SOME PROBLEMS OF THE PERMIAN (ASSELIAN) GLACIATION AND THE SUBSEQUENT CLIMATE IN THE PERMIAN

J. M. Dickins

Despite much study of and widespread evidence from the Asselian, the nature and even the geographical extent and duration of the glaciation in what is broadly the present southern hemisphere (often referred to as Gondwanaland or the Gondwana Region) remains obscure and controversial. In Australia after the Asselian (the lowermost stage of the Permian), a number of strong fluctuations in climate appear to have occurred. In most parts of the world, the Upper Permian was quite warm, and the Triassic appears to have been universally hot and warm. These problems have been considered in Dickins (1985, 1993, in press a). In addition, the nature of Gondwanaland and possibly even its existence in the Lower Permian is unclear (Dickins in press b). Suess (1985) identified Tethys from Mesozoic marine deposits stretching across southern Asia, including the Himalayan Region. To the north he identified a large landmass, Angaraland, and to the south another large landmass, Gondwanaland. A Tethys Ocean in the sense of Suess cannot at present be identified in the Lower Permian, but its existence from the mid-Permian on into the Mesozoic is clear. In the uppermost Carboniferous, land stretched from across Arabia and India to the north with its shore on an ocean across central Asia. No Tethys in the sense of Suess existed. This central Asian ocean closed during the Lower Permian, and the nature of the Lower Permian marine basins remain obscure having been much disrupted by mid-Permian tectonic activity (Dickins 1986, in press c). The occurrence of Lower Permian glacials in Tibet, apparently north of the Upper Permian shore of Gondwanaland, is well established and poses an interesting problem (Dickins et al., 1993).

The climate of the Lower Carboniferous is generally regarded as having been warm and equable although in South America some glaciation may have developed in the Visean. In the latest Lower Carboniferous and perhaps in the beginning of the Upper Carboniferous some glaciation was present in South America and in eastern Australia. Generally this is regarded as limited mountain glaciation but Veevers and Powell (1987) have postulated the existence of ice sheets at this time. A warm period may have intervened before the widespread glaciation of the Asselian (Gonzalez, 1990). The Asselian glaciation is known in Antarctica, South America, southern Africa, Oman (Arabian Peninsula), Tibet, Peninsular and Himalayan India, and western and eastern Australia. These constitute a broadly "southern" cold region. Four interglacials have been recognized in western Australia. The nature of the glaciation remains a matter for debate-how much was a reflection of high altitude and temperate conditions and how much was caused by a continental or polar ice sheet, for example. No corresponding "northern" glacial cold region is known and most of the rest of the world had a warmish climate, judging from the marine faunas, there may have been a somewhat colder region in northeastern Siberia but without established glacial stages. This distribution is quite puzzling and fits neither the present world climate nor any of the popular reconstructions.

The marine fauna accompanying the glacials and occurring immediately afterward in the Tastubian (Lower Sakmarian) associated with a eustatic rise in sea level is a low-diversity fauna known commonly as the *Eurydesma* fauna. This fauna is found in all the regions of glaciation with the exception of Antarctica and Oman, where marine deposits of this age are not known, but, puzzlingly, not in the Paraná Basin of Brazil or, apparently, in western Argentina. This marine fauna and its age and relationships have been reviewed by Archbold (1982) and Archbold et al.

TABLE 14.1. *Occurrence of the Tastubian marine faunas of Gondwanaland followed by the subsequent warmer water faunas*

SERIES/STAGE		WESTERN ARGENTINA	OMAN
LOWER PERMIAN	ARTINSKIAN		
	STERLITAMAKIAN (Substage)	"ORIOCRASSATELLA FAUNA"	HAUSHI
	SAKMARIAN		
	TASTUBIAN (Substage)	" CANCRINELLA FAUNA"	
	ASSELIAN		

(in press). Although the *Glossopteris* flora does occur in all these areas its occurrence seems enigmatic, and it is not confined to these regions.

In the Sterlitamakian (Upper Sakmarian) a distinct change in fauna occurs. The new fauna was first recognized in western Australia and was marked by the entry of many forms with warmer water "Tethyan" relationships. Although it may be recognizable in Queensland, generally the eastern Australian faunas developed from the earlier fauna and do not show the same Tethyan influence. This type of fauna has now been identified in Peninsular and Himalayan India, Tibet, Oman, and western Argentina. It may be present in Timor and Thailand and from largely unpublished information, may also occur in the Karakorums, Afghanistan and the Pamirs. Although in western Australia it was originally thought to suggest a cool temperate climate because of hangovers from the earlier fauna, it may imply warm temperate conditions; in Oman and perhaps India the climate may have even been subtropical. Because of its western Australian occurrence it has usually been considered to be a Gondwana fauna, but perhaps it should be considered as "Tethyan" and its occurrence in other "northern" parts of the world expected. Be that as it may, it does suggest a very widespread warmer climatic change.

In the Lower Permian of western and eastern Australia a number of changes in climate are now apparent, as indicated by changes in the fauna as well as by other features, such as, especially in eastern Australia, the coincidence of dropstones and glendonites, indicating colder periods, although glacial features such as moraine with striated clasts have not been identified. The glendonites are considered to indicate the presence of cold water, probably not above 5°C, and the dropstones might suggest floating ice. The warmer periods are indicated by kaolin and bauxite formation during chemically active weathering and luxuriant vegetation. As many as three such fluctuations have been recorded in the Lower Permian after the Sakmarian and one or more in the Upper Permian.

The climate of the Permian has quite unusual aspects. The assymetry is not easy to understand. It may be related to the distribution of land and sea, but it may also represent an effect of the planetary and galactic systems. The reasons for the fluctuation in climate are equally obscure, taking into account that they occur on a considerably longer time scale than the Milankovitch cycles.

References

Archbold, N.W., 1982. Correlation of the Early Permian faunas of Gondwana: implications for the Gondwanan Carboniferous-Permian boundary. Journal of the Geological Society of Australia, 29: 267–276.

TIBET	HIMALAYAN INDIA	PENINSULAR INDIA	TIMOR	WESTERN AUSTRALIA	EASTERN AUSTRALIA
				HIGH CLIFF (Perth Basin) & ONE GUM (Carnarvon Basin)	
"ORIOCRASSATELLA FAUNA"	UPPER AGGLOMERATIC (Kashmir)	BADHAURA	"BISNAIN FAUNA"	CALLYTHARRA (Carnarvon Basin)	STANLEIGH (Lower part Bowen Basin,Qld)
	GARU (Eastern Himalayas)	BAP UMARIA			
"EURYDESMA FAUNA"	LOWER AGGLOMERATIC (Kashmir)	MANENDRAGARH & RAJHARA		CARRANDIBBY (Carnarvon Basin)	ALLANDALE (Sydney Basin,NSW)

Archbold, N.W., Shah, S.C. and Dickins, J.M., in press. Peninsular Indian Early Permian brachiopod faunas and their Gondwanan relationships. Historical Biology.

Dickins, J.M., 1985. Late Palaeozoic glaciation. BMR Journal of Australian Geology and Geophysics, 9: 163–169.

Dickins, J.M., 1986. Tethys-a geosyncline formed on continental crust? In : K.G.McKenzie (Editor), Shallow Tethys 2. A.A.Balkema, Rotterdam, pp. 149–158.

Dickins, J.M., 1993. Climate of Late Devonian to Triassic. Palaeogeography, Palaeoclimatology, Palaeoecology, 100: 89–94.

Dickins, J.M., in press a. Problems of Upper Palaeozoic glaciation in Australia and subsequent climate in the Permian. Palaeogeography, Palaeoclimatology, Palaeoecology.

Dickins, J.M., in press b. The southern margin of Tethys. Ninth Gondwana Symposium, Hyderabad, India, 1994.

Dickins, J.M., in press c. The mid-Permian: major changes in geology, environment and faunas and some implications. Cambridge University Press, World and Regional Geology Series.

Dickins, J.M., S.C.Shah, Archbold, N.W., Jin Yugan, Liang Dingyi and Liu Benpei, 1993. Some climatic and tectonic implications of the Permian marine faunas of Peninsular India, Himalayas and Tibet. In: R.H. Findlay, R. Unrug, M.R. Banks and J.J. Veevers (Editors), Gondwana Eight: Assembly, evolution and dispersal. A.A. Balkema, Rotterdam, pp. 333–343.

Gonzalez, C.R., 1990. Development of the Late Paleozoic glaciations of the South American Gondwana in western Argentina. Palaeogeography, Palaeoclimatology, Palaeoecology, 79: 275–287.

Suess, E., 1995. Das Antlitz der Erde, V.I. Temsky, Wien, 779 pp.

Veevers, J.J. and Powell, C.McA., 1987. Late Paleozoic glacial episodes in Gondwanaland reflected in transgressive-regressive depositional sequences in Euamerica. Geological Society of America Bulletin, 98: 475–487.

PART IV

AN OLDER PRECAMBRIAN GLACIATION: A Comparative Analysis

15. TECTONIC AND GLACIOEUSTATIC CONTROLS ON POSTGLACIAL STRATIGRAPHY: PROTEROZOIC EXAMPLES

Grant M. Young

The fact that so many major cities, both ancient (Alexandria, London) and modern (New York, San Francisco) have been built at or close to sea level attests to both the importance of maritime commerce and our continuing faith in the permanence of the relationship between land and sea. Such faith is surprising in view of the many mythical and religious allusions to ancient floods (the biblical flood) and drowned lands (Atlantis). Lyell, the great promoter of Hutton's uniformitarian ideas, used borings by marine organisms around the now high and dry pillars at Pozzuoli, near Naples, to imply significant changes in sea level in historic times. It now seems much more likely that there have been changes in elevation of the land in this region related to local volcanic activity, rather than changes in sea level (Ager, 1993) but the message is clear, we should be aware of the possibility of relatively rapid and significant sea-level changes.

Available records, based largely on elevations of dated corals in tectonically stable regions of the world and on radiocarbon dating of peats and other remains of land plants that are now below sea level, clearly indicate that there have been dramatic variations in sea level in the past 18,000 years or so, changes estimated at between about 100 and 150 m. These changes appear to correspond rather closely with isotopic variability (particularly stable isotopes of oxygen) in calcareous microfossils. Such isotopic variations are thought to reflect both ice volume and ambient water temperatures. It is estimated that just over 2 percent of the Earth's water budget is currently locked up in the form of ice and that melting of that ice would result in a sea-level rise of 60 to 80 m. Thus the range of sea level change related to ice volume from the peak of the most recent glaciation to complete meltdown is of the order of 150 to 200 m. Perhaps the most striking aspect of the recent glacioeustatic sea-level rise is its rapidity. A rough figure of about 10 mm/year is estimated for the most recent deglaciation. There are several versions of this curve, reflecting the uncertainties in its construction, but in general the rate of sea-level change appears to have been greatest at the beginning of the collapse of the ice sheets about 15,000 years ago and has gradually slowed down to about 1mm/year for the last century or so. During the last 8,000 years or so, the subsequent effects appear to be very location-specific. For example, when the ice sheets disappeared in northern parts of North America, relative sea level (RSL) was 10–30m higher because of isostatic depression; it has been dropping at an exponentially decreasing rate ever since. To the south of a so-called "hinge line" in central New England, RSL was 10–20m lower than it is today, and it has risen, again at an exponentially decreasing rate. Some have suggested that sea-level rise has actually ceased and that current changes are due to isostatic readjustment to the disappearance of continental glaciers or to tectonic activity quite unrelated to glaciation. In any event, the postglacial sea level rise is among the most dramatic large-scale geological processes and is orders of magnitude faster than other controls on sea level, such as variation in oceanic ridge size. Although there are myriad controls on sea level, such as thermal expansion of sea water, variations in the absolute amount of water present on the surface of the planet, and local tectonic effects, the most important factors in controlling sea level are probably the waxing and the waning of glaciers and global tectonics.

The Permo-Carboniferous glaciation is comparable to that of the Proterozoic in the sense that it appears to have occurred during an episode of supercontinentality. Eyles and Young (1994) suggested that preservation of late Paleozoic glacial sediments in Gondwana may have been the result of development of a series of rift basins. The ultimate cause of this major glacial episode may have been lowering of atmospheric CO_2, possibly related to supercontinentality, but the polar positioning of Gondwana was also an important factor, and the widespread preservation of biomass at this time in the form of coal may have contributed to the drawdown of CO_2 from the atmosphere. There is a significant time gap (almost 100 Ma) between the end of glaciation and the breakup of the Gondwana supercontinent. Evidence from paleomagnetism (Smith et al., 1981) suggests that the supercontinent gradually migrated out of polar latitudes during this period, and this may have played an important role in collapse of the ice sheets. Thus many analogies can be drawn between the Proterozoic and the late Paleozoic glaciations, but there are also differences relating to the paleolatitude.

The late Tertiary to Pleistocene glaciations occurred under a plate-tectonic regime very different from that of most of the preceding major glaciations. These glaciations occurred when much of the continental crust was fragmented and widely dispersed. Clearly the positioning of continental masses at high latitudes (specifically, Antarctica) played an important role, as did special topographic configurations that influenced global oceanic circulation (Broecker and Denton, 1989). Ruddiman et al., (1989) suggested that accelerated elevation of major areas of continental crust such as the Tibetan Plateau, the Himalayas, and the American Cordilleras may have played a major role in triggering the Late Cenozoic glaciations. One feature common to the Neoproterozoic, the late Paleozoic and Tertiary-Recent glaciations, however, is that they all occurred during periods of continental emergence. Such a configuration, enhanced by orogenic uplift, could have contributed to accelerated weathering and reduction in atmospheric CO_2, which may have been the ultimate cause of global cooling. Thus the global climate may be directly linked to plate-tectonic phenomena-specifically, the supercontinental cycle.

The nature of the scientific data available from the Tertiary-Recent glaciation is much more detailed, and interpretations are correspondingly complex. For ancient glaciations, much of the evidence is lost, and the interpretations are correspondingly naive. On the other hand, the ancient glaciations, unlike the present one, have run their course and the record of entire glacial periods is available to us. It seems certain that significant sea-level changes related to changing ice volume must have occurred from time to time throughout geological history. Such rapid changes should have left a diagnostic imprint on the stratigraphic record. The purpose of this chapter is to look for evidence of such changes as recorded in the stratigraphy of some Proterozoic postglacial successions and to attempt to decipher tectonic and glacioeustatic influences.

The Paleoproterozoic (Huronian) Glaciation

The Huronian rocks of the north shore of Lake Huron were among the first studied by the newly formed Geological Survey of Canada in the middle of the ninenteenth century. By the early 1900s a glacial origin was generally accepted for the paraconglomerates (diamictites) and associated rocks of the Gowganda Formation (Coleman, 1914), and considerable effort has been devoted to documenting the stratigraphy and sedimentology of these deposits (Thomson, 1957; Casshyap, 1969; Lindsey 1969; Young, 1970, 1973; Miall, 1985; Young and Nesbitt, 1985; Rainbird and Donaldson, 1988).

It has been suggested that similar Paleoproterozoic glacigenic diamictites in other areas may be contemporaneous and may thus provide a mean of time correlation over a large area of North America and possibly elsewhere (Young, 1970; Ojakangas, 1988; Strand and Laajoki, 1993). The distribution of Paleoproterozoic glacigenic rocks in North America is shown in Figure 15.1. A glacial origin has been proposed for Paleoproterozoic diamictites in southeastern Wyoming, in several areas of Michigan, in the Chibougamau area of northern Quebec, and in the Keewatin District on the west side of Hudson Bay. The age and the glacial interpretation of some of these deposits have been questioned, but the evidence for a widespread Paleoproterozoic glacial event or events in North America remains impressive (Young, 1970).

Outside of North America, Paleoproterozoic glacigenic rocks are known from Karelia, Wes-

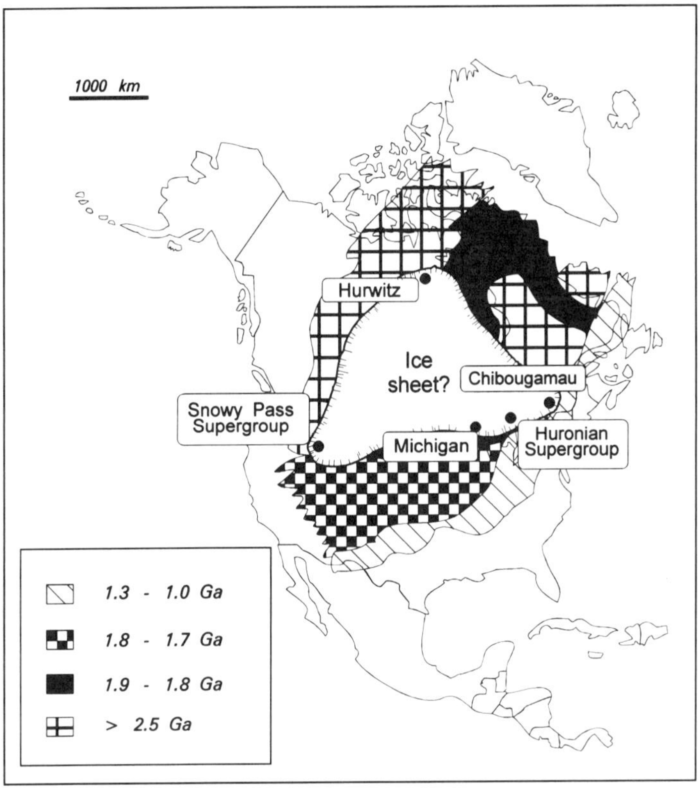

Fig. 15.1. Sketch map showing the location of Paleoproterozoic glacigenic deposits in North America in relation to tectonic provinces (after Hoffman, 1988), the ages of which are shown in the inset. The hypothetical ice sheet would have been less extensive if the trans-Hudson ocean (black) were closed (after Young, 1970).

tern Australia, in the Meteorite Bore Member and in the Griqualand Formation of South Africa.

Gowganda Formation

The Gowganda Formation is part of the Huronian Supergroup on the north shore of Lake Huron (Fig. 15.2). It is the lowest formation of the upper Huronian (Fig. 15.3). The lower part of the Gowganda Formation in the southern part of the Huronian outcrop belt was interpreted in terms of two major glacial advances, separated by an interglacial period when a thick, widespread mudstone was deposited (Lindsey, 1969; Young and Nesbitt, 1985). The second glacial episode was complex, involving several diamictite-rich members separated by argillites and sandstones (Young, 1981; Young and Nesbitt, 1985).

The glacigenic deposits were succeeded by a series of coarsening upward sequences (Fig. 15.3) that have been interpreted as deltaic cycles (Lindsey, 1969; Junnila, 1986; Rainbird and Donaldson, 1988). In the southern part of the Huronian outcrop belt near Whitefish Falls there are four or five such cycles with a composite maximum thickness of almost 1000 m. In this region the contact between the lower and the upper (informal names) Gowganda Formation is taken at the top of the last regionally extensive diamictite. The first coarsening upward sequence, however, contains some evidence of continuing glacial influence in the form of a westerly-thinning diamictite near the base (Young and Nesbitt, 1985).

The individual cycles comprise laminated mudstones, overlain by ripple cross-laminated and wavy bedded siltstone-mudstone units with sporadic intercalations of thicker, faintly graded layers. Some of these massive beds contain rip-up clasts of laminated and wavy bedded mudstones. The upper parts of the coarsening upward sequences are dominated by fine sandstones displaying cross-bedding, ball and pillow structures, and ripple cross-laminations. These vertically stacked facies have been interpreted as

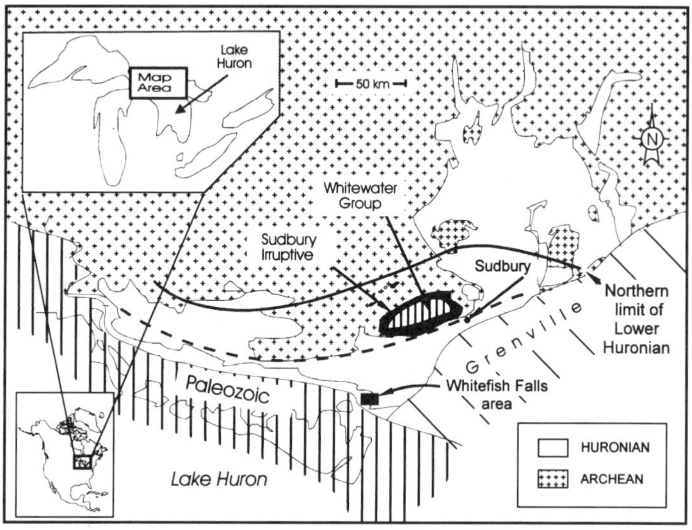

Fig. 15.2. Sketch map showing the distribution of the Huronian Supergroup on the north shore of Lake Huron. Note the line marking the northern limit of the Lower Huronian (pre-Gowganda Formation).

prodelta, delta slope and delta top deposits (Junnila, 1986). Mapping, Junnila (1986) has shown that many of these coarsening upward units are lenticular as displayed in east-striking rocks on the north limb of the McGregor Bay anticline. Paleocurrent measurements suggest dominant transport to the southeast. In the last 5,000 years the Mississippi River has produced several deltaic lobes comparable in size and thickness to those in the upper Gowganda Formation. It has been suggested that the lack of land-plant cover in the Proterozoic could have influenced the nature of the river systems, but the fining upward fluvial cycles overlying the deltaic deposits of the upper Gowganda Formation are similar in scale and in nature to classical meandering river deposits from the Devonian of the United Kingdom.

In the northeast (Cobalt area) and northwest (Bruce Mines-Elliot Lake area), parts of the Huronian outcrop belt, attempts to establish regional subdivisions of the Gowganda Formation have been largely frustrated by a combination of shallow dips and poor exposure. In the Cobalt area Thomson (1957) proposed subdivision of the Gowganda Formation into two units that he called the Coleman and the Firstbrook form-

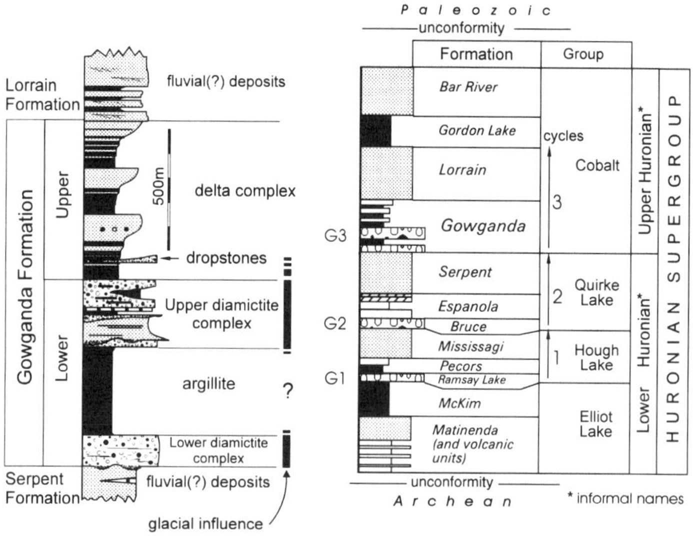

Fig. 15.3. A generalized stratigraphic column for the Huronian Supergroup (maximum thickness about 12 km) in the southern part of the Huronian outcrop belt is shown on the right side of the diagram. The left column is a generalized stratigraphic column for the Gowganda Formation at Whitefish Falls (after Young and Nesbitt, 1985). G1—G3 refers to the three glacial episodes recorded by diamictites in the Huronian Supergroup. For meaning of symbols, see Figure 15.9.

ations. Rainbird and Donaldson (1988), wishing to retain the name Gowganda Formation for these rocks, proposed that the two subdivisions be given member status. The upper Firstbrook Member is highly variable in thickness, ranging from about 75 m to 800 m. It has been interpreted (Rainbird and Donaldson, 1988) as comprising a single coarsening upward sequence comparable to those developed at Whitefish Falls in the south. Rainbird and Donaldson (1988) considered the Firstbrook Member to represent a prograding river-dominated sedimentary wedge built out by a series of coalescing deltas from a braidplain into a shallow marine environment.

Detailed correlations between the upper Gowganda Formation in the Whitefish Falls area and the Coleman Member in the Cobalt area are not possible. If these successions represent a deglaciation sequence, then the accepted paleogeography suggests that the Whitefish Falls succession is, in part at least, somewhat older. As the ice melted and withdrew to the north, outwash deposits would have been transported in a generally southerly direction. One possibility is that the single coarsening upward sequence in the more northerly region is equivalent only to the uppermost sequence in the south. This is possibly true in a time sense, but the presence of dropstones in the lower part of the mudstone succession in the Cobalt area suggests glacial influence at the time of initiation of deltaic sedimentation there. Deglaciation was presumably diachronous, so the last indicator of glaciation could have occurred later in the north than in the south. The reasons for development of several coarsening upward sequences in the south and a single one in the north remain obscure.

Lorrain Formation

The transition from the Upper Gowganda Formation to the Lorrain Formation is gradational in the Whitefish Falls area. The mudstones of the uppermost deltaic sequence pass upward through siltstones and fine sandstones into a coarser grained sandstone-dominated succession characterized by the presence of cross-bedding (Fig. 15.3). The depositional environments of the Lorrain Formation have been the subject of some debate. The sandstones of the Lorrain Formation have been considered to be fluvial or shallow marine.

Results of measurement of a detailed section in the region north of Whitefish Falls are shown in Figures 15.2 and 15.4. The lower part of the Lorrain Formation in this area consists of a series of cross-bedded purple buff and grey sandstones, together with siltstones and mudstones arranged in fining upward cycles, comparable to those of meandering rivers.

The middle portion of the Lorrain Formation consists mainly of coarse grained, commonly cross-bedded buff, pink, and green sandstones with scattered, generally thin quartz pebble conglomerates and rare mudstones. These were described by Hadley (1968) as coarse delta fringe deposits but are considered here as the product of a broad coalescing series of braided rivers.

The upper Lorrain Formation is a thick succession of orthoquartzites with subordinate muddy laminae and beds, some of which are highly aluminous. The depositional environment of these supermature quartzites has also been contentious. Most authors have assumed that they formed in a shallow marine setting. Chandler (1986) considered the supermature mineralogy of the upper Lorrain sandstones to be a secondary, diagenetic phenomenon. In any event, most of the thick arenaceous succession of the Lorrain Formation was probably deposited in fluvial or shallow marine settings.

Plate-Tectonic Setting of the Huronian Supergroup

The distribution of the partially glacigenic Gowganda Formation (Figs. 15.2, 15.5) relative to the underlying Huronian Formations has been of particular significance in interpretations of the tectonic setting of the Huronian Supergroup (Young and Nesbitt, 1985). Throughout a large area north of the main Huronian foldbelt, the Gowganda Formation lies nonconformably on the Archean basement. This stratigraphic relationship is puzzling because a widespread glaciation might be expected to be associated with a regression. The explanation may lie in the dominance of tectonoeustatic over glacioeustatic processes at

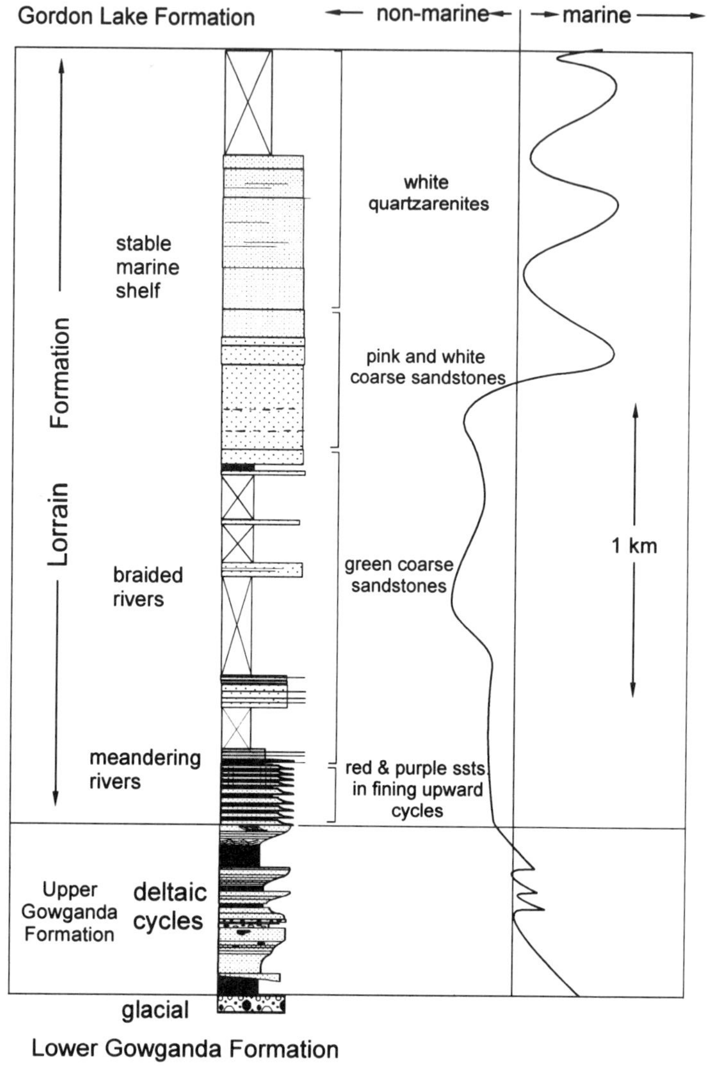

Fig. 15.4. A measured section through the Upper Gowganda and the Lorrain formations near Whitefish Falls at the south end of the Huronian outcrop belt (see Figure 15.2 for location).

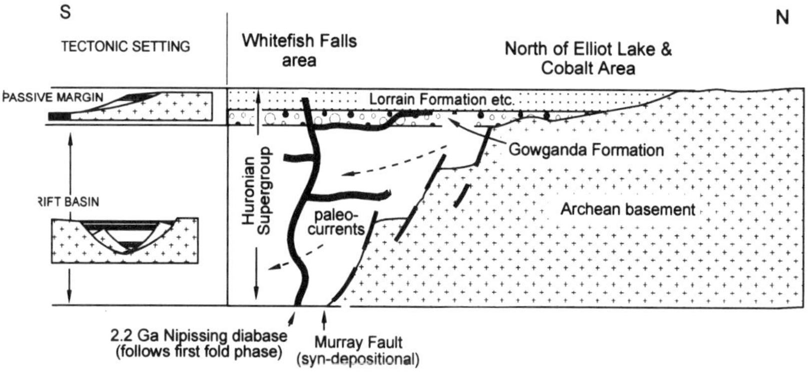

Fig. 15.5. Schematic representation of proposed tectonic setting during deposition of the Huronian Supergroup. The Lower Huronian is interpreted as having accumulated in a rift setting, whereas the Gowganda and the succeeding Huronian formations are considered to have been deposited on a newly formed passive margin.

the time of deposition of the lower part of the Gowganda Formation. Young and Nesbitt (1985) emphasized the role of rifting in deposition of the lower Huronian, and it has been suggested that the Gowganda was deposited during a rift-drift transition. Such an interpretation was based on the presence of rift-type volcanics in the lower Huronian (Card, 1979; Jolly, 1987), the fault-restricted and partially fault-controlled areal distribution of the lower Huronian rocks, and evidence of major subsidence along what is interpreted as a continental margin during deposition of the Gowganda Formation (Young and Nesbitt, 1985).

An interpretation of facies in the upper part of the Gowganda Formation and overlying Lorrain Formation is shown in Figure 15.4. The deltaic cycles of the upper Gowganda Formation are considered to have built out into a shallow marine environment during a brief period of possible glacioeustatic sea level rise. The thick (ca. 2,700 m) arenaceous sequence of the Lorrain Formation records the progradation of fluvial environments over a marine shelf, followed by a gradual return to shallow marine conditions. Deltaic cycles overlain by the thick sandstone-dominated succession of the Lorrain are seen as the products of deglaciation, occurring at some time after breakup of the Paleoproterozoic "supercontinent." The combination of rising sea levels and subsidence of the earlier-formed continental margin provided accommodation for an exceptionally thick pile of sands. Deposition of the dominantly glaciomarine lower Gowganda Formation probably took place during the rift-drift transition (Young and Nesbitt, 1985). The upper Gowganda deltaic apron represents the initial products of glacial meltwaters as the glaciers receded under conditions of glacioeustatic sea-level rise. The deltaic apron was overlain by a succession of felspathic sands as meandering and braided river systems prograded out onto the continental shelf. Eventually, during deposition of the upper part of the Lorrain Formation, a balance between sediment input and basin subsidence was achieved and the thick orthoquartzites of the Lorrain Formation formed. Production of such a thick succession of arenaceous rocks must have been accompanied by production of an even greater thickness of mud-grade sediments (Farey and Roscoe, 1970; Pettijohn, 1970) which were presumably carried southward, beyond the limits of the preserved ancient continental margin. A problem with the alternative suggestion of a diagenetic origin for the orthoquartzites is the removal of fine material derived from the alteration of feldspar. Such material should, theoretically, have had a much greater volume than the quartzites themselves.

The postglacial stratigraphy in the upper Huronian is somewhat similar to that in the Neoproterozoic (Dalradian) succession of Scotland and Ireland but is in marked contrast to the postglacial succession preserved above the Neoproterozoic Rapitan Group in the northern Canadian Cordillera and possible correlative successions (Bell and Jefferson, 1987; Young, 1992) in the Adelaide "Geosyncline" of south Australia.

The Neoproterozoic of the Northern Canadian Cordillera

Neoproterozoic glacial deposits are known throughout the length of the Cordillera from Alaska to Mexico (Fig. 15.6). Three major packages of Proterozoic supracrustal rocks were recognized in this large region (Young et al., 1979) and were informally designated, in ascending order, as Sequences A, B, and C. Glacigenic rocks commonly occur at or near what has traditionally been considered the base of the uppermost Proterozoic succession, Sequence C. Gabrielse and Campbell (1991) and Aitken (1993), however, also included an additional succession (maximum thickness of ca. 1,800 m) of clastics, carbonates and subordinate evaporites, known as the Coates Lake Group (Jefferson and Ruelle, 1986) or Copper cycle of Aitken (1993) in the Windermere Supergroup (Fig. 15.7), because these rocks appear to belong to spatially restricted facies associated with rift faulting attributed to the "Hayhook orogeny" (Young et al., 1979).

The Rapitan Group

The Rapitan Group is one of the most northerly occurrences of glacigenic rocks in the Cordillera.

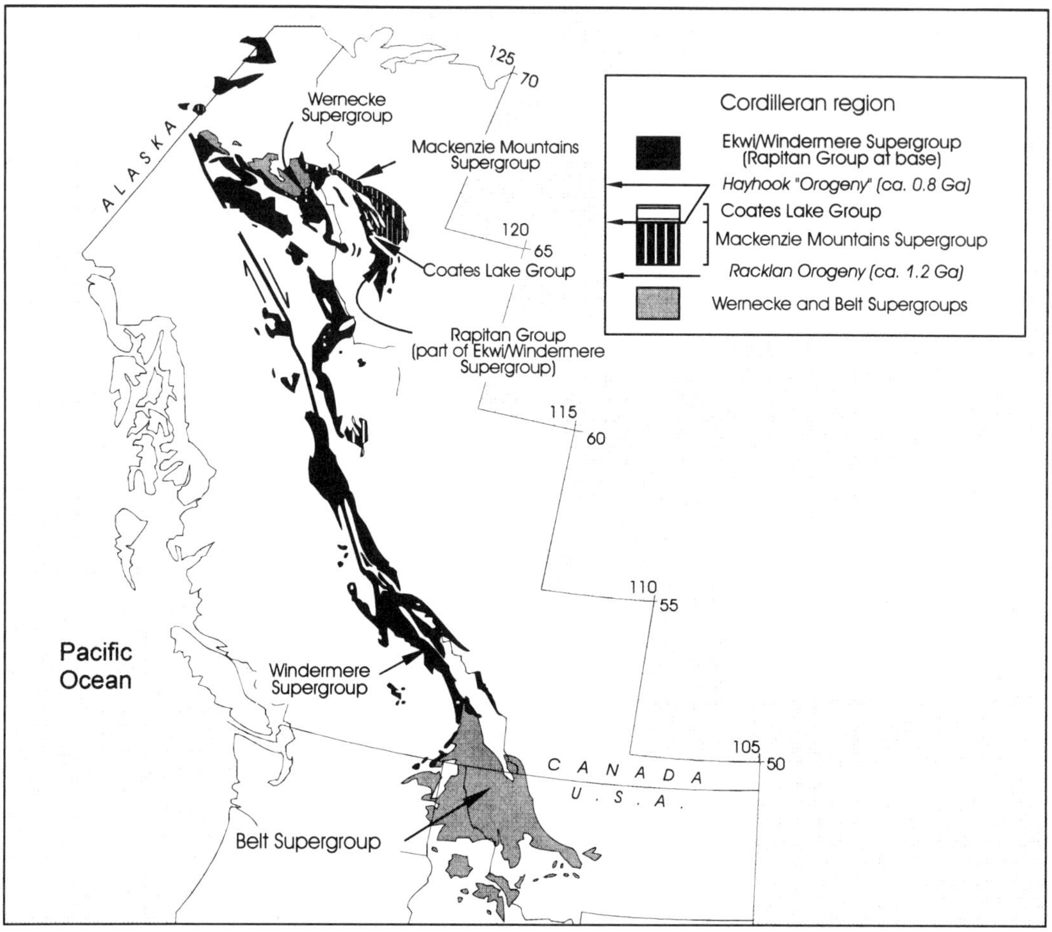

Fig. 15.6. Sketch map of the Proterozoic geology of the northern part of the North American Cordillera, showing the distribution of major stratigraphic units. Note that the Ekwi\Windermere Supergroup is much more widespread than the older units.

It conformably to unconformably overlies rocks ranging from the Little Dal Group to the Coates Lake Group (Fig. 15.7). Geochronological data suggest that the Windermere Supergroup in the northern part of the Cordillera is younger than 770–780 Ma.

The glacial nature of the Rapitan Group has been documented by Eisbacher (1978) and Yeo (1981). A locally developed diamictite known as the Mount Berg Formation is followed by a thick succession of purple and red rhythmites with scattered dropstones, known as the Sayunei Formation. Thc upper part of this unit and the lower part of the succeeding Shezal Formation contain major iron-formations (Yeo, 1981, 1986; Klein and Beukes, 1993). The iron-formation has been ascribed by Yeo (1981) to hydrothermal activity in a rift setting. The majority of the Shezal Formation consists of crudely stratified diamictites with minor intercalations of laminated and ripple cross-laminated silty mudstones. This succession has been interpreted as a series of glaciomarine deposits formed in a rift environment (Eisbacher, 1981; Young, 1984; Aitken, 1993).

Plate-Tectonic Setting of the Rapitan Group

Whereas the Rapitan Group has a restricted distribution, the overlying formations are widespread and have relatively uniform facies. There

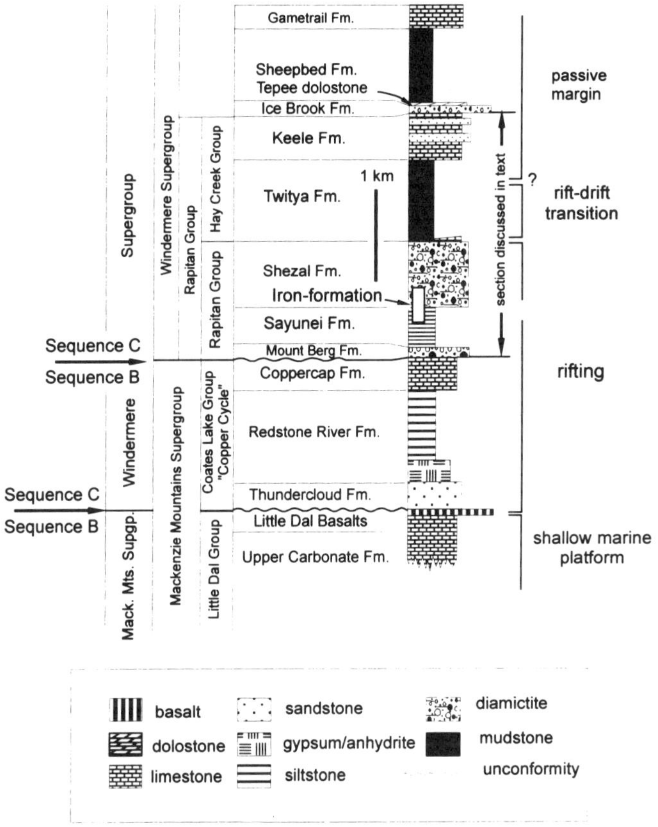

Fig. 15.7. Generalized stratigraphic section showing part of the Neoproterozoic stratigraphy in the Mackenzie Mountains region of the northern Cordillera.

is no evidence of restriction to rift-controlled basins. The Twitya Formation, which overlies the Shezal Formation (Fig. 15.7), is a thick (up to 900 m) succession of dark grey mudstones with minor intercalations of silty and sandy turbidites. The Twitya Formation shoals upward into the carbonate-dominated platformal sequence known as the Keele Formation. In many parts of the world a new style of sedimentation characterized by shallowing upward "grand cycles" (Aitken, 1966) was introduced at this point in Neoproterozoic history and continued into the Cambrian. The widespread and homogeneous nature of these postglacial units has been interpreted to mean the end of the rift phase that characterized deposition of the Coates Lake and the Rapitan groups. The shoaling upward cycles probably developed on a passive margin (Young, 1984; Ross, 1991; Aitken, 1993). These ideas have been summarised by Ross (1991) and by Dalziel et al. (1994), who agreed with the earlier conclusion (Stewart, 1972; Young, 1984) that initial rifting of the Pacific margin of North America took place in Neoproterozoic time (ca. 750 Ma) and that the Cambrian episode possibly represented additional subsidence and/or fragmentation of an already established continental margin.

Figure 15.7 shows a suggested rift-drift transition during deposition of the Twitya Formation immediately above the glacigenic Shezal Formation. The stratigraphic succession in the Neoproterozoic postglacial succession in the northern Cordillera is markedly different from that in the older Huronian Supergroup. There are no deltaic sediments, and in the stratigraphic position occupied by the thick quartzites of the Lorrain Formation there is a thick mudstone-dominated succession displaying evidence of gradual upward shallowing. These features suggest significant deepening of the basin, which became starved of coarse clastics until progradation of the shallow water deposits of the Keele Formation took place and eventually a second cycle of glaciation began (Aitken, 1991).

One possible explanation for the rapid and significant relative sea-level rise is that it is related to the rift-drift transition, involving faulting, thinning and subsidence of the newly formed continental margins of the proto-Pacific Ocean. There is a remarkably similar Neoproterozoic stratigraphy in southeast Australia, which has been considered by a number of workers (Young, 1984; Eisbacher, 1985; Bell and Jefferson, 1987; Dalziel, 1991; Hoffman, 1991; Moores, 1991; Brookfield, 1993) to have been juxtaposed, together with Antarctica and India, against what is now the western margin of North America (Fig. 15.8). In the south Australian Neoproterozoic succession (Young, 1992) the equivalent unit to the deep water shales of the Twitya Formation is the Tapley Hill Formation, which transgresses onto the Gawler craton to the west of the Adelaide "Geosyncline." The effects of postglacial (post-Rapitan in Canada, post-Sturtian in Australia) eustatic sea-level rise would also have been superimposed on these tectonically controlled sea-level changes. The nature of the newly formed continental margin may also have been important. Aitken and Long (1978) suggested, on the basis of shape of isopachs, that a large embayment existed in the Mackenzie Mountains region at the time of deposition of the Mackenzie Mountains Supergroup. This idea was extended by Hansen and others (1993), who proposed that the Mackenzie "embayment" was due to the presence there of a "lower plate segment," which would have been particularly susceptible to subsidence during breakup.

In summary, the presence of a thick mudstone succession above the glacigenic deposits of the Rapitan Group may be due in large part to plate-tectonic phenomena, superimposed on a post-glacial eustatic sea level rise. The timing of the end of glaciation *vis-à-vis* the timing of the rift-drift transition may provide an explanation for the difference between the postglacial stratigraphy in the Neoproterozoic Windermere Supergroup in the northern Cordillera and that of the post-Gowganda Paleoproterozoic succession in Ontario. In the case of the Huronian, major glaciation appears to have largely postdated the rift-drift transition (Young and Nesbitt, 1985), whereas in the case of the Rapitan, the glaciation appears to have terminated with the rift phase, so the effects of tectonic subsidence during breakup were superimposed on postglacial eustatic sea-level rise, resulting in a major transgression and deposition of a thick succession of fine-grained siliciclastic rocks (Twitya and Tapley Hill formations). The significant deepening of the basin may have been exacerbated by positioning of the Mackenzie Mountains "embayment" on an "upper

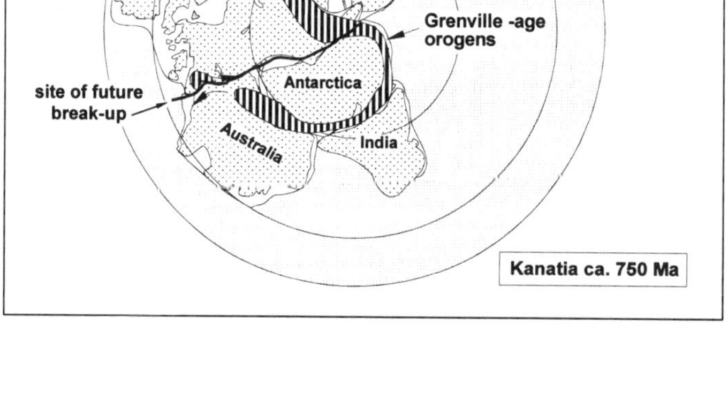

Fig. 15.8. Reconstruction of a Late Proterozoic supercontinent (Kanatia) (reconstruction after Moores, 1991; Dalziel, 1991; Hoffman, 1991). The name "Rodinia" has been used for this reconstruction, but Young (1995), following the original definition of McMenamin and McMenamin (1990), proposed that this name be reserved for the latest Neoproterozoic supercontinent. Lines of latitude are for scale only and are not meant to indicate paleolatitudes. Note the location of future continental breakup, considered to have been contemporaneous with deposition of the Twitya Formation (Fig. 15.7).

plate" (Hansen et al., 1993). If the eustatic component of sea-level rise was important at this time, it opens up the possibility of global correlations based on correlation of postglacial successions in the Neoproterozoic.

The Neoproterozoic Dalradian of Scotland and Ireland

Glacigenic rocks have long been recognized in the Neoproterozoic Dalradian Supergroup of Scotland. The early history of their discovery and description are given in Spencer's (1971) memoir which includes a detailed description and interpretation of the Port Askaig Formation on some small islands (the Garvellachs) on the west coast of Scotland. The Dalradian Supergroup is a thick (> 20 km) succession of mainly sedimentary rocks that has undergone complex folding and metamorphism. A generalized stratigraphy of the Dalradian Supergroup (after Anderton, 1985) is shown in Figure 15.9. Some authors include a still older succession (Grampian Group) with the Dalradian, but Anderton (1985) chose to exclude these rocks. More recently Soper (1994a; 1994b) included the Grampian Group in the lower Dalradian and Soper and England (1995) considered the Dalradian succession, and possible partial time equivalents to the northwest in the Moine and Torridonian, to have been deposited in a long-lived extensional regime that culminated in the opening of the Iapetus Ocean. This section is concerned mainly with the glacial deposits of the Port Askaig Formation and the overlying postglacial succession, but a brief review of the regional setting is necessary to elucidate the tectonic setting.

Port Askaig Formation

The diamictites of the Port Askaig Formation form the base of the Argyll Group. The best exposures of the Port Askaig Formation are in western Scotland on the island of Islay and on a series of small islands known as the Garvellachs. Glacigenic rocks of presumed similar age are known from more easterly outcrops across the width of Scotland (Spencer, 1971), but in most of these areas they are poorly exposed and/or highly deformed. The Port Askaig Tillite also extends west into Ireland (Kilburn et al., 1965; Howarth, 1971; Spencer, 1971), where it is known from an essentially linear (NE-SW) series of exposures extending from northern Donegal to Connemara.

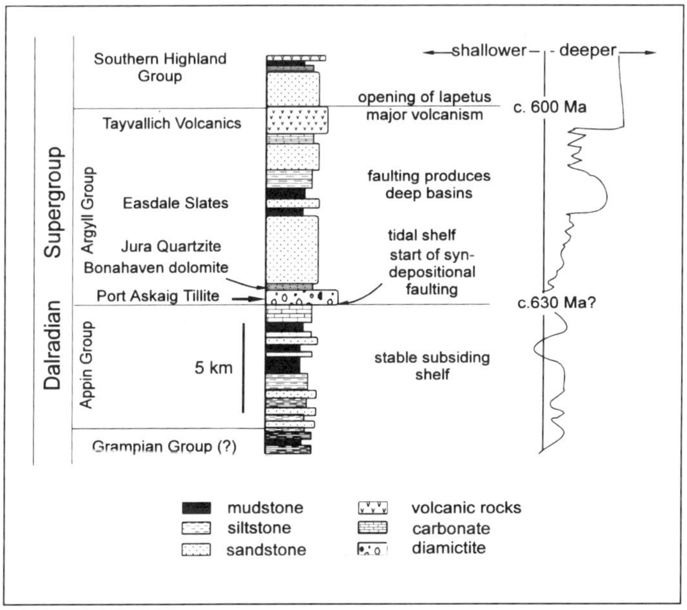

Fig. 15.9. Generalized stratigraphy of Neoproterozoic (Dalradian) stratigraphy in Scotland (after Anderton, 1982, 1985).

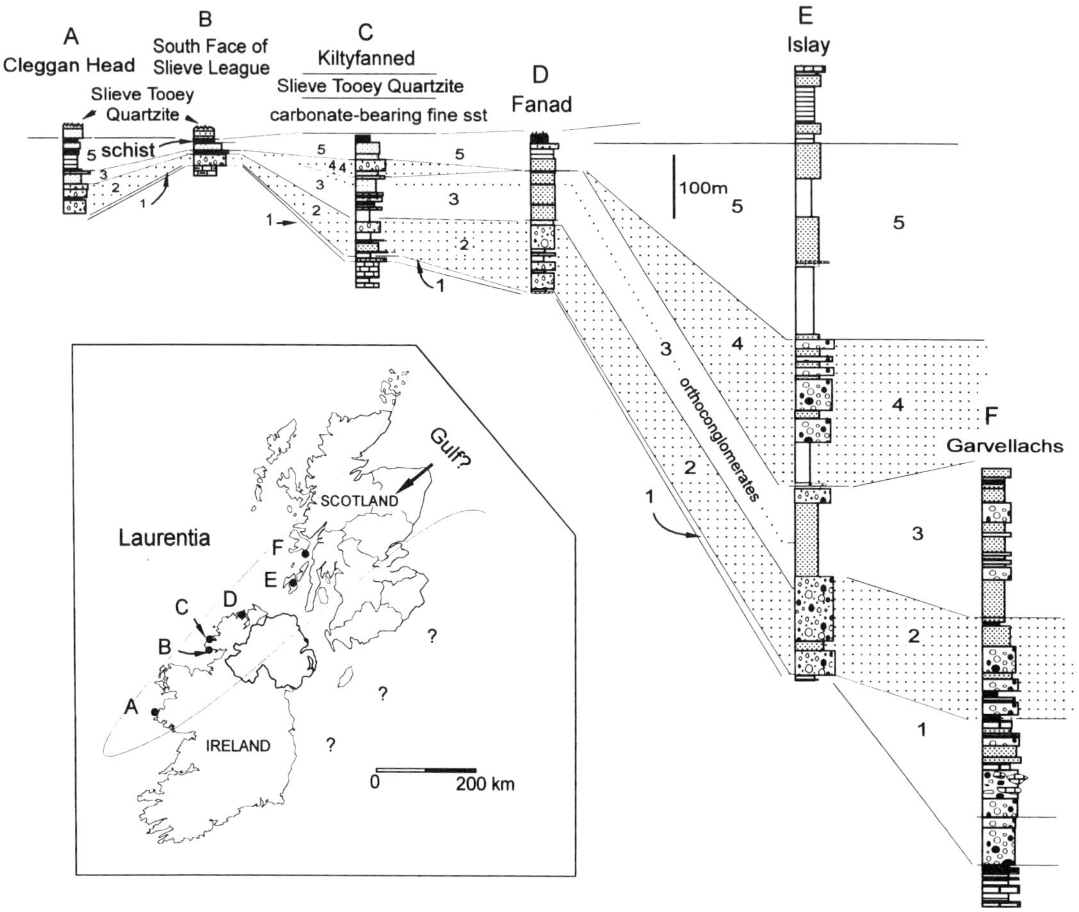

Fig. 15.10. Representative stratigraphic columns for the Port Askaig Formation from the west of Scotland to western Ireland. Columns E and F are after Spencer (1971). Patterns representing different rock types are shown in Figure 15.9. Dotted pattern between columns signifies stratigraphic intervals containing abundant diamictites.

A series of generalized measured sections from these areas (Fig. 15.10) shows a complex stratigraphy. The succession appears to consist of alternating diamictite- and sandstone-rich units on a scale of tens to hundreds of meters (Fig. 15.10).

Sea-Level Changes during Deposition of the Port Askaig Formation

A detailed study of the Garvellachs and the Islay successions (Spencer, 1971) resulted in recognition of cyclical sedimentation involving at least 17 diamictites, with intervening sandstone units. One such cycle is shown in Figure 15.11. Hambrey (1983) suggested that the complex succession of glacial and nonglacial rocks could be an expression of a climatic rhythm or oscillation on the scale of the Milankovitch cycle (tens to hundreds of thousands of years). The diamictites were interpreted by Spencer as being tills formed directly from glacier ice grounded in a shallow marine setting. The upper surfaces of many diamictite units (tillites, according to Spencer, 1971) are intruded by sandstone dykes or wedges that commonly display a polygonal pattern on the bedding surface. These structures were interpreted by Spencer (1971) as periglacial ice wedge structures. They are in turn overlain by orthoconglomerates, attributed to a gradual rise in sea level following deglaciation. The overlying cross-bedded mature sandstones were considered to be

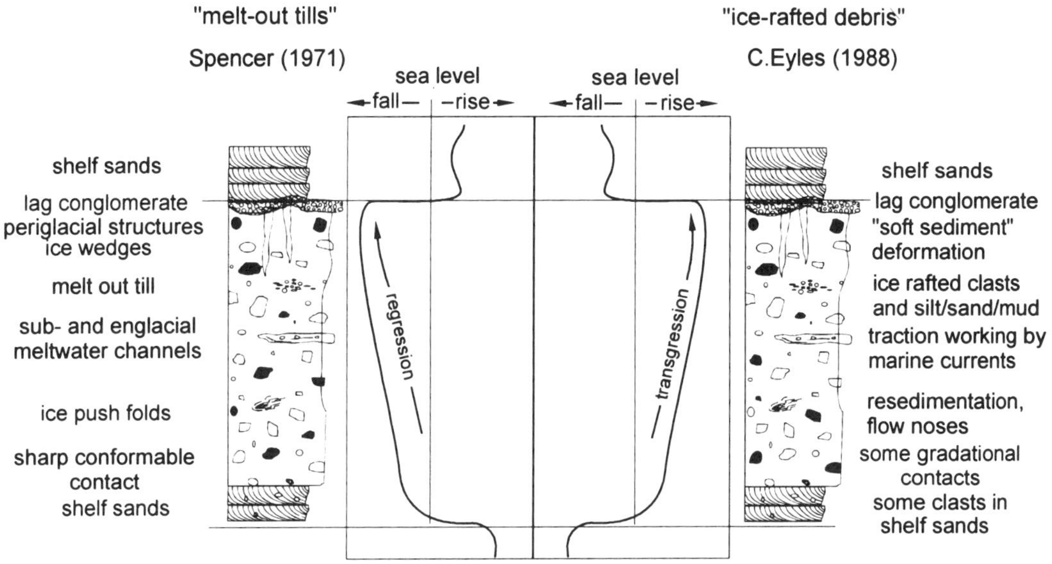

Fig. 15.11. Two contrasting interpretations of the alternating diamictite/sandstone units in the Port Askaig Formation on the Garvellachs Islands in western Scotland.

shallow marine shelf deposits, formed during marine transgression and in turn overlain by diamictite formed during glacial readvance. In 1988 C. Eyles reinterpreted the Garvellachs section as shown in the right side of Figure 15.11. She considered the diamictites to be glacial marine deposits formed by a combination of ice rafting and influx of fine material from meltwater streams. According to this scheme, the diamictites were formed during a rise in sea level. The "ice wedge" structures of Spencer (1971) were considered to be soft-sediment deformation structures formed as a result of "density inversion" due to sands being deposited over muddy diamictons (Eyles and Clark, 1985). The conglomerates capping many diamictites were interpreted as lag conglomerates formed during a fall in sea level, and the sandstones were seen as marine deposits, formed in shallower water than the diamictites. These two radically different interpretations of the Garvellachs sections are shown in Figure 15.11. In 1985 Spencer reaffirmed his interpretation of the Port Askaig diamictites as true tillites. The interpretation hinges to a large extent on the origin of the wedge structures in the upper portions of many diamictites. If these are ice-wedge structures, as suggested by Spencer, then the diamictites must have been exposed to the atmosphere prior to the transgression that produced the succeeding lag conglomerates and sandstones. Eyles's (1988) interpretation, on the other hand, involves a regression that is difficult to reconcile with the termination of a glacial phase unless related to isostatic rebound. Resolution of these differences of interpretation must await further studies.

Plate-Tectonic Setting of the Port Askaig Formation

The glacigenic rocks of the Port Askaig Formation overlie a thick succession of shallow marine aspect (Fig. 15.9). In most areas the Port Askaig Formation is immediately underlain by a thick stable shelf carbonate unit that locally contains stromatolites, flat pebble conglomerates and ooids, suggesting a shallow to intertidal depositional environment (Anderton, 1985). The Appin Group is interpreted as having formed on a stable subsiding shelf, comparable to the setting inferred for the older Mackenzie Mountains

Supergroup in the northern Cordillera. Deposition of the glacigenic rocks of the Port Askaig Formation may have coincided with the onset of faulting. On a regional scale, the Port Askaig Formation varies greatly in thickness, from less than 100 m in parts of western Ireland to 1,200 m in western Scotland. Throughout a strike length of about 500 km the formation shows a consistent "vertical" variation in clast composition. Diamictites in the lower portion of the formation contain abundant carbonate clasts, including orange-weathering dolostones. The matrix material of the lower diamictites is also commonly dolomite-rich. In the upper portion of the formation, the diamictites have a siliceous matrix and contain many quartzite and granitic clasts. Whereas the carbonate clasts could have been derived from the underlying Dalradian units, the source of the extrabasinal granitoid fragments has remained problematic. Some authors (Anderton, 1980), suggested that these clasts may have been derived from a southeasterly source. In any event, their composition suggests that they came from outside the Dalradian basin, probably from a basement complex, as opposed to a recently formed fold belt (Anderton, 1985), which would presumably have yielded a high proportion of deformed sedimentary and volcanic clasts.

The Port Askaig Formation is succeeded by a series of shallow marine rocks. In western Scotland, the glacigenic deposits are overlain by a mixed clastic/carbonate succession, known as the Bonahaven Dolomite (Fairchild, 1980). These rocks include well-preserved stromatolites and have been interpreted as supratidal and subtidal deposits formed in an arid(?) setting. In western Ireland the diamictites are overlain by slightly calcareous siltstones or by a thin argillaceous unit. Throughout the entire region, the succeeding formation is a remarkably thick (several km) quartzite unit, which is dominantly of shallow marine origin (Anderton, 1976; 1985). The homogeneity and thickness of these quartzites attest to a remarkably long-lived balance between subsidence and sedimentation. Deposition of the Jura Quartzite and of correlatives such as the Slieve Tooey Quartzite in Ireland was followed by a distinctive basin-deepening event, during which the Easdale Subgroup was deposited. Outpouring of the Tayvallich Volcanics, higher in the succession, has been linked by Anderton (1985) to the opening of the Iapetus Ocean (Figs. 15.9, 15.12).

The plate-tectonic setting of the Port Askaig Formation is still being debated. For example Bluck and Dempster (1991) proposed that the entire Dalradian block might be allochthonous, having been transferred tectonically from a

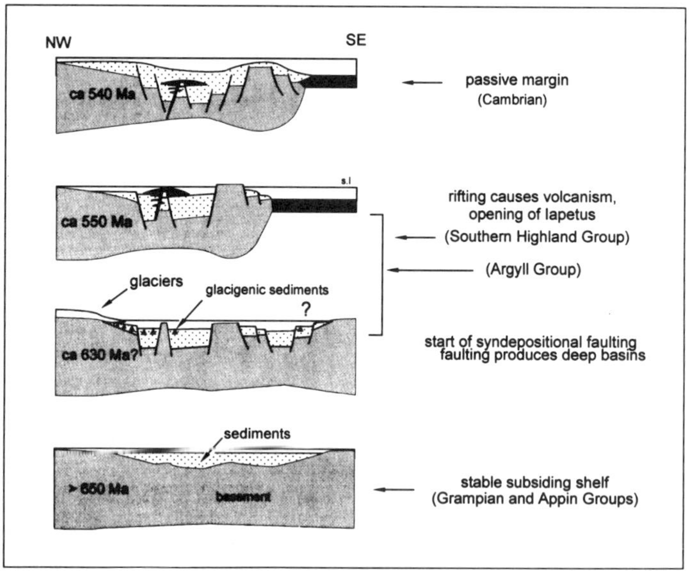

Fig. 15.12. Interpretation of the plate tectonic setting of the Dalradian Supergroup (after Anderton, 1985).

southern plate during a collisional event. Soper (1994a; 1994b) and Soper and England (1995) proposed that, during the Neoproterozoic, Scotland was the site of a triple (RRR) junction and thal the entire Dalradian succession accumulated in an extensional regime. This suggestion is a radical departure from previous interpretations, which involved Precambrian deformation of the Dalradian (Bluck et al., 1995). Anderton (1985) proposed that the Dalradian Supergroup was deposited in a major gulf, several hundred km long, extending southwest into the interior of a rifting "supercontinental" block. It is interesting to note that this supercontinent may be a different assemblage from that on which the Rapitan and the Sturtian glacigenic deposits accumulated. The older supercontinent, which rifted apart at about 750 Ma to produce the proto-Pacific Ocean, is widely known as Rodinia, but Young (1995) pointed out that the original designation and definition of Rodinia are more applicable to a younger assemblage formed by collision of Australia/India/Antarctica with what is now east Africa. The name "Kanatia" was proposed for the older supercontinent. Formation of the second supercontinent (Rodinia) occurred between the two major widespread Neoproterozoic glacial epochs (Dalziel et al., 1994; Young 1995). According to Anderton's (1985) suggestion, most of the clastic component of the Appin and Argyll Groups would have been derived from Laurentia (Fig. 10, insert), or was transported axially along the gulf, but Anderton (1985) speculated that the exotic clasts in the upper part of the diamictite succession could have been "floated in" from the southeastern margin of the basin, which may have included parts of the western South America (Dalziel et al., 1994; Young 1995).

Postglacial stratigraphy of the Neoproterozoic Dalradian differs from that of the Huronian Gowganda-Lorrain transition in lacking the coarsening-upward deltaic cycles. The thick quartzite succession of the Jura Quartzite and correlatives more closely resembles the Lorrain Formation than it does the thick mudstones of the Neoproterozoic post-Rapitan Twitya Formation. The inferred plate tectonic setting of the Port Askaig Formation differs from both the Gowganda Formation and the Rapitan Group. If Anderton's (1985) interpretation is correct, then the Port Askaig Formation was deposited during a rift episode that preceded, perhaps by as much as 100 m.y., the opening of the Iapetus Ocean in this region. The gulf-like setting and the presumed active contemporaneous faulting (Anderton, 1985) may have prevented a geographically widespread postglacial transgression. A shallow marine depositional phase during which siliciclastic sediment influx was diminished (presumably by postglacial transgression) is represented by the Bonahaven Dolomite and correlatives in Ireland, but these carbonates were quickly swamped by sandy siliciclastic debris from the basin margins. This debris was reworked by strong tidal currents, possibly in a Bay of Fundy-type setting, and a remarkable balance between sediment influx and basin subsidence led to the accumulation of a thick succession (several kilometers) of tidally influenced sandstones. This phase was terminated by rapid subsidence, resulting in basinal conditions during deposition of the Easdale Subgroup (Fig. 15.9).

Summary and conclusion

On the basis of the stratigraphic record, glaciations appear to have been relatively rare occurrences throughout geologic time. No significant glaciations are known from the Archean, but two complex major glacial periods are documented in the Paleo- and the Neoproterozoic. Some evidence of glaciation exists in the record of the Late Ordovician and the Devonian, but by far the most significant Phanerozoic glaciation took place in Permo-Carboniferous times. The duration and the severity of the Cenozoic glaciation remain to be seen. All three major glaciations, Paleoproterozoic, Neoproterozoic, and Permo-Carboniferous appear to have taken place during periods of "supercontinentality," and, to variable degrees, the glacial deposits owe their preservation to tectonic processes attendant on the breakup of the supercontinent (Young, 1991). In this chapter an attempt has been made to document and interpret the stratigraphy of postglacial deposits, using three examples from the Proterozoic record.

Termination of a major glaciation would be expected to be accompanied by a eustatic rise in sea level as the ice melted and the water returned to the world ocean. The stratigraphic effects of this rise would be variable, depending on the location of the depositional site. In ice-proximal locations, isostatic rebound would probably outstrip sea-level rise, leaving no sedimentary record. In more distal areas (excluding deep oceanic settings where little or no stratigraphic impact might be registered or preserved), a sea-level rise should be reflected in the stratigraphic record. Available interpretations of the plate-tectonic settings of the Proterozoic glacigenic successions suggest that they were mostly preserved in basins formed at or near the time of breakup of the supercontinents. The glaciations may have been induced by supercontinentality (Young, 1991) or related to uplift of the rift shoulders (Yeo, 1981). Many of the differences in the observed postglacial stratigraphy in the three Proterozoic examples used in this study (Fig. 15.13) may have been controlled by the relative timing of the rift-drift transition and deglaciation. Anderton (1985) and Dalziel et al. (1994) placed the breakup of a supercontinent and the formation of the Iapetus Ocean near the beginning of the Cambrian; that is, long after the end of the Neoproterozoic glaciation in Scotland and Ireland (Fig. 15.13A). Following Anderton (1985), it is suggested that the development of a thick quartzite succession and the lack of extensive muddy deposits are related to deposition in an actively faulting two sided gulf-like trough that developed prior to continental fragmentation.

The Neoproterozoic Rapitan Group, which forms part of the widespread Windermere Supergroup in the Cordilleran region of North America, was deposited on the rifting supercontinent, Rodinia of Dalziel (1991), Hoffman (1991), and Moores (1991), and Kanatia of Young (1995). These rocks have strikingly similar counterparts in the Sturtian of south Australia (Bell and Jefferson, 1987; Young and Gostin, 1991; Young, 1992). The presence of a thick mudstone succession above the glacigenic rocks (Fig. 15.13B) may be explained by a major marine transgression brought about by the coincidence of postglacial eustatic sea-level rise and tectonic subsidence of the newly formed continental margins.

In the case of the Huronian Supergroup (Fig. 15.13C), the rift-drift transition may have occurred during or slightly before deposition of the widespread glacigenic Gowganda Formation (Young and Nesbitt, 1985). Thus the postglacial succession, with its early deltaic cycles and succeeding thick accumulation of fluvial to shallow marine sandstones, is seen as an initial response to eustatic sea level rise, followed by progradation of a clastic wedge out over the earlier established passive continental margin. These three examples of Proterozoic glacial successions are all considered to have formed on supercontinents and to have been preserved in basins formed as a result of breakup. Differences among the postglacial successions are considered to be the result of different relationships between

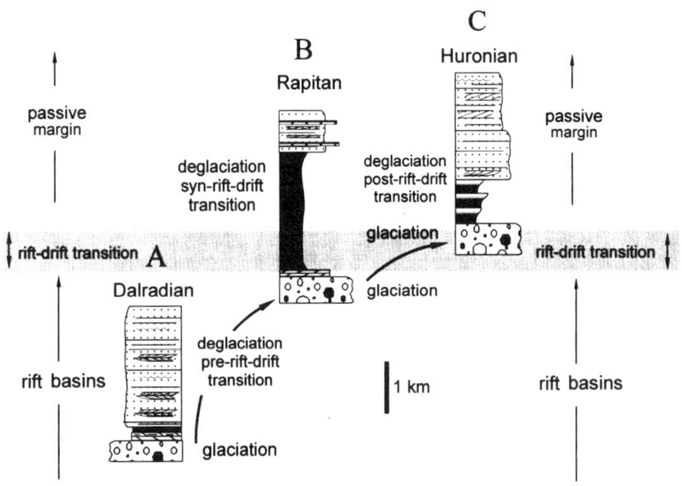

Fig. 15.13. Comparisons of the stratigraphy and the plate-tectonic setting of the three successions described in the text. Note that in each case the rift-drift transition is interpreted to have occurred at different times in relation to the end of glaciation.

the timing of continental fragmentation and deglaciation.

In the case of the Permo-Carboniferous glaciation, breakup of the supercontinent occurred almost 100 Ma later than the end of the glaciation, so the situation differs somewhat from all of the Proterozoic examples discussed in this chapter. The Tertiary to Recent glaciations are also different inasmuch as they do not appear to have been related to the existence and the breakup of a supercontinent. In the absence of rift basins comparable to those in which so many ancient glacial successions were deposited, the preservation potential of these young glacial deposits is not known. Because the Pleistocene glaciation has probably not yet run its course, attempts to explain its postglacial stratigraphy are probably premature.

References

Ager, D.V., 1993. The nature of the stratigraphic record, John Wiley & Son, Chichester, 151 pp.

Aitken, J.D., 1966. Middle Cambrian to Middle Ordovician cyclic sedimentation, southern Rocky Mountains of Alberta: Canadian Petroleum Geology Review, 3: 405–441.

Aitken, J.D., 1991. Two Late Proterozoic glaciations, Mackenzie Mountains, northwestern Canada. Geology, 19: 445–448.

Aitken, J.D., 1993. Proterozoic sedimentary rocks; subchapter 4A. In: D.F. Stott, and J.D. Aitken, (Editors), Sedimentary cover of the craton in Canada. Geology of Canada No. 5, Geological Survey of Canada, pp. 81–95.

Aitken, J.D. and Long, D.G.F., 1978. Mackenzie tectonic arc-reflection of early basin configuration? Geology, 6: 626–629.

Anderton, R., 1976. Tidal-shelf sedimentation: an example from the Scottish Dalradian. Sedimentology, 23: 429–458.

Anderton, R., 1980. Did Iapetus start to open during the Cambrian? Nature, 286: 706–708.

Anderton, R., 1985. Sedimentation and tectonics in the Scottish Dalradian. Scottish Journal of Geology, 21: 407–436.

Bell, R.T. and C.W. Jefferson, 1987. An hypothesis for an Australian-Canadian connection in the Late Proterozoic and the birth of the Pacific Ocean. Proceedings of Pacific Rim Congress of the Geology, Structure, Mineralisation and Economics of the Pacific Rim, Australasian Institute of Mining and Metallurgy, Parkville, Australia, pp. 39–5O.

Bluck, B.J., and T.J. Dempster, 1991. Exotic metamorphic terranes in the Caledonides: tectonic history of the Dalradian block, Scotland. Geology 19: 1133–1136.

Broecker, W.S. and Denton, G.H., 1989, The role of ocean-atmosphere reorganization in glacial cycles. Geochimica et Cosmochimica Acta, 53: 2465–2501.

Brookfield, M.E., 1993. Neoproterozoic Laurentia-Australia fit. Geology, 21: 683–686.

Card, K.D., 1979. Metamorphism of the middle Precambrian (Aphebian) rocks of the eastern Southern Province. Geological Survey of Canada Paper 78-10: pp. 269–282.

Casshyap, S.M., 1969. Petrology of the Bruce and Gowganda Formations and its bearing on the evolution of Huronian sedimentation in the Espanola-Willisville area, Ontario. Palaeogeography, Palaeoclimatology, Palaeoecology, 6: 5–36.

Chandler, F.W., 1986, Sedimentology and paleoclimatology of the Huronian (Early Aphebian) Lorrain and Gordon Lake Formations and their bearing on models for sedimentary copper mineralization. Geological Survey of Canada Paper 86-1A, pp. 121–132.

Coleman, A.P., 1914. The pre-Cambrian rocks north of Lake Huron. Ontario Bureau of Mines, v. 23, Part 1, pp. 204–236.

Dalziel, I.W.D., 1991. Pacific margins of Laurentia and East Antarctica-Australia as a conjugate rift pair: evidence and implications for an Eocambrian supercontinent. Geology, 19: 598–601.

Dalziel, I.W.D., Dalla Salda, L.H. and Gahagan, L.M., 1994. Paleozoic Laurentia-Gondwana interaction and the origin of the Appalachian-Andean mountain system. Geological Society of America Bulletin, 106: 243–252.

Eisbacher, G.H., 1978. Re-definition and subdivisions of the Rapitan Group, Mackenzie Mountains. Geological Survey of Canada, Paper 77–35, 21 pp.

Eisbacher, G.H., 1981. Sedimentary tectonics and glacial record in the Windermere Supergroup, Mackenzie Mountains, northwestern Canada. Geological Survey of Canada, Paper 80-27, 40 pp.

Eisbacher, G.H., 1985. Late Proterozoic rifting, glacial sedimentation and sedimentary cycles in the light of Windermere deposition, western Canada. Palaeogeography, Palaeoclimatology, Palaeoecology, 51: 231–254.

Eyles, C.H., 1988. Glacially- and tidally-influenced shallow marine sedimentation of the Late Precambrian Port Askaig Formation, Scotland. Palaeogeography, Palaeoclimatology, Palaeoecology, 68: 1–25.

Eyles, N. and B.M. Clark, 1985. Gravity-induced soft-sediment deformation in glaciomarine sequences of the Upper Proterozoic Port Askaig Formation, Scotland. Sedimentology, 32: 789–814.

Eyles, N. and Young, G.M., 1994. Geodynamic controls on glaciation in Earth history. In: M. M. Deynoux, J.M.G. Miller, E.W. Domack, N. Eyles, I.J. Fairchild and G.M. Young (Editors), Earth's glacial record. Cambridge University Press, Cambridge, pp. 1–28.

Frarey, M.J. and Roscoe, S.M., 1970. The Huronian Supergroup north of Lake Huron. In: A.J. Baer (Editor), Basins and geosynclines of the Candian Shield. Geological Survey of Canada Paper 70-40, pp. 143–157.

Fairchild, I.J., 1980. Sedimentation and origin of a Late Precambrian "dolomite" from Scotland. Journal of Sedimentary Petrology, 50: 423–446.

Gabrielse, H. and Campbell, R.B., 1991. Upper Proterozoic assemblages. In: H. Gabrielse and R.B. Campbell (Editors), Geology of the Cordilleran orogen in Canada. Geology of Canada No. 4, Geological Survey of Canada, pp. 125–150.

Hadley, D.G., 1968. The sedimentology of the Huronian Lorrain Formation, Ontario and Quebec, Canada. Ph.D. thesis, The Johns Hopkins University, Baltimore MD, 300 pp.

Hambrey, M.J., 1983. Correlation of Late Proterozoic tillites in the North Atlantic region and Europe. Geological Magazine, 120: 209–320.

Hansen, V.L., Goodge, J.W., Keep, M. and Oliver, D.H., 1993. Asymmetric rift interpretation of the western North American margin. Geology, 21: 1067–1070.

Hoffman, P.F., 1988. United Plates of America, the birth of a craton: Early Proterozoic assembly and growth of Laurentia. Annual Reviews of Earth and Planetary Science, 16: 543–603.

Hoffman, P., 1991. Did the breakout of Laurentia turn Gondwanaland inside-out? Science, 252: 1409–1412.

Howarth, R.J., 1971. The Port Askaig Tillite succession (Dalradian) of Co. Donegal. Proceedings of the Royal Irish Academy, 71: 1–35.

Jefferson, C.W. and Ruelle, J.C.L., 1986. The Late Proterozoic Redstone Copper Belt, Mackenzie Mountains, Northwest Territories. In: J.A. Morin (Editor), Mineral deposits of the Northern Cordillera. Canadian Institute of Mining and Metallurgy, Special Volume 37, pp. 154–168.

Jolly, W.T., 1987. Geology and geochemistry of Huronian rhyolites and low-Ti continental tholeiites from the Thessalon region, central Ontario. Canadian Journal of Earth Sciences, 24: 1360–1385.

Junnila, R.M., 1986. Stratigraphy and sedimentology of the upper Gowganda Formation (Early Proterozoic) in the Whitefish Falls area, Ontario. M.Sc. thesis, University of Western Ontario, London, Ontario, 168 pp.

Kilburn, C., Pitcher, W.S. and Shackleton, R.M., 1965. The stratigraphy and origin of the Portaskaig Boulder Bed Series (Dalradian). Geological Journal, 4: 343–360.

Klein, C. and Beukes, N.J., 1993. Sedimentology and Geochemistry of the glaciogenic Late Proterozoic Rapitan Iron-formation in Canada. Economic Geology, 88: 542–565.

Lindsey, D.A., 1969. Glacial sedimentation of the Precambrian Gowganda Formation, Ontario, Canada. Geological Society of America Bulletin, 80: 1685–1704.

McMenamin, M.A.S. and McMenamin, D.L., 1990. The emergence of animals: The Cambrian breakthrough. Columbia University Press, New York, 217 pp.

Miall, A.D., 1985. Sedimentology of an Early Proterozoic continental margin under glacial influence: the Gowganda Formation (Huronian), Elliot Lake area, Ontario, Canada. Journal of Sedimentary Petrology, 23: 763–788.

Moores, E.M. 1991. Southwest U.S.-East Antarctic (SWEAT) connection: a hypothesis. Geology, 19: 425–428.

Ojakangas, R.W., 1988. Glaciation: an uncommon "mega-event" as a key to intracontinental and intercontinental correlation of early Proterozoic basin fill, North America and Baltic cratons. In: K.L. Kleinspehn and C. Paola (Editors), New perspectives in basin analysis. Springer-Verlag, New York. pp. 431–444.

Pettijohn, F.J., 1970. The Canadian shield-a status report. In: A.E. Baer (Editor), Basins and geosynclines of the Canadian Shield. Geological Survey of Canada, Paper 70-40, pp. 239–255.

Rainbird, R.H. and Donaldson, J.A., 1988. Nonglaciogenic deltaic deposits in the early Proterozoic Gowganda Formation, Cobalt Basin, Ontario. Canadian Journal of Earth Sciences, 25: 710–724.

Ross, G.M., 1991. Tectonic setting of the Windermere Supergroup revisited. Geology, 19: 1125–1128.

Ruddiman, W.F., Prell, W.L. and Raymo, M.E., 1989. Late Cenozoic uplift in Southern Asia and the American West: rationale for general circulation modeling experiments. Journal Geophysical Research, 94: 18,379 18,391.

Smith, A.G., Hurley, A.M. and Briden, J.C., 1981. Phanerozoic paleocontinental world maps. Cambridge University Press, Cambridge, 102 pp.

Soper, N.J., 1994a. Was Scotalnd a vendian RRR junction? Journal of the Geological Society, London, 151: 579-582.

Soper, N.J., 1994b. Neoproterozoic sedimentation on the northeast margin of Laurentia and the opening of Iapetus. Geological Magazine, 131: 291-299.

Soper, N.J. and England, R.W., 1995. Vendian and Riphean rifting in NW Scotland. Journal of the Geological Society, London, 152: 11-14.

Spencer, A.M., 1971. Late Pre-Cambrian glaciation in Scotland. Geological Society, Memoir 6, 100 pp.

Spencer, A.M., 1985. Mechanisms and environments of deposition of Late Precambrian geosynclinal tillites: Scotland and East Greenland. Palaeogeography, Palaeoclimatology, Palaeoecology, 51: 143–157.

Stewart, J.H., 1972. Initial deposits in the Cordilleran Geosyncline: evidence of Late Proterozoic (< 850 m.y.) continental separation. Geological Society of America Bulletin, 83: 1345–1360.

Strand, K.O. and Laajoki, K., 1993. Palaeoproterozoic glaciomarine sedimentation in an extensional tectonic setting: the Honkala Formation, Finland. Precambrian Research, 64: 253–271.

Thomson, R., 1957. The Proterozoic of the Cobalt area. Royal Society of Canada, Special Publication 2, pp. 40–45.

Yeo, G.M., 1981. The Late Proterozoic Rapitan glaciation in the Northern Cordillera. In: F.H.A. Campbell (Editor), Proterozoic basins of Canada. Geological Survey of Canada, Paper 81-10, pp. 25–46.

Yeo, G.M., 1986. Iron-formation in the late Proterozoic Rapitan Group, Yukon and Northwest Territories. In: J.A. Morin (Editor), Mineral deposits of the northern Cordillera. Canadian Institute of Mining and Metallurgy, Special Volume 37, pp. 142–153.

Young, G.M., 1970. An extensive Early Proterozoic glaciation in North America? Palaeogeography, Palaeoclimatology, Palaeoecology, 7: 85–101.

Young, G.M., 1973. Tillites and aluminous quartzites as possible time markers for Middle Precambrian (Aphebian) rocks of North America. In: G.M. Young (Editor), Huronian stratigraphy and sedimentation. Geological Association of Canada, Special Paper 12, pp. 97–127.

Young, G.M., 1981. The Early Proterozoic Gowganda Formation. In: M.J. Hambrey and W.B. Harland (Editors), Earth's pre-Pleistocene glacial record. Cambridge University Press, Cambridge, Cambridge, pp. 807–812.

Young, G.M., 1984. Proterozoic plate tectonics in Canada with emphasis on evidence for a Late Proterozoic rifting event. Precambrian Research, 25: 233–256.

Young, G.M., 1991. The geologic record of glaciation; relevance to the climatic history of Earth. Geoscience Canada, 18: 100–108.

Young, G.M., 1992. Late Proterozoic stratigraphy and the Canada-Australia connection. Geology, 20: 215–218.

Young, G.M., 1995. Are Neoproterozoic glacial deposits preserved on the margins of Laurentia related to the fragmentation of two supercontinents? Geology, 23: 153-156.

Young, G.M. and Gostin, V.A., 1991. Late Proterozoic (Sturtian) succession of the North Flinders Basin, South Australia; an example of temperate glaciation in an active rift setting. In: J.B. Anderson and G.M. Ashley (Editors), Glacial marine sedimentation; paleoclimatic significance: Geological Society of America, Special Paper 261, pp. 207–222.

Young, G.M., Jefferson, C.W., Delaney, G.D. and Yeo, G.M., 1979. Late Proterozoic evolution of the northern Canadian Cordillera and Shield. Geology, 7: 329–330.

Young, G.M. and Nesbitt, H.W., 1985. The Gowganda Formation in the southern part of the Huronian outcrop belt, Ontario, Canada: stratigraphy, depositional environments and regional tectonic significance. Precambrian Research, 29: 265–301.

PART V

THE QUATERNARY AND CARBONIFEROUS-PERMIAN: A Few Considerations on Continental Weathering and Organic Deposits

16. INTERPRETING GLACIAL CLIMATE FROM DETRITAL MINERALS IN SEDIMENTS

Prodip Dutta

Sedimentary rock records show that since the middle Precambrian, Earth's climate has fluctuated at least five times between cold glacial and warm tropical periods. Even within a single glacial period, relatively short term glacial and interglacial cycles have been observed.

Evidence of glaciation has come essentially from sedimentary facies such as tillite, dropstone, and varve deposits. Structural features, both erosional and depositional, associated with these sediments added credence to glacial interpretation. Additional evidence from palynology, paleontology, pedology, oxygen isotope, and paleomagnetism has been used to interpret paleoglacial conditions. In spite of this variety of tools, we do not have an unequivocal picture of climates in the past. This prompts us to develop newer tools so that, through an eclectic approach, we can come up with a better and perhaps a high-resolution picture of the past.

The nature and the abundance of silicate minerals in sedimentary rocks have been useful in helping us interpret past climates. However, mineralogical evidence for interpreting cold climates has not been explored fully. The objective of this presentation is to discuss the basis for using mineralogical composition of sediments as a paleoclimatic tool and to demonstrate its usefulness in interpreting Earth's paleoglacial conditions.

The Rationale for Using Minerals in Sediments as Paleoclimatic Indicators

The ultimate source of all sediments is igneous rock. At the interface of Earth's surface and atmosphere, these rocks are subjected to mechanical and chemical weathering. Chemical weathering is volumetrically most significant in disintegrating rocks and can take place only in the presence of liquid water. The effect of chemical weathering in arid climates is reduced because of the lack of precipitation or the lack of water in liquid form. In tropical, warm-humid climates the chemical weathering is most intense. In between these two extremes the products of chemical weathering differ widely as a function of climate (Strakhov, 1967).

Contrary to the common assumption, chemical weathering is widespread even in cold climates (Ugolini, 1986). However, the rate of weathering is extremely slow, and the products are minor amounts of fine-grained secondary minerals: oxides, hydroxides, and clays. Because the disintegration of rocks in a glacial environment is accomplished mostly through mechanical breakage, the mineralogical composition of coarse-grained glacial sediments does not differ much from that of their parents. If the temperature rises above freezing, as during interstadial periods, water in liquid form will initiate chemical decomposition on a very limited scale (Bockheim, 1980), the sediment will be slightly more mature than the sediments deposited during a freezing phase.

By careful documentation of a mineralogical composition of glacial sediments and of their source rocks, it would be possible to reconstruct the climate of a glacial period, including its intra- and interglacial phases. One of the problems in such studies is tracing the source rock and determining its composition. The source rock might have been eroded; quantitative estimation of the composition of a source rock characterized by a mixed bag of different rock types may be difficult, if not impossible. The best results are achieved with first-cycle sediments when the source rock is coarse crystalline and is of uniform character. In such a situation it is possible to determine the composition of both sediments and source rocks. Among sediments, sandstone composition can be determined most precisely in order to identify the source rock. There are also extensive databases on the genesis of sand/sandstone that connect the source rock, process, and products.

Glacial Climate and Sandstone Composition

Glacigenic sediments of Sakmarian age, known as the Talchir Formation, in the Damodar Valley area in India, offer one of the best opportunities to relate sandstone composition and glacial conditions. The Talchir Formation constitutes the basal part of a thick glacial, fluvioglacial, fluviolacustrine, and fluvial deposit, ranging in age from the Sakmarian through the Upper Jurassic and known as the Gondwana Supergroup. Throughout this span the climate changed from polar (Talchir Formation) through cold to temperate humid (Barakar Formation), semihumid (Barren Measures and Raniganj Formation), semiarid to semihumid (Panchet Formation), and warm humid (Mahadeva Formation; Table 16.1).

In the Damodar valley area several isolated basins are aligned from east to west (Fig. 16.1). Not too far in the geological past many of the present-day isolated basins were connected, and a large part of the exposed Precambrian basement was covered by Gondwana sediments. The Precambrian basement rock in the provenance area is mostly coarse-grained crystalline rocks of granitic composition with minor amounts of metasediments and metavolcanics of basic composition (Pascoe, 1959; Bhattacharya and Chatterjee, 1992). The Talchir sediments were not deeply buried, and the diagenetic destruction of framework minerals in sandstones could be estimated to trace the original composition of sand-size sediments (Suttner and Dutta, 1986). In such a geological setting, it is possible to understand the relationship between glacial processes that influence the mineralogical composition of glacigenic sediments.

The Talchir Formation in the Damodar Valley area is about 300m thick (Fig. 16.1). Tillite occurs at the base of the sequence. Overlying tillites are outwash deposits of conglomerate, sandstone, shale, varve, and turbidites. The facies organization indicates a general amelioration from a polar climate to a cool, seasonal climate. However, this unidirectional climatic pattern is reversed in the Karanpura Basin, where resurgence of glaciation is observed higher in the sequence (Fig. 16.1).

Talchir sandstones in the Raniganj Basin are very immature, ranging in QFR composition from 48:49:3 to 59:35:6 (Suttner and Dutta, 1986) (Q = quarz; F = feldspar, R = rock fragments). When QFR compositions are recalculated in terms of QAP (Streckeisen, 1976), all but one sandstone fell well within the granitic field (Fig. 16.2). On the basis of quartz type (Basu et al., 1975), it was observed that the Talchir sandstones were derived primarily from a coarse crystalline granitic source (Suttner and Dutta, 1986). Similar

TABLE 16.1. Stratigraphy of the Gondwana Supergroup in the Damodar Valley area and the climatic vicissitude during Gondwana sedimentation.

Formation Name	Geologic Age	Inferred Climate
Mahadeva	Upper Triassic to Lower Jurassic	Warm Humid
Panchet	Lower Triassic	Semhumid to semiarid
Raniganj	Upper Permian	Semiarid to semihumid
Barren Measures		Temperate humid
Barakar	Permian	Cool/temperate-humid
Talchir		Cold glacial

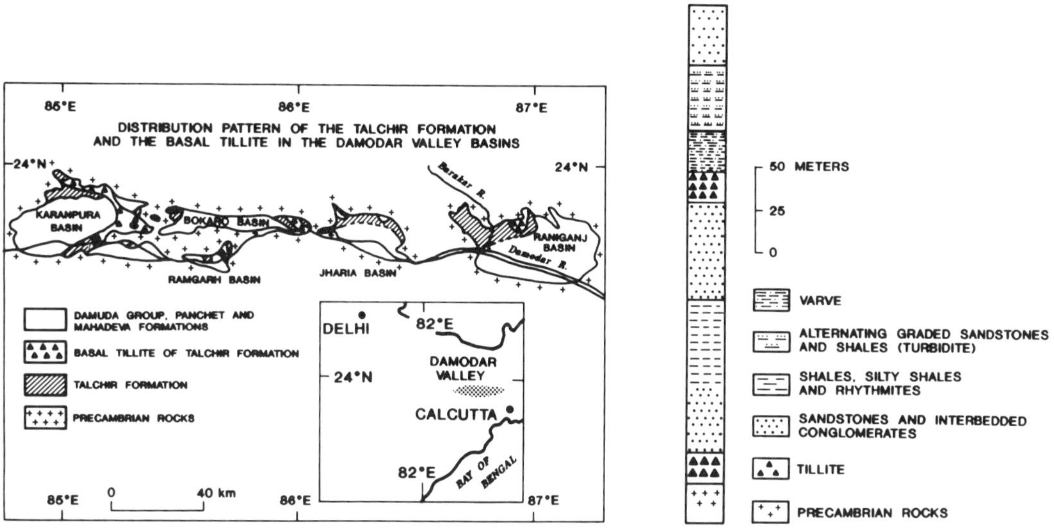

Fig. 16.1. A simplified composite section of the Talchir Formation in the Damodar Valley basins in eastern India. The inset map shows the location of the area. A tillite bed appears near the base of the sequence. Glaciolacustrine and glaciofluvial facies indicate amelioration of the climate, culminating in a plant bed at the top of the sequence. An exception to this generalized section, two tillite beds, has been observed in the westernmost Karanpura Basin (after Ghosh and Mitra, 1975).

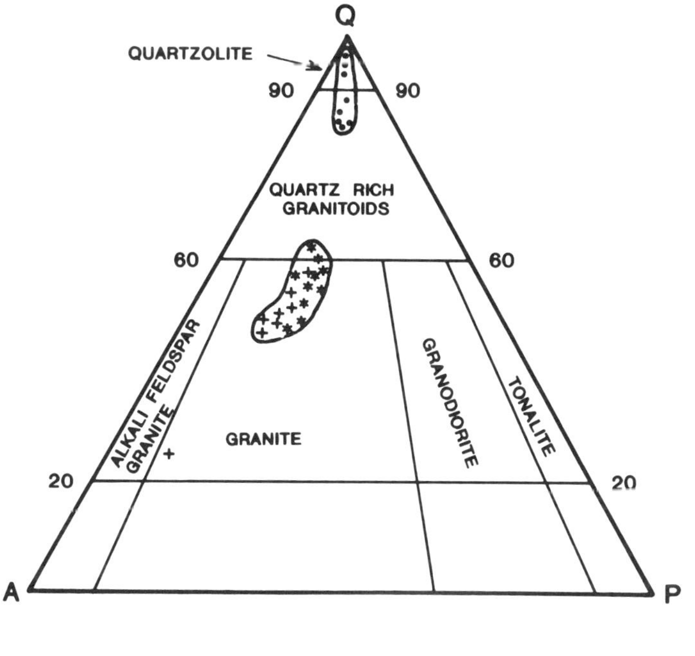

Fig. 16.2. Composition of Talchir and Barakar sandstones from the Raniganj Basin are transposed into a QAP diagram (Streckeisen, 1976) to demonstrate the compositional shift among parent, granite gneiss (data from Bhattacharya and Chatterjee, 1992) and sandstones (data from Suttner and Dutta, 1986) generated during glacial (Talchir) and postglacial (Barakar) climates respectively. Symbols: + = granite; * = Talchir sandstone; o = Barakar sandstone.

observations have also been made by earlier workers (Ghosh and Mitra, 1975). Modal analysis of the granitic source rock located in close proximity to the Raniganj Basin (Bhattacharya and Chatterjee, 1992) shows a composition similar to that of Talchir sandstones in the basin. However, the sandstone contains a little more quartz than the granite (Fig. 16.2). The closeness of the source rock composition and the products of weathering suggest little chemical weathering, and petrographic data confirm that in a glacial environment the chemical destruction is minimal.

When the quartz concentration in sandstone is plotted stratigraphically, an overall maturity trend is observed. Within that trend, a cyclical pattern in sandstone composition from the base of the section to the middle is discernable (Fig. 16.3). This cyclicality reflects the alternation of cold glacial phases and warm, intra- and interglacial phases. However, this cyclicality is not evident from the middle to the top of the section.

The cyclicality of cold (freezing) and warm (thawing) episodes is seen most clearly in the distribution of hornblende in tillites (products of a cold phase) and glaciofluvial sandstone (products of glacial melting). Locally, in a particular section in the Jharia Basin (Fig. 16.1), the hornblende content in the heavy mineral assemblage in tillite is 41.9 percent, the other abundant heavy mineral being garnet. At the same location, immediate above the tillite beds, glaciofluvial sandstone shows no hornblende but abundant garnet. Similar patterns have also been observed elsewhere where hornblende-rich rocks are present (Ghosh and Mitra, 1975). Abundance of hornblende in tillite indicates that even the most unstable minerals have not been chemically decayed while becoming weathered due to the availability of water during a intra- and interglacial warm phase.

The gradual warming trend seen during Talchir deposition occurred over a very short interval of time as indicated by the sharp mineralogical change between the topmost Talchir sandstone and the basal-most Barakar sandstone over the formational boundary. The youngest Talchir sandstone has a QFR composition of 57:37:6 while the succeeding Barakar sandstone composition is 92:6:2.

The development of a coal-bearing unit over

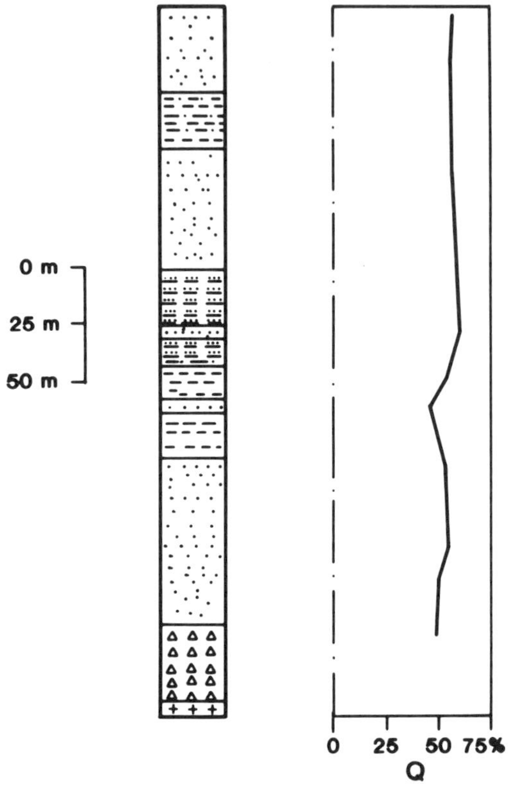

Fig. 16.3. Plot of quartz concentrations of the Talchir sandstones in the Raniganj Basin. Maturity trends are indicated by increasing quartz content. A subtle cyclical pattern between immature and relatively more mature sandstones is observed in the lower half of the section.

the glacigenic Talchir Formation, seemingly without much time lag, suggests proliferation of vegetation. If the immaturity of Talchir sandstones is related to the lack of chemical weathering in a glacial environment, the Barakar sandstones with supermature to submature sandstones (Fig. 16.2) and kaolinitic clay beds represent intense chemical weathering in a cool/temperate humid climate. The differences in the extent of chemical weathering during the glacial Talchir and in the postglacial cool/temperate conditions that prevailed during the Barakar deposition are very significant, as shown by the compositional contrasts.

Conclusion

The petrographic data presented here demonstrate that in a glacial period, chemical weathering is much reduced and is not detectable with petrographic means during the cold phases. During intra- and interglacial warm phases, chemical weathering creates a minor but perceptible change in QFR composition between the source rock and sediments.

Highly unstable minerals are susceptible to chemical decay at much faster rates in the presence of water, even in glacial cold conditions. Careful documentation of the presence and absence of such reactive minerals in glacigenic sediments may reveal a subtle change in climate that can not be detected by any other paleoclimatic tools.

Chemical weathering intense enough to destroy most feldspars in granitic rocks can take place under climatic conditions favorable to vegetation growth. This has occurred in the Barakar sandstones. The change from a glacial to a postglacial (cool/temperate) setting took place in a relatively short period, possibly over thousands of years, a split second on a geological scale. This conforms well with the rapid Pleistocene glacial-postglacial transition.

References

Basu, A., Young, S.W., Suttner, L.J., James, W.C. and Mack, G.H., 1975. Re-evaluation of the use of undulatory extinction and polycrystallinity in detrital quartz for provenance interpretation. Journal of Sedimentary Petrology, 45: 873–882.

Bhattacharya, S. and Chatterjee M., 1992. Evolution of Chotonagpur Gneiss around Kana, Purulia District, W. Bengal. Indian Journal of Geology, 64: 80–87.

Bockheim, J.G., 1980. Properties and classification of some desert soils in coarse textured glacial drift in the Arctic and Antarctica. Geoderma, 24: 45–69.

Ghosh , P.K. and Mitra, N.D., 1975. History of Talchir sedimentation in Damodar Valley basins. Memoir 105, Geological Survey of India. Government of India, Calcutta, 117 pp.

Pascoe, E.W., 1959. A manual of the geology of India and Burma: Vol. 2. Manager of Publications, Government of India, New Delhi, pp. 485–1343.

Strakhov, N.M., 1967. Principles of lithogenesis, Vol. 1. Consultant Bureau, New York, 245 pp.

Streckeisen, A., 1976. Classification of and nomenclature of igneous rocks. Geologische Rundschau, 63: 773–786.

Suttner, L.J. and Dutta, P.K. , 1986. Alluvial sandstone composition and paleoclimate, I. Framework mineralogy. Journal of Sedimentary Petrology, 56: 329–345.

Ugolini, F.C., 1986. Processes and rates of weathering in cold and polar desert environments. In: S.M. Colman and D.P. Detheir (Editors), Rates of chemical weathering of rocks and minerals. Academic Press, Orlando, pp. 193–235.

17. PALEOSOLS OF THE NORTHERN PART OF NORTH AMERICA: THEIR FEATURES AND SIGNIFICANCE AS INDICATORS OF PAST CLIMATES

Charles Tarnocai

Numerous soils occurring in northern North America have been described as paleosols by various researchers. Since most parts of this area were glaciated during the Pleistocene epoch, only those areas that were unglaciated during that time contain paleosols that retain the imprint of pre-Holocene climates. Those paleosols whose properties are useful for determining climates in the Tertiary period and the Pleistocene epoch of the Quaternary period are the main focus of this chapter, since they supply the most complete pedological record; however, Holocene relict soil features that have significance for determining past climates are also considered.

Understanding how a soil forms is essential to understanding how specific soils can be used to determine past climates. Soils are formed as the result of the interaction of soil-forming factors (climate, parent material, landform, drainage, organisms) over time. Paleosols are soils that formed in the past on land surfaces that were sufficiently stable to receive and retain imprints of these soil-forming factors. These paleosols are now generally buried within the geological section, but they can also occur on the surface. Buried paleosols in a geological section can be identified with greatest confidence when they are traceable laterally to show logical changes in the soil profile characteristics corresponding to changes in the landscape (Valentine and Dalrymple, 1975).

The use of paleosols in the reconstruction of paleoenvironments depends on the researcher's ability to recognize them and to ascribe with confidence the origin of some of their properties to specific environmental conditions (Valentine and Dalrymple, 1975; Valentine et al., 1987). In order to develop models to reconstruct the environment under which paleosols developed, researchers study analogous modern soils to determine which principal climatic (including cryogenic), vegetative, and hydrologic influences formed them and then extrapolate these influences to the past. The interpretation is based on the more permanent (irreversible) soil features since these features provide a more reliable indication of the past environment (Yaalon, 1971). This is especially important for paleosols that have a polygenic origin, since some soil properties are direct indicators of the climates under which these paleosols developed. These paleosols are then placed within the geological time frame either by correlating the paleosol with the corresponding geological material or by using dating techniques on the soil material itself.

Thus, the paleosol properties yield information about the climates and paleoenvironments under which the soils formed, and dating these soils, either directly (by ^{14}C dating of buried organic material) or indirectly (by position in the geological section), aids in placing them within a geological time frame. The interpretation of the paleoenvironment can then be further expanded, or verified, by the use of techniques such as pollen analysis and oxygen isotope and paleomagnetic studies.

Soil Classification and Geological Time Frame

Soil Classification

The terminology relating to soil development and classification used in this chapter is that of the Agriculture Canada Expert Committee on Soil Survey (1987). The equivalent U.S. soil classifications (Soil Survey Staff, 1994) are: Inceptisols (Brunisols), Pergelic subgroups (Cryosols), Boralfs and Udalfs (Luvisols), Histosols (Organic soils) and Spodosols (Podzols). Since these classifications refer only to modern soils, the definitions have been slightly modified for use with paleosols. The following are the criteria used for paleosols:

1. Luvisolic paleosols have a paleoargillic (Bt) horizon.
2. Podzolic paleosols have a Bf horizon as determined by morphology and oxalate-extractable Al and Fe.
3. Brunisolic paleosols have a brownish Bm horizon, but lack both Bt and Bf horizons.
4. Cryosolic paleosols have well developed cryogenic features such as ice wedge casts, sand wedges, cryoturbated soil horizons and cryogenic microfabrics.
5. Organic paleosols have developed in peat materials, possibly in a permafrost-free environment.

Geological Time Frame

The geological chronostratigraphy of the Tertiary and the Quaternary periods in North America is shown in Table 17.1. There is no evidence that glaciation occurred during the Tertiary period (approximately 65.0–1.70 Ma). The onset of glaciation at the end of the Tertiary, however, marked the beginning of the Pleistocene epoch (1.70–0.01 Ma) of the Quaternary period. This epoch comprises the major glacial and interglacial periods in Canada, when almost all of the country was covered several times by glacial ice. The only unglaciated areas were parts of the western and northwestern Yukon, the Cypress Hills in southern Alberta and Saskatchewan, and some parts of the Arctic islands. Most of the evidence for early and middle Pleistocene glaciation is found in the Yukon. Bostock (1966) identified the Reid (Illinoian) glaciation and inferred two glacial advances during the pre-Reid (pre-Illinoian) stages, the Nansen (the older) and the Klaza. The Sangamonian stage, which occurred during the early part of the late Pleistocene epoch, was an interglacial period (Fulton and Prest, 1987). The succeeding stage, the Wisconsinan, was glacial (Fulton and Prest, 1987); however, a number of areas of Canada were ice free during this time, permitting biological activity and soil development to take place.

During the Holocene epoch (0.01 Ma to present), the time since the retreat of the last ice sheet, a number of climatic fluctuations have occurred (Ritchie and Hare, 1971; Sorenson et al., 1971; Terasmae, 1972). These fluctuations have produced relict soil features (as defined by Tarnocai and Valentine, 1989) that can be found in contemporary Canadian soils.

Results and Discussion

Descriptions of Paleosols

The climates of the interglacial periods are described in this section according to the associated Tertiary and Pleistocene paleosols. Table 17.1 shows the correlation of these paleosols and their associated environments with the geological timetable. These paleosols occur within three general locations: the High Arctic islands in the northern Northwest Territories, the Yukon and western Northwest Territories, and southern Alberta and Saskatchewan (Figs. 17.1, 17.2).

Eureka Sound Paleosols (Early Tertiary)

These Podzolic, Gleysolic, and Organic paleosols are located in the fossil forest area of Axel Heiberg Island (79° 55' N Lat., 88° 58' W Long.) in the northern Northwest Territories (Basinger, 1991; Tarnocai and Smith, 1991). They occur in fluvial deposits of the Buchanan Lake Formation of the Eureka Sound Group in the Geodetic Hills area (paleosol site 1, Fig. 17.1), and are stacked one above another, with more than 20 being found in some locations. The age of this formation, which developed in the early Tertiary period, is 45 Ma (Ricketts and McIntyre, 1986; Basinger, 1987).

The sola of the Podzolic paleosols are 120 to 200 cm thick, and the Bf horizon is frequently associated with ortstein (cemented) and concretionary layers. The leached Ae horizon is usually very thin, discontinuous, or absent. The surface organic horizons, when present, are also discontinuous and permineralized. The sola of these Podzolic paleosols and the underlying parent materials commonly have faults and cracks

TABLE 17. 1. Geological chronology and the associated paleosols in Canada. Chronology according to Fulton and Prest (1987).

Time (Ma)	Period	Epoch		Stage and substage		Glaciations	Oxygen isotope stages	(ka)	Event	Central Yukon	Old Crow area	Little Bear River area	Other areas
0.01	QUATERNARY	HOLOCENE					1	13	Holocene Interg.	(hb, sa, sp, a)	(sa, a)	(a)	
0.02 0.03		PLEISTOCENE	LATE	WISCONSINAN	LATE	McConnell and Buckland	2	32	Wisconsin Glaciation		Old Crow paleosol (la)		
0.06					MIDDLE		3	65					
0.08					EARLY		4	80					
0.13				SANGAMONIAN			5	130	Sangamon Interglaciation	Diversion Creek paleosols (b)		Paleosol 5 (b)	
0.32			MIDDLE	ILLINOIAN		Reid Thomsen	6 7 8 9	185 235 275 320	Illinois Glaciation				
0.79				PRE-ILLINOIAN		Unnamed Glac. Klaza	: 15	790	Pre-Reid Glac. and Interglacial Periods	Wounded Moose paleosols (dt, op)		Paleosol 4 (b) Paleosol 3 (b)	Mokowan Butte 4–5 (t)
1.10 1.70			EARLY			Nansen, Banks Unnamed Glac.						Paleosol 2 (b) Paleosol 1 (b)	
22.5	TERTIARY	Pl[+] Mi[+]	LATE								Burnt Hill paleosol (t–b) Bluefish River paleosols (t–b)		Cypress Hill paleosols (t) Mokowan Butte 1–3 paleosols (t)
37.5		Ol[+]	MIDDLE										
65.0		Eo[+] Pa[+]	EARLY										Eureka Sound paleosols (t)

* The associated environments, given in brackets, are: temperate (t), dry temperate (dt), open parkland (op), boreal (b), high boreal (hb), subarctic (sa), subalpine (a), low Arctic (la), and high Arctic (ha) (Ecoregions Working Group, 1989).

+ The epochs are: Pliocene (Pl), Miocene (Mi), Oligocene (Ol), Eocene (Eo), and Paleocene (Pa).

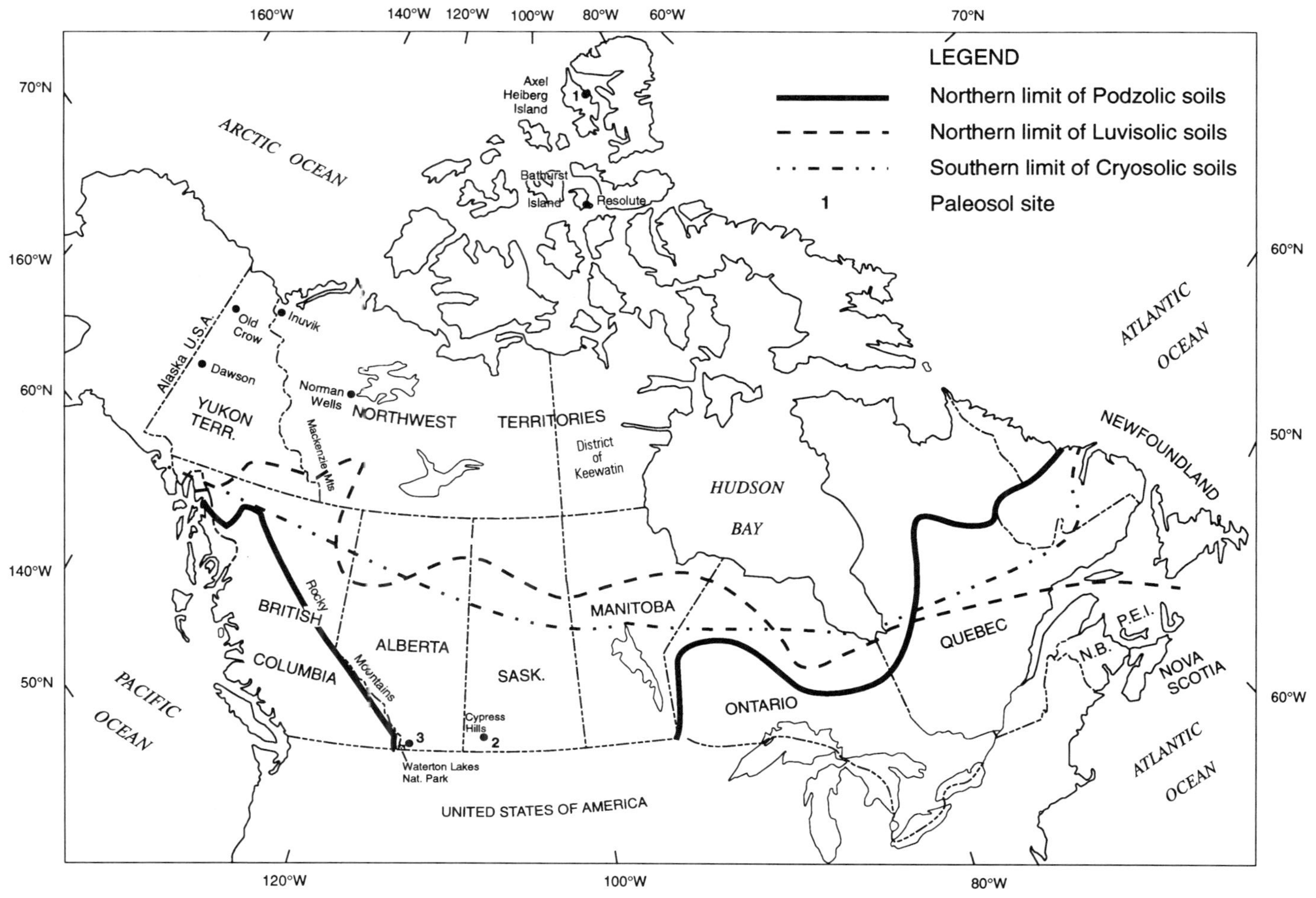

Fig. 17.1. The present northern limit of Luvisolic and Podzolic soils and the southern limit of Cryosolic soils in Canada. The locations of paleosol sites are as follows: 1. Eureka Sound paleosols; 2. Cypress Hills paleosols; and 3. Mokowan Butte paleosols.

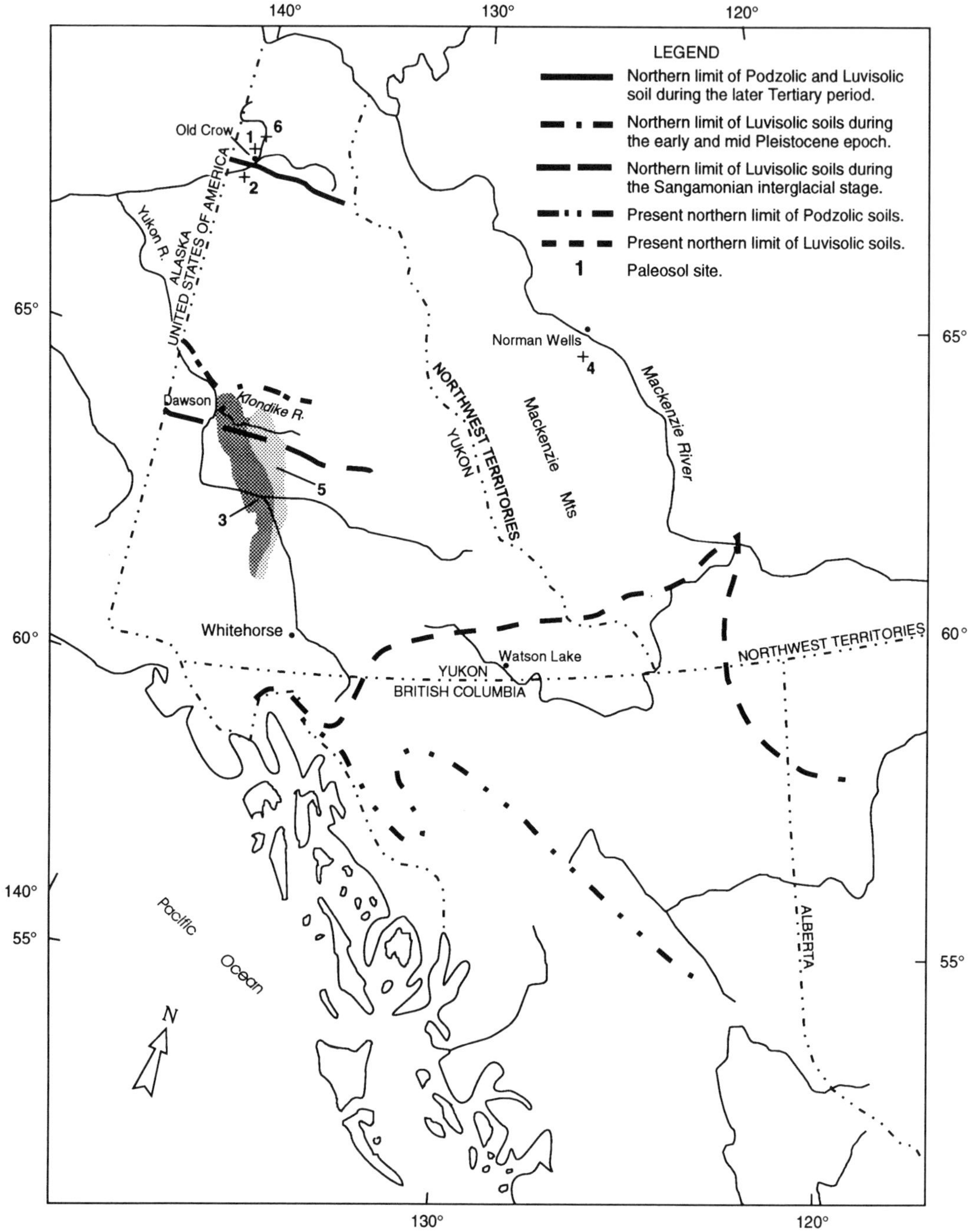

Fig. 17.2. The northern boundaries of some soils during the past 5 Ma. The locations of paleosol sites are as follows: 1. Burnt Hill paleosol; 2. Bluefish River paleosols; 3. Wounded Moose paleosols; 4. Little Bear River paleosols; 5. Diversion Creek paleosols; and 6. Old Crow paleosols.

resulting from compaction and shifting of materials and from tectonic movements.

The Gleysolic paleosols have organic-rich mineral surface horizons (Ah) or well-preserved humic to fibrous organic horizons. The underlying mineral horizons are strongly mottled and reduced.

The Organic paleosols are composed of fibrous woody peat. All coarse materials (twigs, cones, branches, and logs) in these peats are severely flattened because of the pressure resulting from the fluvial overburden or glacial ice. All of these paleosols contain nodules of amber ranging up to 4 cm in size. The wood in this peat material has not been permineralized (petrified), but rather it has been preserved in a mummified or weakly lignified state. This compressed peat is generally 50 to 80 cm thick, but it may be as thick as 200 cm. All of these paleosols are now perennially frozen and perhaps have been frozen since the beginning of the Pleistocene epoch.

Burnt Hill Paleosol (Late Tertiary)

The Burnt Hill paleosol occurs near the mouth of the Old Crow River (67° 35' N Lat., 139° 47.5' W Long.), approximately 10 km from Old Crow (paleosol site 1, Fig. 17.2). This Podzolic paleosol, which was buried by fluvial deposits, formed on gravelly sand of fluvial origin and has a well-developed, reddish (5YR 3/3, moist) Bf horizon overlain by a leached Ae horizon (Tarnocai, 1987a). On the basis of pollen data, the Burnt Hill paleosol is considered to be a soil that developed during the late Pliocene epoch (Tarnocai and Schweger, 1991).

Bluefish River Paleosols (Late Tertiary)

The Bluefish River paleosols occur along the Bluefish River Valley, approximately 60 km south of its confluence with the Porcupine River in the vicinity of Old Crow (paleosol site 2, Fig. 17.2). Some of the Bluefish River paleosols are located in the lower portion of the banks of the Bluefish River (67° 07' 02" N Lat., 140° 48' 24" W Long.), while others occur on the adjacent limestone upland (67° 08' 04" N Lat., 140° 46' 56" W Long.). Some of these paleosols were buried by fluvial deposits, but the upland Bluefish River paleosol lies on the surface, covered by only a thin loess cap.

These Luvisolic paleosols developed from calcareous residual material. Their Bt horizons are dominantly 5YR 4/5, moist, and contain significant clay accumulations (Tarnocai 1987a, 1987c). Soil material from the surface horizon of one of these paleosols had essentially the same pollen assemblage as the sediments near the paleosol at the Burnt Hill exposure (Tarnocai and Schweger, 1991). The soil development and, especially, the pollen data suggest that the Bluefish River paleosols are of late Pliocene age as well.

Cypress Hills Paleosols (Late Tertiary)

Cypress Hills paleosols are located in the Cypress Hills area of southwestern Saskatchewan and southeastern Alberta and lie on the surface, covered by only a thin loess cap (paleosol site 2, Fig. 17.1). The Cypress Hills area is a plateau that was not overrun by Quaternary glaciers. A reddish-yellow weathered surface lies over the bedrock of the Cypress Formation, which is of Oligocene and Miocene age (Storer, 1975). A dark reddish-brown paleo-B horizon is found in this material (Jungerius and Mucher, 1972; Westgate et al., 1972). According to Jungerius and Mucher (1972), the paleo-B horizon formed under forest vegetation in a climate somewhat warmer and moister than that of the Holocene epoch.

Mokowan Butte Paleosols (Late Tertiary; Early and Middle Pleistocene)

Mokowan Butte stands at an elevation of about 1,800 m along the eastern boundary of Waterton Lakes National Park in southwestern Alberta (49°

01' 59.4" N Lat., 107° 38' 38.5" W Long.). A stacked sequence of five paleosols, buried by fluvial deposits, was found in the area (Karlstrom, 1987, 1988) (paleosol site 3, Fig. 17.1).

Late Tertiary. The three oldest paleosols, Paleosols 1–3, occur in late Pliocene (late Tertiary) till or diamict. These three paleosols have strongly developed, rubified, argillic horizons over calcic, petrocalcic, or leached B horizons. In Paleosols 1 and 2 the argillic horizons are 40–150 cm thick while in Paleosol 3 they are 1–5 m thick.

Early and Middle Pleistocene. Paleosols 4 and 5, the more recent paleosols in the Mokowan Butte section (Karlstrom, 1987, 1988), developed during the early Pleistocene. They are characterized by 1–5 m thick, leached, rubified, argillic horizons overlying indurated petrocalcic, calcic, or leached B horizons.

Wounded Moose Paleosols (Early and Middle Pleistocene)

The Wounded Moose paleosols (Smith et al., 1986) occur in central Yukon areas with elevations less than 1,000 m (paleosol site 3, Fig. 17.2). They developed during the pre-Illinoian stage on till and glaciofluvial materials deposited during the pre-Reid glaciations (Tarnocai, 1987d). These soils developed strongly weathered, rubified, paleoargillic Bt horizons with thick clay skins or argillans (Smith et al., 1987; Tarnocai and Smith, 1989). Most Wounded Moose soils display strong cryoturbation in the form of disrupted and displaced soil horizons and oriented stones. As a result of cryoturbation, the clay skins in the upper B horizon are usually fragmented and dispersed. Sand wedges and sand involutions of various sizes are also common. Ventifacts are commonly found at the paleosol surface. These cryogenic features developed during the cold, glacial periods of the Pleistocene epoch. It should be noted that the oxygen isotope record obtained from marine sediments in the North Atlantic Ocean indicates the occurrence of a long cold period, especially between 0.37 and 0.84 Ma (Fig. 17.3B). Perhaps these well-developed and numerous cryogenic features originate mainly from this period.

Foscolos et al. (1977) and Rutter et al. (1978) suggest that the initial stage of Wounded Moose soil development was associated with a climate that was warm and subhumid, with grassland shrub vegetation. This type of climate was favorable for the development of montmorillonite. This was followed by a more temperate and humid climate that induced the degradation of montmorillonite to kaolinite through an intermediate step of mixed-layer montmorillonite-kaolinite.

Little Bear River Paleosols (Pleistocene)

The Little Bear River section lies approximately 80 km southwest of Norman Wells in the Mackenzie Mountains of the western Northwest Territories, on the right bank of the Little Bear River, 62 km southwest of the mouth of the river (64° 27.7' N Lat., 126° 43.3' W Long.). This area lies near the boundary between the Mackenzie Plain and the Mackenzie Mountains, in a zone between the former extent of the local montane glaciers originating in the Canyon Ranges of the Mackenzie Mountains and the maximum westward extent of the Laurentide Ice Sheet. In a section 40–50 m thick, five montane tills separated by paleosols are overlain by late Wisconsinan Laurentide boulder gravel (Hughes et al., 1993) (paleosol site 4, Fig. 17.2). The five paleosols have been numbered 1 to 5 from oldest to youngest. Paleosols 1 to 4 are considered to have developed in the pre-Illinoian interglacial stage of the early and middle Pleistocene, while Paleosol 5 developed in the Sangamonian stage of the late Pleistocene.

Early and Middle Pleistocene. The greater solum and B horizon thicknesses, the increase of redness in the B horizons, and the occurrence of some argillans in Paleosols 1 to 4 indicate stronger soil development with increased age. This

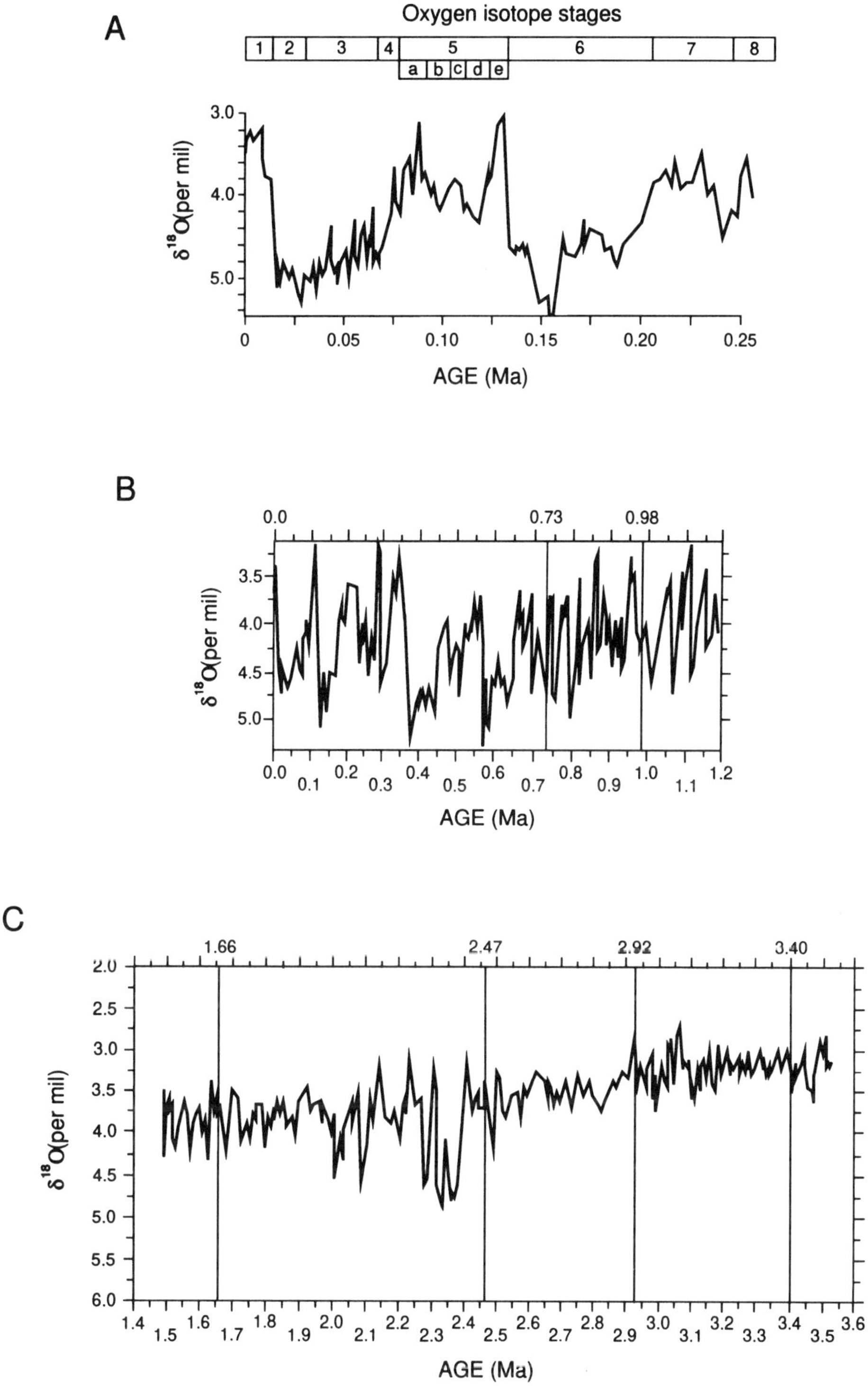

Fig. 17.3. Oxygen isotope records according to Ruddiman and McIntyre (1981) and Shackleton et al. (1984). The vertical lines (with values in Ma) in graphs B and C indicate the time of magnetic reversals.

suggests that, since the formation of the oldest paleosols, the climate has steadily cooled. The lack of evidence for cryoturbation indicates that these paleosols were not associated with permafrost.

An examination of the oxygen isotope record (Fig. 17.3B) indicates that between 1.2 Ma, approximately the time of the Nansen glacial advance, and 0.27 Ma, approximately the beginning of the Reid glaciation, four major warm peaks occurred (at 1.1 Ma, 0.95 Ma, 0.86 Ma, and a double warm peak at approximately 0.3 Ma), possibly representing four interglacial periods. The first warm period was the warmest.

If these paleosols developed during the warm periods identified on this oxygen isotope record (Fig. 17.3B) then, because these paleosols were buried, each is the result of the single interglacial climate under which it developed. Unlike the Wounded Moose paleosols, these paleosols do not show the long-term cumulative effect of several warm interglacial and cold glacial periods. The increased redness and the deeper sola of the older Little Bear River paleosols correlate with the oxygen isotope record and indicate that the climates under which these paleosols developed were much warmer than the present climate.

Late Pleistocene. The redness rating (Torrent et al., 1980) of Paleosol 5 is very similar to that for Paleosol 3 in this section. Paleosol 5 has weakly developed void and grain argillans (Hughes et al., 1993), very similar to those found in the Diversion Creek paleosols (Tarnocai and Smith, 1989).

Diversion Creek Paleosols (Late Pleistocene)

The Diversion Creek paleosols, which are found in the central Yukon (paleosol site 5, Fig. 17.2), began to form during the Sangamonian interglacial stage (Tarnocai et al., 1985; Smith et al., 1986; Tarnocai, 1987a). They developed on till and glaciofluvial materials produced by the Reid (Illinoian) glaciation. These paleosols have moderately well-developed B horizons, with the dominant color of the uppermost B horizon being 7.5YR. The Bt horizons, which are found only in Diversion Creek paleosols developed on till, have weakly developed grain argillans (Tarnocai and Smith, 1989). Sand wedges are present but are not common. Cryoturbated features such as oriented stones, silt cappings, and distorted horizons are more common on the Diversion Creek till paleosols than on the outwash paleosols. Ventifacts occur on all of these Diversion Creek paleosol surfaces. Weathering of coarse fragments in these paleosols is moderate, and frost-shattered coarse fragments commonly occur.

Old Crow Paleosols (Late Pleistocene)

These Cryosolic paleosols were found buried within a dominantly fluvial succession overlain by late Wisconsinan lake sediments in the Old Crow Basin of the northern Yukon (paleosol site 6, Fig. 17.2). They developed in a permafrost environment and are associated with gleyed horizons, mottles, cryoturbated features, patterned ground, and ice wedges. A bone buried in an Old Crow paleosol gave a radiocarbon date of 42±1.2 ka BP (R.E. Morlan, personal communication, 1987). Although Cryosolic soils are common in the Old Crow Basin at the present time, the ice wedge formations found in the Old Crow paleosols indicate the presence of a colder climate than now occurs. This is confirmed by the oxygen isotope record (Fig. 17.3A), which indicates the occurrence at that time of a cold period that was much colder than the present climate.

Buried Organic Layers on Bathurst Island (Late Pleistocene)

Buried organic matter found in Cryosolic soils at Goodsir Inlet on the west central part of Bathurst Island (Fig. 17.1) in the Canadian Arctic Archipelago yielded a radiocarbon date of 38.5±1.4 ka BP (GSC-3048). The morphology and composition of this cryoturbated organic horizon indicate that the organic material was incorporated into the subsoil by cryoturbation and that the soil was a Cryosol 38 ka ago. The soil at that time was most likely very similar to the present-day Cryosolic soils. This suggests that the climate of the area was favorable for biological activity and Cryosolic soil formation. Buried peat layers at Goodsir Inlet yielded a radiocarbon date of 35.0 ka BP (GSC-178; Blake, 1964) while those from the nearby Scoresby Hills area yielded a date of 38.0 ka BP (GSC-1878; Blake, 1964). The presence of these peat layers also indicates that the climate was favorable for peat deposition.

Holocene Relict Soil Features

The variation in climate during the Holocene epoch resulted in the development of soil features that became relict when the climate subsequently changed. After deglaciation the climate gradually improved in all parts of Canada. The climatic optimum, or hypsithermal interval, was reached during the early part of the Holocene epoch, when conditions were warmer and drier than now (Anderson et al., 1989). Holocene relict soils and soil features include relict patterned ground, such as earth hummocks, organic soils from the hypsithermal interval, and forest soil features preserved north of the arctic tree line.

Buried Cryoturbated Organic Materials in Earth Hummocks

Radiocarbon dating was carried out on buried organic materials from earth hummocks in the Arctic and Subarctic regions of Canada (Zoltai et al., 1978; Tarnocai, 1987b). Most of the buried organic materials were younger than 4.8 ka. Cryoturbation, as indicated by Zoltai et al. (1978), began about 4.5 ka and was very intense between 3.0 ka and 3.5 ka. The formation of earth hummocks also began about 4.5 ka. In the Old Crow area of the Yukon, cryoturbated organic material from a Cryosolic soil associated with patterned ground yielded radiocarbon dates between 3.6 ka and 3.7 ka (Tarnocai, 1987b). This period was also considered to have had a cooler climate that initiated extensive cryogenic activities and patterned ground formation. This patterned ground (circles) is now covered with a continuous thick mat of humus and high-organic-content mineral materials.

Relict Earth Hummocks

Relict earth hummocks have been found at a number of locations in Canada. Tarnocai (1973) reported the occurrence of relict earth hummocks with continuous soil horizons in the southern part of the Mackenzie River valley. These soil horizons showed no evidence of cryoturbation, suggesting that when hummock formation and the associated cryoturbation ceased, continuous soil horizons were able to develop in these fossil earth hummocks. Similar fossil earth hummocks were found in the South Indian Lake area of northern Manitoba (H. Veldhuis and G.F. Mills, personal communication, 1987).

Scotter and Zoltai (1982) reported relict earth hummocks and cryoturbated features in alpine soils in the southern Canadian Rocky Mountains. They concluded that these features resulted from the cooler climatic conditions associated with neoglacial activities.

Organic Soil Development during the Hypsithermal Interval

Radiocarbon dates of basal peat materials indicate that peat deposition in continental Canada did not begin until several thousand years after the continental ice sheet melted. A comparison of all of the radiocarbon dates available from the continent indicates that the majority of peat deposits began to develop between 4 and 6 ka BP (Tarnocai, 1978). On the Canadian Arctic islands, however, the basal peat dates are older, ranging from 8.5 to 9 ka BP (Tarnocai, 1978; Tarnocai and Zoltai, 1988). Tarnocai and Zoltai (1988) also indicated that the peat deposits in the High Arctic accumulated at a greater rate between 5 and 10 ka BP than did the Low Arctic peat deposits.

These basal dates indicate that peat development in the High Arctic began shortly after deglaciation and, as was suggested by Tarnocai (1978), probably resulted from the favorable climatic conditions existing there at that time. There is evidence that, after the retreat of glacial ice from the North American continent, there was a relatively warm and dry period lasting several thousand years (Ritchie and Hare, 1971; Terasmae, 1972; Delorme et al., 1977). Tarnocai (1978) suggested that it is possible that the climate was generally too warm and dry in the south during this period for optimum peat development. In the High Arctic areas, however, the climate was assumed to have been cooler and moister (although warmer than at present) after the retreat of the glacial ice, and, consequently, conditions were favorable for peat development. As the climate became colder and peat development virtually ceased in the High Arctic, the Boreal and the Subarctic regions became established as the areas of optimum peat development.

Forest Soils North of the Arctic Tree Line

The distribution of paleosols makes it possible to reconstruct the positions of the arctic tree line and to estimate characteristics of the past climate (Sorenson, 1973, 1977; Sorenson and Knox, 1974). According to these authors, the arctic tree line varied from 280 km north to at least 50 km south of its present position. In the District of Keewatin in the Northwest Territories (Fig. 17.1), the occurrence of buried charcoal and associated Podzolic soils confirms that at some time during the Holocene the tree line was up to 280 km north of its present position (Bryson et al., 1965). It should be noted, however, that the soils Bryson et al. (1965) identified as Podzols are now classified as Brunisols (Agriculture Canada Expert Committee on Soil Survey, 1987). These paleobrunisolic soils are frequently buried by eolian sands on which Arctic Brown soils (Brunisolic Static Cryosols) developed, often in association with polygonal patterned ground (Sorenson et al., 1971). Bryson et al. (1965) and Nichols (1967, 1970) give evidence for southern shifts of the arctic tree line. According to Sorenson (1973, 1977) and Sorenson and Knox (1974) major northward shifts of the arctic tree line occurred about 2.9, 1.8, and 0.8 ka BP and southward movements occurred about 3.5, 2.6, and 1.6 ka BP. The pollen records presented by Ritchie and Hare (1971) and Ritchie (1984) give further evidence of these fluctuations.

Paleoclimatic Reconstruction Using Soil Features

Soil features can be used in paleoclimatic reconstruction as indicated by the following points.

1. Argillic horizons (Bt horizons associated with clay skins on ped surfaces or in pores) are diagnostic of Luvisolic soils.

2. The occurrence of Podzolic horizons (Bf horizons associated with an accumulation of iron and aluminum) indicates a warm or cool, moist temperate climate.

3. Reddening (rubification) through *in situ* hematite formation can take place under temperate climatic conditions with a mean annual temperature of 7°C or greater and total annual precipitation of as little as 500 mm (Torrent et al., 1980; Schwertmann et al., 1982).

4. Soil thickness can indicate the soil's age and give some idea of the climate under which it formed. In most paleosols this thickness is the sum of the thicknesses of all horizons where paleopedological features such as argillans, redness, leaching of carbonates, and soil structure are present.

5. The degree and depth of leaching of carbonates ($CaCO_3$) is controlled by the carbonate content of the parent material, the availability of water (rainfall), and the length of time available.

6. The degree of weathering of both the clay minerals and the coarse fragments is dependent on both the climate (temperature and moisture) and the length of time available.

7. The formation of peat is mainly dependent on climate. If favorable climatic conditions exist during long periods of time, the organic depositions can reach great thicknesses. If the favorable climatic conditions cease to exist, however, peat formation stops.

8. The occurrence of cryogenic macrofeatures (cryoturbation, ice-wedge casts, and sand wedges) and microfeatures (papules, embedded grain argillans, aggregates, oriented fabrics, and silt cappings) indicates the presence of cold soil conditions or permafrost.

Paleoclimatic Reconstruction

The presence of argillic or Podzolic horizons are indicative of advanced soil development. Argillic horizons are usually associated with a moist climate with drier periods, while Podzolic horizons are associated with a warm or cool, moist temperate climate.

The presence of redness, or rubification, is characteristic of late Tertiary and early Pleistocene (pre-Illinoian) paleosols, indicating that the climate at that time was temperate, with a mean annual temperature of 7°C or greater and total annual precipitation of as little as 500 mm.

Smith et al. (1987) studied the pre-Illinoian Wounded Moose paleosols in the central Yukon and found that the redness of the paleoargillic horizons resulted from small amounts of hematite (<3%) concentrated in submicroscopic zones. This rubification was interpreted as a relict feature. This area (Dawson) now has a mean annual temperature of -5.1°C and total annual precipitation of 306 mm (Atmospheric Environment Service, 1982b), which eliminates the possibility of such rubification at the present time and suggests the presence of a much warmer climate during the early Pleistocene epoch.

Thicker sola, which can indicate the soil's age and give some idea of the climate under which it formed, are common in paleosols of the pre-Illinoian interglacial stage. In the central Yukon an average solum thickness of 2 m was associated with the Wounded Moose paleosols (pre-Illinoian), 1 m with the Diversion Creek paleosols (Sangamonian), and 0.5 m with the contemporary (Holocene) soils (Smith et al., 1986).

The depth of leaching of carbonates ($CaCO_3$), which is controlled by the carbonate content of the parent material, the availability of water (rainfall) and the length of time available, was greatest in the pre-Illinoian interglacial paleosols and much less in the recent soils.

In the Yukon the degree of weathering, which is dependent on both the climate (temperature and moisture) and the length of time available, was greatest in the late Tertiary and the pre-Illinoian paleosols, less in the Sangamonian paleosols, and minimal in the Holocene soils.

During the hypsithermal interval in the early Holocene epoch, climatic conditions were more favorable for peat formation on the Arctic islands than in the area that now lies in the Boreal and the Subarctic regions. When the climate cooled following the hypsithermal interval, the climate in the Arctic islands became too cold and dry for peat development. Peat development ceased or became very slow in this area and the optimal area for peat development shifted to more favorable areas, now the Boreal and the Subarctic regions (Tarnocai, 1978).

In modern soils, cryogenic soil features develop under cold climates with mean annual air temperatures of -2°C or less. Some cryogenic features, such as ice-wedge casts, are good indicators that permafrost conditions existed and that ice-wedge formation took place. The presence of sand wedges also indicates an extremely cold environment in which, because of extreme cooling, the frozen soil cracked.

Paleoclimatic Reconstruction Using Northern or Southern Limits of Soils

The northern limit of Luvisols and Podzols and the southern limit of Cryosols are especially sensitive to climate change and can be useful in predicting former climates. It appears that when major changes occur in one of the soil-forming factors such as climate, the system adjusts in a relatively short period of time. This adjustment produces both soil features and soil development that are strongly related to the changed soil-forming factor. Such adjustments occur not only because of a change in climate but also because of changes in other soil-forming factors, such as hydrology and vegetation. If climatic change does take place, however, it has an overriding effect on soil development. In a relatively short period of time the change in climate produces certain soil features or soil development that appear as an overprint on the previous soil development, but features developed before the change of climate often remain as relict features in the soil. In some cases this overprint is visible and long lasting.

On the basis of the northern extent of contemporary soils and an interpretation of their features, a model was developed to reconstruct paleoclimates using soil development and certain soil properties. This model compared the paleosols and the relict features with the present-day soils and soil features and used information concerning the northern limits of Luvisolic and Podzolic soils and the southern limit of Cryosolic soils in western and northwestern Canada.

There are many examples of relict soil features being interpreted in terms of climate change. Bryson and Wendland (1967) and Sorenson (1973) found evidence that the existence of forest soils north of the present arctic tree line can be used to indicate the position of the former tree line. The occurrence of cryogenic features and patterned ground on the northern fringe of the

Brunisols and Luvisols indicates that permafrost extended farther south during the colder periods of the Holocene (Tarnocai, 1973). The presence of collapse scars on the northern fringe of the Organic soils was interpreted as indicating that these scars marked the position of former Organic Cryosols during the Little Ice Age (approximately 1640 to 1870). When the climate warmed during the 1800s, the permafrost melted and produced scars in these areas (Zoltai, 1971).

The northern limit of contemporary Luvisols and Podzols lies in the interior part of western Canada, at approximately 61° and 58° north latitude, respectively (Figs. 17.1, 17.2; Clayton et al., 1977). At latitude 61° north, the Luvisolic limit, the mean annual temperature is -3.5°C and the mean annual precipitation is 375 mm. The corresponding temperature and precipitation at the Podzolic limit are -1.0°C and 450 mm. Luvisols and Podzols are both considered to be forest soils and support various types of forest vegetation. The southern limit of the present Cryosolic soils (Fig. 17.1) coincides approximately with the southern limit of the Discontinuous Permafrost Zone, which corresponds approximately to the -1°C isotherm (Brown, 1967). By using these limits and associated climatic data, it is possible to reconstruct the climate for the northern boundary of the paleosols in the Tertiary period and the various interglacial periods of the Quaternary period.

Tertiary Limits

Early Tertiary Podzolic, Gleysolic, and Organic paleosols have been found on Axel Heiberg Island (fossil forest area), indicating that the climate of the High Arctic at that time was temperate or warm temperate with mild, wet winters and warm, moist summers. Coniferous vegetation was associated with these paleosols on both well to imperfectly drained areas (coniferous forests) and poorly drained areas (forested swamps).

Evidence from the Burnt Hill and the Bluefish River paleosols suggests that the late Tertiary environment in the northern Yukon was associated with Luvisolic and Podzolic soil development. These types of soil development extended to at least 68° north latitude, in the area of Old Crow (Fig. 17.2). At the present time Old Crow has a mean annual temperature of -10.1°C and total annual precipitation of 214 mm (Atmospheric Environment Service, 1982b). The occurrence of Podzolic paleosols suggests that at some point during the late Tertiary (Pliocene), the mean temperature in the Old Crow area was at least 9°C warmer and that the total precipitation was at least twice what it is today. Karlstrom (1987) suggested that the late Tertiary Mokowan Butte Luvisolic paleosols in southwestern Alberta also formed in an environment with mean annual temperatures at least 6° to 10°C warmer than today.

The late Tertiary Luvisolic and Podzolic paleosols are also considered to be forest soils, as are the contemporary soils. The pollen data confirm this, indicating that these paleosols supported a coniferous forest with white pine and hemlock in the tree layer and an understory of alder and *Corylus* (Tarnocai and Schweger, 1991). The northern limit of the *Corylus* sp. now lies in central British Columbia, at approximately 56° north latitude (Drumke, 1964). This indicates that the late Tertiary climate of the Old Crow area was probably comparable to the present climate of the central interior of British Columbia, which has a mean annual temperature of approximately 4°C and total annual precipitation of 550 mm (Atmospheric Environment Service, 1982a). The presence of *Corylus* in the Old Crow area indicates that the temperature in this area was at least 14°C higher than at present. It is thus likely that the northern limits of the Luvisolic and Podzolic soils were even farther north during the late Tertiary.

Pleistocene Limits

The Wounded Moose paleosols represent the northern limit of rubified Luvisolic soils during the early and middle Pleistocene interglacial periods (pre-Illinoian). In the central Yukon these paleosols extend to approximately 65° north latitude and occur at elevations below 1,000 m a.s.l. (Fig. 17.2). If this was the northern limit of the early and middle Pleistocene Luvisolic soils, the temperature was probably 8°C warmer and the precipitation slightly higher than it is now. This limit, however, is arbitrary, since the Wounded-Moose paleosols are associated with pre-Reid glacial deposits. It is possible that Wounded Moose-like paleosols also exist in the unglaciated areas of the Yukon, outside of the pre-Reid limit. Smith et al. (1987) suggested that the mean annual temperature of the central Yukon was 7°C or higher, which is about 10°C warmer than at present. Karlstrom (1987) found that the early Pleistocene Mokowan Butte paleosols in southwestern Alberta were rubified Luvisolic paleosols. He suggested that these paleosols formed under an interglacial climate with a mean annual temperature at least 6°C warmer than the mean temperature today.

The Diversion Creek paleosols mark the northern limit of Luvisolic soils during the Sangamonian stage of the Pleistocene epoch. Although the most northerly occurrence of these soils is in the Chandindu and the Benson Creek moraines in the Klondike Valley (Dawson area, Yukon), the Diversion Creek paleosols in these locations are Brunisols. The most northerly occurrence of the Luvisolic type of Diversion Creek paleosols lies at approximately 65.5° north latitude (Fig. 17.2). These Diversion Creek paleoluvisolic soils are very similar in soil development to the Luvisols now found in the Watson Lake area of the southeastern Yukon and in northeastern British Columbia, at the northern limit of the present Luvisolic soils. This suggests that the Sangamonian climate at the Luvisolic soil boundary was at least 1°C warmer and that the precipitation was only slightly higher than it is at the present time.

The Little Bear River paleosols have no cryogenic features, suggesting that no permafrost was present at that location during the Sangamonian and the pre-Illinoian interglacial stages.

Holocene Limit

No major shifts in soil boundaries have occurred during the Holocene epoch, with the exception of that for the Cryosolic soils. Organic Cryosol development was active in the High Arctic for several thousand years after deglaciation, but then almost ceased because the climate changed, becoming cold and dry. The occurrence of paleoforest soils north of the present tree line suggests that the arctic tree line has fluctuated by at least 200 km a number of times. If the location of the arctic tree line is considered to coincide with the 10°C isotherm for the warmest month (Hopkins, 1959), it indicates the occurrence of alternating periods of slight cooling and warming during the Holocene epoch.

Problems in Interpretation

If the techniques of physical and chemical analysis used for paleosols are the same as those used for contemporary soils, care must be taken in interpreting the data generated. For example, it should be noted that there are indications that the sodium-pyrophosphate-extractable Fe and Al values change over time. Calderoni and Schnitzer (1984) and Schnitzer and Calderoni (1985) found that carbon compounds undergo a conversion with time. The final result is that after 30 ka burial, only the carboxylic and aromatic carbon compounds remain in the paleosols. This suggests that the larger and more soluble fulvic acids, to which the Fe and Al ions are attached and which are determined by the pyrophosphate method, degrade very early and are converted into smaller and more stable carbon compounds. These compounds, identified by Schnitzer and Calderoni (1985), have little or no metallic-ion-holding capability. This indicates that the use of the sodium-pyrophosphate-extractable Fe and Al criteria developed for recent Podzols is questionable when attempting to identify podzolization in paleosols. The identification of these Podzolic paleosols is therefore based on their morphology and the dithionite-citrate- and ammonium-oxalate-extractable Fe and Al values, not on the sodium-pyrophosphate-extractable Fe and Al values, which are used to identify and classify Holocene Podzolic soils (Agriculture Canada Expert Committee on Soil Survey, 1987).

In order to obtain information about when cryogenic processes were active in mineral soils, it is useful to carry out ^{14}C dating. However, the sampling of the soils and the interpretation of such data should also be handled with care. In these soils, cryogenic processes can both translocate organic materials into the mineral soils and mix older and younger organic materials. The ages obtained for the organic material could thus predate the time when these processes were active.

Conclusion

The paleosols that developed during the late Tertiary period and the following interglacial periods in the Quaternary period give evidence of a gradually cooling climate. A summary of the climates, based on the soil development and certain soil features found in various paleosols, follows.

1. *Early Tertiary epoch.* The High Arctic environment was probably characterized by a climate that was temperate to warm temperate, with mild, wet winters and warm, moist summers. Coniferous forest vegetation was present in both upland and wetland areas.

2. *Late Tertiary (Pliocene) epoch.* The northern limit of Luvisolic and Podzolic paleosols was at least 68° north latitude, or perhaps much farther to the north. The prevailing climate probably had a mean annual temperature of at least 4°C (about 14°C higher than at present) and mean annual precipitation at least twice what it is now.

3. *Early and middle Pleistocene interglacial stages.* The northern limit of Luvisols was approximately 65° north latitude. The mean annual temperature was about 8°C higher than the present mean, and the precipitation was slightly higher than it is now. These interglacial periods produced the most highly developed paleosols of the Quaternary period, not only because the climate was favorable for soil development but also because of the long period of time during which these conditions existed.

4. *Sangamonian interglacial stage.* The northern limit of Luvisolic soils was at about 63° north latitude, approximately 3° farther north than at present. The mean annual temperature was probably only 1°C higher and the precipitation only slightly higher that it is now. There is no indication that permafrost was present, at least in some locations. Soil development and the distribution of Luvisols suggests that the climate was similar to or slightly warmer than the present climate.

5. *Ice-free areas during the Wisconsinan stage.* Cryosolic soils with ice-wedge casts at 68° north latitude are evidence of the presence of a permafrost environment with a climate similar to, or slightly cooler than, the present climate. The occurrence of cryoturbated organic layers at 76° north latitude also indicates that Cryosolic soil development and organic matter production took place at that time, further suggesting that the climate was very similar to the present climate.

6. *Holocene interglacial epoch.* Climatic fluctuations have occurred since deglaciation. Shortly after deglaciation, the climate on the Arctic islands was warmer and moister than at present, while in continental Canada the climate was warm and dry. Following this early stage, the climate cooled, and the area of optimum peat development shifted to the south, to the Boreal and Subarctic areas of Canada.

7. A comparison of the northern limits of various soils during the Tertiary and the interglacial periods of the Quaternary indicates that none of the interglacial periods before the Holocene epoch was associated with Cryosolic soils to the extent evident during the Holocene. Even though their boundaries fluctuate, Cryosols are the dominant soils during the Holocene epoch, covering approximately 35 percent of the land area of Canada. The paleosols thus yield evidence of steadily cooling interglacial climates since the end of the Tertiary period.

References

Agriculture Canada Expert Committee on Soil Survey, 1987. The Canadian system of soil classification. Research Branch, Agriculture Canada, Publication No. 164, 2nd ed., 164 pp.

Anderson, T.W., Mathewes, R.W. and Schweger, C.E.,

1989. Holocene climatic trends in Canada with special reference to the Hypsithermal interval. In: R.J. Fulton, J.A. Heginbottom and S. Funder (Editors), Quaternary geology of Canada and Greenland. Geological Survey of Canada, Ottawa. Geological Society of America, Geology of North America, vol. K-1, pp. 520–528.

Atmospheric Environment Service, 1982a. Canadian climate normals, temperature and precipitation: the North-British Columbia. Environment Canada, Atmospheric Environment Service, UDC: 551.582 (711), 268 pp.

Atmospheric Environment Service, 1982b. Canadian climate normals, temperature and precipitation: the North-Yukon Territory and Northwest Territories. Environment Canada, Atmospheric Environment Service, UDC: 551.582(712), 55 pp.

Basinger, F., 1987. Our "tropical" Arctic. Canadian Geographic, 107: 28–37.

Basinger, J.F., 1991. The fossil forests of the Buchanan Lake Formation (early Tertiary), Axel Heiberg Island, Canadian Arctic Archipelago: preliminary floristics and paleoclimate. In: R.L. Christie and N.J. McMillan (Editors), Tertiary fossil forests of the Geodetic Hills, Axel Heiberg Island, Arctic Archipelago. Geological Survey of Canada, Bulletin 403, pp. 39–65.

Blake, W., 1964. Preliminary account of the glacial history of Bathurst Island, Arctic Archipelago. Geological Survey of Canada, Paper 64-30, 8 pp.

Bostock, H.S., 1966. Notes on glaciation in central Yukon Territory. Geological Survey of Canada, Paper 65–36, 18 pp.

Brown, R.J.E., 1967. Permafrost in Canada. Geological Survey of Canada, Map No. 1246A, 1st ed.

Bryson, R.A. and Wendland, W.M., 1967. Tentative climatic patterns for some late-glacial and post-glacial episodes in central North America. In: W.J. Mayer-Oakes (Editor), Life, land and water. University of Manitoba, Winnipeg, Manitoba, pp. 271–298.

Bryson, R.A., Irving, W.N. and Larsen, J.A., 1965. Radiocarbon and soil evidence of former forest in the southern Canadian tundra. Science, 147: 46–48.

Calderoni, G. and Schnitzer, M., 1984. Nitrogen distribution as a function of radiocarbon age in paleosol humic acids. Organic Geochemistry, 5: 203–209.

Clayton, J.S., Ehrlich, W.A., Cann, D.B., Day, J.H. and Marshall, I.B., 1977. Soils of Canada. Research Branch, Canada Department of Agriculture, vol. 1, 243 pp., and vol. 2, 239 pp.

Delorme, L.D., Zoltai, S.C. and Kalas, L.L., 1977. Freshwater shelled invertebrate indicators of paleoclimate in northwestern Canada: the late glacial times. Canadian Journal of Earth Sciences, 14: 2029–2046.

Drumke, J.S., 1964. A systematic survey of *Corylus* in North America. Ph.D. thesis, University of Tennessee, 142 pp.

Ecoregions Working Group. 1989. Ecoclimatic Regions of Canada, first approximation. Ecoregions Working Group of the Canada Committee on Ecological Land Classification. Ecological Land Classification Series, No. 23, Sustainable Development Branch, Canadian Wildlife Service, Conservation and Protection, Environment Canada, Ottawa, Ontario. 119 pp. and map at 1:7 500 000.

Foscolos, A.E., Rutter, N.W. and Hughes, O.L., 1977. The use of pedological studies in interpreting the Quaternary history of central Yukon. Geological Survey of Canada, Bulletin 271, 48 pp.

Fulton, R.J. and Prest, V.K., 1987. Introduction: The Laurentide Ice Sheet and its significance. Géographie physique et Quaternaire, 41: 181–186.

Hopkins, D.M., 1959. Some characteristics of the climate in forest and tundra regions in Alaska. Arctic, 12: 214–220.

Hughes, O.L., Tarnocai, C. and Schweger, C.E., 1993. Pleistocene stratigraphy, paleopedology, and paleoecology of a multiple till sequence exposed on the Little Bear River, Western District of Mackenzie, N.W.T., Canada. Canadian Journal of Earth Sciences, 30: 851–866.

Jungerius, P.D. and Mucher, H.J., 1972. The micromorphology of fossil soils in the Cypress Hills, Alberta, Canada. In: S. Kowalinski (Editor), Soil micromorphology. Panst. Wydawn. Nauk., Warszawa, Poland, pp. 617–627.

Karlstrom, E.T., 1987. Stratigraphy and genesis of five superposed paleosols in pre-Wisconsinan drift on Mokowan Butte, southwestern Alberta. Canadian Journal of Earth Sciences, 24: 2235–2253.

Karlstrom, E.T., 1988. Multiple paleosols in pre-Wisconsinan drift, northwestern Montana and southwestern Alberta. Catena, 15: 147–178.

Nichols, H., 1967. The postglacial history of vegetation and climate at Ennadai Lake, Keewatin, and Lynn Lake, Manitoba. Eiszeitalter und Gegenwart, 18: 176–197.

Nichols, H., 1970. Late Quaternary pollen diagrams from the Canadian Arctic barren grounds at Pelly Lake, northern Keewatin, N.W.T. Arctic and Alpine Research, 2: 43–61.

Ricketts, B.D. and McIntyre, D.J., 1986. The Eureka Sound Group of eastern Axel Heiberg Island: new data on the Eurekan Orogeny. Current Research, Part B, Geological Survey of Canada, Paper 86-1B, pp. 405–410.

Ritchie, J.C., 1984. Past and present vegetation of the

far northwest of Canada. University of Toronto Press, Toronto, Ontario, 251 pp.

Ritchie, J.C. and Hare, F.K., 1971. Late Quaternary vegetation and climate near the arctic tree line of northwestern North America. Quaternary Research, 1: 331–342.

Ruddiman, W.F. and McIntyre, A., 1981. Oceanic mechanism for amplification of the 23,000-year ice volume cycle. Science, 212: 617–627.

Rutter, N.W., Foscolos, A.E. and Hughes, O.L., 1978. Climatic trends during the Quaternary in central Yukon based upon pedological and geological evidence. Proceedings of the Third York Quaternary Symposium on Quaternary Soils. Geo. Abstracts, Norwich, England, pp. 309–395.

Schnitzer, M. and Calderoni, G., 1985. Some chemical characteristics of paleosol humic acids. Chemical Geology, 53: 175–184.

Schwertmann, U., Murad, E. and Schultze, D.G., 1982. Is there Holocene reddening (hematite formation) in soils of axeric temperate areas? Geoderma, 27: 209–223.

Scotter, G.W. and Zoltai, S.C., 1982. Earth hummocks in the Sunshine area of the Rocky Mountains, Alberta and British Columbia. Arctic, 35: 411–416.

Shackleton, N.J., Backman, J., Zimmerman, H., Kent, D.V., Hall, M.A., Roberts, D.G., Schnitker, D., Baldauf, J.G., Desprairies, A., Homrighausen, R., Huddlestun, P., Keene, J.B., Kaltenback, A.J., Krumsiek, K.A.O., Morton, A.C., Murray, J.W. and Westberg-Smith, J., 1984. Oxygen-isotope calibration of the onset of ice-rafting and history of glaciation in the North Atlantic region. Nature, 307: 620–623.

Smith, C.A.S., Spiers, G.A. and Tarnocai, C., 1987. Why are the mid-Pleistocene soils of the central Yukon red? XII INQUA International Congress Abstracts, National Research Council of Canada, Ottawa, p. 266.

Smith, C.A.S., Tarnocai, C. and Hughes, O.L., 1986. Pedological investigations of Pleistocene glacial drift surfaces in the central Yukon. Géographie physique et Quaternaire, 40:29–37.

Soil Survey Staff, 1994. Keys to soil taxonomy. 6th edition. United States Department of Agriculture, Soil Conservation Services, Washington, DC, 306 pp.

Sorenson, C.J., 1973. Interrelationships between soils and climate and between paleosols and paleoclimates: forest/tundra ecotone, north central Canada. Ph.D. thesis, University of Wisconsin, Madison, Wisconsin, 237 pp.

Sorenson, C.J., 1977. Reconstructed Holocene bioclimates. Annals of the Association of American Geography, 67: 214–222.

Sorenson, C.J., Knox, J.C., Larsen, J.A. and Bryson, R.A., 1971. Paleosols and forest border in Keewatin, N.W.T. Quaternary Research, 1: 468–473.

Sorenson, C.J. and Knox, J.C., 1974. Paleosols and paleoclimates related to late Holocene forest/tundra border migration: Mackenzie and Keewatin, N.W.T., Canada. In: S. Raymond and P. Schledermann (Editors), International Conference on the Prehistory and Paleoecology of Western North American Arctic and Subarctic. Archaeological Association, University of Calgary, Calgary, Alberta, pp. 187–203.

Storer, J.E., 1975. Middle Miocene mammals from the Cypress Hills, Canada. Canadian Journal of Earth Sciences, 12: 520–522.

Tarnocai, C., 1973. Soils of the Mackenzie River area. Environmental-Social Committee, Northern Pipelines, Task Force on Northern Oil Development, Report No. 73-26, 136 pp.

Tarnocai, C., 1978. Genesis of organic soils in Manitoba and the Northwest Territories. Proceedings of the Third York Quaternary Symposium on Quaternary Soils, Geo. Abstracts, Norwich, England, pp. 453–470.

Tarnocai, C., 1987a. Quaternary soils. In: S.R. Morison and C.A.S. Smith (Editors), Guidebook to Quaternary research in Yukon. XII INQUA Congress, National Research Council of Canada, Ottawa, pp. 16–21.

Tarnocai, C., 1987b. Stop 29B: Recent, strongly cryoturbated soil. In: S.R. Morison and C.A.S. Smith (Editors), Guidebook to Quaternary research in Yukon. XII INQUA Congress, National Research Council of Canada, Ottawa, pp. 84–87.

Tarnocai, C., 1987c. Stop 29E: Buried paleosol overlain by gravel in the Bluefish River bed. In: S.R. Morison and C.A.S. Smith (Editors), Guidebook to Quaternary research in Yukon. XII INQUA Congress, National Research Council of Canada, Ottawa, pp. 90–95.

Tarnocai, C., 1987d. Stop 22: Wounded Moose paleosol developed on pre-Reid gravel. In: S.R. Morison and C.A.S. Smith (Editors), Guidebook to Quaternary research in Yukon. XII INQUA Congress, Ottawa, National Research Council of Canada, Ottawa, pp. 61–65.

Tarnocai, C. and Schweger, C.E., 1991. Late Tertiary and early Pleistocene paleosols in northwestern Canada. Arctic, 44: 1–11.

Tarnocai, C. and Smith, C.A.S., 1989. Micromorphology and development of some central Yukon paleosols, Canada. Geoderma, 45: 145–162.

Tarnocai, C. and Smith, C.A.S., 1991. Paleosols of the

fossil forest area, Axel Heiberg Island. In: R.L. Christie and N.J. McMillan (Editors), Tertiary fossil forests of the Geodetic Hills, Axel Heiberg Island, Arctic Archipelago. Geological Survey of Canada, Bulletin 403, pp. 171–187.

Tarnocai, C., Smith, S. and Hughes, O.L., 1985. Soil development on Quaternary deposits of various ages in the central Yukon Territory. Current Research, Part A. Geological Survey of Canada, Paper 85-1A, pp. 229–238.

Tarnocai, C. and Valentine, K.W.G., 1989. Relict soil properties of the arctic and subarctic regions of Canada. In: Paleopedology: A. Bronger and J. Catt (Editors), Nature and application of paleosols. Catena Supplement 16, pp. 9–39.

Tarnocai, C. and Zoltai, S.C., 1988. Wetlands of Arctic Canada. In: National Wetlands Working Group, Wetlands of Canada. Ecological Land Classification Series, No. 24. Sustainable Development Branch, Environment Canada, Ottawa, and Polyscience Publications Inc., Montreal, pp. 27–51.

Terasmae, J., 1972. Muskeg as a climate-controlled ecosystem. Proceedings of the 14th Muskeg Research Conference, Kingston, 1971. N.R.C. Technical Memorandum No. 102, pp. 147–158.

Torrent, J., Schwertmann, U. and Schultze, D.G., 1980. Iron oxide mineralogy of some soils of two river terrace sequences in Spain. Geoderma, 23: 191–208.

Valentine, K.W.G. and Dalrymple, J.B., 1975. The identification, lateral variation and chronology of two buried paleocatenas at Woodhall Spa and West Runton, England. Quaternary Research, 4: 551–590.

Valentine, K.W.G., King, R.H., Dormaar, J.F., Vreeken, W.J., Tarnocai, C., De Kimpe, C.R. and Harris, S.A., 1987. Some aspects of Quaternary soils in Canada. Canadian Journal of Soil Science, 67: 221–247.

Westgate, J.A., Bonnischsen, R., Schweger, C. and Dormaar, J.F., 1972. The Cypress Hills. In: N.W. Rutter and E.A. Christiansen (Editors), Guidebook for excursion C-22 of the 24th International Geological Congress, pp. 50–62.

Yaalon, D.H., 1971. Soil forming processes in time and space. In: D.H. Yaalon (Editor), Paleopedology. International Society of Soil Science and Israel Universities Press, pp. 29–38.

Zoltai, S.C., 1971. Southern limit of permafrost features in peat landforms, Manitoba and Saskatchewan. Geological Association of Canada, Special Paper 9, pp. 305–310.

Zoltai, S.C., Tarnocai, C. and Pettapiece, W.W., 1978. Age of cryoturbated organic materials in earth hummocks from the Canadian Arctic. Proceedings of the Third International Conference on Permafrost, Edmonton, Alberta, Canada, National Research Council of Canada, Ottawa, pp. 325–331.

18. COLD-TEMPERATE PEATS AND COALS: THEIR SEDIMENTOLOGY AND COMPOSITION

R. Marc Bustin

Coal geologists have long sought clues to the nature of the original peat-forming vegetation and environmental conditions by studying the composition of coal and the character of the enclosing sediment. This research has been spurred by the importance of coal composition to technological utility and by a general desire to understand the factors that control the distribution and the character of coal in the sedimentary record. Coal seams further provide the opportunity for study of wetland evolution at a time scale not available in modern mires. The ability of coal geologists to interpret the nature of the peatland from coal is closely tied to the understanding of peat-forming environments, and indeed many studies of modern peatlands have been carried out by coal geologists looking for clues to interpret the past geological record (Cohen and Spackman, 1977; Styan and Bustin, 1983a,b; Cameron et al., 1989; Esterle, 1990; Cobb and Cecil, 1993). One area that has received little consideration by geologists is cold-temperate peats. Cold-temperate peats include the majority of peat deposits in the world today (70 percent of the about 5 million km^2; Pfadenhauer et al., 1993) and occur in cold-temperate through subpolar regions (between 40° and 70° N and S Lat.). However, most coals are interpreted to have formed from tropical or subtropical peats (Moore, 1989), with the notable exception of the Gondwana coals. The questions thus arise, why are cold-temperate peats predominant today and yet so under represented in the geological record, and what are the characteristics of these modern peats and how do they compare with ancient coals such as those of Gondwanaland?

The study of cold-temperate peats has recently received a major renewal of interest from ecologists and earth scientists. Peats are excellent repositories of data on past climate and thus are exploited as proxy data for global change (Lottes and Zeigler, 1994). Middle- to high-latitude peats by their sheer volume and decay are a major source of atmospheric methane (a greenhouse gas) and a major carbon sink. In addition, they are important in the horticultural industry and in some northern countries are utilized as an energy resource (Heathwaite and Göttlich, 1993).

The purpose of this chapter is to provide an overview of the characteristics of peatlands, with particular emphasis on those factors that control their geometry, evolution and characteristics; to examine what characteristics, if any, cold-temperate peats possess that reflect their role as progenitors of coal in the geological record; and to describe and compare coals interpreted as cold-temperate to subpolar deposits and to contrast these with coals of tropical to subtropical origin.

Terminology and Background

Study of peat and coal has developed into highly specialized areas of investigation, with terminology that is a constant source of aggravation and confusion even to researchers who work in these fields. Gore (1983) reviews much of the terminology and classification systems that have been utilized. In this chapter I follow Gore (1983) in using the word "mire" to encompass all organic wetland deposits, including those described as swamp, bog, fen, moor, and muskeg. The major types of mire are bog and fen. Bogs are influenced solely by water that falls on them directly as rain or snow (ombrotrophic), whereas fens are influenced by water derived predominately from outside the immediate limits of the mire (groundwater) and are called rheotropic (eutrophic, minerotrophic). The distinction between bog and fen is important because hydrology, in association with climate, is the most important factor giving rise to the shape, chemistry, plant species, and vegetative structure of mires and consequently influencing many characteristics of coal.

Peat is a biochemical sediment that results from the accumulated remains of dead plants (organic matter >90% by dry weight). If the plant material is preserved and subsequently buried in sufficient quantity, it forms coal or, if less abundant, and finely disseminated, carbonaceous sediments. Burial and accompanying diagenesis progressively obscures the initial characteristics of the peat-forming vegetation. Because different vegetation and plant components have different susceptibility to decomposition and diagenesis, the composition of the peat and the subsequent coal may not reflect the floral composition of the original peatland. Recently accumulated peat initially undergoes rapid changes in response to biologically mediated processes. Later, the rate of alteration slows as the more labile compounds are removed as gas or in solution and the peat passes with burial from aerobic to anaerobic conditions (Hoge, 1994). With increasing depth of burial (mainly temperatures reached), the peat evolves (matures) through the progressive coal series from lignite to meta-anthracite. At higher levels of diagenesis (higher coal ranks), evidence for the original character of the peat may be completely lost.

Peat Formation - General Considerations

Mires form as a result of finely balanced environmental and depositional conditions, but at the same time they are dynamic systems that interact with and influence their surroundings. Mires respond to both exogenic factors such as climate and hydrology, and to autogenic forces such as upward growth that modifies hydrology and hydroseral succession. Requirements for the formation of peat can fundamentally be addressed in terms of energy: If total energy of fixation of organic matter (mainly by photosynthesis by green plants) exceeds total energy of respiration, an imbalance exists that results in a net accumulation of organic matter and thus the formation of peat (Moore and Bellamy, 1974). This imbalance occurs only under very restrictive conditions that invariably entail waterlogging, a water table that persists at or near the surface and that serves to inhibit oxygen diffusion and thus respiration and decomposition by aerobic microbes. If decomposition were 100 percent efficient, all organic matter would be recycled. The one feature all mires have in common is that a part of the organic productivity of the ecosystem survives to accumulate as peat (Moore, 1987, 1989), because of either high primary production or a reduced rate of decay. For formation of coal protoliths, however, other qualifications must be met: the peat-forming environment must remain relative free from influx of other sediments, and the peat must accumulate in an environment that has the potential to be preserved in the stratigraphic record. Most environments in the past have been mainly located close to the base level of actively subsiding basins.

The formation of mires may progress via *terrestrialization,* which refers to mires that originate as a result of the infilling of shallow lakes, or by *paludification* (swamping), which refers to mire formation on formerly emergent areas as a result of factors that raise the water table (Ingram, 1983), such as marine transgression in coastal areas or a change to a more humid climate. Paludification and terrestrialization give rise to a succession of sediment and vegetation types. Bogs form in response to paludification or via terrestrialization from underlying fens (Fig. 18.1). Most of the world's mires are raised bogs (ombrotrophic; Ingram, 1983). The elevation of raised bogs is attributed to the low permeability of underlying peat, bedrock, or sediment, which restricts the downward percolation of groundwater. During periods of drought or high precipitation, mires may compensate for vertical oscillations of the water table by contracting, swelling or floating (Tallis, 1983; Eggelsmann et al., 1993; Heathwaite et al., 1993). If the water table drops, plants die, fires are readily started, and surface litter, vegetation, and possibly the peat may be destroyed. The burned material is a nutrient source and leads to a change in vegetation succession and possibly to erosion. Factors leading to the termination of peat accumulation are varied but include both drowning or flooding of the

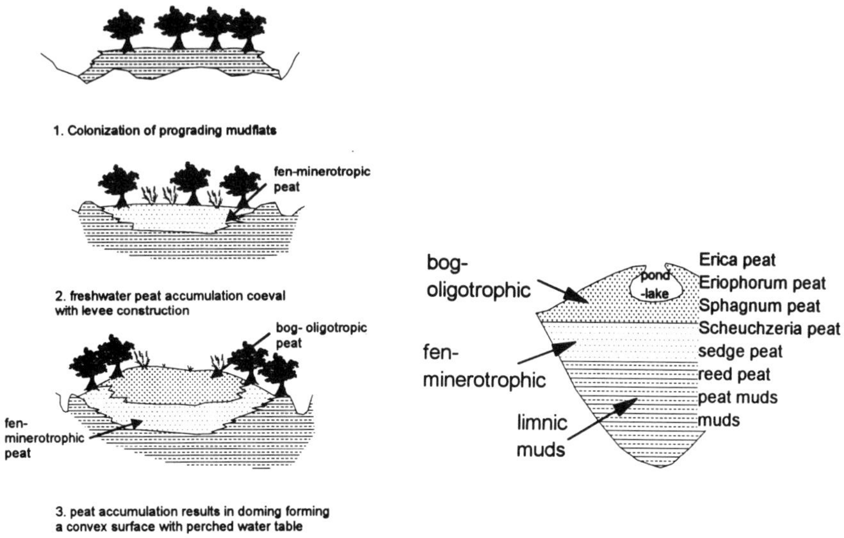

Fig. 18.1. Stages of mire evolution through paludification. Detailed inset is modified from Heathwaite et al.,1993.

wetland in response to a rapid relative rise in the water table and the uplift or slowing of the rate of subsidence so that the peat dries out or is eroded.

In both domed boreal and tropical bogs the vegetation is concentrically zoned away from the dome because of hydrology, which controls nutrient supply, acidity and oxygenation (Fig. 18.1). The species associations that occur from the center to the margin of the domes, referred to as phasic communities by Anderson (1964), exist in response to the degree of doming and oligotrophy. The margins of the deposits supports the most luxuriant and arborescent vegetation as a result of an increased nutrient supply from drainage from the center of the mire, more frequent flooding or presence of thin peat and availability of nutrients from underlying mineral soils. The high, central domed areas, on the other hand, tend to be nutrient poor, highly acidic, and lower in oxygen, giving rise to highly specialized vegetation.

Paludification or terrestrialization results in peat of a thickness limited by concurrent hydrology (Ingram, 1983). Peat growth can not continue indefinitely because with upward growth and thus increased surface gradient, rainfall will eventually be insufficient to maintain the peat in a saturated condition and the peat will dry out and decompose. Ultimately a state of equilibrium will be reached at which the rate of accumulation equals the rate of decay. Thus for formation of thick peats, exogenic factors such as subsidence or a relative rise in sea level must occur, or else substantive accumulations of peat (from the perspective of coal formation) cannot occur. Taking into account compaction of peat with burial and relative rates of peat accumulation, a one-meter coal seam represents on the order of 5 to 10,000 years of accumulation, and thick coals of economic interest, which are commonly 10 m or so thick, represent about 100 m of peat and require at least 50,000 years to accumulate. Such coal progenitors must occupy depositional and tectonic settings amenable to prolonged (or periodic) peat accumulation with little inorganic sedimentation and represent an evolution that invariably represents many episodes of terrestrialization and paludification. Long-term peat accumulation is dependent on available *accommodation space*, and therefore the net long term rates of accumulation do not necessarily reflect climatic or other edaphic controls but rather exogenic factors such as subsidence rates or relative rates of sea-level change.

The effect of climate on peatlands and peat accumulation is well documented on a global scale but in detail the importance of climate is masked by local variability in ecology. A high mean annual rainfall is important for peat formation but as shown by Thompson and Hamilton (1983), Ziegler et al. (1987), Patzowsky et al. (1991), and Lottes and Ziegler (1994), annual precipitation is a poor predictor of peat distribution and frequency. Continuity of rainfall throughout the year appears to be a better predictor of peat distribution for tropical peats. Mid- to high-latitude peats are exceptional, however; cold and dry seasons coincide and biological activity is slowed, so both growth and decay are suspended (Lottes and Ziegler (1994).

With increasing temperature, precipitation and evapotranspiration in all climates, there is an overall increase in species richness and productivity and a relative abundance of arborescent vegetation and microbial activity (Fig. 18.2; Huggett, 1991). The effect of climate on organic productivity and accumulation in mires is pronounced. Accumulation rates of peats from different regions are difficult to compare because of problems in precise age dating and variability in compaction and density of the flora. Typical rates of accumulation of peats are 0.1 to 0.2 mm/a in the high arctic, 0.5 to 0.7 m/a in the subarctic, more than 1 mm/a in cool, wet areas such as the Fraser Delta, British Columbia (49° N. Lat.), and 5 mm/a in topical and subtropical forested swamps (Anderson, 1964; Styan and Bustin, 1983a,b; Martini and Glooschenko, 1985; Heathwaite et al., 1993).

The current widespread distribution of wetlands at high latitudes results from the climate and the physiography of these regions. Interpreting the origin and distribution of the peats is, however, hampered by the difficulty of distinguishing areas of actively accumulating peat and relict peats from interglacial and early postglacial periods. Many temperate peats originated in a geomorphology inherited from earlier glacial and interglacial periods and are relict and moribund and preserved only because of the relatively cool climate, which inhibits decomposition. The physiographic controls on cold-temperate peats are, at the regional scale, related to isostatic and eustatic adjustments to glaciation and, at the local scale, to geomorphology. Eustatic cycles occur on several different scales (Fig. 18.3; Soreghan, 1993). Tectonically driven eustatic sea-level variations due, for example, to changes in rates of sea-floor spreading occur at rates of about 1 cm/ka to 10 cm/ka and have had a profound effect on regional sedimentation (Vail et al., 1977). Climate-driven sea-level changes, on the other hand, occur on the order of 0.1 cm/a to 3 cm/a (Soregham, 1993). In the Pleistocene (and the Permian) climate-driven eustatic cycles had a profound effect on sea level (up to ± 150 m), with glacial periods being associated with lowstands and interglacial periods with highstands. In many low-latitude coastal areas, postglacial sea-level rise outpaced peat accumulation and coastal peats formed during lowstands were transgressed. Only with a slowing in the rate of sea-level rise since about 5,000 yrs BP have coastal wetlands undergone significant expansion (Styan and Bustin, 1983a,b; Cecil et al., 1993; Esterle and Ferm, 1994). In high-latitude regions that have experienced glaciation, postglacial isostatic uplift and postglacial sea-level rise have combined to create a complex and not always predictable response in the mires. In some coastal areas, such as the Mackenzie Delta and the Hudson Bay lowlands of Canada widespread paludification is occurring as a result of a decline in sea level due to postglacial isostatic uplift. In modern mature mires, rates of accumulation in most areas are proportional, up to a point, to rates of subsidence (or eustatic sea-level rise). Where sea-level rise is slow, peat accumulation may keep pace, and net accumulation is therefore directly proportional to sea-level rise.

The geomorphology of high-latitude glaciated terrains is the major cause of the current widespread distribution of high-latitude mires. For peatland formation, saturated surface conditions are a prerequisite, and this condition is commonly met in glaciated terrains where the substrata are relatively impermeable or where drainage has been impeded by deposition of glacial sediments, establishment of permafrost, or deep glacial scouring of bedrock-forming depressions (Ruuhijarvi, 1983). In northern areas, peatlands vary markedly with the presence or absence of permafrost (Terasmae, 1960; Warner, 1992). In the high Canadian Arctic, the peatlands and many other terrain features reflect cryogenic factors. In subarctic regions, permanently ice-cored peaty mounds form distinct mires referred to as palsas.

Fig. 18.2. Comparison of coastal mires from the A. Mackenzie Delta, Arctic Canada (68° N Lat., field of view about 3 km). The vegetation in the Mackenzie Delta is composed predominately of *Sphagnum*, shrubs (*Salix*), heaths, and stunted *Picea* along the nutrient-rich margins. B. Caribbean coast of Panama, field of view about 2 km. Along the Panama coast (8° N Lat.) from the coast inland is a progression from mangrove, *Raphia* and mixed forest to bog plain characterized by *Myrica* and *Cyrilla*.

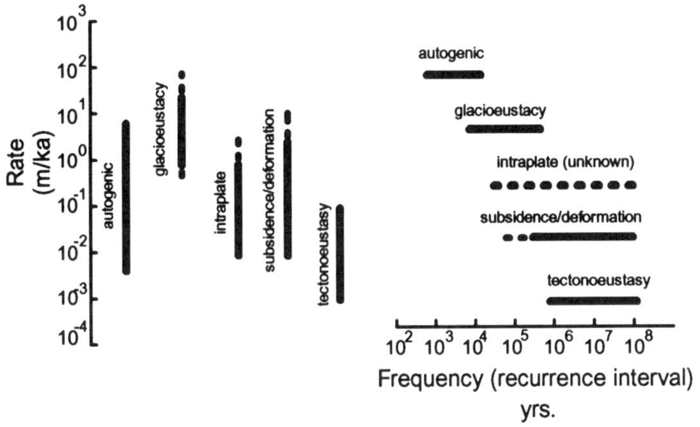

Fig. 18.3. Frequency and rates of autogenic, tectonic (intraplate, structural/basinal), and eustatic (glacio-tectono) processes (after Soreghan, 1993).

In nonpermafrost areas, raised bogs are less pronounced, and string bogs characterized by ribs of peat and troughs filled with water occur.

Comparison between Cold-Temperate Peats and Coals

Interpreting paleomires

To interpret the origin of coals, it is important to identify and evaluate those attributes of cold-temperate peats that are specific and that may be preserved in coal. The distinctiveness of cold-temperate peats, and thus presumably coals, is apparent mainly at the scale of the mire. Even though aspects of the geological setting of cold-temperate peats may be distinct, for the most part such features are seldom definitive in the geological record. Notable exceptions are direct evidence of cold climates such as dropstones, ice-scoured depressions, ice crystal casts, and evidence of past permafrost, such as solifluction lobes. Other possible evidence for cold climates, such as high rates of erosion and sediment transport or climatically related channel morphology (Gibling and Rust, 1990), in isolation rarely provide definitive evidence.

The most distinctive aspect of cold-temperate peatlands is the flora, which is generally stunted and comprises a high proportion of shrubs and mosses and thus contributes significantly less woody material than peats formed in warmer climates. In modern cold-temperate and subpolar climates, ombrotrophic bog vegetation is dominated by *Sphagnum*, shrubs, lichens (*Cladonia*) and sparse *Pinus* and *Picea* trees. Fens support a more luxuriant vegetation, including grasses, sedges, shrubs, and trees. Nutrient-deficient bogs are home to carnivorous plants, such as pitcher plants (*Sarracenia*), butterworts (*Pinguicula*), and sundews (*Drosera*), and heath shrubs characterized by thick cuticles to inhibit evapotranspiration.

Recognition of Paleomires from Coal Composition and Paleoflora

The task of reconstructing the characteristics of an ancient mire environment from a coal deposit is an exercise in interpreting the forest from the debris where the debris has been variably preserved and altered and little is known about the physiology or structure of the original flora. The vegetation and physicochemical changes occurring within a mire are considered by many coal petrologists, recorded in the megascopic appearance, texture, and particularly the microscopic composition (macerals) and inorganic chemistry and mineralogy of the coal (Bustin et al., 1985; Diessel, 1992). A general classification of the microscopic components of coal and their presumed origin is summarized in Table 18.1.

Table 18.1. The main macerals, microlithotypes, lithotypes in which they occur and their presumed origin (after Bustin et al., 1985).

Maceral Group	Major Maceral	Origin	Chemistry
vitrinite	telinite (cell walls) Collinite (cell fillings, structureless)	wood tissues, bark, leaves, etc.	intermediate hydrogen and volatiles
lipinite	resinite sporinite cultinite alginite	resins spores, pollen culticles algae	high hydrogen and volatiles
inertinite	micrinite fusinite semifusinite sclerotintie	plant material charring-fires partially charring-fires fungal sclerotia and hyphae	low carbon and volatiles

Organics

Some macerals are considered facies-diagnostic; they are considered to have a distinctive origin or their depositional setting can be assumed (Diessel, 1986; Lamberson et al., in press). For example, structured vitrinite is generally considered to be derived from the more resistant lignin-rich tissues of trees and shrubs. Most inertinite macerals are formed from the charring (for example, fusinite and macrinite) or partial charring (semifusinite) of preexisting plant tissues and thus are indicative of wildfires. Resinite in cell cavities within wood, suberinite (bark tissues), and cutinite attached to woody stems are mainly derived from arborescent or shrub vegetation. Inertodetrinite has the same origin as fusinite and semifusinite; when associated with detrital minerals such as dispersed clay minerals and quartz, inertodetrinite, vitrodetrinite and sporinite can be interpreted as having been transported. Desmocollinite is the product of a variety of degradation processes and thus suggests relatively enhanced decomposition. Alginite indicates subaqueous conditions. In Tertiary low-rank coals, the relative proportion of angiosperms and gymnosperms, the presence or absence of tree rings, and the relative abundance of wood may be apparent in specific plant tissues. However, in higher-rank coals these attributes may no longer be visible, even with etching, and in older coals our inadequate understanding of their anatomy and physiology makes interpreting plant structure speculative. Even in modern deposits, the plant precursors may be indistinguishable in areas of intense biological decay.

Interpreting the structure of paleomires on the basis of maceral composition or megascopic appearance must be done with caution. As shown by Shearer and Moore (1994), selective preservation of plant tissues may result in misrepresentation of the mire if based on nature of banding. Shearer and Moore (1994) demonstrated that in New Zealand Cretaceous/Tertiary coals, well developed banding derived from secondary xylem of gymnosperms, whereas the lack of banding in other coals is due to their derivation from angiosperms that lacked secondary xylem or, in which xylem, if present, was susceptible to decay. This problem is particularly evident in Carboniferous and Permian coals where the dominant plants were taxonomically, structurally, and compositionally different from modern plants (Collinson and Scott, 1987; DiMichele and Phillips, 1994).

Predicting macerals from cold-temperature peats is a particular problem because of the abundance of nonwoody tissues whose fate on decomposition and burial is unknown. In general, the maceral-precursors documented in cold-temperate peats are similar to maceral precursor materials described from tropical and subtropical peats, although the relative abundance differs (Fig. 18.4). Styan and Bustin (1983a) argue that the overall paucity of arborescent vegetation in cold-temperate peats results in coals relatively

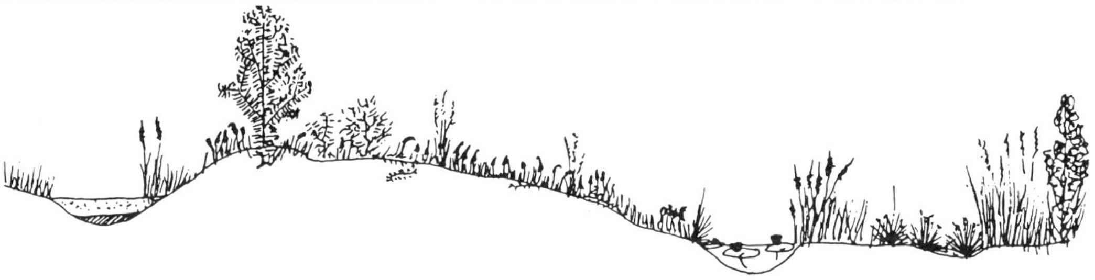

Peat type	Gyttja peats	Ericaceous *Sphagnum* peats	*Sphagnum* peats	*Nuphar* peats	Freshwater sedge-grass peats
dominant coal macerals	liptodetrinite alginite cutinite desmocollinite	suberinite telinite telocollinite pyrofusinite	telinite resinite telinite cutinite	liptodetrinite macrinite sclerotinite cutinite	telocollinite
minor coal macerals	inertodetrinite micrinite	desmocollinite oxyfusinite cutinite sclerotinite resinite sporonite	cutinite telocollinite	suberinite desmocollinite sclerotinite	telinite pyrofusinite desmocollinite cerinite
dominant micro lithotypes	liptite and thin clarite bands	vitrite with bands of clarite lenses of liptite and durite	vitrite	liptite and clarite	clarite with thin interbanded inertinite and vitrite

Fig. 18.4. Peat types and presumed maceral precursors in a freshwater bog of the Fraser River delta, Canada (49º N Lat.) (after Styan and Bustin, 1983b).

impoverished in vitrinite and enriched in liptinite macerals derived from leaf cuticles, spores, pollen, and waxes. They suggest that these peats are precursors of thinly banded coals with a predominance of duller lithotypes reflecting the detrital character of much of the peat as well as the absence of large trees. Styan and Bustin (1983a) point out that the eventual fate of much of the vegetation and its degradation products is speculative.

A direct and less ambiguous interpretation of the flora and climate is supported by direct fossil remains and palynomorphs. Of particular value are "coal balls" which are permineralized peat (generally by silica or carbonate) and that show in unsurpassed detail the flora and composition of the peat. Coals balls are, however, abundant only in certain Carboniferous coals (Cross and Phillips, 1990; DiMichele and Phillips, 1994), and their presence suggests unusual ion rich groundwaters, so interpreting vegetation on the basis of coal balls may yield a flora indicative of only nutrient-rich mire conditions. Plant compressions have also proven useful, but these too are best preserved in the roof or floor rock of coal seams and may not represent the entire mire flora. The relatively abundant of pollen and spores have been extensively used in studies of vegetation of peats and coals (Smith, 1968; Collinson and Scott, 1987). Although the relative abundance of palynomorphs seldom reflects the overall contribution of a particular flora to coal, their near-ubiquitous occurrence in sedimentary rocks makes them extremely valuable. In coal deposits where many of the flora (palynomorphs) are extinct, some with no living relatives, interpreting the climatic record or floral facies (phasic communities) or even determining in which plant the palynomorphs originated can be problematic. Even given these shortcomings, flora evidence, whether derived either directly from palynomorphs or fossil flora or indirectly from the relative abundance and characteristics of coal macerals or lithotypes provides the best evidence for interpreting paleomires.

Inorganics

The inorganic composition of mires in general has no relationship to climate but rather reflects local geomorphology and bedrock chemistry. Mineral matter content of both peats and coals varies markedly, mainly in response to proximity to clastic sediment sources such as rivers or coasts, where periodic flooding can result in inundation of the mire with clastic sediment. Because of their topography, domed peats are the least susceptible to flooding and thus have the lowest mineral matter content (Styan and Bustin, 1983a). Cold-temperate and subpolar mires are more prone to influxes of airborne mineral matter because of the general paucity of surrounding vegetation that might inhibit wind erosion, and the less domed mires are more prone to flooding. There are, however, no quantitative data from coals to support this supposition. Sulphur, which is one of the most deleterious and thus one of the most important elements affecting coal quality, appears to be independent of mire climate and is linked more strongly with the environment of deposition. For example, the total sulphur content of both tropical Panamanian and temperate western Canadian coastal peats averages 3 percent, of which 94 percent is organic sulphur (Phillips and Bustin, in press). High hydrogen and nitrogen content of coals is commonly interpreted to reflect enhanced bacterial degradation at the peat stage and thus indirectly may reflect mire climate, since microbial degradation is overall less important in cool climes (Ross and Ross, 1988). These and other inorganic indices have been utilized in paleomire reconstruction (Diessel, 1992) but have not been shown to be indicative of climate or physiography of the mire.

Characteristics of Cold-Temperate Coals-The Gondwana Coals

Evidence for significant deposits of interglacial, postglacial, and cold-temperate coals (peats) in the geological record are restricted predominately to the Gondwana coals. There are other cold-temperature coals, such as the Late Tertiary lignites of Siberia, Svalbard, and the Canadian Arctic Islands, but these deposits have received little study.

In the late Paleozoic a landmass existed comprising Australia, Antarctica, southern Africa, peninsular India, Madagascar, and part of South America (Fig. 18.5; White, 1990). This large continental mass, referred to as Gondwanaland, was characterized mainly by a deciduous flora that gave rise to major coal deposits in Australia,

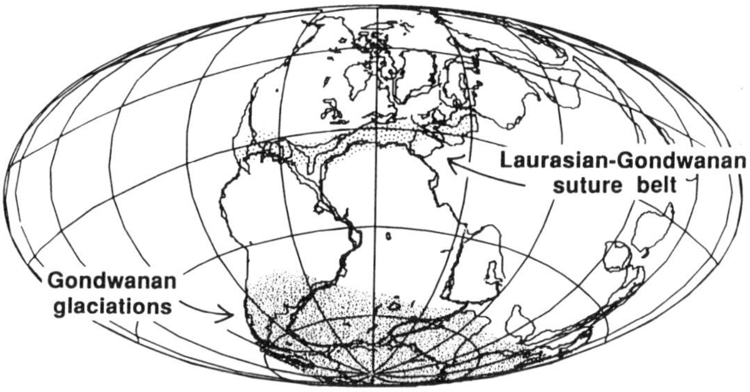

Fig. 18.5. Global paleogeography during the Late Carboniferous (modified after Soreghan, 1993).

India, southern Africa, and, to a lesser extent, South America, Madagascar, and Antarctica (Chandra and Taylor, 1982). The coals are mainly Permian in age but extend into the Carboniferous and Triassic. Paleogeographic reconstructions show that the Gondwana Permian coals accumulated at mostly high latitudes in cold-temperate to subpolar climatic settings, and they therefore have been considered to offer some insight into the compositional effects of climate on coal. Because of their contrasting vegetation and climate, there is considerable literature comparing the nature and composition of Gondwana coals to tropical and subtropical coals of generally similar age in the northern hemisphere.

The Gondwana coals for the most part were derived from what is referred to as *Glossopteris* flora, dominated by stunted broad-leafed deciduous vegetation assigned to *Glossopteris* and *Gangamopteris* (Diessel, 1992). This flora contrasts markedly with the more luxuriant flora of the humid, tropical and subtropical Carboniferous coals of the Northern Hemisphere, which included lycopsides, ferns, sphenopsids, pteridosperms and cordaites. Trees of these five plant groups are considered to comprise about 95 percent of the peat biomass of some Carboniferous coals, a conclusion based on coal ball studies (DiMichele and Phillips, 1994).

The general characteristics and the maceral and floral composition of the major Gondwana coals are summarized here.

Australia

Because of their economic importance, the composition of Australian Permian coals have been extensively studied and compared to tropical and subtropical Carboniferous coals. Gondwana glaciation reached a maximum in the early Permian, and the climate ameliorated from periglacial to cold temperate by the Late Permian. Major coals deposits span the Permian and also occur locally in Triassic strata. Paleoclimate indicators include well-developed tree rings indicative of seasonality (not present in coals of similar age in the Northern Hemisphere), presence of unweathered feldspar, ice-crystal casts, densely packed leaf cuticles reflecting leaf fall (Chandra and Taylor, 1982), dropstones, and evidence of solifluction flows suggesting the presence of permafrost (Hunt, 1989). Waning of the polar climate in the Late Carboniferous and the Early Permian resulted in a succession of impoverished tundra vegetation followed by establishment of a stunted *Gangamopteris* forest taiga. With further climate amelioration, a low-diversity Pteridosperm/Pteridophyte flora dominated by *Glossopteris* flourished through the Late Permian in cold-temperate conditions (Hunt and Hobday, 1982). The Triassic flora in Australia is dominated by conifers.

Early studies of Australian Gondwana coals suggested that they were more thinly banded, finely detrital, and inertinite rich (Fig. 18.6) and thus duller overall than tropical and subtropical coals (Cook, 1975; Teichmüller and Teichmüller, 1982; Teichmüller, 1989). These qualities were attributed to the predominantly herbaceous flora (*Gagamopteris* flora; Cook, 1975), and to seasonal fluctuations in the water table that causes drying out and thus let to a greater abundance of inertinite. Later more detailed studies, however, suggest that, although many Gondwana coals are high in inertinite, there is large variation from seam to seam (Taylor et al., 1989a,b), which is attributed to a change in the overall depositional setting in the Permian foredeep basins from lower delta plain to upper delta plain and alluvial plain coals (Diessel, 1992). Taylor et al. (1989) and Diessel (1992) show that in Carboniferous and Permian foreland basins there is often an overall upward gradational increase in inertinite content and a similar lateral increase across the basin from the craton margin towards the orogen. The increasing inertinite content is attributed to lower subsidence rates, leading to greater oxidation of vitrinite precursors. Hunt and Smyth (1989) also attribute the greater abundance of inertinite in coals of the cratonic Permian Cooper and Galilee basins compared to the foredeep basins of the Sydney and the Bowen basins to the depositional environment. They

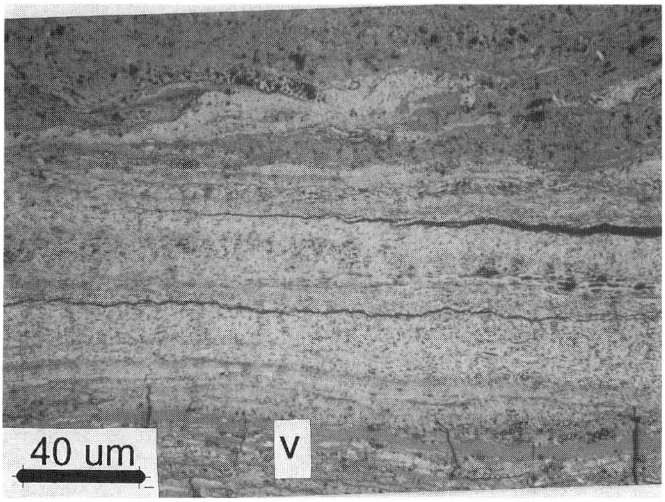

Fig. 18.6. Photomicrograph of Bulli Seam, Sydney Basin, Australia (I= inertinite, V=vitrinite).

argue that the inertinite-rich coals of the cratonic basins formed from freshwater mires where subsidence rates were lower, leading to greater oxidation of the peat.

Diessel (1992) interprets the lower abundance of structured vitrinite and the greater abundance of semifusinite in Australian Gondwana coals compared to Carboniferous coals of the Northern Hemisphere, to indicate that, although the total amount of tissue-derived macerals may be the same, gelification is more pronounced, indicating greater degradation of plant tissues in the Australian Gondwana coals. Diessel (1992) found that the Gondwana coals are poor in sporinite, which he attributes to the coals having been formed from a *Glossopteris* flora dominated by gymnosperms rather than the spore-producing pteridophytes of the Carboniferous of the Northern Hemisphere.

South Africa

Coal measures in the Karoo Basin of South Africa were deposited in the Early Permian, after the Late Carboniferous Dwyka glaciation. The Karoo Basin varied from 80° south latitude during the Late Carboniferous to 40° S latitude during the Early Jurassic (Falcon et al., 1984). During the Dwyka glaciation parts of the hinterland were covered by continental ice sheets. Climate ameliorated during the Late Carboniferous and the Early Permian. Coals commonly directly overlie the Dwyka diamictite in the northern part of the Karoo Basin. The coal measures flora established during the Early Permian comprised a diverse *Glossopteris-Gangamopteris* flora (Falcon et al., 1984; Falcon, 1989). Plumstead (in Falcon, 1989) recognized six stages of plant zonation resulting from vegetation change due to evolution and to climate change (Fig. 18.7). The earliest macrofossils in the Karoo Sequence include rare occurrences of mosses, succeeded by abundant Gangamopterids (reflecting the final stages of the Dwyka glaciation). This flora was followed by a diverse *Glossopteris-Gangamopteris* flora with abundant arborescent and herbaceous Glossopteridophyta with minor lycopods, ferns cordaites, and other early gymnosperms (Falcon, 1989). Climate during the accumulation of the oldest peats (coals) was subpolar to cold-temperate, with wide temperature ranges, short growing seasons, and, possibly, the presence of snow and permafrost for much of the year. Under these conditions, plant growth and accumulation was slow (Falcon, 1989). In the Late Permian a more temperate climate promoted abundant and fast plant growth

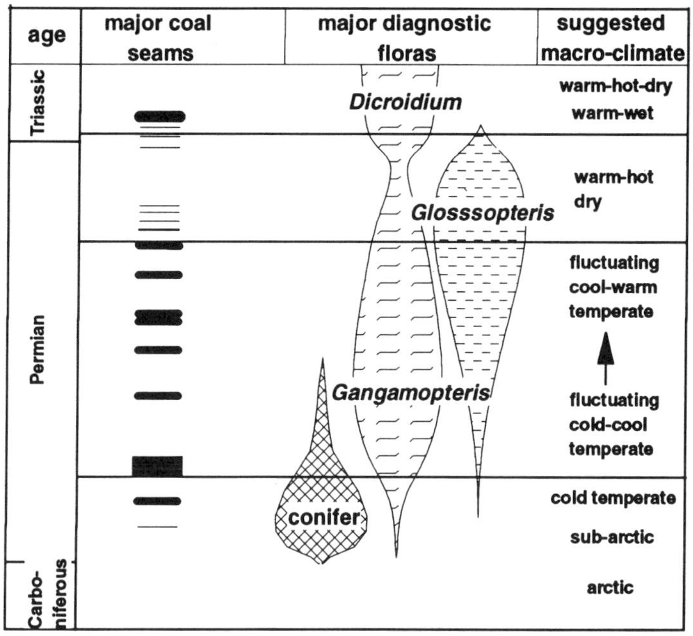

Fig. 18.7. Major flora zones and climate depostion in the Karoo Basin, South Africa (after Falcon, 1989).

and rapid peat accumulation (Falcon, 1989). The presence of more abundant vitrinite in these coals compared to earlier coals is considered consistent with climate amelioration. The oldest coals, those immediately following the Dwyka glaciation, are the thickest in the succession. Falcon et al. (1984) interpret this relationship to reflect slower rates of bacterial and microbial decay in the cooler climate, coupled with a higher water table resulting from glacial meltwater discharge. In general, the coals at the base of the succession are brightest (richest in vitrinite) and grade up to duller coals, which may indicate greater degradation of the younger peats (Falcon et al, 1984).

India

Major Indian coal resources are Permian in age and were formed when India was between 70° and 40° south latitude (Navale and Saxena, 1989). These include the Lower Permian-Carboniferous Talchir and Karharbari formations and the overlying Barakar Formation. The climate during the accumulation of the lower coal deposits was subpolar to cold-temperate, with high precipitation (Bharadwaj, 1976; Navale and Saxena, 1989; Singh and Dwivedi, 1994). The Permian-Carboniferous Talchir Formation locally includes tillite; the overlying Karharbari and Barakar formations include major coal resources. The Talchir Formation is nearly barren of fossils but includes a *Gangamopteris* flora (*Glossopteris* flora in India encompasses *Ganagamopteris* and *Glossopteris* flora phases). Coals of the Karharbari Formation are rich in fusinite, semifusinite, and inertodetrinite and poor in liptinite, which has been interpreted to reflect a paucity of vegetation and rapid "fusinization" due to the shallowness of the basin, slow subsidence, and low humidity (Navale and Saxena, 1989). The coals have been described as microfragmental and include fusinized resins and highly degraded vitrinite (Pareek, 1967). Barakar Formation coals include vitrinite, liptinite and inertinite groups; early Barakar coals are dull, whereas the late Barakar coals are vitrinite-rich, which is attributed to the depositional setting (Pareek, 1967).

Discussion and Conclusion

The formation and accumulation of peat is controlled by a limited number of biological, chemical, and physical controls. All peatlands and peats have characteristics in common irrespective of climate. Cold-temperate peats owe their origin and widespread distribution to a physiography conducive to a high water table and thus to the production and accumulation of organic matter, as well as to a cool climate facilitates preservation. The most distinctive aspect of modern cold-temperate mires is the vegetation that comprises a greater proportion of herbs, shrubs, and mosses but fewer trees and that is less luxuriant than the vegetation of tropical or subtropical mires. Although there is substantial evidence for a cold climate in sediments associated with cold-temperate mires and many cold-temperate mires have a physiography reflecting cold climes (striped bogs, palsas), these features are rarely preserved or recognizable in the rock record and thus only seldom useful in reconstructing paleomires.

Gondwana Coals - Product of Evolution, Climate, or Sedimentology?

Significant differences exist, particularly in the flora, between cold-temperate and tropical peats. Are these differences manifest in coal. and, if so, to what extent can we attribute the contrasting composition of Gondwana and Carboniferous coals of the Northern Hemisphere to climate and flora differences such as exist between modern cold-temperate and tropical peat? Studies of Australian coals, for which the most data exist (Diessel, 1992), show that some Gondwana coals are highly variable in character. This variation exceeds the average variability between Gondwana and Northern Hemisphere Carboniferous coals. At least in these coals, the flora and climatic factors are overshadowed by other factors (mainly depositional environment and tectonic setting) that operated during peat accumulation. As shown by Laskar (1977), Smyth (1989) and others, vitrinite-rich Gondwana coals occur in foredeep basins, whereas inertinite-rich coals tend to occur on continental platforms. This is presumably a result of the slower subsidence rates and the more variable water table affecting platform coals, resulting in a greater propensity for drying out and more rapid oxidation of the peat on more stable platform areas. In contrast, all Carboniferous European coals are vitrinite-rich, irrespective of the depositional environment, suggesting that the Carboniferous flora was overall more amenable to formation of vitrinite and possibly that the climate overall was more humid and less seasonal. It thus appears that depositional environment and tectonic setting played a greater role in controlling the composition of Gondwana coals than that of Carboniferous coals of the Northern Hemisphere.

The occurrence of thinner banded and duller Gondwana coals is consistent with our current understanding of the decomposition and diagenesis of modern flora. Large arboraceous, lignin-rich plants are the main precursor to bright (vitrinite-rich) coals, whereas temperate or cold climate flora composed of proportionately herbs, shrubs, and mosses are protoliths for finely detrital coals (Chandra and Taylor, 1982; Styan and Bustin, 1983b). The high mineral matter content of some coals may be a product in part of windblown material from the surrounding areas that were devoid of vegetation (Plumstead, 1962). The greater abundance of inertinite in Gondwana coals is not, however, consistent with a colder climate. Modern temperate and subpolar mires have a lower wildfire frequency than subtropical or tropical mires (Wein, 1983), so if we extrapolate, Gondwana coals should be relatively impoverished in inertinite. In high-latitude mires, fires occur only during low-rainfall years and burn only shallow humus layers. Wein (1983) estimates a frequency of about 1/100 a for surficial fires, and even less (1/1000 a) for deep-burning peat fires. These values contrast with fire frequencies on the order of 2–3/100 a in the tropics. It is possible, however, that in Gondwanaland rainfall was more seasonal, so periodic drying out and consequent burning of the peat occurred at a greater frequency than in current cold-temperate mires. Enhanced microbial decay would be associated with a dry(er) season resulting in a relative increase in the inertinite macerals, since vitrinite precursors are more susceptible to degradation. Another possibility is that the O_2 content of the atmosphere was higher (Robinson, 1991) during accumulation of Gondwana coals, resulting in more intense burning.

Cold-Temperate Peats as Coal Protoliths?

There has been limited opportunity in the geological record for the accumulation of cold-temperate peats apart from the Permian. Major glaciations in the Phanerozoic are restricted to the late Paleozoic, and prior to the Devonian nonmarine plants were neither evolved nor abundant enough to produce thick peats (and thus coal). Most cold-temperate mires and their deposits today are located on stable landmasses (for example the Canadian Shield) and thus have little opportunity to subside and accumulate thick peats. On currently subsiding continental margins, however, such as the Mackenzie River delta of northern Canada or the Fraser River delta of southwestern Canada, cold-temperate peats have a much higher preservation potential, and, if prolonged peat accumulation can take place, thick coals may form.

It has been assumed (Styan and Bustin, 1983a,b) that cold-temperate peats are good analogues for Gondwana coals because both formed under cold-temperate to subpolar climatic conditions and thus there should be a general similarity in vegetation type and ecological conditions. Although modern mire vegetation is fundamentally different from that of the geological record, with the similarities decreasing in the oldest vegetation, the fundamental principles that control mire development and succession (for example, hydrology and nutrient availability) can, as a first approximation, be considered the same as those operating in modern mires. Two of the most important and complicating factors in interpreting Gondwana coals are the effect of the gross depositional environment, and, indirectly, the effect of tectonic setting on coal composition and character, both of which may overshadow and mask other variables. As is evident from the detailed studies of Australian Gondwana coals reviewed in this chapter, coal composition varies markedly with depositional environment and tectonic setting, independent of climate. Such factors must be subtracted or accounted for if the more subtle effects of climate, ecological setting, and seral succession on coal composition are to be interpreted.

The limitations of our ability to deduce paleomire conditions from coal composition may be seen by attempting to evaluate how well modern mires can be reconstructed from peat stratigraphy. Because the distinctiveness of plant tissues is progressively lost with diagenesis and because the ecology of many plants now long extinct may be unknown, it is reasonable to assume that our interpretation of paleomires on the basis of coal studies can be no better than our interpretation of the succession of phasic communities in modern peats where the ecology of the plants is known. The one conclusion most modern peat studies have in common is a recognition of the complexity of peat development and of the inadequacy of simplistic models of mire evolution (Frenzel, 1983; Heathwaite et al., 1993; Hoge, 1994). Overall, the development of a mire is governed not only by climate change and hydroseral succession but by many other internal and external factors, complicating greatly the interpretation of mire history. In addition, the regulatory processes within an ecosystem may buffer (inertia to change-mires are autoregulatory) even significant climatic or other exogenic changes, whereas in other ecosystems even small exogenic shifts may topple an existing equilibrium, resulting in a marked response in the mire vegetation or physiography. These difficulties in interpreting mires suggest not that there is no merit in attempting reconstruction of paleomires from the study of coal stratigraphy (facies) but rather that the limitations must be recognized and interpretations tempered. Many interpretations of paleomire history that are based on studies of coals far exceed in detail what is possible even in studies of modern mires, calling into question the efficacy of such coal studies.

Additional studies are desperately needed if paleomire reconstruction from coal stratigraphy and composition is to progress. In particular, greater understanding of paleoflora ecology, maceral progenitors, and effects of preferential decay and preservation of different plant types and tissues is needed. Studies with the greatest potential for success, focus on young strata where the flora has the greatest similarity to modern flora and where diagenesis is generally less pronounced. Studies such as those advocated by DiMichele and Phillips (1994), integrating coal petrology, paleobotany, geochemistry, and sedimentology, will contribute most to the understanding of paleomire evolution. Much remains to be learned from the study of modern mires.

References

Anderson, J.A.R., 1964. The structure and development if peat swamps of Sarawak and Brunei. Journal of Tropical Geography, 18: 7–16.

Bharadwaj, D.C., 1976. Paleogeography of India during Gondwana times and its bearing on the climate. Geophytology, 6: 153–161.

Bustin, R.M., Cameron, A.R., Grieve D.A. and Kalkreuth W.D., 1985. Coal petrology its principals, methods and applications. 2nd ed. Geological Association of Canada. 229 pp.

Cameron, C.C., Esterle J.S. and Palmer C.A., 1989. The geology, botany and chemistry of selected peat-forming environments from temperate and tropical latitudes. International Journal of Coal Geology, 12: 105–156.

Cecil, C.B., Dulong, F.T., Cobb J.C. and Supardi, E., 1993. Allogenic and autogenic controls on sedimentation in the central Sumatra Basin as an analogue for Pennsylvanian coal-bearing strata of the Appalachian basin. In: J. Cobb and C.B. Cecil (Editors), Modern and ancient coal-forming environments. Geological Society of America, Special Paper 286, pp. 3–22.

Chandra, D. and Taylor G.H., 1982. Gondwana coals. In: E. Stach, M.Th. Mackowsky, M. Teichmuller, G.H. Taylor, D. Chandra and R. Teichmuller (Editors), Stachs textbook of coal petrology. Borntraeger, Stuttgart, pp. 177–198.

Cobb, J.C. and Cecil, C.B. (Editors), 1993. Modern and ancient coal-forming environments. Geological Society of America, Special Paper 286, 198 pp.

Cohen, A.D. and Spackman, W., 1977. Phytogenic organic sediments and sedimentary environments in the Everglades-mangrove complex of Florida, Part II. Paleontographica, 162-B: 71–114.

Collinson, M.E. and Scott A., 1987. Implication of vegetation change through the geological record on models for coal-formation environments. In: A. C. Scott (Editor), Coal and coal-bearing strata: recent advances. Geological Society, Special Publication 32, pp. 67–85.

Cook, A.C., 1975. The spatial and temporal variation of the type and rank of Australian coals. In: A.C. Cook (Editor), Australian black coal. Australian Institute of Mining and Metallurgy, Illawarra Branch, Melbourne, pp. 63–84.

Cross, A.T. and Phillips T.L., 1990. Coal-forming plants through time in North America. International Journal of Coal Geology, 16: 1–46.

Diessel, C.F.K., 1986. The relationship between coal macerals and sedimentary environments in Australian coals. 12th International Sedimentology Conference, Canberra, Abstracts, p. 82.

Diessel, C.F.K., 1992. Coal-bearing depositional systems. Springer-Verlag, London. 720 pp.

DiMichele, W.A. and Phillips T.L., 1994. Paleobotanical and paleoecological constraints on models of peat formation in the Late Carboniferous of Euramerica. In: J.H. Calder and M.R. Gibling (Editors), The Euramerica coal province: controls on tropical peat accumulation in the Paleozoic. Palaeogeography, Palaeoclimatology, Palaeoecology, 109: 39–90.

Eggelsmann, R., Heathwaite, A.L., Grosse-Brauckmann, G., Kuster, E., Naucke, W., Schuch M. and Schweickle V., 1993. Physical processes and properties of mires. In A.L. Heathwaite and K. Göttlich (Editors), Mires-Process, exploitation and conservation. John Wiles and Sons, Toronto, pp. 171–262.

Esterle, J.S., 1990. Trends in petrographic and chemical characteristics of tropical domed peat deposits in Indonesia and Malaysia as analogues for coal formation. Ph.D. thesis, University of Kentucky, 270 pp.

Esterle, J.S. and Ferm J.C., 1994. Spatial variability in modern tropical peat deposits from Sarawak, Malaysia and Sumatra, Indonesia: analogues for coal. International Journal of Coal Geology, 26: 1–41.

Falcon, R.M.S., 1989. Macro- and micro-factors affecting coal-seam quality and distribution in southern Africa with particular reference to the No. 2 seam, Witbank coalfield, South Africa. International Journal of Coal Geology, 12: 681–731.

Falcon, R.M.S., Pinheiro H.J. and Shepherd P., 1984. The palynostratigraphy of the major coal seams in the Witbank basin with lithostratigraphic, chronostratigraphic and palaeoclimatic implications. In: M.J. Lemos de Sousa (Editor), Symposium on Gondwanaland coals. Lisbon, 70(2): 215–243.

Frenzel, B., 1983. Mire-repositories of climate information or self-perpetuation ecosystems. In A.J.P. Gore (Editor), Ecosystems of the world vol. 4A. Mires: swamp, bog, fen and moor. Elsevier, New York, pp. 35–65.

Gibling, M. and Rust B., 1990. Ribbon sandstones in the Pennsylvanian Waddens Cove Formation, Sydney Basin, Atlantic Canada: the influence of siliceous duricrusts on channel-body geometry. Sedimentology, 37: 45–65.

Gore, A.J.P., 1983. Introduction. In: A.J.P. Gore (Editor), Ecosystems of the world, vol. 4A. Mires: swamp, bog, fen and moor. Elsevier, New York, pp. 1–34.

Heathwaite, A.L. and Göttlich, K. (Editors), 1993. Mires-Process, exploitation and conservation. John Wiley and Sons, New York, 506 pp.

Heathwaite, A.L., Göttlich, K. Burmeister, E.-G. Kaule G. and Grospietsch T., 1993. Mires: Definition and Form. In: A.L. Heathwaite and K. Göttlich (Editors), Mires-Process, exploitation and conservation. John Wiley and Sons, New York, pp. 1–76.

Hoge, B.E., 1994. Wetland ecology and paleoecology: relationships between biogeochemistry and preservable taxa. Current topics in Wetland Biogeochemistry, 1: 48–67.

Huggett, R.J., 1991. Climate, Earth processes and Earth history. Springer-Verlag, London, 281 pp.

Hunt, J.W., 1989. Permian coals of eastern Australia: geological control of petrographic variation. International Journal of Coal Geology, 12: 589–634.

Hunt, J.W. and Hobday D.K., 1982. Petrographic composition and sulphur content of coals associated with alluvial fans in the Permian Sydney Basin and Gunnedah Basin, eastern Australia. In: R. Rahmani and R.M. Flores (Editors), Sedimentology of coal and coal bearing sequences. International Association of Sedimentologists, Special Publication 7, pp. 43–60.

Hunt, J.W. and Smyth M., 1989. Origin of inertinite-rich coals of Australian cratonic basins. International Journal of Coal Geology, 11: 23–46.

Ingram, H.A.P., 1983. Hydrology. In: A.J.P. Gore (Editor), Ecosystems of the world, vol. 4A. Mires: swamp, bog, fen and moor. Elsevier, New York, pp. 67–158.

Lamberson, M.N., Bustin, R.M., Kalkreuth W.D. and Pratt K.C. (in press). The formation of inertinite rich peats in the mid-Cretaceous Gates Formation: implications for the interpretation of mid-Albian history of paleowildfire. Palaeogeography, Palaeoclimatology, Palaeoecology.

Laskar, B., 1977. Evolution of Gondwana coal basins. In: B. Laskar and E.S. Raja Rao (Editors), Proceedings of the fourth international Gondwana symposium, India, pp. 1–21.

Lottes, A.L. and Ziegler A.M., 1994. World peat occurrence and the seasonality of climate and vegetation. Palaeogeography, Palaeoclimatology, Palaeoecology, 106: 23–37.

Martini, I.P. and Glooschenko W.A., 1985. Cold climate peat formation in Canada, and its relevance to Lower Permian coal measures of Australia. Earth-Science Reviews, 22: 107–140.

Moore, P.D., 1987. Ecological and hydrological aspects of peat formation. In: A.C. Scott (Editor), Coal and coal-bearing strata: recent advances. Geological Society, Special Publication 32, pp. 7–15

Moore, P.D., 1989. The ecology of peat-forming processes: a review. In: P.C. Lyons and B. Alpern (Editors), Peat and Coal: origin, facies, and depositional models. Elsevier, Amsterdam, pp. 89–103.

Moore, P.D. and Bellamy, D.J., 1974. Peatlands. Elek Science, London, 220 pp.

Navale, G.K.B. and Saxena, R., 1989. An appraisal of coal petrographic facies in Lower Gondwana (Permian) coal seams of India. International Journal of Coal Geology, 12: 553–588.

Pareek, H.S., 1967. Some observations on the lithology of coal-bearing beds and the nature of coal of the major Gondwana basins of India. Gondwana stratigraphy, Buenos Aires, UNESCO, 2: 883–904.

Patzowsky, M.E., Smith, L.H., Markwick, P.J., Engberts C.J. and Gyllenhaal, E.D., 1991. Application of the Fujita-Zeilger paleoclimate model: Early Permian and Late Cretaceous examples. Palaeogeography, Palaeoclimatology, Palaeoecology, 86: 67–85.

Pfadenhauer, J., Schneekloth, H. and Schneider S., 1993. Mire distribution. In A.L. Heathwaite and K. Göttlich (Editors), Mires-Process, exploitation and conservation. John Wiley and Sons, New York, pp. 77–122.

Phillips, S. and Bustin, R.M., in press. Sulphur distribution as an indicator of environment of deposition of peat and coal. Journal of Sedimentary Research.

Plumstead, E.P., 1962. The Permo-Carboniferous coal

measures of the Transvaal, South Africa- an example of the contrasting stratigraphy in the southern and northern hemispheres. C.r. 4 Congr. Intern. Strat. Geol. Carbonifere, 2: 545–550.

Robinson, J.M., 1991. Phanerozoic atmospheric reconstructions: a terrestrial perspective. Palaeogeography, Palaeoclimatology, Palaeoecology, 97: 51–63.

Ross, C.A. and Ross, J.R.P., 1988. Late Paleozoic transgressive-regressive deposition. In: C.K. Wilgus, C.A. Hastings, H.W. Ross. J. Posamentier, J. Van Wagoner, and C.G. Kendall, (Editors), Sea-level changes: an integrated approach. Society of Economic Mineralogists and Petrologists, Special Publication 42, pp. 227–247.

Ruuhijarvi, R., 1983. The Finnish mire types and their regional distribution. In: A.J.P. Gore (Editor), Ecosystems of the world, vol. 4A. Mires: swamp, bog, fen and moor. Elsevier, New York, pp. 47–67.

Shearer, J.C. and Moore, T., 1994. Botanical control on banding character in two New Zealand coal beds. Palaeogeography, Palaeoclimatology, Palaeoecology, 110: 11–27.

Singh, R.M. and Dwivedi, C.S., 1994. A study of coal facies in the Korba sub-basin, Mahanadi Valley India. International Journal of Coal Geology, 25: 113–132.

Smith, A.H.V., 1968. Seam profiles and seam characteristics. In: D. G. Murchison and T. S. Westoll (Editors), Coal and coal-bearing strata. Oliver Boyd, Edinburgh, pp. 21–44.

Smyth, M., 1989. Organic petrology and clastic depositional environments with special reference to Australian coal basin. In: P.C. Lyons and B. Alpern (Editors), Coal: classification, coalification, mineralogy, trace element geochemistry, and old and gas potential. International Journal of Coal Geology, Special Issue, 12: 635–656.

Soreghan, G.S., 1993. Upper Carboniferous glacio-eustatic cycles in tectonically active basins: implications for global cyclostratigraphy. Douzième congrès international de la stratigraphie et geologie du carbonifere et permiem, Buenos Aires, Comptes Rendus, 1: 337–354.

Styan, W.S. and Bustin, R.M., 1983a. Petrography of some Fraser River peat deposits: coal maceral and microlithotype precursors in temperate-climate peats. International Journal of Coal Geology, 2: 321–370.

Styan, W.S. and Bustin, R.M., 1983b. Sedimentology of Fraser River delta peat deposits: a modern analogue for some deltaic coals. International Journal of Coal Geology, 3: 101–143.

Tallis, J.H., 1983. Changes in wetland communities. In: A.J.P. Gore (Editor), Ecosystems of the world, vol. 4A. Mires: swamp, bog, fen and moor. Elsevier, New York, pp. 311–347.

Taylor, E.L., Taylor, T.N. and Collinson, J.W., 1989a. Depositional setting and paleobotany of Permian and Triassic permineralized peat from the central Transantarctic Mountains, Antarctica. International Journal of Coal Geology, 12: 657–679.

Taylor, G.H., Liu, S.Y. and Diessel, F.K., 1989b. The cold climate origin of inertinite-rich Gondwana coals. International Journal of Coal Geology, 11: 1–22.

Teichmüller, M., 1989. The genesis of coal from the viewpoint of coal petrology. In: P.C. Lyons and B. Alpern (Editors), Coal: classification, coalification, mineralogy, trace element geochemistry, and old and gas potential. International Journal of Coal Geology, Special Issue, 12: 1–87.

Teichmüller, M. and Teichmüller, R., 1982. Fundamentals of coal petrology. In: E. Stach, M. T. Mackowsky, M. Teichmüller, G.H. Taylor, D. Chandra and R. Teichmuller (Editors) ,Textbook of coal petrology. Borntraeger, Stuttgart, pp. 6–86.

Terasmae, J., 1960. Post-glacial deposits in the St. Lawrence Lowlands. Geological Survey of Canada Bulletin 56: 1–22.

Thompson, K. and Hamilton, A.C., 1983. Peatlands and swamps of the African continent. In: A.J.P. Gore (Editor), Ecosystems of the World, vol. 4B. Mires: swamp, bog, fen and moor. Elsevier, Amsterdam, pp. 331–373.

Vail, P.R., Mitchum, R.M. and Thompson, S., 1977. Seismic stratigraphy and global changes in sea level, Part 3. Relative changes of sea level from coastal onlap. In: C.W. Payton (Editor), Seismic stratigraphy-applications to hydrocarbon exploration. American Association of Petroleum Geologists, Memoir 26: 83–97.

Warner, B., 1992. Peat: nature's compost. Earth, 1: 44–49.

Wein, R.W., 1983. Fire behaviour and ecological effects in organic terrain. In: R.W. Wein and D. A. MacLean (Editors), The role of fire in northern circumpolar ecosystems. John Wiley and Sons, Toronto, pp. 81–89.

White, M.E., 1990. Greening of Gondwana. Princeton University Press, Princeton, N.J , 256 pp.

Ziegler, A.M., Raymond, A.L., Gierlowski, T.C., Horrell, M.A., Rowley, D.B. and Lottes, A.L., 1987. Coal, climate and terrestrial productivity: the present and early Cretaceous compared. In: A.C. Scott (Editor), Coal and coal-bearing strata: recent advances. Geological Society, Special Publication 32, pp. 25–49.

19. RECENT COLD CLIMATE PEATS OF CENTRAL CANADA AND PERMIAN COALS OF GONDWANA

I. Peter Martini

Holocene settings resemble those of the early Permian times more than those of any other period. Such similarities include: (1) postglacial settings; (2) development of extensive peatlands in early postglacial times contemporaneously in both cold to cold-temperate zones and in tropical zones; (3) and significant changes in climate, sea level, and environments.

Differences exist as well. They include: (1) differences in time frame. Pleistocene-Holocene changes were of short duration, whereas those detectable in the Permian were of long duration; (2) differences in fauna and flora, although one would presume that analogous niches were colonized by species with similar requirements (Collinson and Scott, 1987); (3) differences in the distribution of land and sea, as well as different trends in continental drift, with a predominant east-west movement in the Quaternary and one with a north-south component in the Permian.

Although no perfect analogy can be made between the Pleistocene-Holocene and the Carboniferous-Permian for any particular single feature (Styan and Bustin, 1983, 1984; Bustin, this volume), the comparative analysis of deposits from these two geological times can provide a conceptual feedback system leading to a better understanding of what has happened in the past and what may happen in the future (Galloway and Hobday, 1983). The longer geological record of the Paleozoic case poses questions better analyzed in the more recent, well-exposed and well-preserved Quaternary systems; conversely features developing or present at the surface of Earth today can be tested for preservability in the ancient rock record.

Recent peatlands can be used as analogs for ancient coal settings (Cowell et al., 1989; Esterle et al., 1989). Peatlands are dynamic systems that react rapidly to climatic and environmental changes, and the information with which they are imprinted can be recognized in derived coal layers. Hence, a comparative analysis of selected peat and coal deposits should lead to a good understanding of local and global postglacial environmental changes.

Objectives

In this chapter I compare and contrast the distribution and occurrence of some modern peats and those of some Permian coals of Gondwana, with emphasis on cold-climate organic deposits. Among the questions to be addressed are what we know of the records of true cold-climate coals, whether there are interglacial coals, and whether we know enough about the coals' diagnostic characters to be able to recognize them. Indeed, there appears to be a discrepancy between the data obtainable from the recent geological record and those of the ancient record; the most extensive, albeit not necessarily the thickest, modern peats occur in the subarctic to cold-temperate regions, whereas most Carboniferous-Permian coals are believed to be derived from warm to tropical peats (Galloway and Hobday, 1983; Calder and Gibling, 1994).

I first offer a brief analysis of the occurrence of modern peats primarily along a central, north-south corridor of North America; second, I describe examples of Lower Permian coals from selected areas of Gondwana, Australia, South Africa, Brazil, and India; third, I provide a comparative analysis of peat and coal.

Modern Peat

Terminology

It is important to use proper, simple terminology to simplify communication. A good set of defined terms is put forward in the Canadian Wetlands

Classification (Zoltai and Pollett, 1983). Terms like wetlands, peatlands, marsh, fen, bog, and swamp have been given specific and unambiguous meanings (Fig. 19.1). Often, however, more general terms are used in the literature, where at times one term can have different meanings. For instance, "mire" refers loosely to both wetlands and peatlands, disregarding the fact that a peatland is always a wetland, but a wetland is not always a peat-producing environment; according to the Canadian Wetlands Classification, a wetland is not a peatland unless it has at least a surficial 40cm-thick peat layer. Descriptive terms like "raised forest" are used as well, some of which have local significance (Esterle and Fern, 1994; Staub and Esterle, 1994). The "raised forests" of Malaysia, for example, include true swamps, (more that 25 percent of the surface area covered by trees) with treed , minerotrophic peatlands at the margins of peat domes and treed bogs in the central parts of those domes (Anderson, 1964, 1983).

Distribution

Modern peatlands cover great parts of the boreal to cold temperate regions of the Northern Hemisphere and of the southern tip of South America, and local areas of the temperate regions of southern Australia, mainly Tasmania, and the warm to tropical areas in the southeastern United States and, more remarkably, of Indonesia (Fig. 19.2).

In terms of types of occurrence, peats show reasonably good zonal variations, modified locally by maritime and groundwater conditions (Damman, 1979; Martini and Glooschenko, 1985). Vast unconfined peatlands are best developed in the subarctic and high boreal zones, such as the vast fens and bogs of Siberia and of the Hudson Bay Lowland in Canada (Sjörs, 1959; Martini, 1989). Farther south, in the middle and low boreal zones, the peats become more localized in valleys surrounded by nonpeat-bearing forest (taiga). In the case of the Canadian Shield in the Great Lakes region, the valleys are structurally controlled but modified by glacial erosion and sedimentation. In such settings, thick fen and raised-bog peat successions develop that may locally exceed 12 m in thickness. Still farther south, in the cold-temperate and temperate regions, the conditions of development become even more restricted, and peat is found in relatively small basins, in many cases associated with kettle depressions on glacial deposits. Under such conditions, the peat is frequently part of terrestrialization successions (the filling of shallow lakes) (Dachnowski, 1912).

Except for special cases on continents and in maritime areas where moisture is abundant, peat does not develop well in the temperate areas due to seasonality of rainfall and an excess of evapotranspiration in relation to precipitation during several months of the year. Peat starts forming again and is more common in rainy subtropical and tropical humid areas, but thick deposits are somewhat restricted, as in the Okefenokee Swamp in Florida (Cohen, 1974) and the Snuggedy Swamp of South Carolina (Staub and Cohen, 1979).

In every case the peat formation is independent of the lithological composition of the substratum, although the initial stages of development may be affected by topography. In the case of the unconfined peatlands of the Hudson Bay Lowland, peat has developed on Holocene emersed coastal successions, tidal flats, beach ridges, and estuarine terrains. Farther south in Canada, peats have formed on Precambrian metamorphic and igneous terrains. To be sure, in most cases thin (less than 0.5 m in places) glacial deposits or lacustrine fine material occur between the Precambrian substratum and the peat itself. Farther south still, peat has formed in depressions floored by Paleozoic carbonates or shales or, especially in Canada, in depressions within Pleistocene glacial material such as till and glaciofluvial, glaciolacustrine, or glaciomarine deposits. Peat is thus an opportunistic organic deposit that forms under specific climatic and hydrologic conditions irrespective of the substratum type, although some mean of initially reducing the permeability of underlying deposits is required (such as deposition of clays or formation of pedogenetic crusts). Furthermore, peat develops most frequently but not exclusively in relatively flat areas where impeded drainage is more likely. However, where much moisture is present throughout the year, such as in the Queen Charlotte Islands in western Canada, thick peats can

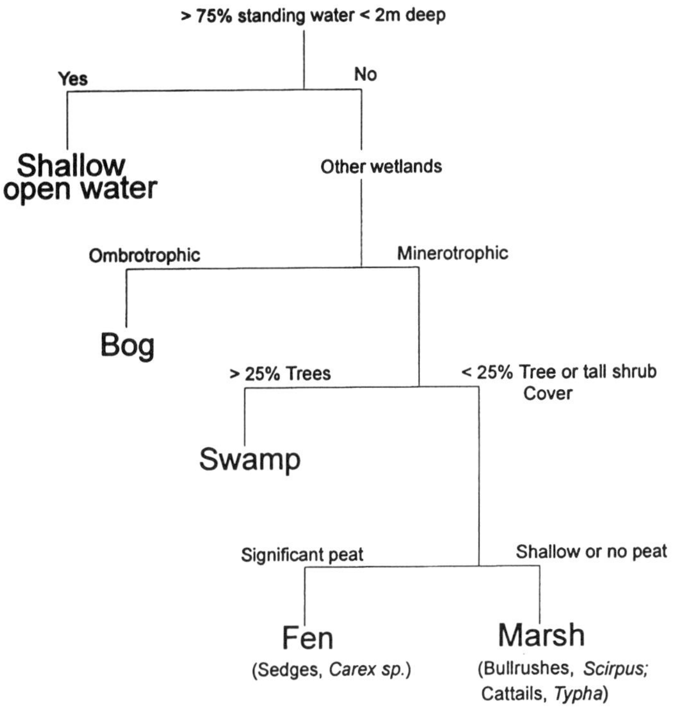

Fig. 19.1. Simplified version of the Canadian Wetlands Classification (after Zoltai and Pollet, 1983).

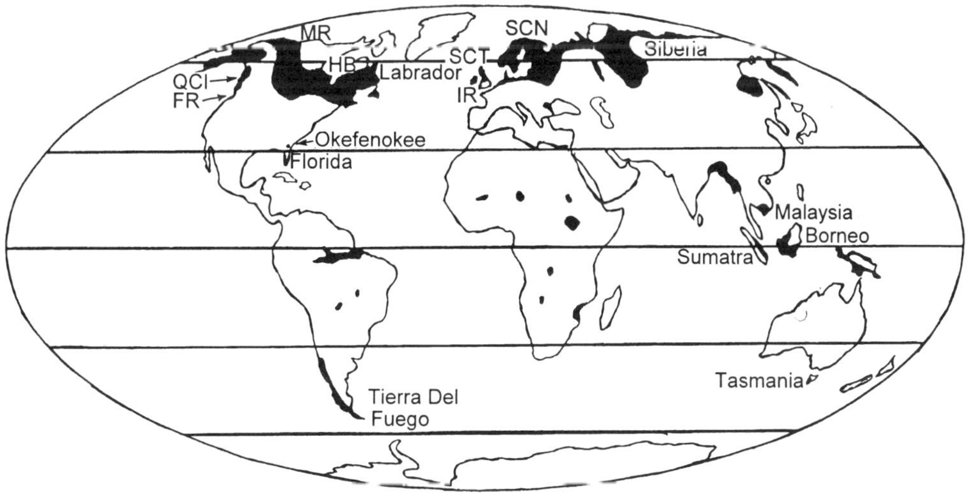

Fig. 19.2. Worldwide distribution of major wetlands (after McCabe and Parrish, 1992). Major peatlands have been identified. HB= Hudson Bay; IR= Ireland; FR= Fraser River; MR= Mackenzie River; QCI= Queen Charlotte Islands; SCN= Scandinavian countries; SCT= Scotland.

develop on relatively steep mountainous slopes (Clague et al., 1982; Mathewes, 1989).

Age

Recent peats have formed in the last 12–13,000 years; most of those in North America irrespective of the area and the climate, formed in the last 8–6,000 years (Heinselman, 1975). In glaciated areas, peat formed only after the glaciers had retreated and, in areas submerged during deglaciation, after emersion occurred due to isostatic rebound or sedimentation. In non-glaciated areas peats formed in coastal areas after the major eustatic rise, when sea level stabilized about 5–6,000 years ago and platforms for the growth of peat, such as deltas and coastal plains, had developed (Galloway and Hobday, 1983). Even in nonglaciated inland areas, such as the Okefenokee Swamp, most of the peat started forming at about this time because of the changing climatic conditions during the Late Pleistocene-Holocene.

Peat formation has responded rapidly to the changing climates during the Holocene (Wells and Pollet, 1983), and in different ways, depending on the latitude of the peatlands. The effect of climatic change is very strong in areas near the northern transition zones.

(1) In the far north (Mackenzie River area), peat formed during the postglacial warming period, and the majority of peat has an age of about 8,000–5,000 years BP (Mackay and Terasmae, 1963; Terasmae, 1967; Tarnocai, this volume). Afterward, cold climate returned, permafrost developed, and peat ceased to grow in any significant amount.

(2) In the subarctic and boreal areas, peat started forming about 5,000 years ago as the climate was deteriorating after the hypsithermal interval (Terasmae, 1977). During the hypsithermal interval, either suitable platforms were not available or the climate was too warm and too dry (too much evapotranspiration in respect to precipitation) for substantial amount of peat to form.

(3) At mid-high latitudes, since the later part of the hypsithermal, cool temperatures, high levels of precipitation, and low rates of evaporation led to high ground-water levels and to the development of much *Sphagnum* peat, hence of bogs. Bog deposition continues today in those regions. Indeed, since the inception of the Holocene, the changing climatic conditions have had, on the whole, the effect of reducing the extent of the tropical peats and increasing the extent of the boreal ones (Klinger, 1990).

Thickness

The rate of deposition of modern peats varies widely, from 0.01 mm/yr to 4 mm/yr (Anderson, 1964; Cowell et al., 1983). Variations also occur in the degree of compaction undergone by various types of peats, on the order of 10 to 1. Assuming, for simplicity, an average rate of growth of 1 mm/year and a peat-coal average thickness ratio of 10:1, development of a 15m coal layer would require a 150m-thick peat deposit, which in turn would require 150,000 years of stability in the wetland environment in a basin that was undergoing progressive subsidence. Because the Pleistocene–Holocene has been a period of significant and rapid changes we have no modern analog for such an occurrence. The study of modern peatlands can, however, add to our understanding of the behavior of ancient peatlands and of their response to changing environmental conditions.

The thickness of peats is constrained by climatic conditions as well. The thickest peat is found in tropical raised forests in Malaysia, but comparable thicknesses, up to 14–15 m, are found in bogs in the maritime boreal regions of Canada. However, peat in inland boreal regions is generally thinner (2–4 m thick) than that in tropical regions, where thickness can exceed 20 m. This is related to the balance between peat organic productivity and organic preservation. In tropical areas, much organic matter is produced, but much is decomposed (Moore, 1987). In subarctic, boreal, and cold-temperate regions, the number of vegetation growth-days per year is greatly reduced by the occurrence of six months of frozen ground conditions and the persistence of a cold climate for several additional months. During the summer, however, an explosive productivity occurs during short periods of warm climate. The key factor here is that although the amount of material produced is not large when compared to that of tropical areas, most of what is produced is preserved because of the cold climate, which inhibits the decomposition of the organic matter by micro-organisms.

Finally, where peat-bearing basins do not undergo subsidence, the maximum thickness that peat can reach depends on the depth of the valleys into which it is constrained. Outside the valleys, in open peatlands, the domed bogs and raised forests reach maximum thickness as a result of a balance among the rate of decomposition and water retention capability of peat, piovosity, and slope stability. In both tropical and boreal areas, domed treed peatlands can maintain steeper slopes and hence rise higher than the grassy and *Sphagnum* peatlands.

Makeup

Modern peatlands contain a large variety of plants, apparently much larger than the correspondent Carboniferous communities (Collinson and Scott, 1987). Some researchers believe that in the evolution of modern peatland the tendency toward the formation of raised bogs or raised forests represents climactic stage (Klinger, 1990). The trends in cold and cold-temperate areas are:

Terrestrialization: from basal lacustrine deposits (locally marls), to aquatic peat, fen, swamp, and finally bog.

Paludification : from basal grassy peat, to graminaceous fen peat, fen-forest or swamp peat, to finally bog.

Heinselman (1963, 1970, 1975) believes that a forestation stage occurred in Minnesota after an initial postglacial grassy-prairie stage and before paludification.

Animals can affect the peatlands as well. In marginal areas of the peatlands, for example, beavers build dams and impound waters. In places, humankind has been involved not only in destroying wetlands but also in fostering the growth of some peatlands. In Ireland, for example, paludification was initiated on fields cultivated by ancient agriculturalists. The shallow scraping of the organic-rich surfaces for planting crops allowed the mobilization of iron and the formation of an impermeable iron pan, which impeded drainage and initiated the formation of the wetland.

As a result of climatic conditions, northern peatlands are characterized primarily by grassy and mossy vegetation, but treed peatlands develop as well in subarctic and boreal (treed fens and treed bogs, with tamarack and spruce primarily) and cold-temperate (wide-leaf hardwood swamps) regions. Contemporaneous tropical wetlands are most frequently covered by trees, ranging from the cyprus swamps of Florida to the extensive coastal mangrove swamps of tropical areas to the hardwood raised forests of Malaysia (Gore, 1983).

Decomposition versus Humification

Peat deposits represent the net accumulation of organic matter after it has undergone partial decomposition, humification and compaction. Organic matter undergoes varying rates and amounts of decomposition and humification, depending on the wetness, nutrients, micro-organisms, type of plants and climate.

Bogs, for instance, show various decomposition trends. Raised bogs undergo progressive drying out, at least for part of the year, as they are greatly affected by summers with low piovosity. The upper part of the bog therefore becomes more aerobic, and oxidation and activity of aerobic micro-organisms are enhanced. This seems to be the case in boreal to subarctic regions. In fact, in the Hudson Bay Lowland the bog areas are euphemistically called the dry regions by Coombs (1954). Furthermore, Cowell et al. (1983) found that the free water table in bogs can be as deep as 35 to 75 cm in certain years, hence the surface of these peatlands is subject to aerobic conditions during the brief summer periods.

Whereas oxidation of the surficial layer of the bog may exceed that of fens, the total amount of decomposition of the peat depends also on other variables. For instance, the low pH of bogs inhibits the growth of micro-organisms, so the rate of peat decomposition is relatively low. In contrast, the minerotrophic waters that flow slowly through the less acid fen carry more nutrients; they therefore may feed micro-organisms, which effectively decompose these peats (Staub and Esterle, 1994).

Furthermore, according to Moore (1987), decomposition affects differently the various types of plants and their various components; for instance, leaves decompose faster than woody material. Consequently, the composition of the peat does not accurately reflect the nature of the original peat-forming vegetation. *Sphagnum* material decomposes more slowly than grassy and graminaceous materials; in this respect, as well, bogs experience slower transformation.

Humification starts as soon as the organic matter is subjected to reducing conditions. This is the first stage of the organic metamorphism that transforms vegetable matter into coal. Independent of other possible variables, such as plant composition, peats that have been decomposed to some degree under oxidized conditions before being buried and humified will result in coals with higher levels of inertinite, whereas those that have not been decomposed form coals with higher levels of vitrinite (Hunt and Smyth, 1989).

Main Impurities in Peat

Peat have variable geochemistry (Shotyk, 1988) and contain various mineral impurities, either transported from outside or generated by the plants themselves (phytoliths). Of the various components, the amounts of ash and pyrite are the most important when peat and coals are considered as energy resources. It is assumed that because of their relative position in respect to active clastic sedimentary environments, fens are richer in ash content and sulfur than bogs. This is correct, in the sense that the highest level of impurities is found in some fens. However, it is also true that many fens contain low levels of ash and pyrite and that some bogs are rich in both, having acquired them as windblown impurities such as volcanic ashes or as silt from unvegetated coastal and stream areas.

Burial and Preservation

Peat is not an easily eroded material. Pleistocene lignite and Holocene peats are found preserved in various conditions. Essentially, peat requires a stable or slowly changing setting to grow, but a relatively rapid inundation and burial are necessary for it to be preserved and changed into coal. Burial can be accomplished in many ways. Peat and coals are found under glacial, marine, deltaic, lacustrine, and alluvial fan deposits. The tectonic style of the basins does not seem to matter, either; coals are found associated with every terrestrial humid deposit ranging from those of ancient cratonic areas to extensional basins of various types and molassic basins. This does not exclude the fact that some subsiding tectonic settings may facilitate formation and preservation of thick peats.

Hobday (1987) stresses tectonic subsidence as the regulator of groundwater regime and marine transgressions and thus of the type of coal (inertinite-rich or inertinite-poor). However, subsidence is not the only variable. For example, particularly in postglacial times, climatic changes may occur that lead to a change in the moisture regime, which in turn causes changes in the water table in peatlands and therefore in the rate of peat decomposition, independent of tectonic settings. The development of the very thick peat deposits that can form thick coals requires long-lasting, specific balanced conditions. Tectonic subsidence is generally a slow process, usually much slower than the rate of growth of peat. On the other hand, postglacial climatic changes may trigger a rise in water levels in the sea or lakes rapid enough to drown the peatlands. In addition, Falini (1965) suggests that the rate and the amount of local variation in sea or lake level and the variation in rates of transgression may be due not only to general tectonism but also to other variables, such as the geometry of the basin.

Permian Coals

Ancient cold-to-temperate climate coals have been studied to varying degrees in Australia, Brazil, South Africa, India, and Antarctica. A few characteristics of the first three areas are emphasized here. Other information has been presented in various chapters of this book and in other recent reviews (Rahamnai and Flores, 1984; Scott, 1987; Lyons and Alpern, 1989a, b; Cobb and Cecil, 1993; Calder and Gibling, 1994).

Australia

In most cases the easternmost Australian coals are associated with deltaic or other coastal successions or fluvial and alluvial fan successions (Hunt, 1988; Diessel, 1992). These are part of clastic wedges that prograded westward from the rising New England Fold Belt into the Sydney and the Bowen basins (Fig. 19.3A). Tectonism was somewhat subdued in the Tasmanian Basin to the south; the major Permian deposits developed as infilling of deep fjord-like embayments, with steep banks and perhaps occasional local reactivation of the bounding faults, and do contain some coal-bearing successions. Different from these are the deposits of the west side of the Bowen Basin and the associated basins of central-eastern Australia, as well as some of the smaller basins of western Australia. In these locations, cratonic conditions prevailed, and the coal measures developed on coastal or alluvial-lacustrine plains on glacially eroded paleosurfaces.

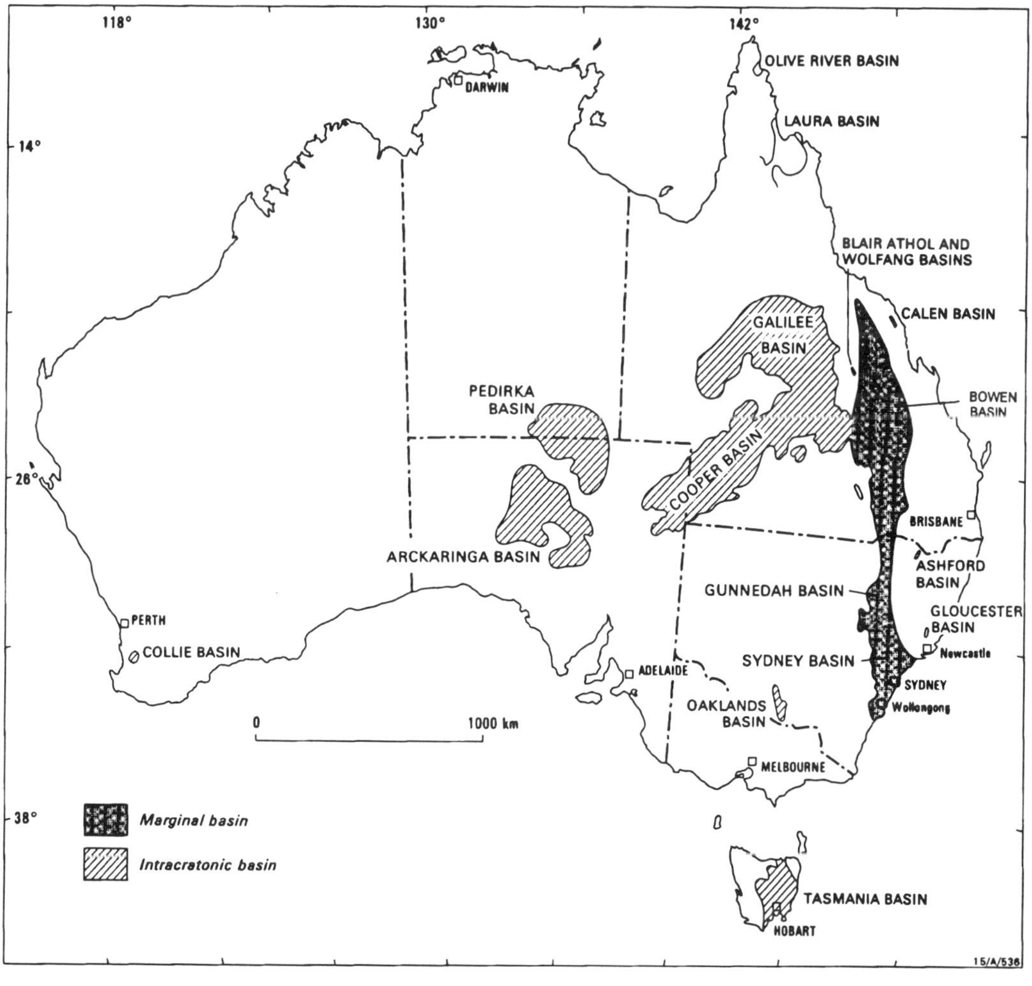

Fig. 19.3A. Coal basins of Australia (after Lindsay, this volume).

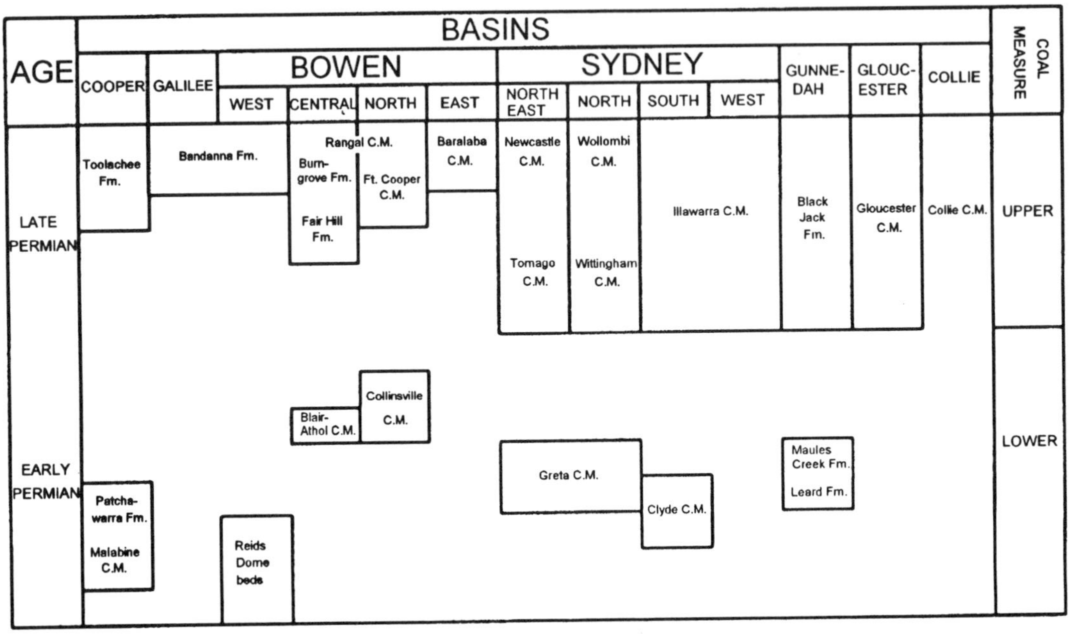

Fig. 19.3B. Stratigraphic distribution of major Permian coals in Australia (after Lindsay, this volume).

The lower coal measures (Fig. 19.3B) of eastern Australian are laterally equivalent to cold, lonestone-ridden marine successions, indicating that adjacent seas received ice floes and icebergs from southern glaciers. The lowermost coals of the Sydney Basin (Greta and equivalent coals) were formed under cold, postglacial conditions, as indicated also by associated permafrost features (casts of ice wedges, imprints of ice crystals, convolute deformations). However, the climatic conditions changed rapidly, and the subsequent Permian economic coals derived from peats of temperate and warm settings.

Brazil

Brazilian coals are found in the Paraná Basin (Fig. 19.4A). This basin has been glaciated and retains thick glaciomarine and minor glacioterrestrial deposits, as well as exhumed glacially eroded surfaces, such as on the peneplaned Uruguayan Precambrian Shield. Thin (<0.5 m), noneconomic coals are interbedded in glacial deposits in São Paulo State and have been interpreted to have formed during interglacial times (Fig. 19.4; Martini and Rocha-Campos, 1991). Temperate, postglacial coals have developed in glacially modified valleys of the Uruguayan Shield in Rio Grande Do Sul (Silva, 1991). Only a very thin (less than 30 cm thick) layer of lacustrine clay separates these coals from the underlying, glacially eroded Precambrian surface. All other coals in Brazil are found in the postglacial Rio Bonito Formation (Fig. 19.4B). These coals developed on coastal and deltaic platforms. They derived from already temperate to warm climate peats, and no indication exists in correlative marine sediments of lonestones or coldwater fauna.

South Africa

The basal coals of South Africa have varying platforms, ranging from Archean to Precambrian bedrock to Paleozoic tillites, to diamictites of various sorts to fluvial deposits (conglomerates and sandstones) to floodplain and lacustrine fine sediments (Fig. 19.5; Hobday, 1987). The coals are confined to paleovalleys, and reflect the influence of substratum inhomogenieties in their variable thicknesses (Cairncross and Cadle, 1988; Holland et al., 1989). In South Africa the controlling factors for the growth of the peatlands were the rate of basinal subsidence and of shifting of

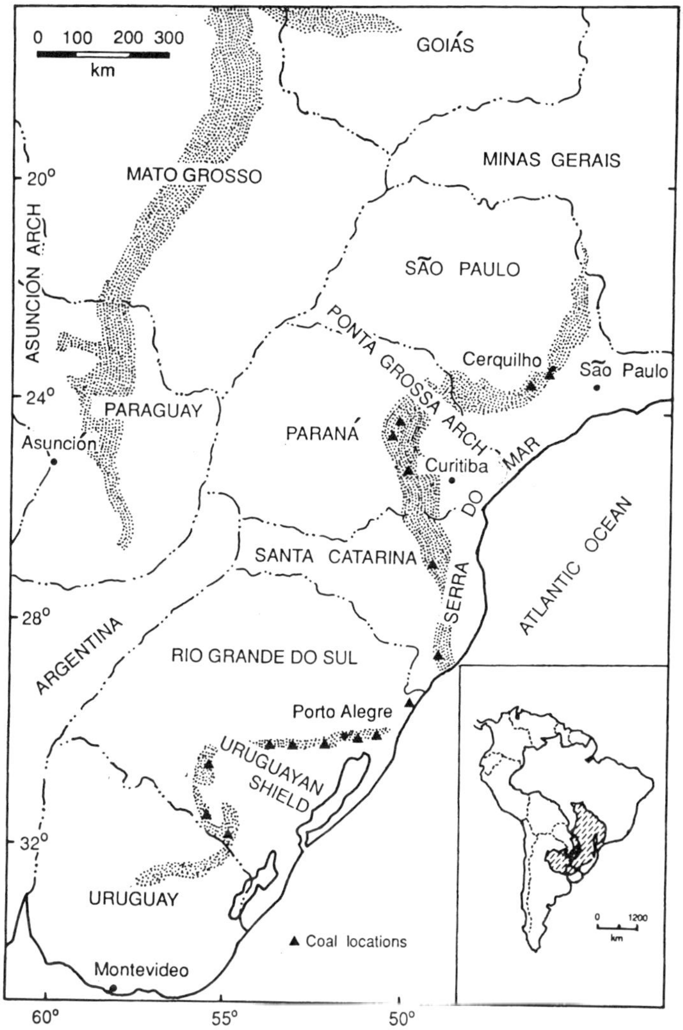

Fig. 19.4A. Coal deposits of the Paraná Basin (after Martini and Rocha-Campos, 1991).

streams. In contrast, the major factor controlling the development of the equivalent peatlands of the lower coals of Brazil were the changing lacustrine and marine conditions.

In Africa, as in Brazil, there are indications that coal is present in interstadial glacial successions, as suggested by Stavrakis (1986) and Stavrakis and Smyth (1991) for the Dwyka coal #1, in the Orange Free State (Fig. 19.5A). A good indication of changing climatic conditions during the formation of coals has been identified by Falcon (1989). She found cold-climate flora in coal no. 1 and the transition to a warmer-climate flora in the upper part of coal no. 2 of the Transvaal region (Fig. 19.5B). It is reported in the literature that coal no. 1 is associated with glaciofluvial deposits, and that perhaps the same periglacial to paraglacial conditions existed for part of coal no. 2. Consistent with to the Quaternary case, it is unlikely that the Permian thick and vast cold peatlands developed during full glacial times. Rather, the lower Permian peatlands of South Africa may have formed under postglacial cold-climate conditions, when mountain glaciers may have still been present in the highlands, but had no direct relationship with the development of the peat itself.

Younger, South African Permian coals formed

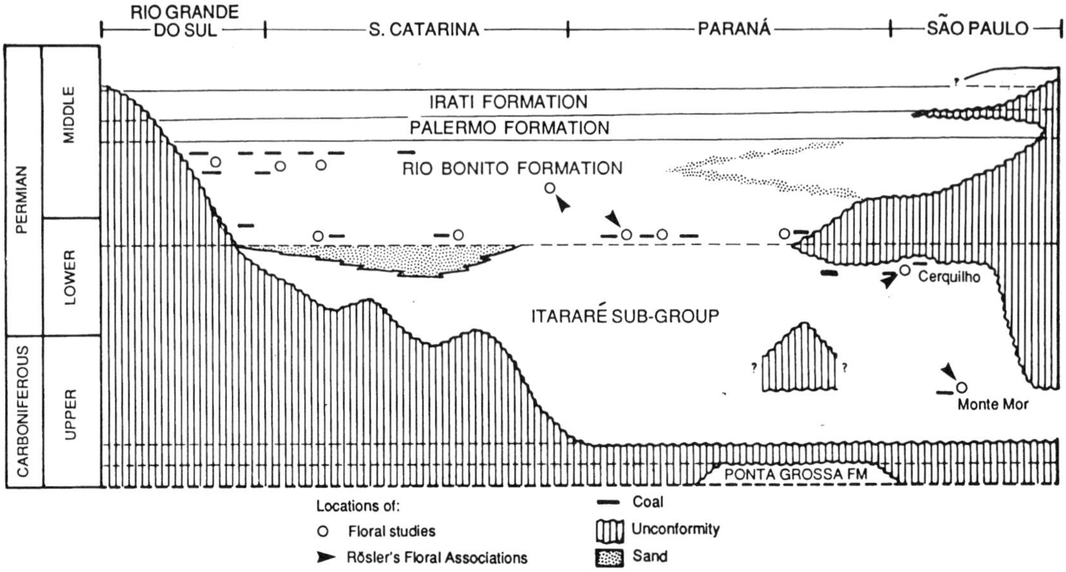

Fig. 19.4B. Cross-section of the Paraná Basin showing major coal deposits (after Martini and Rocha-Campos, 1991).

from warm-climate peats that developed on sedimentary platforms, such as delta lobes and paralic settings, and in river floodplains, particularly in Transvaal and Natal provinces.

India

The Gondwana coals of India are preserved in rift basins (Fig. 19.6A; Navale and Saxena, 1989; Mita, 1991). Their characters are similar to those of other deposits described earlier, except that several seams have clay-rich seat earths (Mitra, 1991) not common in other Gondwana coals. The lowermost Permian coals rest directly on diamictites or on marine deposits that, in turn, rest on diamictites. All coals derived from postglacial peats; the lowermost ones derived from peats of cold-climate areas (Fig. 19.6B,C), and subsequent ones derived from peats of temperate to warm settings, as indicated by floral and petrographic evidence (Tewari and Tripathi, 1987; Dutta, this volume).

Discussion

Peat can be considered both as soil and sediment.

Soil

Peat can form on surfaces that are stable from the point of view of erosion, sedimentation, and drainage. As a soil (histosol) it carries specific information on the climatic conditions under which it developed (Shotyk, 1992). For peat to form, sustained, strong growth of plants and anaerobic conditions for the preservation of dead organic material are necessary. These are achieved either by the maintenance of waterlogged conditions in tropical areas by continuous rain throughout the year to compensate for the strong evapotranspiration, or, under cold climatic conditions, by the presence of frozen ground for many months of the year. Both conditions - either drowned surfaces or frozen surfaces - inhibit oxidation of plant matter and the disruptive action of microbial activities; as a result, peat can form and persist. Like any other soil, peat achieves various degrees of development. The thickness and the degree of horizonization vary according the to length of time the peat has to form. In very stable surfaces peat's ability to achieve a given thickness is determined by the soil's ability to remain waterlogged. Peat, like any other soil, marks unconformable surfaces.

Unlike some other soils, however, peat can

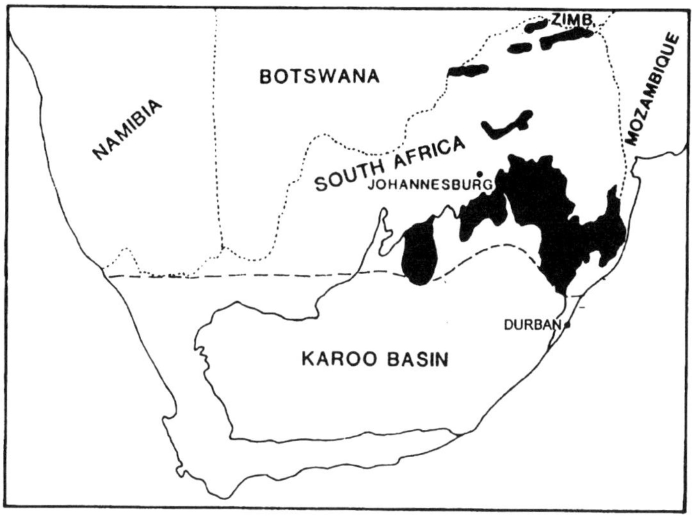

Fig. 19.5A. Permian coal deposits of South Africa (after Falcon, 1989).

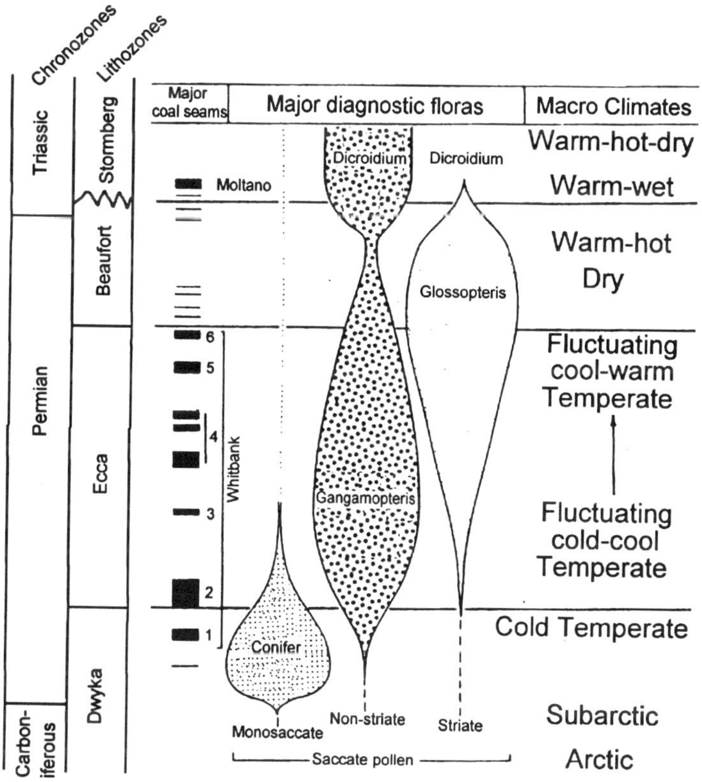

Fig. 19.5B. Stratigraphic distribution of major coals of the Transvaal with climatic interpretations (after Falcon, 1989).

Fig. 19.6A. Permian coals of peninsular India (after Mishra, 1991).

STRATIGRAPHY			LITHOTYPES	MICROFLORAL ASSEMBLAGES	CLIMATE
Age	Fm.	Subdiv.			
Late Permian	Raniganj	Upper	Sandstone, Siltstone, Shale, Coal seam	Striated Disaccates Trilete Group	Warm and humid
		Lower			
	Barren Measures		Ironstone, Shale Sandstone	Striated Disaccates Densipollenites Group	Hot and dry
	Barakar	Upper	Feldspathic Sandstone, Carbonaceous Shale, Fire clay, Coal seam	Striated Disaccates Group	Warm and humid
		Lower		Non-striated Disaccates Group	
Early Permian	Kar-harbari	Upper	Carbonaceous Sandstone Conglomerate, Coal seam	Callumispora Radial Monosaccate Group	Cold and warm humid
		Lower			
	Talchir		Diamictite, Shales Sandstone	Radial Monosaccates Group	Cold and dry

Fig. 19.6B. Stratigraphic distribution of Permian coals in India (after Navale and Saxena, 1989).

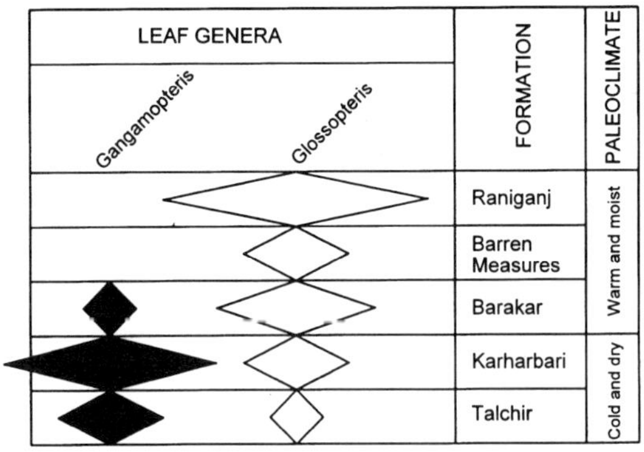

Fig. 19.6C. Distributin of leaf species in the coal-bearing units of India (after Navale and Saxena, 1989).

continue to develop and thicken indefinitely if the right environmental conditions are maintained and accommodation space is provided by subsidence of the basin. Under such conditions, it behaves like a sedimentary deposit.

Sediment

Peat is an organic sediment that forms in special settings known as peatlands. Given certain environmental conditions, peatlands, once established, form a self-sufficient community of plants and animals. They can be considered the terrestrial equivalent of marine reefs. Peatlands and reefs have obvious differences, but they show strong similarities as well. Among these similarities are the following:

1. Peatlands can develop positive growth forms over any substratum if the environmental conditions are suitable. Unlike reefs, peats develop preferentially in low-lying areas, valleys, basins, but they can also build up above the local minerotrophic water table by generating their own wetland ecosystem, and, like reef, they form self-sufficient systems: the bogs.
2. Peatlands and reefs are both strongly affected by changing climatic and environmental conditions.
3. Both peatlands and reef can be damaged or destroyed by input of clastic sediments or by drowning in fast-subsiding basins. To develop fully, they require stable conditions with no erosion or active clastic sedimentation.
4. Both peatlands and reef usually have a great variety of organisms (flora and fauna).
5. Peatlands and reefs both form binding deposits and stable structures. Like reefs, peatlands are resistant to erosion, although they export particulate matter in addition to gaseous substances (predominantly methane and carbon dioxide) to the atmosphere (Shotyk, 1989; Klinger et al., 1994; Roulet et al., 1994)) and ions and organic acids in solution.
6. Both form in warm- and cold-climate conditions, although peatlands are more widely distributed in modern cold settings. Both cold and warm deposits (peat and reef) are preserved in the ancient geological record, with the warm types predominating.
7. Both peatlands and reefs have developed since Paleozoic times. For both, modern examples can be only approximate analogs to ancient cases because their component organisms have changed drastically. Nevertheless, analogies in environmental settings can be made, and it can be assumed that similar niches have been occupied by organisms with similar requirements. Thus, most of the processes and the complex interrelations we observe in modern ecosystems can serve as guide to ancient ones.

Like any other soil or sediment, peat, once formed, is affected by weathering (oxidation and microbial transformation), and when buried it is transformed into a rock: coal. This transformation occurs by stages, and as the pressure and temperature of the deposit increase, the material experiences diagenesis and metamorphism. Diagenesis provides a sedimentary rock-equivalent, lignite or brown coal, where much of the original plant makeup is still recognizable. During subsequent organic metamorphism the original organic matter is transformed into new compounds (macerals), which are equivalent to the formation of metamorphic minerals of other types of rocks. The composition of the buried deposits changes as the grade of metamorphism increases, with progressive reorganization of the texture and structure and loss of hydrogen and concentration of carbon. It is therefore not a surprising that, although some paleoenviromental reconstruction can be made from coal, it may be difficult to impossible to reconstruct the exact conditions of the original peatlands from certain coal deposits (Bustin, this volume). Our success in interpreting paleoenvironment depends on making reasonable demands of the material we work with. As much evidence as possible should be brought to bear on the solution of the problem, because no single character of the rock record, even of peat-coal, is usually sufficient to reconstruct a paleoenvironment with certainty.

Another problem that peat and coal have in common with other soils and sedimentary deposits is that they have been affected by changing climates and by sedimentary and tectonic settings. It is often difficult to disentangle the various effects, particularly when some precision is required. Carbonate deposits, for example, usually indicate subtropical to tropical conditions, but they can develop also by allochemical (such as shells) material in cold seas; evaporites are associated with an excess evaporation over precipitation and usually form under warm climatic conditions, but they can also form, albeit in thinner deposits, under cold conditions; red-beds are formed primarily in warm, semiarid settings, but they develop as well in cold settings, albeit less extensively; thick peat develops in tropical wetlands, but the most extensive modern peatlands occur under subarctic to boreal conditions.

Tectonism affects the accommodation space for sediments within a basin as well as the persistence of certain environments of sedimentation. Rising lands are subject to erosion; thus, their materials and soils are not likely to be preserved in the geological record. Subsiding basins, on the other hand, provide space for sediments to accumulate, and, depending on the rate of subsidence and on the surrounding environmental conditions, the sediments may either be variable or rather uniform. It is the interaction between the tectonic factor, which affects the boundary conditions of the basin, and the climatic factor, which affect the type of materials transported and deposited, that in many cases becomes difficult for paleogeographs to disentangle. This difficulty is reflected in the geological literature, which shows cyclic thinking. A classical example is the interpretation of diamictites, which were initially considered to be tectonically controlled sediment gravity flows, but later they were reinterpreted as glacial deposits; today an effort is being made to accommodate both ideas, since both the climatic (glacier) and tectonic effects may have acted on their formation. Another example of the interaction between tectonic forces and climate is the presence of readily weatherable minerals, such as fresh feldspar in first-cycle sedimentary rocks. The presence of a large amount of feldspar indicates, first of all, that plutonic rocks were abundant in the source area; second that tectonic uplift, erosion, and sedimentation were rapid enough or that the climate was cold enough to reduce the rate of chemical weathering reactions sufficiently for feldspar to be preserved (Dutta, this volume).

Coal apparently shows similar climatic and tectonic influences. Coal macerals can form from different plant materials (primarily climatic control) or from plant materials that were variously affected by weathering and decomposition prior to burial (both climatic and tectonic control). In the first instance, the presence of woody material should lead to a greater abundance of vitrinite, and graminaceous peat, to inertinite. The development of these types of plants depends in part on climate, the warm peatlands having a predominance of trees, the cold climate ones having a predominance of graminaceous and mossy plants. However, plants that could have produced vitrinite may not do so if their remains have been oxidized or otherwise decomposed prior to removal from contact with the atmosphere.

It has been observed in Carboniferous-Permian deposits that the coals of the Northern Hemisphere are invariably rich in vitrinite, independent of the tectonic settings. The climatic effect that led to the development of subtropical to tropical swamps was stable enough not to be obscured by the tectonic effect. The coals of the Southern Hemisphere, in contrast, although generally rich in inertinite, show variations in inertinite-vitrinite ratios between and within basins. Thus, in general, the intracratonic basins of Australia contain more inertinite-rich coals than do the more tectonically active foredeep basins. Within the foredeep basins, both vitrinite-rich and inertinite-rich coals are present. The intracratonic basins are more stable and experience lower rates of subsidence. Therefore, the peatlands are likely to grow to their water-table upper limit and be subjected to oxidation and decomposition; they are more likely to generate precursors for inertinite, independent of climate and plant composition. The foredeep basins are more tectonically active; they subside more rapidly, and more clastic sediments are injected into them. Peatlands that develop and can keep pace with subsidence without being drowned experience wetter condition; dead plant material is stored under anoxic waters, and the peat has a high likelihood of being buried by other sediments and being preserved in the geological record. Under these conditions, the probability of formation of vitrinite is greater. However, the formation of inertinite in these basins is not excluded because of the composition of the original plant material, the tectonic subsidence is not a regularly continuous phenomenon, and the rates of subsidence and the rate of water-table variation and weathering of plant material differ by several orders of magnitude.

Although all of this is true, it should also be considered that the continents experienced drastic climatic changes during the Carboniferous-Permian (Frakes, 1979; Rowley et al., 1985). As a result, different peats and derived coals were formed. This is true from a petrographic point of view; for instance, the Greta coals differ from all the other younger Australian coals. Palynologically, the lowermost coals of every Gondwana continent differ from the higher coals. The enclosing rocks differ as well, the lower coals being associated with true cold-climate sediments. Indeed, only the lowermost Permian coals of Australia (Greta and correlatives), South Africa (coal seam no. 1 and part of no. 2 and correlatives) and Brazil (Cherquilho and correlatives) derived from cold to cold-temperate peats. All other overlying coals derived from cool-temperate to warm peats. Thus, the Gondwana coals do not form a homogenous group. The differences observed between the lowermost truly cold coals and the younger warm coals conform to a large extent with the overall expectation associated with the different composition of the original peatlands: the graminaceous-equivalent cold peatlands led preferentially to the formation of inertinite-rich coals, whereas the warm, tree-rich peatlands led preferentially to inertinite-rich coals. Variations that occur among coals of different horizons are associated with peatlands environmental and tectonic conditions, much the same as indicated nowadays on Earth's surface.

Conclusion

In both Holocene and Permian postglacial times, vast peatlands developed in both cold and warm areas. Few coals, however, from true cold peatlands have been preserved. What has been preferentially preserved are deposits that formed under postglacial, warm conditions.

The relative low preservability of cold peats may be an effect of the type of peatlands that formed under different climatic conditions - for example thicker peat formed in tropical areas - or the result of different tectono-sedimentary conditions. Glaciated, cold-climate areas of both Quaternary and Permian times, for example, were subjected to uplift rather than to subsidence, due to postglacial isostatic rebound. Furthermore, after the strong pulse of sediment dispersal associated with the glacial events, the rate of sedimentation in glaciated lands was greatly reduced in comparison to warmer, more rapidly weatherable areas. Therefore, although more widespread peatlands could develop in the colder, glaciated settings, the likelihood of burial and preservation of such peats was less than for those formed in sedimentologically more active environments.

Because of all this and because of the difference in flora, it is not possible to consider the modern settings as very accurate analogs for the ancient peatlands. However, as it has happened for reefs, similarity in environmental conditions may have fostered colonization of similar niches by organisms with similar requirements. The overall successions in climatic and peatland conditions we have experienced in the recent past can therefore foster a good understanding of what went on in ancient times. Conversely, analysis of ancient settings, such as the unique Permian coal-bearing successions of Gondwana, provides us with an understanding of long-range changes that have occurred and may yet occur on Earth.

References

Anderson, J.A.R., 1964. Structure and development of the peat swamps of Saravak and Brunei. Journal of Tropical Geography, 18: 7–16.

Anderson , J.A.R., 1983. The tropical swamps of western Malesia. In: A.P.J. Gore (Editor), Mires: Swamp, Bog, Fen and Moor, B. Regional studies. Ecosystems of the World, vol. 4B, Elsevier, Amsterdam, pp. 181–200.

Cairncross B. and Cadle, A.B., 1988. Depositional palaeoenvironments of the coal-bearing Permian Vryheid Formation in the east Witbank Coalfield, South Africa. South African Journal of Geology, 91: 1–17.

Calder J.H. and Gibling M.R. (Editor), 1994. The Euroamerican coal province: controls on tropical peat accumulation in the Paleozoic. Palaeogeography, Palaeoclimatology and Palaeoenvironment, Special Issue, vol. 106, 351 pp.

Cameron, C.C., Esterle, J.S. and Palmerm C.A., 1989. The geology, botany and chemistry of selected peat-forming environments from temperate and tropical latitudes. International Journal of Coal Geology, 12: 105–156.

Clague, J.J., Mathewes, R.W. and Warner, B.G., 1982. Late Quaternary geology of eastern Graham Island, Queen Charlotte Islands, British Columbia. Canadian Journal of Earth Science, 19: 1786–1795.

Cobb, J.C. and Cecil, C.B. (Editors), 1993. Modern and ancient coal-forming environments. Geological Society of America, Special Paper 286, 198 pp.

Cohen, A.D., 1974. Petrology and paleoecology of Holocene peats from the Okefenokee swamp-marsh complex of Georgia. Journal of Sedimentary Petrology, 44: 716–726.

Collinson, M.E. and Scott, A.C., 1987. Implications of vegetational change through the geological record on models for coal-forming environments. In: A.C. Scott (Editor), Coal and coal-bearing strata: recent advances. Geological Society, Special Publication 32, pp. 67–85.

Coombs, D.B., 1954. The physiographic subdivisions of the Hudson Bay Lowlands south of 60 degrees north. Geographical Bulletin, 6: 1–16.

Cowell, D.W., Wickware, G.M., Boissoneau, A.W., Jeglum, J.,K. and Sims, R.A., 1983. Hudson Bay Lowland peatland inventory. Proceedings of the Peatland Inventory Methodology Workshop, Land Resource Institute, Agriculture Canada, Ottawa, pp. 88–102.

Dachnowski, A., 1912. Peat deposits in Ohio: their origin, formation, and uses. Soil Science, 17: 107–133.

Damman, A.W.H., 1979. Geographic patterns in peatland development in eastern North America. Classification of peat and peatlands. In: E. Kivinen, L. Heikurainen and P. Pakarinen (Editors), Proceedings of the International Symposium, Hyytiala, Finland. International Peat Society, Helsinski, pp. 213–228.

Esterle, J.S., Ferm, J.C. and Yiu-Liong, T., 1989. A test for the analogy of tropical domed peat deposits to "dulling-up" sequences in coal beds - preliminary results. Organic Geochemistry, 14: 333–342.

Esterle, J.S. and Ferm, J.C., 1994. Spatial variability in modern tropical peat deposits from Sarawak, Malaysia and Sumatra, Indonesia: analogues for coal. International Journal of Coal Geology, 26: 1–41.

Falcon, R.M.S., 1989. Macro- and micro-factors affecting coal-seam quality and distribution in southern Africa with particular reference to the No. 2 seam, Witbank coalfield, South Africa. International Journal of Coal Geology, 12: 681–731.

Falini, F., 1965. On the formation of coal deposits of lacustrine origin. Geological Society of America Bulletin, 76: 1317–1346.

Frakes, L.A., 1979. Climates throughout geologic time. Elsevier, New York, 310 pp.

Galloway, W.E. and Hobday, D.K., 1983. Terrigenous clastic depositional systems: applications to petroleum, coal, and uranium exploration. Springer-Verlag, New York, 433 pp.

Gore, A.J.P. (Editor), 1983. Mires: swamp, bog, fen and moor - General studies. Ecosystems of the World, vol. 4A. Elsevier, Amsterdam, 440 pp.

Heinselman, M.L., 1963. Forest sites, bog processes, and peatland types in the glacial Lake Agassiz region, Minnesota. Ecological Monograph 33, pp. 327–374.

Heinselman, M.L., 1970. Landscape evolution, peatland types, and the environment in the Lake Agassiz peatlands natural area, Minnesota. Ecological Monograph 40, pp. 235–261.

Heinselman, M.L., 1975. Boreal peatlands in relation to environment. In: A.D. Hasler (Editor), Coupling of land and water systems. Ecological Studies, vol. 10. Springer-Verlag, New York, pp. 93–103.

Hobday, D.K., 1987. Gondwana coal basins of Australia and South Africa: tectonic setting, depositional systems and resources. In: A.C. Scott (Editor), Coal and coal-bearing strata: problems and perspectives. Geological Society, Special Publication 32, pp. 219–233.

Holland, M.J, Cadle, A.B., Pinheiro, R. and Falcon, R.M.S., 1989. Depositional environments and coal petrography of the Permian Karoo sequence: Witbank Coalfield, South Africa. International Journal of Coal Geology, 11: 143–169.

Hunt, J.W., 1988. Sedimentation rates and coal formation in the Permian basin of eastern Australia. Australian Journal of Earth Science, 35: 259–274.

Hunt, J.W. and Smyth, M., 1989. Origin of inertite-rich coals of Australian cratonic basins. International Journal of Coal Geology, 11: 23–46.

Klinger, L.F., Elias, S.A., Behan-Pelletier, V.M. and Williams, N.E., 1990. The bog climax hypothesis: fossil arthropod and stratigraphic evidence in peat sections form southeast Alaska, USA. Holarctic Ecology, 13: 72–80.

Klinger, L.F., Zimmerman, P.R., Greenberg, J.P., Heidt, L.E. and Guether, A.B., 1994. Carbon trace gas fluxes along a succession gradient in the Hudson Bay lowland. Journal of Geophysical Research, 99: 1469–1494.

Lyons P.C. and Alpern B. (Editors), 1989a. Peat and coal: origin, facies, and depositional models. Elsevier, Amsterdam.

Lyons, P.C. and Alpern, B. (Editors), 1989b. Coal: classification, coalification, mineralogy, trace element geochemistry, and old and gas potential. International Journal of Coal Geology, Special Issue 12, 626 pp.

Mackay, J.R. and Terasmae, J., 1963. Pollen diagrams in the Mackenzie Delta area, N.W.T. Arctic, 16: 228–238.

Martini, I.P., 1989. The Hudson Bay Lowland: Major geologic features and assets. Geologie en Mijnbouw, 68: 25–34.

Martini, I.P. and Glooschenko, W., 1985. Cold climate peat formation in Canada, and its relevance to Lower Permian Coal Measures of Australia. Earth-Science Reviews, 22: 107–140

Martini, I.P. and Rocha-Campos, A.C., 1991. Interglacial and early post-glacial, lower Gondwana coal sequences in the Paraná Basin, Brazil. In: H. Ulbrich and A.C. Rocha-Campos (Editors), Proceedings of the seventh International Gondwana Symposium, Universidade de São Paulo, São Paulo, p. 317–336.

Mathewes, R.W., 1989. The Queen Charlotte Islands refugium: a paleoecological perspective. In: R.J. Fulton (Editor), Quaternary geology of Canada and Greenland. Geological Survey of Canada, Geology of Canada No. 1, pp. 486–491.

McCabe, P.J. and Parrish, J.T., 1992. Tectonic and climatic controls on the distribution and quality of coals. In: P. J. McCabe and J.T. Parrish (Editors), Controls on the distribution and quality of Cretaceous coals. Geological Society of America, Special Paper 267, pp. 1-15.

Mishra, H.K., 1991. A comparison of the petrology of some Permian coals of India with those of Western Australia. In: H. Ulbrich and A.C. Rocha-Campos (Editors), Proceedings of the seventh International Gondwana Symposium. Universidade de São Paulo, São Paulo, p. 260–271.

Mitra, N.D., 1991. The sedimentary history of lower Gondwana coal basins of peninsular India. In: H. Ulbrich and A.C. Rocha-Campos (Editors), Proceedings of the seventh International Gondwana Symposium. Universidade de São Paulo, São Paulo, p. 273–288.

Moore, P.D., 1987. Ecological and hydrological aspects of peat formation. In: A.C. Scott (Editor), Coal and coal-bearing strata: recent advances. Geological Society, Special Publication 32, pp. 7–15.

Navale, G.K.B. and Saxena, R., 1989. An appraisal of coal petrographic facies in Lower Gondwana (Permian) coal seams in India. International Journal of Coal Geology, 12: 553–588.

Rahmani, R.A. and Flores, R.M. (Editors), 1984. Sedimentology of coal and coal-bearing sequences. International Association Sedimentology, Special Publication 7, 412 pp.

Roulet, N.T., Jano, A., Kelly C.A., Klinger, L.F., Moore, T.R., Protz, R., Ritter, J.A. and Rouse, W.R., 1994. Role of the Hudson Bay lowland as a source of atmospheric methane. Journal of Geophysical Research, 99: 1439–1454.

Rowley, D.B., Raymond, A., Parish J.T., Lottes, A.L., Scotese, C.R. and Ziegler, A.M., 1985. Carboniferous paleogeographic, phytogeographic, and paleoclimatic reconstructions. International Journal of Coal Geology, 5: 7–42.

Scott, A.C., 1987. Coal and coal-bearing strata: recent advances. Geological Society, Special Publication 32, 332 pp.

Shotyk, W., 1988. Review of the inorganic geochemistry of peats and peatland waters. Earth-Science Reviews, 25: 95–176.

Shotyk, W., 1989. An overview of the geochemistry of methane dynamics in mires. International Peat Journal, 3: 25–44.

Shotyk, W., 1992. Organic soils. In: I.P. Martini and W. Chesworth (Editors), Weathering, soils and paleosols. Elsevier, Amsterdam, pp. 203–224.

Silva Correa da, Z.C., 1991. The formation of coal deposits in south Brazil. In: H. Ulbrich and A.C. Rocha-Campos (Editors), Proceedings of the seventh International Gondwana Symposium. Universidade de São Paulo, São Paulo, pp. 233–247.

Sjors, H., 1959. Bogs and fens in the Hudson Bay Lowlands. Arctic, 12: 3–19.

Smyth, M., 1970. Type seam sequence for some Permian-Australian coals. Proceedings Australian Institute Min. Metall., vol. 233, pp. 7–16.

Smyth, M., 1980. Thick coal members: products of an inflammatory environment? Australian Coal Geology, 2: 53–76.

Stavrakis, N., 1986. Sedimentary environments and facies of the Orange Free State coalfield. In: C.R. Anhaeusser and S. Maske (Editors), Mineral deposits of Southern Africa, vol. 2. Geological Society of South Africa, pp. 1939–1952.

Stavrakis, N. and Smyth, M., 1991. Clastic sedimentary environments and organic petrology of coals in Orange Free State, South Africa. International Journal of Coal Geology, 18: 1–16.

Staub, J.R., Cohen, A.D., 1979. The Snuggedy swamp of South Carolina: a back-barrier estuarine coal-forming environment. Journal of Sedimentary Petrology, 49: 133–144.

Staub, J.R. and Esterle, J.S., 1994. Peat-accumulating depositional systems of Sarawak, east Malaysia. Sedimentary Geology, 89: 91–106.

Styan, W.B. and Bustin, R.M., 1983. Petrography of

some Fraser River delta peat deposits: coal maceral and microlithotype precursors in temperate-climate peats. International Journal of Coal Geology, 2: 321–370.

Styan, W.B. and Bustin, R.M., 1984. Sedimentology of Fraser River delta peat deposits: a modern analogue for some deltaic coals. In: R.A. Rahmani and R.M. Flores (Editors), Sedimentology of coal and coal-bearing sequences. International Association of Sedimentology, Special Publication 7, pp. 241–271.

Terasmae, J., 1967. Recent pollen deposition in the northeastern district of Mackenzie (Northwest Territories, Canada). Palaeogeography, Palaeoclimatology, Palaeoecology, 3: 17–27.

Terasmae, J., 1977. Postglacial history of Canadian muskeg. Muskeg and the northern environment in Canada. In: N.W. Radforth and C.O. Brawner (Editors), Muskeg Research Conference, Edmonton. University of Toronto Press, Toronto, pp. 9–30.

Wells, E.D. and Pollet, F.C., 1983. Peatlands. In: G.R. South (Editor), Biogeography and ecology of the island of Newfoundland. Junk, The Hague, pp. 207–265.

Zoltai, S.C. and Pollett, F.C., 1983. Wetlands in Canada, their classification, distribution, and use. In: A.J.P. Gore (Editor), Mires: swamp, bog, fen and moor. Regional studies. Ecosystems of the World, vol. 4B. Elsevier, Amsterdam, pp. 245–269.

INDEX

accommodation space, 296, 323-324
accumulation rates, 297
Adelaide, 216, 255, 258
Adelaide Fold Belt, 125
Adrar, 120
Adrar des Iforas, 120
Adria Turkish Plate, 122
aerobic, 295, 315, 320
Afghanistan, 244
Agglomerate Slate, 203
Agulhas Plateaus, 117
Air Massif, 123
Aker moraines, 42, 44
Akhdar, 122, 145
Al Khlata diamictites, 121
Al Khlata Formation, 198
Aland Archipelago, 44
Alaska, 119, 255
Alberta, 277, 281, 288-289
Alexander Island, 72
Alexandria, 249
Algeria, 120, 145
Algiers, 192
Alice Springs Orogen, 125
Allerod, 22-23, 37-38, 57
allochemical, 324
Altaids, 118
Altay Mountain, 100
Altiplano plateau, 70
Amazon River, 103-104
Amazonas, 147, 150, 162
Amb, 198
Amb Formations, 198
Ambikapur, 200
Amery Group, 121
Amur River, 118
anaerobic, 295, 320
Andes, 61, 63-66, 68, 70-75, 116
Angara, 54
Angaran, 118, 131
Angaran floral province, 130
Angola, 170, 177, 181-182, 185
Antarctic
  continent, 21-22
  Convergence, 66
  East Antarctic Ice Sheet, 16, 61-62, 69, 82, 235
  Ice Sheet, 11, 69, 171, 216
  Peninsula, 63, 72
  West Antarctic Ice Sheet, 9, 11, 61, 63, 84, 235
Antarctica, 4, 13, 22, 61-63, 66-69, 73, 75, 117, 119, 121, 127, 129, 133, 145-147, 150, 152, 169, 172, 183-185, 213-214, 216, 230-232, 234-237, 239, 243, 250, 258, 263, 302, 303, 317
Antler Belt, 124, 143
Appalachian
  basin, 115
  chain, 129
  foldbelt, 120
  Mountains, 111, 115, 123, 128-129
  Orogeny, 120
  peneplain, 111
Appin Group, 261, 263
Aquidauana Formation, 156
Arabia, 4, 122, 145, 170, 183-185, 193, 198, 204-208, 243
  Afro-Arabia, 192, 198, 204-205, 208,
Arabian, 204
  Basin, 132, 145
  Peninsula, 121, 169, 197, 243
  platform, 198,
Arafura Basin, 213
Aral Sea, 53, 100
  Basin, 99
Archean, 253, 263, 315
Arckaringa Basin, 217-219, 223
Arctic
  area, 53,
  Brown, 286
  Canadian Arctic, 13, 297
  Canadian Arctic Archipelago, 284
  Canadian Arctic Channels, 13
  Canadian Arctic islands, 285, 302
  coast, 13
  flora, 57
  high, 277, 285-286, 288 290, 297
  islands, 277, 287, 290
  Ocean, 99
  region, 285
  Sea, 44, 53
  tree line, 285-287, 289

Arctic *(continued)*
  tundra, 56-57,
  Urals, 54-55,
Arequipa Massif, 150-151
Argentina, 65-66, 70, 72, 117, 133, 144, 150, 152-153, 156, 160, 172, 184, 243-244
argillic horizons, 277, 281, 286-287
Argyll Group, 259, 263
Artinskian, 112, 119, 131, 162, 169, 171, 175-176, 179, 180, 182-188, 198, 203, 205, 209,219
Arunachal Pradesh, 193, 201, 207
ash, 62, 71, 157, 313
Asselian, 131, 171, 176, 180, 183-186, 198, 203, 205, 208, 218, 222, 243
Asuncion Arch, 120, 144, 158
Atacora Mountains, 126, 145
Atlantic Basin, 111
Atlantic
  North Atlantic, 3, 12, 15-16, 19, 21, 38, 44, 68, 92, 104-105, 120, 282
  Ocean, 12, 55, 62, 98, 279
  South Atlantic, 126
Atlantis, 249
Atlas, 115-116, 145
Aussenkjer, 174, 178
Australia, 4, 66, 75, 112, 117, 121-122, 125, 127, 129, 132-133, 146-148, 171, 194, 198, 205, 213,226, 231, 240, 243-244, 251, 255, 258, 263, 264, 301-303, 305-306, 311-312, 317-318, 324-325
Axel Heiberg Island, 277-288

back-arc basin, 118, 124, 174
Baffin Shelf, 15, 21-23
Baffin Island, 15, 18, 21
Bagra/Jabalpur sequence , 194
Bahia Choiseul Formation, 164
Baikalian rift, 123
Bajada, 122
Baltic
  Basin, 33, 37, 39, 40
  Depression, 34-35, 37
  Ice Lake, 39-40, 42
  Ice Margin, 33
  readvance, 37
  Sea, 34, 36-37, 39, 42, 44-45
  Shield, 32
Bansjhol, 200
Baoshan, 204
Baoshan terrane, 193, 203
Barakar, 201, 273-274, 305
Barakar Formation, 199, 201, 208, 272, 305
Barbados, 13, 19-21
Barents, 13, 18
Barents Sea, 32-34, 38-39, 45, 131, 143
Barents/Spitsergen, 21
Barren Measures, 199, 201, 208, 271
Bathurst Island, 284-285
Beacon Supergroup, 183
Beardmore
  area, 231, 234, 236, 238-240
  Glacier, 230-232, 234-238, 240
  Range, 236
Bechar Basin, 120
beetles, 70
Belt Lines, 37
Belt, 29
Benson Creek Moraines, 289
Berghem, 29
Berghem Moraine, 39
Bhaduara fauna, 201
Bhutan, 193, 201, 207
Billingen, 29.38-40
Bjornoya Trough, 32-33, 38
Black Sea, 53, 100, 115
Bluefish River, 280-281, 288
Boam Canyon, 100
bog, 73, 294-296, 298-299, 301, 306, 312, 314-316, 323
  peat, 57, 69-70, 129, 222
  string, 299
  treed, 312, 315
Bolivia, 65, 68, 70, 72, 147, 150-151, 160-162
Bolling, 22-23, 37, 56
Bonahaven Dolomite, 262-263
Bonaparte
  Basin, 213, 215-222, 225
  Gulf Basin, 194
Bonete Formation, 158
Boralf, 276
boreal, 5, 44, 57, 118, 205, 208, 286-287, 290, 296, 312, 314-315,324
Bothnian Bay, 31, 34, 44
Botswana, 117, 170, 177
Bouguer anomaly, 124
Bowen
  Basin, 117, 146, 214, 218-223, 225, 303, 317
  Trough, 221
Boyup, 213
Brandenburg moraines, 33, 36
Brandenburg-Leszno moraines, 36
Brazil, 72, 76, 117, 120, 126-127, 132, 144, 147, 150-151, 156-158, 172, 179, 182, 185, 192, 243, 311, 317-319, 325
British Columbia, 117, 124, 288-289, 297
British Isles, 33
British ice sheets, 33
Broome Arch, 216
brown coal, 155, 323

Browse Basin, 213, 215, 217
Bruce Mines, 252
brunisol, 276, 286, 288-289
  brunisolic, 277, 286
  brunisolic Static Cryosol, 286
Buckeye Formation, 233-234
Buckley Formation, 232, 237-239
Burt Hill
  exposure, 281
  paleosol, 280-281, 288
Burundi, 170
Bylot Island, 92
Byrd
  area, 232, 234, 236
  -Beardmore area, 231-232, 234, 236, 239-240
  Glacier, 230-232, 235-237, 240
  Marie Byrd Land, 236
  -Nimrod area, 236
Byrranga, 53
Byrranga Mountains, 53-54
Byrranga Range, 116, 143

Caledonian, 123-124
  Mountatins, 32
  Orogeny, 124
  root, 123
  tectonic-magmatic reactivation, 124
  thrust belt, 124
Caledonides, 122
Calingasta-Uspallata, 151-155, 161-162
calving, 10, 15, 17, 36, 44, 63, 69, 205
Cambrian, 126, 193, 198, 257, 264
Campo Mourao, 156
Canada, 131, 143, 161, 250, 258, 276-278, 285-290, 297, 298, 301, 308, 311-312, 314
Canadian
  Cordillera, 255
  Shield, 125, 308, 312
canal, 84, 88
Canning Basin, 146, 213, 215-216, 218-222, 225
Canyon Ranges, 282
Cape Foldbelt, 117
Cape York, 222
Cape Orogen, 127
carbon dioxide, 4, 5, 206, 208-209, 226, 250, 323
carbonate, 13, 15-16, 18, 22-23, 85, 123, 131, 133, 177, 180, 194, 196, 198, 207-209, 219-220, 255,257, 261-263, 286-287, 301, 312, 323
Carboniferous, 3-5, 75, 103, 115-134, 147-162, 169-188, 192, 194-195, 197-198, 203-204, 206, 213-216, 226, 231-232, 239, 243, 249, 263, 265, 300-306, 311, 315, 324-325
Cargonian Highlands, 114-115, 126-127, 145, 174, 177
Caribbean, 19, 104, 298
Caribou Terrane, 124
Carnarvon Basin, 213, 215-216, 218-222
Caspian Sea, 3, 100
cataclysmic flood, 98-106
catastrophic, 3, 11, 39, 40, 42, 62-63, 76, 103-104, 106, 200
Cathaysian
  flora, 118, 132-133
  floral affinities, 119
  forms, 203
  tropical rain forest province, 129
Caucasian Arc, 118, 143
Caucasus, 118
Cenozoic, 11, 117, 121-122, 126, 235, 250, 263
Cenozoic Ice Age, 11, 23
Central African Republic, 170, 181
Central Finland Ice-Marginal Formation, 29, 44
Central Swedish Ice-Marginal Line, 29
Chad, 170
Chandindu, 289
Changxinain Stage, 118
Chanic, 150
Channeled Scabland, 101-102, 106
Chapeu do Sol Member, 157
Charata, 156
charcoal, 286
Cherquilho, 325
Chert, 112, 132, 159, 162, 181-182
Chesapeake Bay, 111
Chhidru Formations, 198
Chibougamau, 250
Chikiala Formation, 194
Chile, 65, 68, 70-71, 144
Chilean Lakes Region, 70
Chimborazo, 65, 70
China, 115-116, 118-119, 123-125, 129, 132, 144, 202
Chios Arc, 118
Chorro Formation, 151
clast pavement, 87-88, 93
climate forcing, 13
climatic optimum, 59, 285
coal balls, 301, 303
Coates Lake Group, 255-257
Cobalt, 252-253
Coleman
  formation, 252
  Member, 253
Collapse, 15, 38, 42, 52, 55, 61, 63, 76, 104, 156, 162, 169, 171, 184-185,187, 195, 234-235, 249-250, 288
Collie Basin, 213, 216, 219-221, 225
Colombia, 65, 68, 70, 71, 144
Colorado, 119, 177
Columbia River, 102
Conglomerate, 150-151, 154-155, 157-158, 160-161, 172, 174, 178-179, 181-183, 197, 200-201, 206, 222, 233, 238, 250, 253, 260-261, 272, 318

Congo, 145, 170
  Basin, 126, 170, 180-186
  craton, 126
Connemara, 259
convolute, 318
Cooper Basin, 213, 215-216, 218-220, 222-223, 225-226, 303
Cordillera
  American Cordilleras, 250
  Blanca, 63, 65
  Canadian Cordillera, 255-256
  Ice Sheet, 9, 13, 23
  North American Cordillera, 256, 264
  northern, 257-258, 262
Coronel Oviedo Formation, 157
cratonic, 102, 127, 152, 162, 201, 214, 236, 238, 240, 303-304, 316-317
creep, 82-83, 89
Cretaceous, 111, 116, 119-120, 125-127, 194, 300
crevasse fillings, 30
cryogenic, 276
  activities, 285
  factors, 297
  feature, 30, 277, 282, 286-287, 289
  microfabric, 277
  process, 290
cryosol, 276, 284, 286-290
  cryosolic, 277, 284-285, 287-290
  cryosphere, 9, 23
crystalline bedrock, 30, 32, 85-87, 90-94
Cumana Formation, 150,
cycles, 4, 5, 71, 76, 123, 171, 189, 195, 201, 225, 244, 251-253, 255, 257, 263-264, 271, 297
Cypress Hills, 277, 281

Dalradian Supergroup, 259, 262-263
Damara orogen, 126
Damodar valley, 272-273
Damuda Group, 200-201
Dandot Formation, 198
Dansgaard-Oeschger events, 68
Darjeeling, 193, 201
Datta Formation, 198
Dawson area, 287, 289
De Geer moraines, 44
Deccan
  Lavas, 194
  Traps, 127
deforming sediment, 81-82, 84-85, 87-90, 93-94
Dekese borehole, 181-182
Delaware Basin, 131, 143
Denison Trough, 218-219, 225
Denmark, 32-33, 35-37, 39, 143
depression, 10, 28, 34-35, 37, 46, 100, 119, 127, 131, 143, 161-162, 180-183, 207, 213, 235, 249, 297, 299, 312
Des Moines Lobe, 92
Devonian, 4, 119-126, 132, 147, 150-151, 153, 158, 192, 208, 214, 252, 263, 307
Diamantina Ranges, 144
diamictite, 98, 121, 131, 150-151, 153-158, 160-161, 172, 174, 176-184, 188, 196, 198-201, 203, 207, 231, 233-234, 240, 250-252, 256, 259-263, 304, 318, 320, 324
Diego (Mahajanga) Basin, 197
discontinuous Permafrost Zone, 288
Discovery Ridge Formation, 234, 236
District of Keewatin, 250, 286
Diversion Creek Paleosols, 280, 284, 287, 289
Djeffara Basin, 132, 145
Djulfian Zewan Formation, 203
dolomite, 22, 23, 182, 198, 206-207, 262-263
Donegal, 259
Donets Basin, 119
Drakensberg volcanics, 194
Dronning Maud Land, 172, 183, 230
dropstone, 30, 131, 150-151, 153-155, 157-158, 160, 192, 196, 199-201, 205, 207, 233-234, 236, 244, 253, 256, 271, 299, 303
drumlin, 34, 37, 46-47, 103-106
drumlinoid, 46-47
Dryas
  Early, 56
  Late, 56-57
  Middle, 57
  Older, 37
  Younger, 11, 15, 21-23, 30, 36-46, 71
Dry Valleys area, 69
Dwyka
  coal, 319
  Diamictite, 304
  Formation, 156
  glaciation, 304-305
  Group, 169, 171, 174, 177-178, 184, 194, 204
  tillite, 127
Dzulfian, 114

Easdale Subgroup, 262-263
East Jutland, 29, 37
East Siberian Shelf, 23
Ecuador, 65, 68, 70-71, 74
Edenburg, 174
edoma, 56
Eemian Interglacial, 31
effective pressure, 83, 85
Egga moraines, 33, 38
El Imperial Formation, 155
El Niño, 71
Elandsvlei, 174

Elliot Lake, 252
Ellsworth Mountains, 117, 183, 230
Emei Mountains, 123
Emei Shan basalts, 119
Enisei, 54
Ephrata Fan, 102
equilibrium profile, 11, 22
Erie, 92
erosional
grooves, 105-106
marks, 103
eskers, 3, 30, 44, 46, 54, 90-91, 93, 103-106, 163, 200
Eskimonia High, 122, 143
Ethiopia, 170, 183
Eurasia, 98-99, 102, 122
Ice Sheet, 99-100
shelf, 22
Eureka Sound
Group, 277
Paleosols, 277
Euroasian Ice Sheet, 53
Europe, 3, 11, 16, 32, 53, 58, 71, 115, 122, 131-132, 143, 306
European
Alps, 74
continent, 33, 46
Ice Sheet, 17, 30
Eurydesma, 156, 158, 161-162, 171, 177-178, 180, 186-187, 198, 201, 205, 207, 243
eustatic, 33, 39, 44, 58, 61, 111, 127, 160-162, 198, 205, 207-208, 221, 225, 243, 258-259, 263-264, 297, 299, 314
eustatic cycles, 297
evaporites, 128-129, 131-132, 185, 198, 206-208, 255, 324

fabrics, 87, 127, 286
Fairchild Formation, 232, 236-237
Falkland, 72, 75, 117, 148, 155-156, 158, 161-163
Famennian, 150, 162
feldspar, 4, 125, 128, 197, 206, 208, 255, 272, 275, 303, 324
fen, 294-295, 299, 312, 315-316
Fenestella Beds, 203
Fenno-
scandanavian ice sheets, 18
scandian Ice Sheet, 10
Finland, 31-33, 40, 42-44, 46
Finnish Karelia, 41
Firstbrook
Formations, 252
Member, 253
Fitzroy Tough, 216, 221-222, 225
fjord, 40, 63, 66, 174, 178, 185, 205, 317
Flinders Range, 125, 146
floating terminus, 30-31
Florida, 312, 315
Foraminifera, 15, 18-21, 23, 55, 177
foredeep, 151, 303, 306, 324
foreswell, 120
Fox glacier, 66
Frankfurt, 29
moraines, 33, 36
Franz Josef
glacier, 66, 71, 75
Land, 34
Fraser River, 313
Delta, 297, 301, 307
Front Range, 119
frost, 30, 46, 133, 239, 284
action, 30, 46, 65

Gabon, 126, 145, 170, 181-182, 204, 208
Basin, 132, 145, 170, 181, 184
Galilee Basin, 213, 215, 217-219, 222-223, 225, 303
Gamburtsev Subglacial Mountains, 119
Gangamopteris, 198, 203, 205, 207, 223, 239, 303-305
Ganges/Brahmaputra, 193
Gansu Province, 124
Gansu-Qinhai region, 124
Garu Formation, 202
Garvellachs, 259-61
Gawler Craton, 216, 258
Gawlor, 126-127
Geodetic Hills, 277
George VI Sound, 72
Germany, 33-34, 36, 39, 122
Ghana, 126,
Gharif Formation, 198
Ghats
Eastern, 127, 145
Western, 127
giant ripple, 106
Gilbert delta, 236
glacial
erosion, 11, 45, 67, 312
isostatic, 9-10, 19, 297, 325
pavement, 103, 179
striae, 42, 172
glacioeustatic, 58, 160-162, 249-250, 253, 255
glaciofluvial, 30, 42, 44, 46, 199, 216, 218, 273-274, 282, 312, 319
glacioisostatic, 160
glaciolacustrine, 30, 37, 54, 105-106, 154, 198, 273, 312
glaciomarine, 3, 30, 45, 105, 119, 150, 153, 155, 157-158, 161, 202, 216, 231-234, 236, 239-240, 255-256, 312, 318

glaciotectonic, 35-37, 45-46, 172, 174, 178, 299
glendonite, 244
global sea level, 19, 21, 62, 68-70
global warming, 53, 55, 61-62, 72, 75, 192
Glossopteris, 133, 156-159, 161-162, 197-198, 203, 205, 208, 223, 231, 239, 244, 303, 305
Godavari Basin, 199
Godavari Valley, 194
Golconda Belt, 117-118, 143
Gondwana, 4, 112, 114-115, 117, 119-122, 125, 132-133, 150, 163, 169, 171-172, 180, 183, 186-188, 192-194, 198-209, 213-216, 222-223, 225-226, 232, 234, 243-244, 249-250, 311, 325
  Basin, 207
  coals, 5-6, 161, 294, 302-304, 306-307, 320, 325
  craton, 193
  glaciation, 148, 150, 160, 162, 234, 303
  Ice Sheet, 150, 152, 156, 162, 169, 171, 183-188
  sediments, 193, 198, 272
  subcontinent, 147, 160
  supercontinent, 4, 162, 169, 215, 250
  Supergroup, 194, 199, 201, 272
Gondwanaland, 75-76, 127, 147, 150-152, 160-162, 236, 239, 243-244, 294, 302, 306
Gondwanan
  graben, 122
  sequences, 129, 202-203
Goodsir Inlet, 284
Gothenburg, 29
Gotland, 29
Gowganda Formation, 250-255, 258, 263-264
  Upper Gowganda Formation, 252-254
Grampian Group, 259
Great Lakes, 16, 312
Greece, 143-144
Green Bay Lobe, 90, 92
Greenland, 122, 143
  ice-core record, 104
  Ice Sheet, 9, 11, 13, 16-17, 19, 68
Grenville Mountains, 125, 143
Greta, 318, 325
Griqualand Formation, 251
Grounded ice sheet, 10, 38
Guandacol embayment, 154
Guinea, 170
Gulf of Bothnia, 31, 34, 44
Gulf of Carpentaria, 217
Gulf of Mexico, 98
Gulf of St. Lawrence, 92
Gulf Stream, 53
Gunnedah, 214-215, 220, 222-223
Gymnosperms, 155, 223, 300, 304

Hado River, 200
Halland Coastal Moraines, 29-37
Hammersley Range, 127, 146
Hasdo River, 201
Haushi Group, 198, 204, 206
Havallah Basin, 117-118
Hayhook orogeny, 255
Heinrich-events, 13, 15-17, 21, 68, 92, 104
herbaceous, 222, 303-304
Hercynian, 115, 143, 150
heterolithic, 157
high Arctic,
Himalaya, 111, 127, 133, 193, 199, 201, 203, 208, 243-244, 250
  Greater Himalaya, 193
  Lesser Himalaya, 193, 201-202, 208
  Tethyan Himalayan, 193, 201-203, 208
hinge line, 111, 249
histosol, 279, 320
Hoggar Massif, 120
Holocene, 22, 42, 57, 59, 61-63, 69-74, 161, 209, 276,277, 281, 285-290, 311-312, 314, 316, 325
hornblende, 274
Hoyada Verde Formation, 161
Huab Basin, 170, 177-179
Hudson Bay, 15, 22, 250, 313
Hudson Bay Lowland, 297, 312, 315
Hudson Strait, 14-17, 19, 21-22, 92, 104
humification, 315-316
Huqf Axis, 121, 145
Huron, 92, 251-252,
Huronian, 251-252, 254-255, 258, 263
  foldbelt, 253
  formations, 253-254
  glaciation, 250
  outcrop belt, 251-252, 254
  rocks, 250, 255
  Supergroup, 251-254, 257, 264
hydraulic conductivity, 86
hypsithermal, 57
hypsithermal Interval, 285, 287, 314

Iapetus Ocean, 259, 262-264
ice
  age, 10-11, 23, 75-76, 82
  crystal, 318
  crystal cast, 299, 303
  dammed lake 42, 45, 54, 58, 98, 100-101
  flux, 84
  house, 4, 192, 209
  rafted 28, 33, 101, 104, 154-155, 157, 162, 177, 179-181, 192
  rafting, 151, 153-154, 162, 217
  sheet collapse, 15, 63, 104, 169
  shelf, 22, 34, 36, 72, 216-217, 234-235, 240

stream, 15, 17, 21, 34-35, 62-63, 75, 84, 127, 172, 177, 184, 186
wedge 260-261, 277, 284, 286-287, 290, 318
icebergs, 15-17, 19, 21, 33, 63, 69, 92, 100, 104, 172, 216, 318
Idaho, 124
Idusi Formation, 196-198, 204-205
Iforas Mountains, 120, 145
Illinoian, 277, 282, 284, 286-289
inceptisol, 276
India, 4, 75, 121-122, 127, 132-133, 145, 171, 185, 192-194, 198-202, 204-205, 207-208, 213, 216,218, 243-244, 258, 263, 272-273, 285, 302-303, 305, 311, 317, 320, 322
Indo-Antarctic Metamorphic Terrain, 127
Indochina, 119, 132
Indochinese Arcs, 119, 144
Indonesia, 312
inertinite, 5, 157, 300, 303-307, 316, 324-325
Ines, 66
Intertropical Convergence Zone, 129
intraclast, 88-89, 155
intracratonic, 119, 156, 181, 193-194, 198, 213, 223, 324
Iran, 122, 144
Irati Shale, 127
Ireland, 255, 259-260, 262-264, 313, 315
Irian Jaya, 213
Irtysh, 55, 57, 99, 116
Isalo Group, 197
Isla de Chiloe, 70
Islay, 259-260
Itacuami Formation, 151
Itararé, 156-157

Jamal-Gydan, 54-55
James Lobe, 92
Jebel Akhdar, 122
Jharia basin, 274
jokulhlaups, 98
Jos Plateau, 123
Jotunheimen, 29, 42
Juan de Fuca, 102
Jura Quartzite, 262-263
Jurassic, 118-119, 124, 158, 180, 194, 198, 206, 221, 272, 304

Kaapvaal
craton, 126
Province, 127
Kalahari Basin, 114, 117, 127, 156, 161-162, 169-171, 174, 177-180, 183-185
Kalevala moraine, 40
kame terraces, 105
Kandalaksha, 29
Bay, 40
Kaokoland, 177
Kaokoveld, 170, 178-179
kaolinitic, 196, 205, 274,
Kara Sea, 23, 38, 55, 99, 116
ice dome, 34
ice lobe, 38
Shelf, 3, 53-55, 58
Karakum, 116
Karelia, 32, 40-42, 250
Karharbari
coalfields, 201
Formation, 199, 201, 208, 305
sedimentation, 201
Karoo, 121, 182, 183, 193, 304
Basin, 114, 117, 127, 145, 147, 152, 156, 158, 161-162, 169-172, 174-177, 179-181, 183-186, 188, 193, 195, 304-305
rift systems, 193
rocks, 121, 183
sequence, 304
strata, 182
succession, 183
Supergroup, 180, 194, 197
Karpinsky Swell, 114, 119, 143
Kasa Formation, 150
Kashmir, 193-194, 202-203
Kaspian, 53
Kattegat, 29, 36-37
Kazakhstania, 115-116
Kazanian, 114, 118, 128-129, 131, 203, 207, 221
Keele Formation, 257
Keewatin District, 250, 286
Keiva moraines, 38
Kenya, 66, 170, 182-183, 193
Khaduttei, 54-55
Kharpari Creek, 200
Khuff
Formation, 198
Group, 198, 206-207
Khvalyn Sea, 99
Kilifi area, 182
Kimberley
Block, 215-216
highlands, 219
Plateau, 146
Range, 127
uplands, 222
Kingori Sandstone, 197, 207
Kingriali Dolomite, 198
Kivu, 181-182

Klaarstroom, 174
Klamath terranes, 117
Klaza, 277
Klondike Valley, 289
Koel-Damodar, 199-200
Kola Peninsula, 32, 34, 38, 40, 42
Kolpashevo, 55-56
Korba, 200
Kulshill Group, 194
Kunene Valley, 126
Kungurian, 156, 207, 219, 221
Kuznetsk, 131, 143
Kygyzstan, 100
Kyzylkum Upland, 116, 143

Labrador, 87, 91-92
  Sea, 3, 12, 15-16, 18-19, 92, 104, 125
Lachlan Highlands, 125, 146, 222
Lafonian diamictites, 158, 161
Lafonia Formation, 156, 158
Lagoa Azul, 156
Laingsburg/Floriskraal, 174
Lake Agassiz, 98
Lake Baikal, 124
Lake Issyk-Kul, 100
Lake Mentz Subgroup, 194
Lake Michigan, 89, 92
Lake Michigan Lobe, 92-93
Lake Missoula, 100
Lake Nyasa, 180, 193
Lake Tanganyika, 126, 204
Lake Titicaca, 72, 147, 150
Lake Vattern, 29, 37
Lambert Rift, 121-122, 145
Lambert Glacier, 121
laminite, 201, 204-205, 207
Las Peñas (= Escarpment) Formation, 151
Late Wisconsin, 53, 58, 235, 282, 284
Late-glacial climatic change, 71
Laurentia, 9, 13, 263
Laurentide Ice Sheet, 3, 10-13, 15-16, 19-21, 35, 45, 62, 81-82, 85-94, 98, 103-104, 282
Laurussia, 115
Lena, 54
Lesotho, 170, 174
Leszno
  moraines, 33, 36
  Zone, 34
Levene, 29
  line, 37
  moraine, 37
Lhasa Block, 122, 133, 144, 193
Libya, 132, 145, 204
lignin, 300, 306
lignite, 295, 302, 316, 323
Lilangu Member, 196, 205
Limpopo, 121, 170, 179-180
Little Ice Age, 74-75, 288
Little Bear River, 280, 282, 284, 289
lodgement, 81, 87-88
  till, 28, 37, 86, 157, 199, 204, 207, 216, 233
Loeriesfontein, 174
Lomagundi
  Highlands, 121, 145
  uplands, 179-180
London, 249
Lord Howe Rise, 117, 213-214
Lorrain Formation, 253-255, 257, 263
Low Baltic Ice, 37
Lower Subgroup, 181
Luangwa rift valley, 121
Lukuga
  Glacials, 181
  Group, 181
  tillite, 126
Lulangu Basin, 193
lutite, 196-198, 200, 205, 207, 219
Lutoe Series, 182
luvisol, 276, 287-290
  luvisolic, 277, 281, 286-290

MacDonnell Ranges, 125, 146
maceral, 5, 299-301, 303-304, 307, 308, 323-324
Mackellar Formation, 232, 234-236
Mackenzie
  Delta, 297-298, 307
  Mountains, 257-258, 282
  Mountains Supergroup, 258, 261
  Plain, 282, 313-314
  River, 285
Madagascar, 121, 127, 169-170, 172, 182-183, 192-194, 197-199, 204-208, 302-303
Madumabisa Mudstone, 121
Magellan Strait, 68, 70, 72
Mahadeva Formation, 272
Mahanadi
  Basin, 145, 201
  Valley, 121
main stationary line, 29, 37
Makutu Mountains, 121, 145
Malagasy Basin, 170, 180-184, 186
Malagasy Trough, 193, 197-198
Malawi, 170
Malaysia, 312, 314-315
Mali, 120, 145
Mamal Formation, 203
Manda Beds, 197, 207
Manitoba, 285

Mansy Lake, 54, 99
Manych, 99
Maranhao Basin, 126, 132, 144
Marie Byrd Land, 236
Marine Isotope Stage, 11-12, 33, 68
Morocco, 145, 192
Mars, 3, 98, 104-106
  Martian, 104-105
marsh, 222, 312
Mato Grosso, 120
Mauritanide Belt, 120, 144
Mayombe Range, 126, 145
Mbuyura Formation, 196-197, 206
McCloud Arc, 117-118, 124, 143
McGregor Bay, 252
Mchuchuma Formation, 196-197, 205
Mediterranean
  ocean basin, 122
  Sea, 3, 53, 115
Melo Formation, 157
melt-out, 28, 81, 86-88, 103
  till, 28, 200, 233
Merida Basin, 144
Mesocenzoic, 126
Mesozoic, 116-118, 120-123, 126-128, 172, 182, 192, 225, 232, 243
Metangula Basin, 193
Meteorite Bore Member, 251
methane, 5, 294, 323
Mexico, 115, 255
Mhukuru Formation, 196-197, 207
Michigan, 250
micro-organisms, 315-316
microfabric, 277
microlithotype, 300
Middle Swedish End Moraines, 38-39
Mikumi
  Basin, 121
  rift valleys, 121
Milankovitch
  cycle, 244, 260
  insolation rhythms, 10
  theory, 55
  time-scale, 9
Milky Way, 4, 209
Minnesota, 315
Miocene, 61-62, 281
mire, 70, 161, 294-302, 304, 306-308, 312
Mississippi, 98, 252
Mississippian, 119
Missoula flood, 101-102, 104
Mitumba Range, 126, 145
mixed bed, 85
Mojon de Hierro Formation, 155, 161
Mokowan Butte, 281-282, 289
Mokowan Butte Luvisolic, 288
Mongolia, 116, 118, 123, 144
Mongolian Arc, 116, 118, 144
Mongolo-Okhotsk Seaway, 118
moor, 294
moraine, 29-30, 33-42, 44-47, 54-55, 58, 66, 70-72, 74, 90, 103-105, 244, 289
Morondava Basin, 197, 204-205
Moscovian Stage, 120
Mount Berg Formation, 256
Mount Glossopteris Formation, 236-237
Mozambique, 145, 170, 179-180
Mpera Sandstone Member, 196, 205
Mt. Cook National Park, 66
Mt. Billingen, 39-40
Mt. Kilimanjaro, 66
Muchinga Scarp, 121, 145
mudrock, 172, 174, 177-181, 183
mudstone, 12, 121, 150-151, 154-158, 160-162, 177-179, 182, 197, 199, 201, 207, 238, 251, 253, 256-258, 263-264
Mueller Glacier, 74
Murmansk, 40
Murray Basin, 125, 215-219
Musgrave Ranges, 125, 127, 146
muskeg, 294

Namibia, 126, 145, 169-170, 177, 179, 185
Namurian, 150-152, 155, 161-162, 171, 175, 194
Nansen, 277
  glacial advance, 284
Napier Mountains, 127
Naples, 249
Natal, 320
neoglacial, 74-75, 285
Neoproterozoic, 250, 255, 257-259, 263-264
Nepal, 193, 201
Nevada, 117, 124
Nevado Ausangate-Quelccaya ice-cap, 70
New England, 94, 219, 222, 249
  Arc, 117
  Fold Belt, 317
  highlands, 221
  Mountains, 146
  Orogen, 214-215
New South Wales, 66, 218, 219
New Zealand, 61, 66, 68-74, 213-214, 300
New Zealand Alps, 72-74
New York, 249
Newfoundland, 15
Niger, 123, 145, 170, 204-205
Nigeria, 123, 170
Nimrod Glaciers, 233, 235-236

Nishatbag Formation, 203
Norman Wells, 282
North America, 3, 9-11, 28, 45, 53, 58, 85, 98-99, 101-102, 117, 119-120, 129, 131-132, 143, 249-251, 256-258, 264, 276-277, 285, 311, 314
North Caspian Depression, 119
North Land, 55
North Sea, 33-34, 36, 46, 122, 143
North Urals, 53
North West Cape, 213
Northern Hemisphere, 9, 11, 17, 19, 21-22, 61-62, 68, 71-72, 75, 82, 93-94, 129, 303-304, 306, 312, 324
Northwest Territories, 277, 282, 286
Norway, 21, 32-34, 38, 40-42, 44, 123-124, 143
Norwegian Trench, 32-33
Novaya Semlja, 34, 38
Nyasa
  Rift, 194
  rise, 179

Ob', 55, 57, 99
Odderade Interstadial, 32
Officer Basin, 215-216, 218
Ogaden Basin, 183
Ohio Range, 183, 230, 232-240
Okefeenokee Swamp, 312, 314
öland, 29
Old Crow River, 281
Oman, 122, 145, 193, 197-198, 204-205, 207-208, 243-244
Omeishan, 123, 144
Omolon, 131
  Block, 130-131, 144
  ice-marine deposit, 131
Orange Free State, 319
Ordos Basin, 132, 144
Ordovician, 4, 120, 123, 183, 192, 208-209, 214, 263
organic soil, 68, 70, 276, 285, 288
  paleosol, 277, 281, 288
Orkneys, 63
Oslo
  Dome, 122-124, 143
  Fjord, 40, 42
Oslofjord, 37, 39-40, 42
Osorno, 66
Otira stage
  Early, 69
  Late, 69
Ouachita-Marathon Orogeny, 119
Ougarta
  Chain, 120
  Mountains, 120, 145
outwash, 33, 40, 45-46, 71, 98, 100, 105, 150, 155-156, 160-161, 172, 174, 178-182, 184, 200-202, 204, 216-217, 233, 253, 272, 284

Ovamboland, 177
oxygen isotope, 32, 61

Pacific
  Islands, 117
  Ocean, 66, 71, 132
Paganzo, 147, 151, 153-155, 160, 162
Pagoda
  Formation, 232, 234
  strata, 233
paleobathymetry, 111
Paleoproterozoic, 250-251, 255, 258, 263
paleosol, 4-5, 276-278, 280-282, 284, 286-290
paleotopograhy, 111, 120, 125, 128, 153, 177, 179, 181-183
paleo-Pacific, 150, 162, 185, 188, 213, 216, 232, 236-239
Paleozoic, 5, 15, 32, 98, 106, 111, 115-116, 118-120, 123-128, 131-132, 147, 150-151, 154, 156, 158-161, 169, 172, 180, 182-183, 187-188, 192, 194, 198, 201, 204-205, 208-209, 213, 216, 222, 230-237, 239, 249, 250, 302, 307, 311-312, 318, 323
palsas, 297, 306
paludification, 295-297, 315
Pamir, 116, 244
Pampa de Tepuel Formation, 155
pampas, 72
Pan Xi Mountains, 124, 144
Panafrican, 114, 120
Panamanian, 302
Panchet Formation, 208, 272
Pangea, 112, 115, 129, 131-132, 134, 193-194, 197-198, 201, 203-204, 209, 214
Pangea Mega-Sequence, 194-196, 199
Panjal
  area, 122
  High, 145
  Traps, 203
  Rift, 122
  Volcanics, 203
Papua-New Guinea, 66
Paraguay, 120, 144, 156
Paraná Basin, 114, 117, 120, 127, 133, 144, 148, 152, 155-157, 160, 162, 169, 171, 177, 179, 184-185, 243, 318-320
Patagonia, 68-70, 72
Patom Uplands, 124, 144
peaty mounds, 297
pebbly mudstones, 12, 155, 177, 182
Pedirka Basin, 215-216, 218-219
Pelagonian Block, 118, 122, 144
Pensacola, 183, 230
Perito Moreno glacier, 66-67
permafrost, 30, 56, 105, 192, 223, 277, 284, 286-290, 297, 299, 303-304, 314, 318

Permian, 3-5, 75-76, 111-134, 143, 148, 152, 155-156, 158, 162, 169, 174-175, 179-180, 182-183, 186, 193-194, 197-200, 203-209, 213-226, 231-232, 238-239, 243-244, 272, 297, 300, 303-305, 307, 311, 317-322, 324-325
Permo-Carboniferous, 76, 103, 169, 171-172, 186-187,192, 194, 197-198, 249, 263, 265
Perth, 222
  Basin, 122, 146, 215-216, 218-222, 225-226
Peru, 63, 65, 70-72
Phanerozoic, 125, 127, 263, 307
phosphate, 181, 289
phytoliths, 316
Piedra Azul Formation, 158-159
Pilbara, 215-216
Pleistocene, 3-4, 28, 31, 58, 61-62, 75-76, 98-100, 102, 104, 111, 132-134, 160, 171, 188, 209, 250, 265, 275-277, 281-282, 284, 286-290, 297, 311-312, 314, 316
Pliocene, 281-282, 288, 290
Podzol, 276, 286-289
  Podzolic, 277, 281, 286-290
  Podzolic horizon, 286
Poland, 33-34, 36, 39, 122
Polar Plateau, 62, 232, 237-238
polygenic, 276
Pomeranian moraines, 29, 36
Ponoy Ice Cap, 38, 40
Porcupine River, 281
Port Askaig Formation, 259-263
Port Askaig Tillite, 259
Port Sussex Formation, 159
Poznan, 29, 33
Pozzuoli, 249
preboreal, 44, 57
Pre-Illinoian, 277, 282, 286-289
Pre-Reid, 277, 282, 289
Precambrian, 3-4, 32, 35, 85, 98, 105, 111-112, 116, 121, 124-128, 131, 192, 199, 216, 230, 232, 263, 271, 272, 312, 318
Precaspian, 131, 143
Precordillera, 153
pressure melting point, 15, 82-83
Prieska, 174
Prince Albert Formation, 156, 161
Prince Charles Mountains, 230
Productus Limestone, 198
proglacial lakes, 3, 15-16, 69-70, 98, 205
Proterozoic, 4, 105, 119-120, 125, 127, 186, 209, 249-250, 252, 255, 256, 258, 263-265
Proto-Pacific, 258, 263
Prydz Bay, 122
Pulju moraine, 46
Putorana, 53-55
pyrite, 181-182, 205, 316

quartz, 112, 197, 207, 253, 272, 274, 300
Quaternary, 3-5, 9-11, 13, 21, 23, 28, 30, 32, 35, 45, 47, 53, 106, 129, 159, 209, 276-277, 281, 288, 290, 311, 319, 325
Quebec, 125, 250
Queen Charlotte Island, 312, 313
Queen Maud Land, 127
Queensland, 221, 244
Quelccaya ice-cap, 65, 70-71
Quesnel terrane, 117
Quincy Basin, 102

R-channels, 84, 90, 93
Ra Moraine, 39-41
radar, 11
radiocarbon dates, 12-15, 21, 23, 29, 33, 67, 72, 74, 92, 249, 284-285
Rahab Shale, 198, 205, 208
raised forest, 312, 314-315
Rangit
  Pebble Slate, 204
  tectonic window, 204
Raniganj, 199, 201, 208, 272-274
Rapitan Group, 255-258, 263-264
Recent, 250, 265
recession, 29, 36-37, 44, 54-55, 58, 61, 69-72, 74-75, 93, 161
red beds, 201, 206
reef, 5, 19, 104, 115, 131, 323, 325
Reid glaciation, 277, 284
relict soil, 276-277, 285, 287
Reschreiter glacier, 71
Resolution Basin, 21-23
rhythmite, 30, 47, 101, 106, 157, 161, 172, 174, 177-183, 196, 256
rigid bed, 81-82, 84-86, 90, 93
Rio Bonito Formation, 157-158, 318
Rio de Sul Member, 157-158
Rio Grande Do Sul, 318
Rio Segredo Member, 157
roches moutonnées, 126, 157, 207
Rockies, 119
Rocky Mountains, 119, 143, 285
Rogen moraine, 46-47, 103-104
Roma Shelf, 225
Ross
  Embayment, 63, 69, 235
  High, 120, 146
  Orogenic Belt, 232
  Sea, 117, 232, 235, 237-238
Ruhuhu
  Basin, 121, 145, 180, 194, 197, 204-205

Ruhuhu *(continued)*
  Formation, 196-197, 207
Rukwa rift, 121
Russia, 32, 34, 118-119, 123, 143-144
Russian
  Karelia, 42
  Plain, 39
  Platform, 115, 131
Ruwenzori, 66
Rwanda, 170

Sabi-Limpopo, 179-180
Sakamena Group, 197
Sakmarian, 4, 112, 119, 128, 131-132, 156, 159, 169, 171, 175-180, 182-184, 186-187, 205-206, 208, 218-220, 223, 243-244, 272
Sakoa Group, 182, 197
Salar de Uyuni, 70
Salpausselkas, 29, 40-44
Salt Range, 193, 197-198, 203, 205, 207-208
Samfrau Orogenic Zone, 117
San Gregorio Formation, 157
San Miguel Formation, 158
San Rafael, 117, 151, 155, 162
sandstone, 85-86, 150-151, 154-161, 172, 174, 177-183, 194, 196-202,205-208, 219, 226, 231, 233, 236, 237-238, 251, 253, 255, 260-261, 263-264, 271-275, 318
Sangamonian, 277, 282, 284, 287, 289-290
Santa Catarina, 120
São Francisco craton, 126
São Paulo State, 318
Sartan, 53-56, 58
Sarthai Formation, 198
Saskatchewan, 88, 277, 281
Satpura Basin, 199-200
Sauce Grande, 158-162
  Basin, 117, 148, 152, 155-156, 158-162
  deposits, 158
  Formation, 156
Sayunei Formation, 256
scabland, 100-102, 106
Scandes Mountains, 123-124, 143
Scandinavia, 28, 31-36, 38-40, 42, 44-46, 74, 124
Scandinavian
  countries, 313
  Ice-Sheets, 13, 28, 30, 33-35, 38,42, 45, 47, 62
  inland ice, 28, 32
  margin, 21
  Peninsula, 31-32, 35, 37, 42, 44, 46
Scotland Shetland Isles, 33
Scott Glaciers, 234, 240
sediment
  deformation, 11, 81-87, 89, 92-94, 155, 157-158, 261
  gravity flow, 86, 153, 172, 178, 240, 324
Selous
  Basin, 197
  rift valley, 121
Senegal Microplate, 120
Sergipe-Algohas Basin, 182
Severnaya Zemlya islands, 116
Shaba, 181-182
Shackleton Glacier, 232, 234, 240
Shackleton-Scott
  area, 234,
  High, 233-234, 236, 240
shale, 85, 127, 132, 150-151, 153-155, 157-162, 172, 174, 177-183, 198-201, 205, 207-208, 226, 231, 233-240, 258, 272, 312
shear strength, 34, 83, 85
shear stress, 11-12, 16, 35, 81, 83-85, 90, 93, 103
Shetlands, 63, 66
Shezal Formation, 256-257
Siberia, 3, 31, 53-58, 115-116, 123, 130, 143-144, 243, 302, 312
Siberian, 23, 34, 53-59, 99, 116, 131
  West Siberia, 54, 57-58
Siberian
  Craton, 116, 123
  Flood Basalt Province, 122
  Platform, 124, 130-131
  West Siberian Lowlands, 116
  West Siberian Plain, 53, 55-58, 99
Sibumasu, 119, 133, 143, 204, 213
  Arc, 119, 144
  Block, 193
  microcontinent, 119
Sichuan Basin, 132, 178
Sierra de la Ventana, 117
Sierras Australes, 158-159
Sikkim, 193, 201, 207
Silurian, 124-125, 158
Simrishamn, 29
Siripaca Formation, 150
Skagerrak, 29, 32, 36, 40
Skovde, 29, 38, 40
slate, 85, 120, 201-203
sliding, 81-83, 86, 94, 104, 112
Slieve Tooey Quartzite, 262
slump, 35, 47, 204-205
  slumped, 150-151, 157, 196, 200
  slumping, 30, 36
smectite, 15-16
Snowy Range, 66
solar radiation, 4, 14, 61, 68, 75, 214
solifluction, 30, 299, 303
Solimoes, 147, 150, 162
Somalia, 170, 193

Son-Mahanadi, 199-200, 207
Songea
 Group, 195-196
 Series, 195
 Upper Songea Group, 206
Sorkapp Troughs, 38
South America, 3-4, 61, 63, 66, 72-73, 75, 116-117, 132, 144, 147-148, 150-151, 156, 160-162, 169, 172, 177, 183-186, 208, 243, 263, 302-303, 312
South Africa 4, 114, 117, 145, 148, 150, 152, 156, 158, 161, 163, 170, 174, 177, 183, 185, 192-194, 204, 251, 304-305, 311, 317-319, 321, 325
South Georgia, 66, 69, 71-72
South Indian Lake, 285
South Patagonian Ice-field, 66-67, 71, 74
southern Africa, 75, 126, 132-133, 147, 162, 169, 174, 177, 180, 182, 185-188, 194, 243, 302-303
Southern Alps, 61, 66, 70, 74
Southern Hemisphere, 4, 61-62, 66-72, 74-75, 114, 206, 324
Southern Ocean, 61-62, 66, 69-70
Spain, 15, 132, 143
sphagnum, 298-299, 314-316
spodosol, 276
Springsure Shelf, 222
St. Lawrence, 98
steady-state, 19, 81, 89
steppe, 54, 57, 74
Stephanian, 148, 152, 155, 162, 171, 175-176, 178, 183-184, 186, 194, 205, 208
Sterlitamakian, 244
sticky spots, 83, 85, 94
Stockholm, 42, 44
striated clast, 47, 88, 150-151, 155, 198, 244
Stuart Blocks, 216
Sturtian, 258, 263-264
Sub-Antarctic islands, 61, 63, 67-69
subaqueous fans, 154
subarctic, 57, 59, 285-287, 290, 297, 311-312, 314-315, 324
subglacial
 drainage, 81, 84, 90
 hydrology, 81, 83, 85, 90, 93-94
 lakes, 11, 104
 sediment, 11, 81-86, 89-90, 92-94
Sudair Formation, 198, 207
Sudan, 170, 205
sulphur, 75, 155, 157, 236, 302, 316
Sumatra, 119
Sur Riograndense Arch, 157
Svalbard, 33 34, 38, 302
Sverdrup Basin, 143
swamp, 131-133, 178, 183, 196, 206, 209, 220-221, 223, 263, 288, 294-295, 297, 312, 314-315, 324
Swaziland, 170
Sweden, 31-33, 37, 39, 44-45, 123, 143
Sydney, 221
Sydney
 Basin, 117, 146, 213-215, 217-223, 303-304, 317-318
 Bowen trough, 221

Taberg, 29
Taciba Formation, 156-157
Tacuary Formation, 158
Taiguati Formation, 151, 161
Taimyr Peninsula, 53-54, 116
Talchir, 207-208, 272-274
 deposition, 274
 Formation, 199-201, 207-208, 272-274, 305
 glaciation, 194, 198, 273-274
 glaciers, 207
 rocks, 199
 sandstone, 272-274
 sediment, 208, 272
 Tillite, 194
Tan River, 207
Tanama, 54
Tanzania, 133, 145, 170, 180, 183, 194-197, 204, 206
Tanzanian
 basins, 170, 180-181, 184
 Craton, 121
 Uplands, 121, 145
Tapley Hill Formation, 258
Tarija, 151, 161
Tarija Formation, 151
Tarim, 115-116, 144
Tasman Geosyncline, 214
Tasmania, 66, 214-222, 312
Tasmanian Basin, 317
Tastubian, 243-244
Tatarian, 114, 131, 203, 221-222
Tauca, 70
Tayvallich Volcanics, 262
Temi Basin, 218
Tengchong Terranes, 193
Tepuel, 147, 151, 155, 161
Termination, 9, 18, 61, 263
Terras Motas Formation, 159
terrestrialization, 295-296, 312, 315
Tertiary, 4-5, 15, 32, 111-112, 123-124, 126, 128, 250, 265, 276-277, 281-282, 286-288, 290, 300, 302
Tethyan, 207, 244
 belt, 193, 198
 coast, 132
 environment, 198
 fauna, 118,
 Himalayan, 193, 201-203, 208

Tethyan *(continued)*
  margin, 198
  microcontinent, 133,
  ocean, 183
  region, 120, 122
  Rift, 122
  Sea, 185-187
  seaway, 117
Tethys, 112, 115, 118, 129, 193-194, 202-203, 243
  Ocean, 243
Thailand, 119, 144, 244
thermal
  optimum, 72
  regime, 10-11, 30, 83
Theron, 230
Thurston Island, 236
Tibet, 144, 192, 202-204, 208, 243-244
Tibetan Plateau, 250
Tierra del Fuego, 66, 70
Timor, 213, 244
Tindouf, 115
Titicaca Lake, 147, 150
Torngat Mountains, 91
Tournaisian, 150, 162
Trans-Indus Ranges, 198
Transantarctic
  Basin, 117,
  Mountians, 69, 120, 230-240
Transbaykalia, 123
Transbaikalian Uplands, 123-124, 144
Transvaal, 320, 322
tree line, 285-287, 289
Tres Islas Formation, 157
Triassic, 114-116, 118-120, 122-123, 133, 192, 194-195, 197-199, 201, 207-209, 214, 230, 237-239, 243, 272, 303
Troll, 21
Trollhattan, 29
Tronador, 65-66
tundra, 54, 56-58, 222, 303
Tunguska, 54, 131, 143
Tunisia, 132, 145
tunnel channels, 103, 105
turbidites, 153-155, 157, 172, 179, 196, 198, 204-205, 216, 257, 272
Turgay, 54, 99
  drainage area, 54
  Hollow, 99
  Pass, 53-54, 58
  spillway, 99
Turkey, 122, 144
Turkmenia, 116
Tuupovaara end moraine, 41
Twitya Formation, 257-258, 263

udalf, 276
ufimian, 221
Uganda, 145, 170
Ukraina, 119, 143
Ulan Bator, 118
Uluguru horst, 121
Uluguru Uplands, 121, 145
Unayzah Group, 198, 204
Uncompahgre Uplifts, 119
United States, 119, 312
Upper Izegouanane Formation, 123
Upsala glacier, 66, 74
Urals, 53-55, 114-116, 123, 131
Uruguayan
  Precambrian Shield, 318
  Shield, 318
Usili Formation, 197, 207

Vardar, 118
Variscan, 115, 122
varve, 37, 42-44, 157, 205, 271-272
varved clays, 30
Veadeiros, 144
Veiki moraine, 46
Venezuela, 65, 68, 115, 132, 144
ventifact, 282, 284
Verkhoyansk Complex, 123, 130
Victoria, 216, 218-219
  Land, 216, 230, 236-237, 240
Villarica, 66
Vilyui Basin, 130-131, 144
Visean, 150, 243
Vistula, 33
vitrinite, 5, 157, 300-301, 303-307, 316, 324
volcano, 70-71, 123
Volta Basin, 126
Vryheid Formation, 161

Warchha Formation, 198
Wargal, 198
Warmbad Basin, 170, 177, 179
Washington, 102
waterlaid
  moraine, 55
  sediment, 28, 30
Waterton Lakes National Park, 281
wave-cut cliffs, 105
Weddell
  embayments, 235
  Sea, 63, 117, 172, 183, 235, 237-238
Wegener-Mawson Belt, 119, 127
Weichselian, 31-34, 46, 53, 57
West African Shield, 120

Western Interior Seaway, 119
Westphalian-Stephanian, 155, 171, 184
wetlands classification, 312-313
White Sea, 38-40
Whitefish Falls, 251-254
Wilga, 213
Williston Basin, 131, 143
Windermere Supergroup, 255-256, 258, 264
Windhoek Highlands, 126, 145, 177, 179
Wolwefontein, 174
Wordian, 114
Wounded Moose, 280, 282, 284, 287-289

Xi, 124

Yangtze platform, 123
Yarrol, 217, 219-221
Yeneisei rivers, 99
Yenesei Ridge, 116, 143
Yilgarn, 216, 225
  Block, 215-216, 218-219
  center, 216
  Ice Cap, 225
Yin
  Arc, 118
  Mountains, 118
  Shan, 118
  Shan Arc, 116, 124, 144
Yoldia Sea, 42
Yukon, 277, 282, 284-285, 287-289
Yunnan, 118-119, 192, 203, 208

Zaire, 145, 170, 181, 183
  Basin, 126
Zaisan, 116, 143
Zambesi, 161, 170, 179-180, 184, 193
Zambia, 121, 145, 170, 204, 206
Zechstein flora, 132
Zimbabwe, 121, 145, 170, 179, 185
Zinder, 123
  Hot Spot, 123, 145

(Reference lists, Table 5.1 and names drafted on figures have not been indexed.)